FUNCTIONAL INTEGRATION

Functional integration successfully entered physics as path integrals in the 1942 Ph.D. dissertation of Richard P. Feynman, but it made no sense at all as a mathematical definition. Cartier and DeWitt-Morette have created, in this book, a new approach to functional integration. The close collaboration between a mathematician and a physicist brings a unique perspective to this topic. The book is self-contained: mathematical ideas are introduced, developed, generalized, and applied. In the authors' hands, functional integration is shown to be a robust, user-friendly, and multi-purpose tool that can be applied to a great variety of situations, for example systems of indistinguishable particles, caustics-analysis, superanalysis, and non-gaussian integrals. Problems in quantum field theory are also considered. In the final part the authors outline topics that can profitably be pursued using material already presented.

PIERRE CARTIER is a mathematician with an extraordinarily wide range of interests and expertise. He has been called "un homme de la Renaissance." He is Emeritus Director of Research at the Centre National de la Recherche Scientifique, France, and a long-term visitor of the Institut des Hautes Etudes Scientifiques. From 1981 to 1989, he was a senior researcher at the Ecole Polytechnique de Paris, and, between 1988 and 1997, held a professorship at the Ecole Normale Supérieure. He is a member of the Société Mathématique de France, the American Mathematical Society, and the Vietnamese Mathematical Society.

CÉCILE DEWITT-MORETTE is the Jane and Roland Blumberg Centennial Professor in Physics, Emerita, at the University of Texas at Austin. She is a member of the American and European Physical Societies, and a Membre d'Honneur de la Société Française de Physique. DeWitt-Morette's interest in functional integration began in 1948. In F. J. Dyson's words, "she was the first of the younger generation to grasp the full scope and power of the Feynman path integral approach in physics." She is co-author with Yvonne Choquet-Bruhat of the two-volume book *Analysis, Manifolds and Physics*, a standard text first published in 1977, which is now in its seventh edition. She is the author of 100 publications in various areas of theoretical physics and has edited 28 books. She has lectured, worldwide, in many institutions and summer schools on topics related to functional integration.

CAMBRIDGE MONOGRAPHS ON MATHEMATICAL PHYSICS

General editors: P. V. Landshoff, D. R. Nelson, S. Weinberg

S. J. Aarseth *Gravitational N-Body Simulations*
J. Ambjørn, B. Durhuus and T. Jonsson *Quantum Geometry: A Statistical Field Theory Approach*[†]
A. M. Anile *Relativistic Fluids and Magneto-Fluids*[†]
J. A. de Azcárrage and J. M. Izquierdo *Lie Groups, Lie Algebras, Cohomology and Some Applications in Physics*[†]
O. Babelon, D. Bernard and M. Talon *Introduction to Classical Integrable Systems*
F. Bastianelli and P. van Nieuwenhuizen *Path Integrals and Anomalies in Curved Space*
V. Belinkski and E. Verdaguer *Gravitational Solitons*[†]
J. Bernstein *Kinetic Theory in the Expanding Universe*[†]
G. F. Bertsch and R. A. Broglia *Oscillations in Finite Quantum Systems*[†]
N. D. Birrell and P. C. W. Davies *Quantum Fields in Curved Space*[†]
M. Burgess *Classical Covariant Fields*[†]
S. Carlip *Quantum Gravity in 2 + 1 Dimensions*[†]
J. C. Collins *Renormalization*[†]
M. Creutz *Quarks, Gluons and Lattices*[†]
P. D. D'Eath *Supersymmetric Quantum Cosmology*[†]
F. de Felice and C. J. S. Clarke *Relativity on Curved Manifolds*[†]
B. S. deWitt *Supermanifolds*, 2nd edition[†]
P. G. O. Freund *Introduction to Supersymmetry*[†]
J. Fuchs *Affine Lie Algebras and Quantum Groups*[†]
J. Fuchs and C. Schweigert *Symmetries, Lie Algebras and Representations: A Graduate Course for Physicists*[†]
Y. Fujii and K. Maeda *The Scalar-Tensor Theory of Gravitation*
A. S. Galperin, E. A. Ivanov, V. I. Orievetsky and E. S. Sokatchev *Harmonic Superspace*[†]
R. Gambini and J. Pullin *Loops, Knots, Gauge Theories and Quantum Gravity*[†]
M. Göckeler and T. Schücker *Differential Geometry, Gauge Theories and Gravity*[†]
C. Gómez, M. Ruiz Altaba and G. Sierra *Quantum Groups in Two-Dimensional Physics*[†]
M. B. Green, J. H. Schwarz and E. Witten *Superstring Theory, volume 1: Introduction*[†]
M. B. Green, J. H. Schwarz and E. Witten *Superstring Theory, volume 2: Loop Amplitudes, Anomalies and Phenomenology*[†]
V. N. Gribov *The Theory of Complex Angular Momenta*
S. W. Hawking and G. F. R. Ellis *The Large Scale Structure of Space-Time*[†]
F. Iachello and A. Arima *The Interacting Boson Model*
F. Iachello and P. van Isacker *The Interacting Boson–Fermion Model*[†]
C. Itzykson and J.-M. Drouffe *Statistical Field Theory, volume 1: From Brownian Motion to Renormalization and Lattice Gauge Theory*[†]
C. Itzykson and J.-M. Drouffe *Statistical Field Theory, volume 2: Strong Coupling, Monte Carlo Methods, Conformal Field Theory and Random Systems*[†]
C. Johnson *D-Branes*
J. I. Kapusta and C. Gale *Finite Temperature Field Theory*, 2nd edition
V. E. Korepin, N. M. Boguliubov and A. G. Izergin *The Quantum Inverse Scattering Method and Correlation Functions*[†]
M. Le Bellac *Thermal Field Theory*[†]
Y. Makeenko *Methods of Contemporary Gauge Theory*[†]
N. Manton and P. Sutcliffe *Topological Solitons*
N. H. March *Liquid Metals: Concepts and Theory*[†]
I. M. Montvay and G. Münster *Quantum Fields on a Lattice*[†]
L. O'Raifeartaigh *Group Structure of Gauge Theories*[†]
T. Ortín *Gravity and Strings*
A. Ozorio de Almeida *Hamiltonian Systems: Chaos and Quantization*[†]
R. Penrose and W. Rindler *Spinors and Space-Time, volume 1: Two-Spinor Calculus and Relativistic Fields*[†]
R. Penrose and W. Rindler *Spinors and Space-Time, volume 2: Spinor and Twistor Methods in Space-Time Geometry*[†]
S. Pokorski *Gauge Field Theories*, 2nd edition[†]
J. Polchinski *String Theory, volume 1: An Introduction to the Bosonic String*[†]
J. Polchinski *String Theory, volume 2: Superstring Theory and Beyond*[†]
V. N. Popov *Functional Integrals and Collective Excitations*[†]
R. J. Rivers *Path Integral Methods in Quantum Field Theory*[†]
R. G. Roberts *The Structure of the Proton*[†]
C. Roveli *Quantum Gravity*
W. C. Saslaw *Gravitational Physics of Stellar Galactic Systems*[†]
H. Stephani, D. Kramer, M. A. H. MacCallum, C. Hoenselaers and E. Herlt *Exact Solutions of Einstein's Field Equations*, 2nd edition
J. M. Stewart *Advanced General Relativity*[†]
A. Vilenkin and E. P. S. Shellard *Cosmic Strings and Other Topological Defects*[†]
R. S. Ward and R. O. Wells Jr *Twister Geometry and Field Theory*[†]
J. R. Wilson and G. J. Mathews *Relativistic Numerical Hydrodynamics*

[†] Issued as a paperback

Functional Integration: Action and Symmetries

P. CARTIER AND C. DEWITT-MORETTE

CAMBRIDGE UNIVERSITY PRESS

CAMBRIDGE UNIVERSITY PRESS
Cambridge, New York, Melbourne, Madrid, Cape Town, Singapore, São Paulo

Cambridge University Press
The Edinburgh Building, Cambridge CB2 2RU, UK

Published in the United States of America by Cambridge University Press, New York

www.cambridge.org
Information on this title: www.cambridge.org/9780521866965

© P. Cartier and C. DeWitt-Morette 2006

This publication is in copyright. Subject to statutory exception
and to the provisions of relevant collective licensing agreements,
no reproduction of any part may take place without
the written permission of Cambridge University Press.

First published 2006

Printed in the United Kingdom at the University Press, Cambridge

A catalog record for this publication is available from the British Library

ISBN-13 978-0-521-86696-5 hardback
ISBN-10 0-521-86696-0 hardback

Cambridge University Press has no responsibility for the persistence or accuracy of URLs for
external or third-party internet websites referred to in this publication, and does not
guarantee that any content on such websites is, or will remain, accurate or appropriate.

Contents

Acknowledgements   *page* xi
List of symbols, conventions, and formulary   xv

PART I   THE PHYSICAL AND MATHEMATICAL ENVIRONMENT

1   The physical and mathematical environment   **3**
A: An inheritance from physics   3
1.1   The beginning   3
1.2   Integrals over function spaces   6
1.3   The operator formalism   6
1.4   A few titles   7
B: A toolkit from analysis   9
1.5   A tutorial in Lebesgue integration   9
1.6   Stochastic processes and promeasures   15
1.7   Fourier transformation and prodistributions   19
C: Feynman's integral versus Kac's integral   23
1.8   Planck's blackbody radiation law   23
1.9   Imaginary time and inverse temperature   26
1.10   Feynman's integral versus Kac's integral   27
1.11   Hamiltonian versus lagrangian   29
     References   31

PART II   QUANTUM MECHANICS

2   First lesson: gaussian integrals   **35**
2.1   Gaussians in $\mathbb{R}$ 35
2.2   Gaussians in $\mathbb{R}^D$ 35
2.3   Gaussians on a Banach space   38

2.4	Variances and covariances	42
2.5	Scaling and coarse-graining	46
	References	55

3 Selected examples — 56

3.1	The Wiener measure and brownian paths	57
3.2	Canonical gaussians in L^2 and $L^{2,1}$	59
3.3	The forced harmonic oscillator	63
3.4	Phase-space path integrals	73
	References	76

4 Semiclassical expansion; WKB — 78

4.1	Introduction	78
4.2	The WKB approximation	80
4.3	An example: the anharmonic oscillator	88
4.4	Incompatibility with analytic continuation	92
4.5	Physical interpretation of the WKB approximation	93
	References	94

5 Semiclassical expansion; beyond WKB — 96

5.1	Introduction	96
5.2	Constants of the motion	100
5.3	Caustics	101
5.4	Glory scattering	104
5.5	Tunneling	106
	References	111

6 Quantum dynamics: path integrals and the operator formalism — 114

6.1	Physical dimensions and expansions	114
6.2	A free particle	115
6.3	Particles in a scalar potential V	118
6.4	Particles in a vector potential $\vec{A}$	126
6.5	Matrix elements and kernels	129
	References	130

PART III METHODS FROM DIFFERENTIAL GEOMETRY

7 Symmetries — 135

7.1	Groups of transformations. Dynamical vector fields	135
7.2	A basic theorem	137

7.3	The group of transformations on a frame bundle	139
7.4	Symplectic manifolds	141
	References	144

8 Homotopy — 146
8.1	An example: quantizing a spinning top	146
8.2	Propagators on SO(3) and SU(2)	147
8.3	The homotopy theorem for path integration	150
8.4	Systems of indistinguishable particles. Anyons	151
8.5	A simple model of the Aharanov–Bohm effect	152
	References	156

9 Grassmann analysis: basics — 157
9.1	Introduction	157
9.2	A compendium of Grassmann analysis	158
9.3	Berezin integration	164
9.4	Forms and densities	168
	References	173

10 Grassmann analysis: applications — 175
10.1	The Euler–Poincaré characteristic	175
10.2	Supersymmetric quantum field theory	183
10.3	The Dirac operator and Dirac matrices	186
	References	189

11 Volume elements, divergences, gradients — 191
11.1	Introduction. Divergences	191
11.2	Comparing volume elements	197
11.3	Integration by parts	202
	References	210

PART IV NON-GAUSSIAN APPLICATIONS

12 Poisson processes in physics — 215
12.1	The telegraph equation	215
12.2	Klein–Gordon and Dirac equations	220
12.3	Two-state systems interacting with their environment	225
	References	231

13 A mathematical theory of Poisson processes — 233
13.1	Poisson stochastic processes	234
13.2	Spaces of Poisson paths	241

13.3	Stochastic solutions of differential equations	251
13.4	Differential equations: explicit solutions	262
	References	266

14 The first exit time; energy problems — 268

14.1	Introduction: fixed-energy Green's function	268
14.2	The path integral for a fixed-energy amplitude	272
14.3	Periodic and quasiperiodic orbits	276
14.4	Intrinsic and tuned times of a process	281
	References	284

PART V PROBLEMS IN QUANTUM FIELD THEORY

15 Renormalization 1: an introduction — 289

15.1	Introduction	289
15.2	From paths to fields	291
15.3	Green's example	297
15.4	Dimensional regularization	300
	References	307

16 Renormalization 2: scaling — 308

16.1	The renormalization group	308
16.2	The $\lambda\phi^4$ system	314
	References	323

17 Renormalization 3: combinatorics, contributed by Markus Berg — 324

17.1	Introduction	324
17.2	Background	325
17.3	Graph summary	327
17.4	The grafting operator	328
17.5	Lie algebra	331
17.6	Other operations	338
17.7	Renormalization	339
17.8	A three-loop example	342
17.9	Renormalization-group flows and nonrenormalizable theories	344
17.10	Conclusion	345
	References	351

18 Volume elements in quantum field theory, contributed by Bryce DeWitt — 355

18.1	Introduction	355

18.2 Cases in which equation (18.3) is exact	357
18.3 Loop expansions	358
References	364

PART VI PROJECTS

19 Projects	**367**
19.1 Gaussian integrals	367
19.2 Semiclassical expansions	370
19.3 Homotopy	371
19.4 Grassmann analysis	373
19.5 Volume elements, divergences, gradients	376
19.6 Poisson processes	379
19.7 Renormalization	380

APPENDICES

Appendix A Forward and backward integrals. Spaces of pointed paths	**387**
Appendix B Product integrals	**391**
Appendix C A compendium of gaussian integrals	**395**
Appendix D Wick calculus, contributed by Alexander Wurm	**399**
Appendix E The Jacobi operator	**404**
Appendix F Change of variables of integration	**415**
Appendix G Analytic properties of covariances	**422**
Appendix H Feynman's checkerboard	**432**
Bibliography	437
Index	451

Acknowledgements

Throughout the years, several institutions and their directors have provided the support necessary for the research and completion of this book.

From the inception of our collaboration in the late seventies to the conclusion of this book, the Institut des Hautes Etudes Scientifiques (IHES) at Bures-sur-Yvette has provided "La paix nécessaire à un travail intellectuel intense et la stimulation d'un auditoire d'élite."[1] We have received much help from the Director, J. P. Bourguignon, and the intelligent and always helpful supportive staff of the IHES. Thanks to a grant from the Lounsbery Foundation in 2003 C. DeW. has spent three months at the IHES.

Among several institutions that have given us blocks of uninterrupted time, the Mathematical Institute of the University of Warwick played a special role thanks to K. David Elworthy and his mentoring of one of us (C. DeW.).

In the Fall of 2002, one of us (C. DeW.) was privileged to teach a course at the Sharif University of Technology (Tehran), jointly with Neda Sadooghi. C. DeW. created the course from the first draft of this book; the quality, the motivation, and the contributions of the students (16 men, 14 women) made teaching this course the experience that we all dream of.

The Department of Physics and the Center for Relativity of the University of Texas at Austin have been home to one of us and a welcoming retreat to the other. Thanks to Alfred Schild, founder and director of the Center for Relativity, one of us (C. DeW.) resumed a full scientific career after sixteen years cramped by rules regarding alleged nepotism.

This book has been so long on the drawing board that many friends have contributed to its preparation. One of them, Alex Wurm, has helped

---

[1] An expression of L. Rosenfeld.

C. DeW. in all aspects of the preparation from critical comments to typing the final version.

Cécile thanks her graduate students

My career began on October 1, 1944. My gratitude encompasses many teachers and colleagues. The list would be an exercise in name-dropping. For this book I wish to bring forth the names of those who have been my graduate students. Working with graduate students has been the most rewarding experience of my professional life. In a few years the relationship evolves from guiding a student to being guided by a promising young colleague.

Dissertations often begin with a challenging statement. When completed, a good dissertation is a wonderful document, understandable, carefully crafted, well referenced, presenting new results in a broad context.

I am proud and humble to thank the following.

Michael G. G. Laidlaw (Ph.D. 1971, UNC Chapel Hill) *Quantum Mechanics in Multiply Connected Spaces.*

Maurice M. Mizrahi (Ph.D. 1975, UT Austin) *An Investigation of the Feynman Path Integral Formulation of Quantum Mechanics.*

Bruce L. Nelson (Ph.D. 1978, UT Austin) *Relativity, Topology, and Path Integration.*

Benny Sheeks (Ph.D. 1979, UT Austin) *Some Applications of Path Integration Using Prodistributions.*

Theodore A. Jacobson (Ph.D. 1983, UT Austin) *Spinor Chain Path Integral for the Dirac Electron.*

Tian Rong Zhang (Ph.D. 1985, UT Austin) *Path Integral Formulation of Scattering Theory With Application to Scattering by Black Holes.*

Alice Mae Young (Ph.D. 1985, UT Austin) *Change of Variable in the Path Integral Using Stochastic Calculus.*

Charles Rogers Doering (Ph.D. 1985, UT Austin) *Functional Stochastic Differential Equations: Mathematical Theory of Nonlinear Parabolic Systems with Applications in Field Theory and Statistical Mechanics.*

Stephen Low (Ph.D. 1985, UT Austin) *Path Integration on Spacetimes with Symmetry.*

John LaChapelle (Ph.D. 1995, UT Austin) *Functional Integration on Symplectic Manifolds.*

Clemens S. Utzny (Master 1995, UT Austin) *Application of a New Approach to Functional Integration to Mesoscopic Systems.*

Alexander Wurm (Master 1995, UT Austin) *Angular Momentum-to-Angular Momentum Transition in the DeWitt/Cartier Path Integral Formalism.*

Xiao-Rong Wu-Morrow (Ph.D. 1996, UT Austin) *Topological Invariants and Green's Functions on a Lattice.*

Alexander Wurm (Diplomarbeit, 1997, Julius-Maximilians-Universität, Würzburg) *The Cartier/DeWitt Path Integral Formalism and its Extension to Fixed Energy Green's Functions.*

David Collins (Ph.D. 1997, UT Austin) *Two-State Quantum Systems Interacting with their Environments: A Functional Integral Approach.*

Christian Sämann (Master 2001, UT Austin) *A New Representation of Creation/Annihilation Operators for Supersymmetric Systems.*

Matthias Ihl (Master 2001, UT Austin) *The Bose/Fermi Oscillators in a New Supersymmetric Representation.*

Gustav Markus Berg (Ph.D. 2001, UT Austin) *Geometry, Renormalization, and Supersymmetry.*

Alexander Wurm (Ph.D. 2002, UT Austin) *Renormalization Group Applications in Area-Preserving Nontwist Maps and Relativistic Quantum Field Theory.*

Marie E. Bell (Master 2002, UT Austin) *Introduction to Supersymmetry.*

List of symbols, conventions, and formulary

Symbols

$A := B$	A is defined by B
$A \stackrel{\int}{=} B$	the two sides are equal only after they have been integrated
θ	step function
$B \succ A$	B is inside the lightcone of A
$d^\times l = dl/l$	multiplicative differential
$\partial^\times/\partial l = l\partial/\partial l$	multiplicative derivative
$\mathbb{R}^D, \mathbb{R}_D$	are dual of each other; $\mathbb{R}^D$ is a space of contravariant vectors, $\mathbb{R}_D$ is a space of covariant vectors
$\mathbb{R}^{D \times D}$	space of D by D matrices
$\mathbb{X}, \mathbb{X}'$	$\mathbb{X}'$ is dual to $\mathbb{X}$
$\langle x', x \rangle$	dual product of $x \in \mathbb{X}$ and $x' \in \mathbb{X}'$
$(x\|y)$	scalar product of $x, y \in \mathbb{X}$, assuming a metric
$(\mathbb{M}^D, g)$	D-dimensional riemannian space with metric g
$T\mathbb{M}$	tangent bundle over $\mathbb{M}$
$T^*\mathbb{M}$	cotangent bundle over $\mathbb{M}$
$\mathcal{L}_X$	Lie derivative in the X-direction
$U^{2D}(S), U^{2D}$	space of critical points of the action functional S (Chapter 4)
$\mathcal{P}_{\mu,\nu}(\mathbb{M}^D)$	space of paths with values in $\mathbb{M}^D$, satisfying μ initial conditions and ν final conditions
$U_{\mu,\nu} := U^{2D}(S) \cap \mathcal{P}_{\mu,\nu}(\mathbb{M}^D)$	arena for WKB

$\hbar = h/(2\pi)$	Planck's constant
$[h] = \mathrm{ML}^2\mathrm{T}^{-1}$	physical (engineering) dimension of h
$\omega = 2\pi\nu$	ν frequency, ω pulsation
$t_\mathrm{B} = -\mathrm{i}\hbar\beta = -\mathrm{i}\hbar/(k_\mathrm{B}T)$	(1.70)
$\tau = \mathrm{i}t$	(1.100)

Superanalysis
(Chapter 9)

$\tilde{A}$	parity of $A \in \{0,1\}$
$AB = (-1)^{\tilde{A}\tilde{B}} BA$	graded commutativity
$[A,B]$	graded commutator (9.5)
$\{A,B\}$	graded anticommutator (9.6)
$A \wedge B = -(-1)^{\tilde{A}\tilde{B}} B \wedge A$	graded exterior algebra
$\xi^\mu \xi^\sigma = -\xi^\sigma \xi^\mu$	Grassmann generators (9.11)
$z = u + v$	supernumber, u even $\in \mathbb{C}_c$, v odd $\in \mathbb{C}_a$ (9.12)
$\mathbb{R}_c \subset \mathbb{C}_c$	real elements of $\mathbb{C}_c$ (9.16)
$\mathbb{R}_a \subset \mathbb{C}_a$	real elements of $\mathbb{C}_a$ (9.16)
$z = z_\mathrm{B} + z_\mathrm{S}$	supernumber; z_B body, z_S soul (9.12)
$x^A = (x^a, \xi^\alpha) \in \mathbb{R}^{n\vert\nu}$	superpoints (9.17)
$z = c_0 + c_i \xi^i$ $\quad + \frac{1}{2!} c_{ij} \xi^i \xi^j + \cdots$	
$= \rho + \mathrm{i}\sigma,$	where both ρ and σ have real coefficients
$z^* := \rho - \mathrm{i}\sigma,$	imaginary conjugate, $(zz')^* = z^* z'^*$ (9.13)
$z^\dagger$	hermitian conjugate, $(zz')^\dagger = z'^\dagger z^\dagger$

Conventions

We use traditional conventions unless there is a compelling reason for using a different one. If a sign is hard to remember, we recall its origin.

Metric signature on pseudoriemannian spaces

$$\eta_{\mu\nu} = \mathrm{diag}(+,-,-,-)$$
$$p_\mu p^\mu = (p^0)^2 - |\vec{p}|^2 = m^2 c^2, \ p^0 = E/c$$
$$p_\mu x^\mu = Et - \vec{p}\cdot\vec{x}, \ x^0 = ct$$
$$E = \hbar\omega = h\nu, \ \vec{p} = \hbar\vec{k}, \ \text{plane wave } \omega = \vec{v}\cdot\vec{k}$$

Positive-energy plane wave $\exp(-\mathrm{i}p_\mu x^\mu/\hbar)$
Clifford algebra

$$\gamma_\mu \gamma_\nu + \gamma_\nu \gamma_\mu = 2\eta_{\mu\nu}$$

Quantum operators
$$[p_\mu, x^\nu] = -i\hbar \delta_\mu^\nu \qquad \Rightarrow p_\mu = -i\hbar \partial_\mu$$

Quantum physics (time t) and statistical mechanics (parameter τ)
$$\tau = it \qquad \text{(see (1.100))}$$

Physical dimension
$$[\hbar] = ML^2T^{-1}$$
$$\hbar^{-1}\langle p, x \rangle = \frac{2\pi}{h}\langle p, x \rangle \text{ is dimensionless}$$

Fourier transforms
$$(\mathcal{F}f)(x') := \int_{\mathbb{R}^D} d^D x \, \exp(-2\pi i \langle x', x \rangle) f(x) \qquad x \in \mathbb{R}^D, x' \in \mathbb{R}_D$$

For Grassmann variables
$$(\mathcal{F}f)(\kappa) := \int \delta\xi \, \exp(-2\pi i \kappa \xi) f(\xi)$$

In both cases
$$\langle \delta, f \rangle = f(0) \qquad \text{i.e. } \delta(\xi) = C^{-1}\xi$$
$$\mathcal{F}\delta = 1 \qquad \text{i.e. } C^2 = (2\pi i)^{-1}$$
$$\int \delta\xi \, \xi = C, \qquad \text{here } C^2 = (2\pi i)^{-1}$$

Formulary (giving a context to symbols)

- Wiener integral
$$\mathbb{E}\left[\exp\left(-\int_{\tau_a}^{\tau_b} d\tau \, V(q(\tau))\right)\right] \qquad (1.1)$$

- Peierls bracket
$$(A, B) := \mathcal{D}_A^- B - (-1)^{\tilde{A}\tilde{B}} \mathcal{D}_B^- A \qquad (1.9)$$

- Schwinger variational principle
$$\delta \langle A | B \rangle = i \langle A | \delta \mathbf{S}/\hbar | B \rangle \qquad (1.11)$$

- Quantum partition function
$$Z(\beta) = \text{Tr}(e^{-\beta \hat{H}}) \qquad (1.71)$$

- Schrödinger equation

$$\begin{cases} i\partial_t \psi(x,t) = \left(-\tfrac{1}{2}\mu^2 \Delta_x + \hbar^{-1} V(x)\right)\psi(x,t) \\ \psi(x,t_a) = \phi(x) \end{cases} \quad (1.77)$$

$$\mu^2 = \hbar/m$$

- Gaussian integral

$$\int_{\mathbb{X}} d\Gamma_{s,Q}(x) \exp(-2\pi i \langle x', x \rangle) := \exp(-s\pi W(x')) \quad (2.29)_s$$

$$d\Gamma_{s,Q} x \stackrel{f}{=} \mathcal{D}_{s,Q}(x) \exp\left(-\frac{\pi}{s} Q(x)\right) \quad (2.30)_s$$

$$Q(x) = \langle Dx, x \rangle, \qquad W(x') = \langle x', Gx' \rangle \quad (2.28)$$

$$\int_{\mathbb{X}} d\Gamma_{s,Q}(x) \langle x'_1, x \rangle \ldots \langle x'_{2n}, x \rangle = \left(\frac{s}{2\pi}\right)^n {\sum}' W(x'_{i_1}, x'_{i_2}) \ldots W(x'_{i_{2n-1}}, x'_{i_{2n}})$$

sum without repetition
- Linear maps

$$\langle \tilde{L} y', x \rangle = \langle y', Lx \rangle \quad (2.58)$$

$$W_{\mathbb{Y}'} = W_{\mathbb{X}'} \circ \tilde{L}, \qquad Q_{\mathbb{X}} = Q_{\mathbb{Y}} \circ L \quad \text{(Chapter 3, box)}$$

- Scaling and coarse graining (Section 2.5)

$$S_l u(x) = l^{[u]} u\left(\frac{x}{l}\right)$$

$$S_l [a, b[= \left[\frac{a}{l}, \frac{b}{l}\right[$$

$$P_l := S_{l/l_0} \cdot \mu_{[l_0, l[*} \quad (2.94)$$

- Jacobi operator

$$S''(q) \cdot \xi\xi = \langle \mathcal{J}(q) \cdot \xi, \xi \rangle \quad (5.7)$$

- Operator formalism

$$\langle b|\hat{O}|a \rangle = \int_{\mathcal{P}_{a,b}} O(\gamma) \exp(iS(\gamma)/\hbar) \mu(\gamma) \mathcal{D}\gamma \quad \text{(Chapter 6, box)}$$

- Time-ordered exponential

$$T \exp\left(\int_{t_0}^{t} ds\, A(s)\right) \quad (6.38)$$

List of symbols, conventions and formulary   xix

- Dynamical vector fields

$$\mathrm{d}x(t,z) = X_{(A)}(x(t,z))\mathrm{d}z^A(t) + Y(x(t,z))\mathrm{d}t \qquad (7.14)$$

$$\Psi(t,\mathbf{x}_0) := \int_{\mathcal{P}_0\mathbb{R}^D} \mathcal{D}_{s,Q_0}z \cdot \exp\left(-\frac{\pi}{s}Q_0(z)\right)\phi(\mathbf{x}_0 \cdot \Sigma(t,z)) \qquad (7.12)$$

$$Q_0(z) := \int_{\mathbb{T}} \mathrm{d}t\, h_{AB}\dot{z}^A(t)\dot{z}^B(t) \qquad (7.8)$$

$$\begin{cases} \dfrac{\partial \Psi}{\partial t} = \dfrac{s}{4\pi} h^{AB}\mathcal{L}_{X_{(A)}}\mathcal{L}_{X_{(B)}}\Psi + \mathcal{L}_Y\Psi \\ \Psi(t_0,\mathbf{x}) = \phi(\mathbf{x}) \end{cases} \qquad (7.15)$$

- Homotopy

$$|K(b,t_b;a,t_a)| = \left|\sum_\alpha \chi(\alpha) K^\alpha(b,t_b;a,t_a)\right| \qquad \text{(Chapter 8, box)}$$

- Koszul formula

$$\mathcal{L}_X\omega = \mathrm{Div}_\omega(X)\cdot\omega \qquad (11.1)$$

- Miscellaneous

$$\det \exp A = \exp \mathrm{tr}\, A \qquad (11.48)$$
$$\mathrm{d}\ln\det A = \mathrm{tr}(A^{-1}\,\mathrm{d}A) \qquad (11.47)$$
$$\nabla^i f \equiv \nabla^i_{g^{-1}} f := g^{ij}\,\partial f/\partial x^j \qquad \text{gradient} \qquad (11.73)$$
$$(\nabla_{g^{-1}}|V)_g = V^j{}_{,j} \qquad \text{divergence} \qquad (11.74)$$
$$(V|\nabla f) = -(\mathrm{div}\,V|f) \qquad \text{gradient/divergence} \qquad (11.79)$$

- Poisson processes

$$N(t) := \sum_{k=1}^\infty \theta(t-T_k) \qquad \text{counting process} \qquad (13.17)$$

- Density of energy states

$$\rho(E) = \sum_n \delta(E-E_n), \qquad H\psi_n = E_n\psi_n$$

- Time ordering

$$T(\phi(x_j)\phi(x_i)) = \begin{cases} \phi(x_j)\phi(x_i) & \text{for } j > i \\ \phi(x_i)\phi(x_j) & \text{for } i > j \end{cases} \qquad (15.7)$$

xx   *List of symbols, conventions and formulary*

- The "measure" (Chapter 18)

$$\mu[\phi] \approx \left(\text{sdet}\, G^+[\phi]\right)^{-1/2} \tag{18.3}$$

$$_{i,}S_{,k}[\phi] G^{+kj}[\phi] = -_i\delta^j \tag{18.4}$$

$$G^{+ij}[\phi] = 0 \qquad \text{when } i \geqslant j \tag{18.5}$$

$$\phi^i = \begin{cases} u^i_{\text{in}\,A}\mathbf{a}^A_{\text{in}} + u^{i\,*}_{\text{in}\,A}\mathbf{a}^{A\,*}_{\text{in}} \\ u^i_{\text{out}\,X}\mathbf{a}^X_{\text{out}} + u^{i\,*}_{\text{out}\,X}\mathbf{a}^{X\,*}_{\text{out}} \end{cases} \tag{18.18}$$

- Wick (normal ordering)
operator normal ordering

$$(a + a^\dagger)(a + a^\dagger) = :(a + a^\dagger)^2: +1 \tag{D.1}$$

functional normal ordering

$$:F(\phi):_G := \exp\left(-\frac{1}{2}\Delta_G\right) F(\phi) \tag{D.4}$$

functional laplacian defined by the covariance G

$$\Delta_G := \int_{\mathbb{M}^D} \mathrm{d}^D x \int_{\mathbb{M}^D} \mathrm{d}^D y\, G(x,y) \frac{\delta^2}{\delta\phi(x)\delta\phi(y)} \tag{2.63}$$

Part I
The physical and mathematical environment

1
The physical and mathematical environment

> A physicist needs that his equations should be mathematically sound[1]
> Dirac [1]

A: An inheritance from physics

1.1 The beginning

In 1933, Dirac [2] laid the foundation stone for what was destined to become in the hands of Feynman a new formulation of quantum mechanics. Dirac introduced the action principle in quantum physics [3], and Feynman, in his doctoral thesis [4] "The principle of least action in quantum physics," introduced path integrals driven by the classical action *functional*, the integral of the lagrangian. The power of this formalism was vindicated [5] when Feynman developed techniques for computing functional integrals and applied them to the relativistic computation of the Lamb shift.

In 1923, after some preliminary work by Paul Lévy [6], Norbert Wiener published "Differential space" [7], a landmark article in the development of functional integration. Wiener uses the term "differential space" to emphasize the advantage of working not with the values of a function but with the difference of two consecutive values. He constructed a measure in "differential space." Mark Kac remarked that the Wiener integral

$$\mathbb{E}\left[\exp\left(-\int_{\tau_a}^{\tau_b} d\tau\, V(q(\tau))\right)\right], \tag{1.1}$$

---

[1] Because, says N. G. Van Kampen, "When dealing with less simple and concrete equations, physical intuition is less reliable and often borders on wishful thinking."

where $\mathbb{E}$ denotes the expectation value relative to the Wiener measure, becomes the Feynman integral

$$\int \mathcal{D}q \cdot \exp(\mathrm{i}S(q)) \qquad \text{for } S(q) = \int_{t_a}^{t_b} \mathrm{d}t \left(\frac{1}{2}\dot{q}(t)^2 - V(q(t)) \right) \qquad (1.2)$$

if one sets $\tau = \mathrm{i}t$. Kac concluded that, because of i in the exponent, Feynman's theory is not easily made rigorous. Indeed, one needs an integration theory more general than Lebesgue theory for making sense of Kac's integral for $\tau = \mathrm{i}t$, and such a theory has been proposed in [8, 9]. The usefulness of this theory and its further developments can be seen in this book. For a thorough discussion of the relation $\tau = \mathrm{i}t$, see Sections 1.9–1.11. Feynman, however, objected to a Kac-type integral because "it spoils the physical unification of kinetic and potential parts of the action" [10]. The kinetic contribution is hidden from sight.

The arguments of the functionals considered above are functions on a line. The line need not be a time line; it can be a scale line, a one-parameter subgroup, etc. In all cases, the line can be used to "time" order products of values of the function at different times.[2] Given a product of factors $U_1, \ldots, U_N$, each attached to a point t_i on an oriented line, one denotes by $T(U_1 \ldots U_N)$ the product $U_{i_N} \ldots U_{i_1}$, where the permutation $i_1 \ldots i_N$ of $1 \ldots N$ is such that $t_{i_1} < \ldots < t_{i_N}$. Hence in the rearranged product the times attached to the factors increase from right to left.

The evolution of a system is dictated by the "time" line. Thus Dirac and Feynman expressed the time evolution of a system by the following N-tuple integral over the variables of integration $\{q'_i\}$, where q'_i is the "position" of the system at "time" t_i; in Feynman's notation the probability amplitude $(q'_t|q'_0)$ for finding at time t in position q'_t a system known to be in position q'_0 at time t_0 is given by

$$(q'_t|q'_0) = \iint \ldots \int (q'_t|q'_N)\mathrm{d}q'_N(q'_N|q'_{N-1})\mathrm{d}q'_{N-1} \ldots (q'_2|q'_1)\mathrm{d}q'_1(q'_1|q'_0). \qquad (1.3)$$

The continuum limit, if it exists, is a "path integral" with its domain of integration consisting of functions on $[t_0, t]$. Dirac showed that $(q'_t|q'_0)$ defines the exponential of a function $\mathcal{S}$,

$$\exp(\mathrm{i}\mathcal{S}(q'_t, q'_0, t)/\hbar) := (q'_t|q'_0). \qquad (1.4)$$

The function $\mathcal{S}$ is called by Dirac [3] the "quantum analogue of the classical action function (a.k.a. Hamilton's principal function)" because its real

---

[2] There are many presentations of time-ordering. A convenient one for our purpose can be found in "Mathemagics" [11]. In early publications, there are sometimes two different time-ordering symbols: T^*, which commutes with both differentiation and integration, and T, which does not. T^* is the good time-ordering and simply called T nowadays. The reverse time-ordering defined by (1.3) is sometimes labeled $\overleftarrow{T}$.

part[3] is equal to the classical action function $\mathcal{S}$ and its imaginary part is of order $\hbar$. Feynman remarked that $(q'_{t+\delta t}|q'_t)$ is "often equal to

$$\exp \frac{i}{\hbar} L\left(\frac{q'_{t+\delta t} - q'_t}{\delta t}, q'_{t+\delta t}\right) \delta t \tag{1.5}$$

within a normalization constant in the limit as δt approaches zero"[4]. Feynman expressed the finite probability amplitude $(q'_t|q'_0)$ as a path integral

$$(q'_t|q'_0) = \int \mathcal{D}q \cdot \exp\left(\frac{i}{\hbar} S(q)\right), \tag{1.6}$$

where the action functional $S(q)$ is

$$S(q) = \int_{t_o}^{t} ds\, L(\dot{q}(s), q(s)). \tag{1.7}$$

The path integral (1.6) constructed from the infinitesimals (1.5) is a product integral (see Appendix B for the definition and properties of, and references on, product integrals). *The action functional, broken up into time steps, is a key building block of the path integral.*

Notation. The Dirac quantum analog of the classical action, labeled $\mathcal{S}$, will not be used. The action function, namely the solution of the Hamilton–Jacobi equation, is labeled $\mathcal{S}$, and the action functional, namely the integral of the lagrangian, is labeled S. The letters $\mathcal{S}$ and S are not strikingly different but are clearly identified by the context. See Appendix E.

Two phrases give a foretaste of the rich and complex issue of path integrals: "the imaginary part [of $\mathcal{S}$] is of order $\hbar$," says Dirac; "within a normalization constant,"[4] says Feynman, who summarizes the issue in the symbol $\mathcal{D}q$.

Feynman rules for computing asymptotic expansions of path integrals, order by order in perturbation theory, are stated in terms of graphs, which are also called diagrams [12]. The Feynman-diagram technique is widely used because the diagrams are not only a powerful mathematical aid to the calculation but also provide easy bookkeeping for the physical process of interest. Moreover, the diagram expansion of functional integrals in quantum field theory proceeds like the diagram expansion of path integrals in quantum mechanics. The time-ordering (1.3) becomes a chronological

---

[3] More precisely, the real part is the classical action, up to order $\hbar$.
[4] See the remark at the end of Section 2.2 for a brief comment on the early calculations of the normalization constant.

ordering, dictated by lightcones: if the point x_j is in the future lightcone of x_i, one writes $j \gg i$, and defines the chronological ordering T by the symmetric function

$$T(U_j U_i) = T(U_i U_j) := U_j U_i, \qquad (1.8)$$

where $U_i := U(x_i)$, $U_j := U(x_j)$, and $j \gg i$.

The Feynman diagrams are a graphic expression of gaussian integrals of polynomials. The first step for computing the diagram expansion of a given functional integral is the expansion into polynomials of the exponential in the integrand. We shall give an explicit diagram expansion as an application of gaussian path integrals in Section 2.4.

1.2 Integrals over function spaces

Wiener and Feynman introduced path integrals as the limit for $N = \infty$ of an N-tuple integral. Feynman noted that the limit of N-tuple integrals is at best a crude way of defining path integrals. Indeed, the drawbacks are several.

- How does one choose the short-time probability amplitude $(q'_{t+\delta t}|q'_t)$ and the undefined normalization constant?
- How does one compute the N-tuple integral?
- How does one know whether the N-tuple integral has a unique limit for $N = \infty$?

The answer is to do away with N-tuple integrals and to identify the function spaces which serve as domains of integration for functional integrals. The theory of promeasures [13] (projective systems of measures on topological vector spaces, which are locally convex, but not necessarily locally compact), combined with Schwartz distributions, yields a practical method for integrating on function spaces. The step from promeasures to prodistributions (which were first introduced as pseudomeasures [8]) is straightforward [14]. It is presented in Section 1.7. Already in their original form, prodistributions have been used for computing nontrivial examples, e.g. the explicit cross section for glory scattering of waves by Schwarzschild black holes [15]. A useful byproduct of prodistributions is the definition of complex gaussians in Banach spaces presented in Section 2.3.

1.3 The operator formalism

A functional integral is a mathematical object, but historically its use in physics is intimately connected with quantum physics. Matrix elements of an operator on Hilbert spaces or on Fock spaces have been used for defining their functional integral representation.

Bryce DeWitt [16] constructs the operator formalism of quantum physics from the Peierls bracket. This formalism leads to the Schwinger variational principle and to functional integral representations.

The bracket invented by Peierls [17] in 1952 is a beautiful, but often neglected, covariant replacement for the canonical Poisson bracket, or its generalizations, used in canonical quantization. Let A and B be any two physical observables. Their *Peierls bracket* (A, B) is by definition

$$(A, B) := \mathcal{D}_A^- B - (-1)^{\tilde{A}\tilde{B}} \mathcal{D}_B^- A, \tag{1.9}$$

where the symbol $\tilde{A} \in \{0, 1\}$ is the Grassmann parity of A and $\mathcal{D}_A^- B$ ($\mathcal{D}_A^+ B$) is known as the retarded (advanced) effect of A on B. The precise definition follows from the theory of measurement, and can be found in [16].

The *operator quantization rule* associates an operator $\mathbf{A}$ with an observable A; the supercommutator $[\mathbf{A}, \mathbf{B}]$ is given by the Peierls bracket:

$$[\mathbf{A}, \mathbf{B}] = -i\hbar(A, B) + O(\hbar^2). \tag{1.10}$$

Let $|A\rangle$ be an eigenvector of the operator $\mathbf{A}$ for the eigenvalue A. The *Schwinger variational principle* states that the variation of the transition amplitude $\langle A|B\rangle$ generated by the variation $\delta \mathbf{S}$ of an action $\mathbf{S}$, which is a functional of field operators, acting on a space of state vectors is

$$\delta\langle A|B\rangle = i\langle A|\delta \mathbf{S}/\hbar|B\rangle. \tag{1.11}$$

The variations of matrix elements have led to their functional integral representation. The solution of this equation, obtained by Bryce DeWitt, is the Feynman functional integral representation of $\langle A|B\rangle$. It brings out explicitly the exponential of the classical action functional in the integrand, and the "measure" on the space of paths, or the space of histories, as the case may be. The domain of validity of this solution encompasses many different functional integrals needed in quantum field theory and quantum mechanics. The measure, called $\mu(\phi)$, is an important contribution in the applications of functional integrals over fields ϕ. (See Chapter 18.)

1.4 A few titles

By now functional integration has proved itself useful. It is no longer a "secret weapon used by a small group of mathematical physicists"[5] but

---

[5] Namely "an extremely powerful tool used as a kind of secret weapon by a small group of mathematical physicists," B. Simon (1979).

is still not infrequently a "sacramental formula."[6] Compiling a bibliography of functional integration in physics, other than a computer-generated list of references, would be an enormous task, best undertaken by a historian of science. We shall mention only a few books, which together give an idea of the scope of the subject. In chronological order:

- R. P. Feynman and A. R. Hibbs (1965). *Quantum Mechanics and Path Integrals* (New York, McGraw-Hill);
- S. A. Albeverio and R. J. Høegh-Krohn (1976). *Mathematical Theory of Feynman Path Integrals* (Berlin, Springer);
- B. Simon (1979). *Functional Integration and Quantum Physics* (New York, Academic Press);
- J. Glimm and A. Jaffe (1981). *Quantum Physics*, 2nd edn. (New York, Springer);
- L. S. Schulman (1981). *Techniques and Applications of Path Integration* (New York, John Wiley);
- K. D. Elworthy (1982). *Stochastic Differential Equations on Manifolds* (Cambridge, Cambridge University Press);
- A. Das (1993). *Field Theory, A Path Integral Approach* (Singapore, World Scientific);
- G. Roepstorff (1994). *Path Integral Approach to Quantum Physics: An Introduction* (Berlin, Springer); (original German edition: *Pfadintegrale in der Quantenphysik*, Braunschweig, Friedrich Vieweg & Sohn, 1991);
- C. Grosche and F. Steiner (1998). *Handbook of Feynman Path Integrals* (Berlin, Springer);
- A *Festschrift* dedicated to Hagen Kleinert (2001): *Fluctuating Paths and Fields*, eds. W. Janke, A. Pelster, H.-J. Schmidt, and M. Bachmann (Singapore, World Scientific);
- M. Chaichian and A. Demichev (2001). *Path Integrals in Physics*, Vols. I and II (Bristol, Institute of Physics);
- G. W. Johnson and M. L. Lapidus (2000). *The Feynman Integral and Feynman's Operational Calculus* (Oxford, Oxford University Press; paperback 2002);
- B. DeWitt (2003). *The Global Approach to Quantum Field Theory* (Oxford, Oxford University Press; with corrections, 2004).
- J. Zinn-Justin (2003). *Intégrale de chemin en mécanique quantique: introduction* (Les Ulis, EDP Sciences and Paris, CNRS);

---

[6] "A starting point of many modern works in various areas of theoretical physics is the path integral $\int \mathcal{D}q \exp \mathrm{i} S(q)/\hbar$. What is the meaning of this sacramental formula?" M. Marinov [18].

- J. Zinn-Justin (2004). *Path Integrals in Quantum Mechanics* (Oxford, Oxford University Press); and
- H. Kleinert (2004). *Path Integrals in Quantum Mechanics, Statistics, and Polymer Physics*, 3rd edn. (Singapore, World Scientific).

Many books on quantum field theory include several chapters on functional integration. See also [19].

These few books, together with their bibliographies, give a good picture of functional integration in physics at the beginning of the twenty-first century. We apologize for having given an incomplete list of our inheritance.

B: A toolkit from analysis
1.5 A tutorial in Lebesgue integration

It is now fully understood that Feynman's path integrals are *not* integrals in the sense of Lebesgue. Nevertheless, Lebesgue integration is a useful device, and the Kac variant of Feynman's path integrals is a genuine Lebesgue integral (see Part C). Lebesgue integration introduces concepts useful in the study of stochastic processes presented in Section 1.6.

Polish spaces

This class of spaces is named after Bourbaki [13]. A Polish space is a *metric space*, hence a set $\mathbb{X}$ endowed with a *distance* $d(x,y)$ defined for the pairs of points in $\mathbb{X}$, satisfying the following axioms ($d(x,y)$ is a real number):

- $d(x,y) > 0$ for $x \neq y$ and $d(x,y) = 0$ for $x = y$;
- *symmetry* $d(x,y) = d(y,x)$; and
- *triangle inequality* $d(x,z) \leq d(x,y) + d(y,z)$.

Furthermore, a Polish space should be *complete*: if $x_1, x_2, \ldots$ is a sequence of points in $\mathbb{X}$ such that $\lim_{m=\infty, n=\infty} d(x_m, x_n) = 0$, then there exists a (unique) point x in $\mathbb{X}$ such that $\lim_{n=\infty} d(x_n, x) = 0$. Finally, we assume that *separability* holds: there exists a countable subset D in $\mathbb{X}$ such that any point in $\mathbb{X}$ is a limit of a sequence of points of D.

Note. Any separable Banach space is a Polish space.

We give the basic examples:

- a countable set D, with $d(x,y) = 1$ for $x \neq y$;
- the real line $\mathbb{R}$, with $d(x,y) = |x - y|$;

- the euclidean n-space $\mathbb{R}^n$ with

$$d(x,y) = \left(\sum_{i=1}^{n}(x_i - y_i)^2\right)^{1/2} \tag{1.12}$$

for $x = (x_1, \ldots, x_n)$ and $y = (y_1, \ldots, y_n)$;
- the Hilbert space ℓ^2 of infinite sequences $x = (x_1, x_2, \ldots)$ of real numbers such that $\sum_{n=1}^{\infty} x_n^2 < +\infty$, with

$$d(x,y) = \left(\sum_{n=1}^{\infty}(x_n - y_n)^2\right)^{1/2}; \tag{1.13}$$

- the space $\mathbb{R}^\infty$ of all unrestricted sequences $x = (x_1, x_2, \ldots)$ of real numbers with

$$d(x,y) = \sum_{n=1}^{\infty} \min(2^{-n}, |x_n - y_n|); \tag{1.14}$$

- let $\mathbb{T} = [t_a, t_b]$ be a real interval. The space $\mathcal{P}(\mathbb{T})$ (also denoted by $C^0(\mathbb{T}; \mathbb{R})$) of paths consists of the continuous functions $f : \mathbb{T} \to \mathbb{R}$; the distance is defined by

$$d(f,g) = \sup_{t \in \mathbb{T}} |f(t) - g(t)|. \tag{1.15}$$

In a Polish space $\mathbb{X}$, a subset U is *open*[7] if, for every x_0 in U, there exists an $\epsilon > 0$ such that U contains the ϵ-neighborhood of x_0, namely the set of points x with $d(x_0, x) < \epsilon$. The complement $F = \mathbb{X} \setminus U$ of an open set is *closed*. A set $A \subset \mathbb{X}$ is called a G_δ if it is of the form $\cap_{n \geq 1} U_n$, where $U_1 \supset U_2 \supset \ldots \supset U_n \supset \ldots$ are open. Dually, we define an F_σ set $A = \cup_{n \geq 1} F_n$ with $F_1 \subset F_2 \subset \ldots$ closed. Any open set is an F_σ; any closed set is a G_δ.

Next we define the *Baire hierarchy* of subsets of $\mathbb{X}$:

- class 1: all open or closed sets;
- class 2: all limits of an increasing or decreasing sequence of sets of class 1 (in particular the F_σ sets and the G_δ sets).

In general, if α is any ordinal with predecessor β, the class α consists of the limits of monotonic sequences of sets of class β; if α is a limit ordinal, a set of class α is any set belonging to a class β with $\beta < \alpha$.

We stop at the first uncountable ordinal ϵ_0. The corresponding class ϵ_0 consists of the *Borel subsets* of $\mathbb{X}$. Hence, if $B_1, B_2, \ldots, B_n, \ldots$ are Borel subsets, so are $\cap_{n \geq 1} B_n$ and $\cup_{n \geq 1} B_n$ as well as the differences $\mathbb{X} \setminus B_n$. Any open (or closed) set is a Borel set.

---

[7] A closed (or open) set is itself a Polish space.

Measures in a Polish space

A *measure* in a Polish space $\mathbb{X}$ associates a real number $\mu(B)$ with any Borel subset B of $\mathbb{X}$ in such a way that

- $0 \leq \mu(B) \leq +\infty$;
- if B is a disjoint union of Borel sets $B_1, B_2, \ldots, B_n, \ldots$ then $\mu(B) = \sum_{n=1}^{\infty} \mu(B_n)$; and
- the space $\mathbb{X}$ can be covered[8] by a family of open sets U_n (for $n = 1, 2, \ldots$) such that $\mu(U_n) < +\infty$ for all n.

The measure μ is *finite* (or bounded) if $\mu(\mathbb{X})$ is finite. The previous definition defines the so-called positive measures. A *complex measure* μ assigns to any Borel set B a complex number, such that $\mu(B) = \sum_{n=1}^{\infty} \mu(B_n)$ for a disjoint union B of $B_1, B_2, \ldots$ (the series is then absolutely convergent). Such a complex measure is of the form $\mu(B) = c_1 \mu_1(B) + \cdots + c_p \mu_p(B)$ where $c_1, \ldots, c_p$ are complex constants, and $\mu_1, \ldots, \mu_p$ are bounded (positive) measures. The *variation* of μ is the bounded (positive) measure $|\mu|$ defined by

$$|\mu|(B) = \text{l.u.b.} \left\{ \sum_{i=1}^{q} |\mu(B_i)| : B = B_1 \cup \ldots \cup B_q \quad \text{disjoint union} \right\}. \tag{1.16}$$

The number $|\mu|(\mathbb{X})$ is called the *total variation* of μ, and l.u.b stands for least upper bound.

A measure μ on a Polish space is *regular*. That is,

(a) if K is a compact[9] set in $\mathbb{X}$, then $\mu(K)$ is finite; and
(b) if B is a Borel set with $\mu(B)$ finite, and $\epsilon > 0$ is arbitrary, then there exist two sets K and U, with K compact, U open, $K \subset B \subset U$ and $\mu(U \setminus K) < \epsilon$.

As a consequence, if B is a Borel set with $\mu(B) = +\infty$, there exists,[10] for each $n \geq 1$, a compact subset K_n of B such that $\mu(K_n) > n$. Since the measure μ is regular, the knowledge of the numbers $\mu(K)$ for K compact enables us to reconstruct the measure $\mu(B)$ of the Borel sets. It is possible

---

[8] This condition can be expressed by saying that the measures are "locally bounded."
[9] A subset K in $\mathbb{X}$ is compact if it is closed and if, from any sequence of points in K, we can extract a convergent subsequence.
[10] From the regularity of the measures, we can deduce the following statement: the Polish space $\mathbb{X}$ can be decomposed as a disjoint union

$$\mathbb{X} = N \cup K_1 \cup K_2 \cup \ldots,$$

where each K_n is compact, and $\mu(N) = 0$. After discarding the null set N, the space $\mathbb{X}$ is replaced by a locally compact space.

to characterize the functionals $\mu(K)$ of compact subsets giving rise to a measure [13].

Construction of measures

To construct a measure, we can use *Caratheodory's extension theorem*, whose statement follows:

let $\mathcal{C}$ be a class of Borel subsets, stable under finite union, finite intersection and set difference;
assume that all Borel sets can be obtained via a Baire hierarchy, starting with the sets in $\mathcal{C}$ as class 1; and
let $I : \mathcal{C} \to \mathbb{R}$ be a functional with (finite) positive values, which is *additive*,

$$I(C \cup C') = I(C) + I(C') \qquad \text{if } C \cap C' = \emptyset, \qquad (1.17)$$

and *continuous at $\emptyset$*,

$$\lim_{n=\infty} I(C_n) = 0 \qquad \text{for } C_1 \supset C_2 \supset \ldots \text{ and } \cap_n C_n = \emptyset. \qquad (1.18)$$

We can then extend uniquely the functional $I(C)$ to a functional $\mu(B)$ of Borel subsets such that $\mu(B) = \sum_{n \geq 1} \mu(B_n)$ when B is the disjoint union of $B_1, B_2, \ldots, B_n, \ldots$

As an application, consider the case in which $\mathbb{X} = \mathbb{R}$ and $\mathcal{C}$ consists of the sets of the form

$$C =]a_1, b_1] \cup \cdots \cup]a_n, b_n].$$

Let $F : \mathbb{R} \to \mathbb{R}$ be a function that is increasing[11] and right-continuous.[12] Then there exists a unique measure μ on $\mathbb{R}$ such that $\mu(]a, b]) = F(b) - F(a)$. The special case $F(x) = x$ leads to the so-called Lebesgue measure λ on $\mathbb{R}$ satisfying $\lambda(]a, b]) = b - a$.

Product measures

Another corollary of Caratheodory's result is about the construction of a *product measure*. Suppose that $\mathbb{X}$ is the cartesian product $\mathbb{X}_1 \times \mathbb{X}_2$ of two Polish spaces; we can define on $\mathbb{X}$ the distance

$$d(x, y) = d(x_1, y_1) + d(x_2, y_2) \qquad (1.19)$$

for $x = (x_1, x_2)$ and $y = (y_1, y_2)$ and $\mathbb{X}$ becomes a Polish space. Let μ_i be

---

[11] That is $F(a) \leq F(b)$ for $a \leq b$.
[12] That is $F(a) = \lim_{n=\infty} F(a + 1/n)$ for every a.

a measure on $\mathbb{X}_i$ for $i = 1, 2$. Then there exists a unique measure μ on $\mathbb{X}$ such that

$$\mu(B_1 \times B_2) = \mu_1(B_1) \cdot \mu_2(B_2), \tag{1.20}$$

where $B_1 \subset \mathbb{X}_1$, $B_2 \subset \mathbb{X}_2$ are Borel subsets. This measure is usually denoted $\mu_1 \otimes \mu_2$. This construction can be easily generalized to $\mu_1 \otimes \cdots \otimes \mu_p$, where μ_i is a measure on $\mathbb{X}_i$ for $i = 1, \ldots, p$. The same construction applies to complex measures.

For probabilistic applications, it is useful to consider infinite products $\bigotimes_{i=1}^\infty \mu_i$ of measures. For simplicity, we consider only the case of $\mathbb{R}^\infty = \mathbb{R} \times \mathbb{R} \times \cdots$. Suppose given a sequence of measures $\mu_1, \mu_2, \mu_3, \ldots$ on $\mathbb{R}$; assume that each μ_i is positive and normalized $\mu_i(\mathbb{R}) = 1$, or else that each μ_i is a complex measure and that the infinite product $\prod_{i=1}^\infty |\mu_i|(\mathbb{R})$ is convergent. Then there exists a unique measure $\mu = \bigotimes_{i=1}^\infty \mu_i$ on $\mathbb{R}^\infty$ such that

$$\mu(B_1 \times \cdots \times B_p \times \mathbb{R} \times \mathbb{R} \times \cdots) = \mu_1(B_1) \cdots \mu_p(B_p) \tag{1.21}$$

for $p \geq 1$ arbitrary and Borel sets $B_1, \ldots, B_p$ in $\mathbb{R}$.

Integration in a Polish space

We fix a Polish space $\mathbb{X}$ and a positive measure μ, possibly unbounded. A function $f : \mathbb{X} \to \overline{\mathbb{R}} = [-\infty, +\infty]$ is called Borel if the set

$$\{a < f < b\} = \{x \in \mathbb{X} : a < f(x) < b\} \tag{1.22}$$

is a Borel set in $\mathbb{X}$ for arbitrary numbers a, b with $a < b$. The standard operations (sum, product, pointwise limit, series, ...) applied to Borel functions yield Borel functions. Note also that continuous functions are Borel. We denote by $\mathcal{F}_+(\mathbb{X})$ the set of Borel functions on $\mathbb{X}$ with values in $\overline{\mathbb{R}}_+ = [0, +\infty]$.

An *integral on* $\mathbb{X}$ is a functional $I : \mathcal{F}_+(\mathbb{X}) \to \overline{\mathbb{R}}_+$ with the property

$$I(f) = \sum_{n=1}^\infty I(f_n)$$

if

$$f(x) = \sum_{n=1}^\infty f_n(x) \qquad \text{for all } x \text{ in } \mathbb{X}.$$

The integral is *locally bounded* if there exists an increasing sequence of continuous functions f_n such that

- $0 \leq f_n(x)$, $\lim_{n=\infty} f_n(x) = 1$ for all x in $\mathbb{X}$; and
- $I(f_n) < +\infty$ for every $n \geq 1$.

Given a locally bounded integral on $\mathbb{X}$, we define as follows a measure μ on $\mathbb{X}$: if B is any Borel set in $\mathbb{X}$, let χ_B be the corresponding characteristic function[13] and let $\mu(B) = I(\chi_B)$. It is trivial that μ is a measure, but the converse is also true – and a little more difficult to prove – that is, given any measure μ on $\mathbb{X}$, there exists a unique locally bounded integral I such that $I(\chi_B) = \mu(B)$ for every Borel set B in $\mathbb{X}$. In this case, we use the notation[14] $\int f \, d\mu$ instead of $I(f)$ for f in $\mathcal{F}_+(\mathbb{X})$.

We denote by $L^1 = L^1(\mathbb{X}, \mu)$ the class of Borel functions f on $\mathbb{X}$ such that $\int |f| \, d\mu < +\infty$. The integral $\int f \, d\mu$ for the nonnegative elements f of L^1 extends to a linear form on the vector space L^1. Then L^1 is a separable Banach space[15] for the norm

$$\|f\|_1 := \int |f| \, d\mu. \tag{1.23}$$

The familiar convergence theorems are valid. We quote only the "dominated convergence theorem."

Theorem. If $f_1, f_2, \ldots, f_n, \ldots$ are functions in L^1, and if there exists a Borel function $\phi \geq 0$ with $\int \phi \, d\mu < \infty$ and $|f_n(x)| \leq \phi(x)$ for all x and all n, then the pointwise convergence

$$\lim_{n=\infty} f_n(x) = f(x) \tag{1.24}$$

for all x entails that f is in L^1 and that

$$\lim_{n=\infty} \|f_n - f\|_1 = 0, \quad \lim_{n=\infty} \int f_n \, d\mu = \int f \, d\mu. \tag{1.25}$$

For $p \geq 1$, the space L^p consists of the Borel functions f such that $\int |f|^p \, d\mu < \infty$, and the norm $\|f\|_p$ is then defined as

$$\|f\|_p := \left(\int |f|^p \, d\mu \right)^{1/p}. \tag{1.26}$$

Then L^p is a separable Banach space. The two more important cases are $p = 1$ and $p = 2$. The space L^2 is a (real) Hilbert space with scalar product $(f|g) = \int fg \, d\mu$. We leave to the reader the extension to complex-valued functions.

Assume that $\mathbb{X}$ is a compact metric space. Then it is complete and separable, and hence a Polish space. Denote by $C^0(\mathbb{X})$ the Banach space

---

[13] That is $\chi_B(x) = 1$ for x in B and $\chi_B(x) = 0$ for x in $\mathbb{X} \setminus B$.
[14] Or any of the standard aliases $\int_\mathbb{X} f \, d\mu$, $\int_\mathbb{X} f(x) d\mu(x), \ldots$
[15] This norm $\|f\|_1$ is zero iff the set $\{f \neq 0\}$ is of measure 0. We therefore have to identify, as usual, two functions f and g such that the set $\{f \neq g\}$ be of measure 0.

of continuous functions $f : \mathbb{X} \to \mathbb{R}$ with the norm

$$\|f\|_\infty = \text{l.u.b.} \{|f(x)| : x \in \mathbb{X}\}. \tag{1.27}$$

If μ is a (positive) measure on $\mathbb{X}$, then it is bounded, and one defines a linear form I on $C^0(\mathbb{X})$ by the formula[16]

$$I(f) = \int_\mathbb{X} f \, \mathrm{d}\mu. \tag{1.28}$$

Then I is positive, that is $f(x) \geq 0$ for all x in $\mathbb{X}$ entails $I(f) \geq 0$. Conversely, any positive linear form on $C^0(\mathbb{X})$ comes from a unique[17] positive[18] measure μ on $\mathbb{X}$ (the "Riesz–Markoff theorem").

A variation on the previous theme: $\mathbb{X}$ is a locally compact Polish space (that is, every point has a compact neighborhood), $C^0_c(\mathbb{X})$ is the space of continuous functions $f : \mathbb{X} \to \mathbb{R}$ vanishing off a compact subset K of $\mathbb{X}$. Then the measures on $\mathbb{X}$ can be identified with the positive linear forms on the vector space $C^0_c(\mathbb{X})$. Again $C^0_c(\mathbb{X})$ is dense in the Banach space $L^1(\mathbb{X}, \mu)$ for every measure μ on $\mathbb{X}$.

1.6 Stochastic processes and promeasures

The language of probability

Suppose that we are measuring the blood parameters of a sample: numbers of leukocytes, platelets... The record is a sequence of numbers (after choosing proper units) subject eventually to certain restrictions. The record of a sample is then a point in a certain region Ω of a numerical space $\mathbb{R}^D$ (D is the number of parameters measured). We could imagine an infinite sequence of measurements, and we should replace Ω by a suitable subset of $\mathbb{R}^\infty$.

In such situations, the record of a sample is a point in a certain space, the *sample space* Ω. We make the mathematical assumption that Ω is a Polish space, for instance the space $C^0(\mathbb{T}; \mathbb{R})$ of continuous paths $(q(t))_{t \in \mathbb{T}}$, for the observation of a sample path of the brownian motion during the time interval $\mathbb{T} = [t_a, t_b]$. The outcome is subject to random fluctuations, which are modeled by specifying a measure $\mathbb{P}$ on Ω, normalized by $\mathbb{P}[\Omega] = 1$ (a so-called *probability measure*, or *probability law*). Any Borel subset B of Ω corresponds to a (random) event, whose *probability* is the measure

---

[16] Notice the inclusion $C^0(\mathbb{X}) \subset L^1(\mathbb{X}, \mu)$.

[17] The uniqueness comes from the fact that $C^0(\mathbb{X})$ is dense in $L^1(\mathbb{X}, \mu)$ for every positive measure μ on $\mathbb{X}$.

[18] More generally, the elements of the dual space $C^0(\mathbb{X})'$ of the Banach space $C^0(\mathbb{X})$ can be identified with the (real) measures on $\mathbb{X}$.

$\mathbb{P}[B]$; hence $0 \leq \mathbb{P}[B] \leq 1$. Any numerical quantity that can be measured on the sample corresponds to a Borel function $F : \Omega \to \mathbb{R}$, called a *random variable*. The mean value of F is by definition

$$\mathbb{E}[F] := \int_\Omega F \, d\mathbb{P}.$$

A basic notion is that of (stochastic) *independence*. Suppose that we perform independently the selection of samples ω_1 in Ω_1 and ω_2 in Ω_2. The joint outcome is a point $\omega = (\omega_1, \omega_2)$ in the Polish space $\Omega = \Omega_1 \times \Omega_2$. Let $\mathbb{P}_1$ ($\mathbb{P}_2$) be the probability law of Ω_1 (Ω_2). Then $\mathbb{P} = \mathbb{P}_1 \otimes \mathbb{P}_2$ is a probability law on Ω characterized by the rule

$$\mathbb{P}[A_1 \times A_2] = \mathbb{P}_1[A_1] \cdot \mathbb{P}_2[A_2]. \tag{1.29}$$

Stated otherwise, the probability of finding jointly ω_1 in A_1 and ω_2 in A_2 is the product of the probabilities of these two events E_1 and E_2,

$$\mathbb{P}[E_1 \cap E_2] = \mathbb{P}[E_1] \cdot \mathbb{P}[E_2]. \tag{1.30}$$

This is the definition of the *stochastic independence of the events E_1 and E_2*.

A general measurement on the sample modeled by $(\Omega, \mathbb{P})$ is a function $F : \Omega \to \mathbb{X}$, where $\mathbb{X}$ is a Polish space, such that the inverse image $F^{-1}(B)$ of a Borel set B in $\mathbb{X}$ is a Borel set in Ω. Two measurements are stochastically independent if the corresponding maps $F_i : \Omega \to \mathbb{X}_i$ satisfy

$$\mathbb{P}[E_1 \cap E_2] = \mathbb{P}[E_1] \cdot \mathbb{P}[E_2], \tag{1.31a}$$

where E_i is the event that $F_i(\omega)$ belongs to B_i, that is $E_i = F_i^{-1}(B_i)$, where B_i is a Borel set in $\mathbb{X}_i$ for $i = 1, 2$. In terms of mean values, we obtain the equivalent condition

$$\mathbb{E}[\xi_1 \cdot \xi_2] = \mathbb{E}[\xi_1] \cdot \mathbb{E}[\xi_2], \tag{1.31b}$$

where ξ_i is of the form $u_i \circ F_i$ for $i = 1, 2$, and u_i is a (bounded) Borel function on $\mathbb{X}_i$.

Marginals

To be specific, let the sample space Ω be $C^0(\mathbb{T}; \mathbb{R})$; that is we record continuous curves depicting the evolution of a particle on a line during an interval of time $\mathbb{T} = [t_a, t_b]$ (figure 1.1).

In general, we don't record the whole curve, but we make measurements at certain selected times $t_1, \ldots, t_n$ such that

$$t_a \leq t_1 < \cdots < t_n \leq t_b. \tag{1.32}$$

1.6 Stochastic processes and promeasures

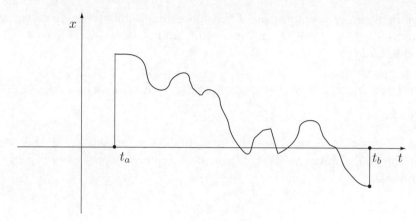

Fig. 1.1

We can approximate the probability law $\mathbb{P}$ on Ω as follows. Given $t_1, \ldots, t_n$ as above,[19] consider the map

$$\Pi_{t_1,\ldots,t_n} : \Omega \to \mathbb{R}^n$$

associating with a path $x = (x(t))_{t_a \leq t \leq t_b}$ the collection $(x(t_1), \ldots, x(t_n))$. Then there exists a measure $\mu_{t_1,\ldots,t_n}$ on $\mathbb{R}^n$ given by

$$\mu_{t_1,\ldots,t_n}([a_1, b_1] \times \cdots \times [a_n, b_n]) = \mathbb{P}[a_1 \leq x(t_1) \leq b_1, \ldots, a_n \leq x(t_n) \leq b_n]. \tag{1.33}$$

In most cases, $\mu_{t_1,\ldots,t_n}$ is given by a probability density $p(t_1, \ldots, t_n; x_1, \ldots, x_n)$ that is

$$\mu_{t_1,\ldots,t_n}(A) = \int_A d^n x \, p(t_1, \ldots, t_n; x_1, \ldots, x_n) \tag{1.34}$$

for any Borel subset A in $\mathbb{R}^n$. The measures $\mu_{t_1,\ldots,t_n}$ or the corresponding probability densities are called the *marginals* of the process $(x(t))_{t_a \leq t \leq t_b}$ modeled by the probability law $\mathbb{P}$ on $\Omega = C^0(\mathbb{T}; \mathbb{R})$.

The marginals satisfy certain coherence rules. We state them in terms of probability density:[20]

- *probability density*:

$$p(t_1, \ldots, t_n; x_1, \ldots, x_n) \geq 0, \tag{1.35}$$

$$\int_{\mathbb{R}^n} d^n x \, p(t_1, \ldots, t_n; x_1, \ldots, x_n) = 1; \tag{1.36}$$

---

[19] The restriction $t_1 < \cdots < t_n$ is irrelevant, since any system of *distinct* epochs $t_1, \ldots, t_n$ can always be rearranged.

[20] The general case is formally the same if we interpret the probability density in terms of generalized functions, such as Dirac δ-functions.

- *symmetry*: any permutation of the spacetime points $(t_1, x_1), \ldots, (t_n, x_n)$ leaves $p(t_1, \ldots, t_n; x_1, \ldots, x_n)$ invariant; and
- *compatibility*:

$$p(t_1, \ldots, t_n; x_1, \ldots, x_n) = \int_{\mathbb{R}} \mathrm{d}x_{n+1} p(t_1, \ldots, t_n, t_{n+1}; x_1, \ldots, x_n, x_{n+1}). \tag{1.37}$$

The information carried by the marginals is exhaustive, in the sense that the probability law $\mathbb{P}$ on Ω can be uniquely reconstructed from the marginals.

Promeasures

The converse of the previous construction is fundamental. Assume given for each sequence $t_1, \ldots, t_n$ of distinct epochs in the time interval $\mathbb{T}$ a probability measure $\mu_{t_1,\ldots,t_n}$ on $\mathbb{R}^n$ satisfying the previous rules of symmetry and compatibility. Such a system is called (by Bourbaki) a *promeasure*, or (by other authors) a *cylindrical measure*. Such a promeasure enables us to define the mean value $\mathbb{E}[F]$ for a certain class of functionals of the process. Namely, if F is given by

$$F(x) = f(x(t_1), \ldots, x(t_n)) \tag{1.38}$$

for a suitable Borel function f in $\mathbb{R}^n$ (let it be bounded in order to get a finite result), then $\mathbb{E}[F]$ is unambiguously defined by

$$\mathbb{E}[F] = \int_{\mathbb{R}^n} \mathrm{d}^n x \, f(x_1, \ldots, x_n) \, p(t_1, \ldots, t_n; x_1, \ldots, x_n). \tag{1.39}$$

The problem is that of how to construct a probability measure $\mathbb{P}$ on Ω such that $\mathbb{E}[F] = \int_\Omega F \, \mathrm{d}\mathbb{P}$ for F as above. We already know that $\mathbb{P}$ is unique; the question is that of whether it exists.

For the existence, the best result is given by the following criterion, due to Prokhorov [13]:

Criterion. In order that there exists a probability measure $\mathbb{P}$ on Ω with marginals $\mu_{t_1,\ldots,t_n}$, it is necessary and sufficient that, for any given $\epsilon > 0$, there exists a compact set K in the Polish space Ω such that, for all sequences $t_1, \ldots, t_n$ in $\mathbb{T}$, the measure under $\mu_{t_1,\ldots,t_n}$ of the projected set $\Pi_{t_1,\ldots,t_n}(K)$ in $\mathbb{R}^n$ be at least $1 - \epsilon$.

Notice that, by regularity of the measures, given $t_1, \ldots, t_n$ there exists a compact set $K_{t_1,\ldots,t_n}$ in $\mathbb{R}^n$ such that

$$\mu_{t_1,\ldots,t_n}(K_{t_1,\ldots,t_n}) \geq 1 - \epsilon. \tag{1.40}$$

The important point is the *uniformity*, that is the various sets $K_{t_1,\ldots,t_n}$ are obtained as the projection of a unique compact set K in Ω.

Prokhorov's criterion was distilled from the many known particular cases. The most conspicuous one is the brownian motion. The main assumptions about this process can be formulated as follows:

- the successive differences $x(t_1)-x(t_a), x(t_2)-x(t_1), \ldots, x(t_n)-x(t_{n-1})$ (for $t_a \leq t_1 < \cdots < t_n \leq t_b$) are stochastically independent; and
- a given difference $x(t_2) - x(t_1)$ obeys a gaussian law with variance $D \cdot (t_2 - t_1)$, where $D > 0$ is a given constant. This gives immediately

$$p(t_1,\ldots,t_n;x_1,\ldots,x_n) = (2\pi D)^{-n/2} \prod_{i=1}^{n}(t_i - t_{i-1})^{-1/2}$$

$$\times \exp\left(-\sum_{i=1}^{n} \frac{(x_i - x_{i-1})^2}{2D \cdot (t_i - t_{i-1})}\right) \quad (1.41)$$

(where $t_0 = t_a$ and $x_0 = x_a$, the initial position at time t_a).

The coherence rules are easily checked. To use Prokhorov's criterion, we need Arzela's theorem, which implies that, for given constants $C > 0$ and $\alpha > 0$, the set $K_{C,\alpha}$ of functions satisfying a Lipschitz condition

$$|x(t_2) - x(t_1)| \leq C|t_2 - t_1|^\alpha \quad \text{for } t_1 < t_2 \quad (1.42)$$

is compact in the Banach space $C^0(\mathbb{T};\mathbb{R})$. Then we need to estimate $\mu_{t_1,\ldots,t_n}(\Pi_{t_1,\ldots,t_n}(K_{C,\alpha}))$, a problem in the geometry of the euclidean space $\mathbb{R}^n$, but with constants independent of n (and of $t_1,\ldots,t_n$). The uniformity is crucial, and typical of such problems of integration in infinite-dimensional spaces!

1.7 Fourier transformation and prodistributions[21]

Characteristic functions

Let Ω be a Polish space, endowed with a probability measure $\mathbb{P}$. If X is a random variable, that is a Borel function $X : \Omega \to \mathbb{R}$, the *probability law of X* is the measure μ_X on $\mathbb{R}$ such that

$$\mu_X([a,b]) = \mathbb{P}[a \leq X \leq b] \quad (1.43)$$

for real numbers a, b such that $a \leq b$, where the right-hand side denotes the probability of the event that $a \leq X \leq b$. Very often, there exists a

---

[21] An informal presentation of prodistributions (which were originally introduced as pseudomeasures [8]) accessible to nonspecialists can be found in [14].

probability density p_X such that

$$\mathbb{P}[a \leq X \leq b] = \int_a^b \mathrm{d}x\, p_X(x). \tag{1.44}$$

Then, for any function $f(x)$ of a real variable, we have

$$\mathbb{E}[f(X)] = \int_\mathbb{R} \mathrm{d}x\, p_X(x) f(x). \tag{1.45}$$

In particular, for $f(x) = \mathrm{e}^{\mathrm{i}px}$ we get

$$\mathbb{E}[\mathrm{e}^{\mathrm{i}pX}] = \int_\mathbb{R} \mathrm{d}x\, \mathrm{e}^{\mathrm{i}px} p_X(x). \tag{1.46}$$

This is the so-called *characteristic function* of X, or else the Fourier transform of the probability law μ_X (or the probability density p_X).

Given a collection $(X^1, \ldots, X^n)$ of random variables, or, better said, a random vector $\vec{X}$, we define the probability law $\mu_{\vec{X}}$ as a probability measure in $\mathbb{R}^n$, such that

$$\mathbb{P}[\vec{X} \in A] = \mu_{\vec{X}}(A) \tag{1.47}$$

for every Borel subset A of $\mathbb{R}^n$, or equivalently

$$\mathbb{E}[f(\vec{X})] = \int_{\mathbb{R}^n} \mathrm{d}\mu_{\vec{X}}(\vec{x}) f(\vec{x}) \tag{1.48}$$

for every (bounded) Borel function $f(\vec{x})$ on $\mathbb{R}^n$. In particular, the Fourier transform of $\mu_{\vec{X}}$ is given by

$$\chi_{\vec{X}}(\vec{p}) := \int_{\mathbb{R}^n} \mathrm{d}\mu_{\vec{X}}(\vec{x}) \mathrm{e}^{\mathrm{i}\vec{p}\cdot\vec{x}} \tag{1.49}$$

for $\vec{p} = (p_1, \ldots, p_n)$, $\vec{x} = (x^1, \ldots, x^n)$, and $\vec{p} \cdot \vec{x} = \sum_{j=1}^n p_j x^j$. In probabilistic terms, we obtain

$$\chi_{\vec{X}}(\vec{p}) = \mathbb{E}[\exp(\mathrm{i}\vec{p} \cdot \vec{X})], \tag{1.50}$$

where, as expected, we define $\vec{p} \cdot \vec{X}$ as the random variable $\sum_{j=1}^n p_j X^j$.

Hence the conclusion that *knowing the characteristic function of all linear combinations $p_1 X^1 + \cdots + p_n X^n$ with nonrandom coefficients $p_1, \ldots, p_n$ is equivalent to knowledge of the probability law $\mu_{\vec{X}}$ of the random vector $\vec{X}$*.

The characteristic functional

The idea of the characteristic functional was introduced by Bochner [20] in 1955. As in Section 1.6, we consider a stochastic process $(X(t))_{t_a \leq t \leq t_b}$.

1.7 Fourier transformation and prodistributions

Assuming that the trajectories are continuous (at least with probability unity!), our process is modeled by a probability measure $\mathbb{P}$ on the Banach space[22] $C^0(\mathbb{T};\mathbb{R})$, or equivalently by the marginals $\mu_{t_1,\ldots,t_n}$ for $t_a \leq t_1 < \cdots < t_n \leq t_b$. These marginals are just the probability laws of the random vectors $(X(t_1),\ldots,X(t_n))$ obtained by sampling the path at the given times. It is therefore advisable to consider their characteristic functions

$$\chi_{t_1,\ldots,t_n}(p_1,\ldots,p_n) = \int_{\mathbb{R}^n} d\mu_{t_1,\ldots,t_n}(x^1,\ldots,x^n) e^{i\vec{p}\cdot\vec{x}}. \tag{1.51}$$

Knowledge of the marginals, that is of the promeasure corresponding to the process, is therefore equivalent to knowledge of the characteristic functions $\chi_{t_1,\ldots,t_n}(p_1,\ldots,p_n)$. But we have the obvious relations

$$\chi_{t_1,\ldots,t_n}(p_1,\ldots,p_n) = \mathbb{E}\left[\exp\left(i\sum_{j=1}^n p_j X(t_j)\right)\right], \tag{1.52}$$

$$\sum_{j=1}^n p_j X(t_j) = \int_\mathbb{T} d\lambda(t) X(t), \tag{1.53}$$

where λ is the measure $\sum_{j=1}^n p_j \delta(t - t_j)$ on $\mathbb{T}$. We are led to introduce the characteristic functional

$$\Phi(\lambda) = \mathbb{E}\left[\exp\left(i\int_\mathbb{T} d\lambda(t) X(t)\right)\right], \tag{1.54}$$

where λ runs over the (real) measures on the compact space $\mathbb{T}$, that is (by the Riesz–Markoff theorem) over the continuous linear forms over the Banach space $C^0(\mathbb{T};\mathbb{R})$. On going backwards from equation (1.54) to equation (1.51), we see how to *define* the marginals $\mu_{t_1,\ldots,t_n}$ starting from the characteristic functional $\Phi(\lambda)$ over the dual space $C^0(\mathbb{T};\mathbb{R})'$ of $C^0(\mathbb{T};\mathbb{R})$. But then the coherence rules (see equations (1.35)–(1.37)) are tautologically satisfied. Hence the probability law $\mathbb{P}$ of the process is completely characterized by the characteristic functional. Notice that we have a kind of infinite-dimensional Fourier transform,[23] namely

$$\Phi(\lambda) = \int_{C^0(\mathbb{T};\mathbb{R})} d\mathbb{P}(X) e^{i\langle\lambda,X\rangle} \tag{1.55}$$

where $\langle \lambda, X \rangle = \int_\mathbb{T} d\lambda\, X$ defines the duality on the Banach space $C^0(\mathbb{T};\mathbb{R})$.

---

[22] Of course $\mathbb{T}$ is the interval $[t_a, t_b]$.
[23] Later in this book, we shall modify the normalization of the Fourier transform, by putting $2\pi i$ instead of i in the exponential.

As an example, the characteristic functional of the brownian motion is given by[24]

$$\Phi(\lambda) = \exp\left(-\frac{D}{2} \int_\mathbb{T} \int_\mathbb{T} d\lambda(t) d\lambda(t') \inf(t,t')\right). \tag{1.56}$$

Prodistributions

As we shall explain in Part C, going from Kac's formula to Feynman's formula requires analytic continuation. One way to do it is for instance to replace D by iD in the last formula.

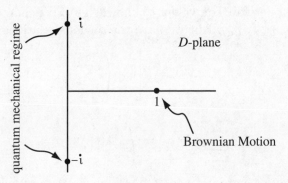

Fig. 1.2

Changing D into iD in the definition (1.41) of the marginals for the brownian motion causes no difficulty, but now $p(t_1,\ldots,t_n;x_1,\ldots,x_n)$ is of constant modulus (for fixed $t_1,\ldots,t_n$) and hence cannot be integrated over the whole space. In order to define the Fourier transformation, we need to resort to the theory of tempered distributions. So we come to the definition of a *prodistribution*[25] (see [8] and [14]) as a collection of marginals $\mu_{t_1,\ldots,t_n}(x_1,\ldots,x_n)$ that are now (tempered) distributions on the spaces $\mathbb{R}^n$. In order to formulate the compatibility condition (1.37), we restrict our distributions to those whose Fourier transform is a continuous function (necessarily of polynomial growth at infinity). As in the case of promeasures, it is convenient to summarize the marginals into the characteristic functional $\Phi(\lambda)$. The correct analytic assumption is an estimate

$$|\Phi(\lambda)| \leq C(\|\lambda\|+1)^N \tag{1.57}$$

with constants $C > 0$ and $N \geq 0$, and the continuity of the functional Φ in the dual of the Banach space $C^0(\mathbb{T};\mathbb{R})$. These restrictions are fulfilled

---

[24] See Section 3.1.
[25] Called "pseudomeasures" in [8], where they were originally introduced.

if we make D purely imaginary in (1.56). This is one way to make sense of the analytic continuation from Wiener–Kac to Feynman. We shall develop it in Part C.

C: Feynman's integral versus Kac's integral

Kac's integral is an application of brownian motion, and is nowadays used in problems of statistical mechanics, especially in euclidean quantum field theories, and in the study of bound states. Feynman's integral is mostly used to study the dynamics of quantum systems. Formally, *one goes from Kac's integral to Feynman's integral by using analytic continuation* with respect to suitable parameters. We would like to argue in favor of the "equation"

$$\text{inverse temperature} \sim \text{imaginary time}.$$

1.8 Planck's blackbody radiation law

The methods of thermodynamics were used in the derivation of the (pre-quantum) laws for the blackbody radiation.

Stefan's law: in a given volume V, the total energy E carried by the blackbody radiation is proportional to the fourth power of the absolute temperature T. (The homogeneity of radiation assumes also that E is proportional to V.)

Wien's displacement law: by reference to the spectral structure of the radiation, it states that the frequency ν_m of maximum energy density is proportional to T.

To put these two laws into proper perspective, we need some *dimensional analysis*. Ever since the pioneering work of Gabor in the 1940s, it has been customary to analyze oscillations in a *time–frequency diagram*[26] (figure 1.3).

Here t is the time variable, and $\omega = 2\pi\nu$ is the *pulsation* (ν being the *frequency*). The product ωt is a phase, hence it is a pure number using the radian as the natural phase unit. Hence the area element $dt \cdot d\omega$ is without physical dimensions, and it makes sense to speak of the energy in a given cell with $\Delta t \cdot \Delta \omega = 1$. Similarly, for a moving wave with wave-vector $\vec{k}$, the dot product $\vec{k} \cdot \vec{x}$, where $\vec{x}$ is the position vector, is a phase; hence there is a dimensionless volume element $d^3\vec{x} \cdot d^3\vec{k}$ in the phase diagram with coordinates $x^1, x^2, x^3, k^1, k^2, k^3$. This is plainly visible in the Fourier

---

[26] Compare this with the standard representation of musical scores, as well as the modern wavelet analysis.

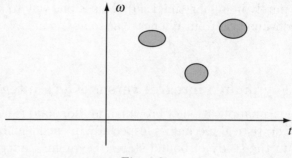

Fig. 1.3

inversion formula, which reads

$$f(0) = (2\pi)^{-3} \int\int d^3\vec{x} \cdot d^3\vec{k}\, e^{-i\vec{k}\cdot\vec{x}} f(\vec{x}). \qquad (1.58)$$

Hence, in the spectral analysis of the blackbody radiation, one can speak of the energy per unit cell $\Delta^3\vec{x}\cdot\Delta^3\vec{k} = 1$ around the point $\vec{x}$, $\vec{k}$ in the phase diagram. Since it depends also on the temperature T, we write it as $E(\vec{x},\vec{k},T)$. It obeys the following laws:

homogeneity: the spectral energy density $E(\vec{x},\vec{k},T)$ is independent of the space position $\vec{x}$; and
isotropy: the spectral energy density $E(\vec{x},\vec{k},T)$ depends only on the length $|\vec{k}|$ of the wave-vector, or rather on the pulsation $\omega = c\cdot|\vec{k}|$ (where c is the speed of light).

Hence the spectral energy density is of the form

$$E(\vec{x},\vec{k},T) = E(\omega,T).$$

Stefan's and Wien's law can be reformulated as scaling:

$$E(\vec{x},\lambda\vec{k},\lambda T) = \lambda E(\vec{x},\vec{k},T) \qquad (1.59)$$

for an arbitrary scalar $\lambda > 0$ (or $E(\lambda\omega,\lambda T) = \lambda E(\omega,T)$).

To conclude, the spectral energy density is given by

$$E(\vec{x},\vec{k},T) = \omega f(\omega/T), \qquad (1.60)$$

where f is a universal function. What one measures in the observations is the energy per unit volume in the small pulsation interval $[\omega,\omega+\Delta\omega]$ of the form

$$\Delta\mathcal{E} = E(\omega,T)\frac{4\pi\omega^2}{c^3}\Delta\omega. \qquad (1.61)$$

For a given temperature, we get a unimodal curve (figure 1.4).

1.8 Planck's blackbody radiation law

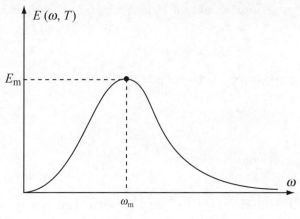

Fig. 1.4

Hence, the coordinates of the maximum define a pulsation $\omega_{\mathrm{m}} = \omega_{\mathrm{m}}(T)$ and an energy

$$E_{\mathrm{m}}(T) = \frac{c^3}{4\pi\omega_{\mathrm{m}}^2} \left(\frac{\Delta\mathcal{E}}{\Delta\omega}\right)_{\mathrm{m}}.$$

The scaling law is rewritten as[27]

$$\omega_{\mathrm{m}}(\lambda T) = \lambda\omega_{\mathrm{m}}(T), \qquad E_{\mathrm{m}}(\lambda T) = \lambda E_{\mathrm{m}}(T)$$

and amounts to an *identification of the three scales* temperature, pulsation, and energy.

This identification is made explicit in *Planck's Ansatz*

$$\Delta\mathcal{E} = \frac{A\nu^3}{e^{B\nu/T} - 1} \cdot \Delta\nu \qquad (1.62)$$

using the frequency $\nu = \omega/(2\pi)$ and *two* universal constants A and B. With a little algebra, we rewrite this in the standard form (using $A = 8\pi h/c^3$ and $B = h/k_{\mathrm{B}}$)

$$\Delta\mathcal{E} = \frac{8\pi h}{c^3} \frac{\nu^3}{e^{h\nu/(k_{\mathrm{B}} T)} - 1} \cdot \Delta\nu. \qquad (1.63)$$

We have now *two* Planck constants,[28] h and k_{B}, and two laws

$$E = h\nu, \qquad E = k_{\mathrm{B}} T, \qquad (1.64)$$

identifying energies E, frequencies ν, and temperatures T.

---

[27] The first formula is Wien's law, the second Stefan's law.
[28] The constant k_{B} was first considered by Planck and called "Boltzmann's constant" by him in his reformulation of the Boltzmann–Gibbs laws of statistical mechanics.

Here is a better formulation of Planck's law: in the phase diagram, the elementary spectral energy of the blackbody radiation is given by[29]

$$\langle E \rangle \times (2\pi)^{-3} \mathrm{d}^3 \vec{x} \cdot \mathrm{d}^3 \vec{k} \times 2, \tag{1.65a}$$

where the last factor of 2 corresponds to the two states of polarization of light and

$$\langle E \rangle = \frac{h\nu}{\mathrm{e}^{h\nu/(k_\mathrm{B} T)} - 1} \tag{1.65b}$$

is the average thermal energy.

1.9 Imaginary time and inverse temperature

As usual, we associate with a temperature the inverse $\beta = 1/(k_\mathrm{B} T)$. According to (1.64), βE is dimensionless, and the laws of statistical mechanics are summarized as follows.

In a mechanical system with energy levels $E_0, E_1, \ldots$, the thermal equilibrium distribution at temperature T is given by weights proportional to $\mathrm{e}^{-\beta E_0}, \mathrm{e}^{-\beta E_1}, \ldots$ and hence the average thermal energy is given by

$$\langle E \rangle_\beta = \frac{\sum_n E_n \mathrm{e}^{-\beta E_n}}{\sum_n \mathrm{e}^{-\beta E_n}}, \tag{1.66}$$

or by the equivalent form

$$\langle E \rangle_\beta = -\frac{\mathrm{d}}{\mathrm{d}\beta} \ln Z(\beta), \tag{1.67}$$

with the *partition function*

$$Z(\beta) = \sum_n \mathrm{e}^{-\beta E_n}. \tag{1.68}$$

The thermodynamical explanation of Planck's law (1.65) is then that $\langle E \rangle$ is the average thermal energy of an oscillator of frequency ν, the energy being quantized as

$$E_n = nh\nu, \quad \text{where } n = 0, 1, \ldots \tag{1.69}$$

Let's go back to the two fundamental laws $E = h\nu = k_\mathrm{B} T$. We have now a physical picture: in a thermal bath at temperature T, an oscillator of frequency ν corresponds to a Boltzmann weight $\mathrm{e}^{-h\nu/(k_\mathrm{B} T)}$ and hence to a phase factor $\mathrm{e}^{-2\pi \mathrm{i} \nu t_\mathrm{B}}$ upon introducing an *imaginary Boltzmann time*

$$t_\mathrm{B} = -\mathrm{i}\hbar\beta = -\mathrm{i}\hbar/(k_\mathrm{B} T) \tag{1.70}$$

---

[29] It follows from formula (1.58) that the natural volume element in the phase diagram is $(2\pi)^{-3} \mathrm{d}^3 \vec{x} \cdot \mathrm{d}^3 \vec{k} = \prod_{i=1}^{3} \mathrm{d}x^i \mathrm{d}k_i/(2\pi)$, not $\mathrm{d}^3 \vec{x} \cdot \mathrm{d}^3 \vec{k}$. We refer the reader to the conventions given earlier for normalization constants in the Fourier transformation.

1.9 Imaginary time and inverse temperature

(with the standard notation $\hbar = h/(2\pi)$) proportional to the inverse temperature.

The meaning of the constants k_B and $\hbar$ is further elucidated as follows. In quantum mechanics, the hamiltonian H is quantized by an operator $\hat{H}$ acting on some Hilbert space. According to von Neumann, the quantum partition function is given by

$$Z(\beta) = \text{Tr}(e^{-\beta \hat{H}}), \tag{1.71}$$

and the thermal average of an observable A, quantized as an operator $\hat{A}$, is given by

$$\langle A \rangle_\beta = \frac{\text{Tr}(\hat{A} e^{-\beta \hat{H}})}{\text{Tr}(e^{-\beta \hat{H}})}. \tag{1.72}$$

On the other hand, according to Heisenberg, the quantum dynamics is given by the differential equation

$$\frac{d\hat{A}}{dt} = \frac{i}{\hbar}[\hat{H}, \hat{A}] \tag{1.73}$$

with solution

$$\hat{A}(t) = U_{-t}\hat{A}(0)U_t \tag{1.74}$$

by using the evolution operator

$$U_t = e^{-it\hat{H}/\hbar}. \tag{1.75}$$

Notice the relation

$$e^{-\beta \hat{H}} = U_{t_B}. \tag{1.76}$$

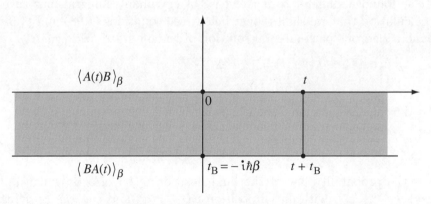

Fig. 1.5 The domain of existence of $F(z)$

From these relations, it is easy to derive the so-called KMS relation, named after Kubo, Martin, and Schwinger. Suppose given two observables, $A(t)$ evolving in time, and B fixed. There exists a function $F(z)$

holomorphic in the strip $-\hbar\beta \leq \operatorname{Im} z \leq 0$ with limiting values
$$F(t) = \langle A(t)B \rangle_\beta, \qquad F(t+t_\mathrm{B}) = \langle BA(t) \rangle_\beta.$$
This is the best illustration of the following principle:

inverse temperature = imaginary time.

1.10 Feynman's integral versus Kac's integral

The Feynman formula is the path-integral solution of the *Schrödinger equation*
$$\mathrm{i}\partial_t \psi(x,t) = \left(-\tfrac{1}{2}\mu^2 \Delta_x + \hbar^{-1} V(x)\right)\psi(x,t), \qquad (1.77)$$
$$\psi(x,t_a) = \phi(x),$$
where $\mu^2 = \hbar/m$. It is given by
$$\psi(x_b,t_b) = \int \mathcal{D}q \cdot \phi(q(t_a)) \exp\left(\frac{\mathrm{i}}{\hbar} S(q)\right), \qquad (1.78)$$
where the integral is over the space of paths $q = (q(t))_{t_a \leq t \leq t_b}$ with fixed final point $q(t_b) = x_b$. Here $S(q)$ is the action, the integral of the lagrangian
$$S(q) = \int_{t_a}^{t_b} \mathrm{d}t\, L(\dot{q}(t), q(t)), \qquad (1.79)$$
with
$$L(\dot{q}, q) = \frac{m}{2} |\dot{q}|^2 - V(q). \qquad (1.80)$$

Kac found it difficult to make sense of Feynman's integral, but (using a calculation that we shall repeat below, see equations (1.98) and (1.99)) found a rigorous path-integral solution of the modified *heat equation*
$$\partial_t \psi(x,t) = \left(\tfrac{1}{2}\mu^2 \Delta_x - W(x)\right)\psi(x,t),$$
$$\psi(x,t_a) = \delta(x - x_a). \qquad (1.81)$$
For $W = 0$, the solution is well known,[30]
$$\psi(x_b, t_b) = (\mu\sqrt{2\pi})^{-D} \exp\left(-\frac{|x_b - x_a|^2}{2\mu^2(t_b - t_a)}\right). \qquad (1.82)$$
It is the probability law of the final position $w(t_b)$ of a brownian path $w = (w(t))_{t_a \leq t \leq t_b}$ with initial position $w(t_a) = x_a$ and *covariance* μ^2: the

---
[30] We denote by D the dimension of the configuration space, that is the number of coordinates $x^1, \ldots, x^D$ of x. In general $D = 3$.

mean value of the square of the increment
$$\Delta w = w(t + \Delta t) - w(t) \tag{1.83}$$
is $D\mu^2 \Delta t$, that is[31]
$$\mathbb{E}_a\big[|w(t + \Delta t) - w(t)|^2\big] = D\mu^2\, \Delta t. \tag{1.84}$$
Kac's formula expresses the solution of equation (1.81) for general W as follows:
$$\psi(x_b, t_b) = \mathbb{E}_{a,b}\bigg[\exp\bigg(-\int_{t_a}^{t_b} W(w(t))\mathrm{d}t\bigg)\bigg], \tag{1.85}$$
where $\mathbb{E}_{a,b}$ is the integral over the sample paths of the brownian motion conditioned at both ends (see Appendix A)
$$w(t_a) = x_a, \qquad w(t_b) = x_b. \tag{1.86}$$

1.11 Hamiltonian versus lagrangian

Kac's formula has interesting consequences in quantum statistical mechanics. As we saw, the quantum partition function is given by
$$Z(\beta) = \mathrm{Tr}(\mathrm{e}^{-\beta \hat{H}}), \tag{1.87}$$
where $\hat{H}$ is the quantum hamiltonian. If $\hat{H}$ has a discrete spectrum $\{E_n\}$, the formula $Z(\beta) = \sum_n \mathrm{e}^{-\beta E_n}$ allows one to calculate the energy levels E_n from the partition function $Z(\beta)$. For a particle of mass m moving under the influence of a potential V, the hamiltonian operator is
$$\hat{H} = -\frac{\hbar^2}{2m}\Delta_x + V(x). \tag{1.88}$$
The operator $\mathrm{e}^{-\beta \hat{H}}$ is given by a kernel $K_\beta(x_b; x_a)$, hence the partition function is the integral
$$Z(\beta) = \int \mathrm{d}^3\vec{x}\, K_\beta(x; x). \tag{1.89}$$
Using Kac's formula (1.85), we obtain
$$K_\beta(x_b; x_a) = \mathbb{E}_{a,b}\bigg[\exp\bigg(-\int_0^\beta \mathrm{d}\tau\, V(q(\tau))\bigg)\bigg] \tag{1.90}$$
(in the normalized case $\hbar = m = 1$). The integral $\mathbb{E}_{a,b}$ is taken over the sample paths of a brownian motion $q = (q(\tau))_{0 \le \tau \le \beta}$ with boundary

---

[31] We denote by $\mathbb{E}_a$ the mean value with respect to the Wiener measure for paths beginning at $a = (x_a, t_a)$.

conditions

$$q(0) = x_a, \qquad q(\beta) = x_b. \tag{1.91}$$

Hence the integral (1.89) is over all *loops*

$$q(0) = q(\beta) \qquad (=x). \tag{1.92}$$

Formally, Wiener's integral is given by

$$\mathbb{E}_{a,b}[F] = \int \mathcal{D}q \cdot F(q)\exp\left(-\frac{1}{2}\int_0^\beta d\tau\, |\dot{q}(\tau)|^2\right). \tag{1.93}$$

Restoring the constants $\hbar$ and m, we rewrite equation (1.90) as

$$K_\beta(x_b; x_a) = \int \mathcal{D}q \cdot \exp\left(-\frac{1}{\hbar}\int_0^{\beta\hbar} d\tau\, H(q(\tau), \dot{q}(\tau))\right) \tag{1.94}$$

with the *energy function*

$$H(q, \dot{q}) = \frac{m}{2}|\dot{q}|^2 + V(q). \tag{1.95}$$

By comparison, we can rewrite Feynman's formula (1.78) as an expression for the *propagator* $L_u(x_b; x_a)$, that is the kernel of the operator $e^{-iu\hat{H}/\hbar}$:

$$L_u(x_b; x_a) = \int \mathcal{D}q \cdot \exp\left(\frac{i}{\hbar}\int_0^u dt\, L(\dot{q}(t), q(t))\right) \tag{1.96}$$

with the *lagrangian*

$$L(\dot{q}, q) = \frac{m}{2}|\dot{q}|^2 - V(q). \tag{1.97}$$

The riddle is that of how to understand the change from the energy function to the lagrangian. Here is the calculation:

$$-\frac{1}{\hbar}\int_0^{\beta\hbar} d\tau\, H(q, \dot{q}) = \int_0^{\beta\hbar} \left\{-\frac{m}{2\hbar}\frac{|dq|^2}{d\tau} - \frac{1}{\hbar}V(q)d\tau\right\}, \tag{1.98}$$

$$\frac{i}{\hbar}\int_0^u dt\, L(\dot{q}, q) = \int_0^u \left\{\frac{im}{2\hbar}\frac{|dq|^2}{dt} - \frac{i}{\hbar}V(q)dt\right\}. \tag{1.99}$$

The two expressions match if we put formally

$$\tau = it, \qquad \beta\hbar = iu; \tag{1.100}$$

that is, $u = -i\beta\hbar$ is what we called the imaginary Boltzmann time (see equation (1.70)). As we shall see in Chapter 6, these conventions give the right i factor if a magnetic field is available.

Conclusion

To *unify* the formulas, we shall introduce a parameter s as follows:

$s = 1$ for real gaussians, i.e. Wiener–Kac integrals;

$s = i$ for imaginary gaussians, i.e. Feynman's integrals.

Introducing a quadratic form $Q(x)$ corresponding to the action for the path x of a free particle (that is, the time integral of the kinetic energy), in Chapter 2 we define the two kinds of gaussians by their Fourier transforms:

$$\int_{\mathbb{X}} \mathcal{D}_{s,Q}x \cdot \exp\left(-\frac{\pi}{s}Q(x)\right)\exp(-2\pi i\langle x', x\rangle) = \exp(-\pi s W(x')). \quad (1.101)$$

References

[1] P. A. M. Dirac (1977). "The relativistic electron wave equation," *Europhys. News* **8**, 1–4. This is abbreviated from "The relativistic wave equation of electron," *Fiz. Szle* **27**, 443–445 (1977).

[2] P. A. M. Dirac (1933). "The Lagrangian in quantum mechanics," *Phys. Z. Sowjetunion* **3**, 64–72. Also in the collected works of P. A. M. Dirac, ed. R. H. Dalitz (Cambridge, Cambridge University Press, 1995).

[3] P. A. M. Dirac (1947). *The Principles of Quantum Mechanics* (Oxford, Clarendon Press).

[4] R. P. Feynman (1942). *The Principle of Least Action in Quantum Mechanics*, Princeton University Publication No. 2948. Doctoral Dissertation Series, Ann Arbor, MI.

[5] C. DeWitt-Morette (1995). "Functional integration; a semihistorical perspective," in *Symposia Gaussiana*, eds. M. Behara, R. Fritsch, and R. Lintz (Berlin, W. de Gruyter and Co.), pp. 17–24.

[6] P. Lévy (1952). *Problèmes d'analyse fonctionnelle*, 2nd edn. (Paris, Gauthier-Villars).

[7] N. Wiener (1923). "Differential space," *J. Math. Phys.* **2**, 131–174.

[8] C. DeWitt-Morette (1972). "Feynman's path integral; definition without limiting procedure," *Commun. Math. Phys.* **28**, 47–67.
C. DeWitt-Morette (1974). "Feynman path integrals; I. Linear and affine techniques; II. The Feynman–Green function," *Commun. Math. Phys.* **37**, 63–81.

[9] S. A. Albeverio and R. J. Høegh-Krohn (1976). *Mathematical Theory of Feynman Path Integrals* (Berlin, Springer).

[10] Letter from Feynman to Cécile DeWitt, December 10, 1971.

[11] P. Cartier (2000). "Mathemagics" (a tribute to L. Euler and R. Feynman), *Sém. Lothar. Combin.* **44** B44d. 71 pp.

[12] G. 't Hooft and M. J. G. Veltmann (1973). "Diagrammar," CERN Preprint 73–9, pp. 1–114 (Geneva, CERN).

[13] N. Bourbaki (1969). *Eléments de mathématiques, intégration* (Paris, Hermann), Chapter 9. See in particular "Note historique" pp. 113–125. English translation by S. K. Berberian, *Integration II* (Berlin, Springer, 2004).

[14] C. DeWitt-Morette, A. Masheshwari, and B. Nelson (1979). "Path integration in non relativistic quantum mechanics," *Phys. Rep.* **50**, 266–372.

[15] C. DeWitt-Morette (1984). "Feynman path integrals. From the prodistribution definition to the calculation of glory scattering," in *Stochastic Methods and Computer Techniques in Quantum Dynamics*, eds. H. Mitter and L. Pittner, *Acta Phys. Austriaca Supp.* **26**, 101–170. Reviewed in *Zentralblatt für Mathematik* 1985.
C. DeWitt-Morette and B. Nelson (1984). "Glories – and other degenerate critical points of the action," *Phys. Rev.* **D29**, 1663–1668.
C. DeWitt-Morette and T.-R. Zhang (1984). "WKB cross section for polarized glories," *Phys. Rev. Lett.* **52**, 2313–2316.

[16] B. DeWitt (2003). *The Global Approach to Quantum Field Theory* (Oxford, Oxford University Press; with corrections 2004).

[17] R. E. Peierls (1952). "The commutation laws of relativistic field theory," *Proc. Roy. Soc. (London)* **A214**, 143–157.

[18] M. S. Marinov (1993). "Path integrals in phase space," in *Lectures on Path Integration, Trieste 1991*, eds. H. A. Cerdeira, S. Lundqvist, D. Mugnai et al. (Singapore, World Scientific), pp. 84–108.

[19] C. DeWitt-Morette (1993). "Stochastic processes on fibre bundles; their uses in path integration," in *Lectures on Path Integration, Trieste 1991*, eds. H. A. Cerdeira, S. Lundqvist, D. Mugnai et al. (Singapore, World Scientific) (includes a bibliography by topics).

[20] S. Bochner (1955). *Harmonic Analysis and the Theory of Probability* (Berkeley, CA, University of California Press).

Part II
Quantum mechanics

2
First lesson: gaussian integrals

$$\int \mathcal{D}_{s,Q} x \, \exp\left(-\frac{\pi}{s} Q(x)\right) \exp(-2\pi \mathrm{i} \langle x', x \rangle) = \exp(-\pi s W(x'))$$

Given the experience accumulated since Feynman's doctoral thesis, the time has come to extract simple and robust axioms for functional integration from the body of work done during the past sixty years, and to investigate approaches other than those dictated by an action functional.

Here, "simple and robust" means easy and reliable techniques for computing integrals by integration by parts, change of variable of integration, expansions, approximations, etc.

We begin with gaussian integrals in $\mathbb{R}$ and $\mathbb{R}^D$, defined in such a way that their definitions can readily be extended to gaussians in Banach spaces $\mathbb{X}$.

2.1 Gaussians in $\mathbb{R}$

A gaussian random variable and its concomitant the gaussian volume element are marvelous multifaceted tools. We summarize their properties in Appendix C. In the following we focus on properties particularly relevant to functional integrals.

2.2 Gaussians in $\mathbb{R}^D$

Let

$$I_D(a) := \int_{\mathbb{R}^D} \mathrm{d}^D x \, \exp\left(-\frac{\pi}{a} |x|^2\right) \quad \text{for } a > 0, \tag{2.1}$$

with $d^D x := dx^1 \cdots dx^D$ and $|x|^2 = \sum_{j=1}^{D}(x^j)^2 = \delta_{ij}x^i x^j$. From elementary calculus, one gets

$$I_D(a) = a^{D/2}. \tag{2.2}$$

Therefore, when $D = \infty$,

$$I_\infty(a) = \begin{cases} 0 & \text{if } 0 < a < 1, \\ 1 & \text{if } a = 1, \\ \infty & \text{if } 1 < a, \end{cases} \tag{2.3}$$

which is clearly an unsatisfactory situation, but it can be corrected by introducing a volume element $D_a x$ scaled by the parameter a as follows:

$$D_a x := \frac{1}{a^{D/2}} dx^1 \cdots dx^D. \tag{2.4}$$

The volume element $D_a x$ can be characterized by the integral

$$\int_{\mathbb{R}^D} D_a x \, \exp\left(-\frac{\pi}{a}|x|^2 - 2\pi i \langle x', x \rangle\right) := \exp(-a\pi |x'|^2), \tag{2.5}$$

where x' is in the dual $\mathbb{R}_D$ of $\mathbb{R}^D$. A point $x \in \mathbb{R}^D$ is a contravariant (or column) vector. A point $x' \in \mathbb{R}_D$ is a covariant (or row) vector.

The integral (2.5) suggests the following definition of a volume element $d\Gamma_a(x)$:

$$\int_{\mathbb{R}^D} d\Gamma_a(x) \exp(-2\pi i \langle x', x \rangle) := \exp(-a\pi |x'|^2). \tag{2.6}$$

Here we can write

$$d\Gamma_a(x) = D_a x \, \exp\left(-\frac{\pi}{a}|x|^2\right). \tag{2.7}$$

This equality is meaningless in infinite dimensions; however, the integrals (2.5) and (2.6) remain meaningful. We introduce a different equality symbol:

$$\overset{\int}{=}, \tag{2.8}$$

which is a qualified equality in integration theory; e.g. the expression

$$d\Gamma_a(x) \overset{\int}{=} D_a x \, \exp\left(-\frac{\pi}{a}|x|^2\right) \tag{2.9}$$

indicates that both sides of the equality are equal after integration (compare (2.5) and (2.6)).

A linear map $A : \mathbb{R}^D \to \mathbb{R}^D$ given by

$$y = Ax, \quad \text{i.e. } y^j = A^j{}_i \, x^i, \tag{2.10}$$

2.2 Gaussians in $\mathbb{R}^D$

transforms the quadratic form $\delta_{ij}y^iy^j$ into a general positive quadratic form

$$Q(x) = \delta_{ij}A^i{}_k A^j{}_\ell x^k x^\ell =: Q_{k\ell}x^k x^\ell. \qquad (2.11)$$

Consequently a linear change of variable in the integral (2.5) can be used for defining the gaussian volume element $\mathrm{d}\Gamma_{a,Q}$ with respect to the quadratic form (2.11). We begin with the definition

$$\int_{\mathbb{R}^D} D_a y \exp\left(-\frac{\pi}{a}|y|^2 - 2\pi i \langle y', y \rangle\right) := \exp(-a\pi |y'|^2). \qquad (2.12)$$

Under the change of variable $y = Ax$, the volume element

$$D_a y = a^{-D/2}\, \mathrm{d}y^1 \cdots \mathrm{d}y^D$$

becomes

$$\begin{aligned} D_{a,Q}x &= a^{-D/2}|\det A|\, \mathrm{d}x^1 \cdots \mathrm{d}x^D \\ &= |\det(Q/a)|^{1/2}\, \mathrm{d}x^1 \cdots \mathrm{d}x^D. \end{aligned} \qquad (2.13)$$

The change of variable

$$y'_j = x'_i B^i{}_j \qquad (2.14)$$

(shorthand $y' = x'B$) defined by transposition

$$\langle y', y \rangle = \langle x', x \rangle, \qquad \text{i.e. } y'_j y^j = x'_i x^i, \qquad (2.15)$$

implies

$$B^i{}_j A^j{}_k = \delta^i{}_k. \qquad (2.16)$$

Equation (2.12) now reads

$$\int_{\mathbb{R}^D} D_{a,Q}x \exp\left(-\frac{\pi}{a}Q(x) - 2\pi i \langle x', x \rangle\right) := \exp(-a\pi W(x')), \qquad (2.17)$$

where

$$\begin{aligned} W(x') &= \delta^{ij} x'_k x'_\ell B^k{}_i B^\ell{}_j, \\ &=: x'_k x'_\ell W^{k\ell}. \end{aligned} \qquad (2.18)$$

The quadratic form $W(x') = x'_k x'_\ell W^{k\ell}$ on $\mathbb{R}_D$ can be said to be the inverse of $Q(x) = Q_{k\ell}x^k x^\ell$ on $\mathbb{R}^D$ since the matrices $(W^{k\ell})$ and $(Q_{k\ell})$ are inverse to each other.

In conclusion, in $\mathbb{R}^D$, the gaussian volume element defined in (2.17) by the quadratic form aW is

$$\mathrm{d}\Gamma_{a,Q}(x) = D_{a,Q}x \exp\left(-\frac{\pi}{a}Q(x)\right) \qquad (2.19)$$

$$= \mathrm{d}x^1 \ldots \mathrm{d}x^D \left(\det_{k,\ell} \frac{Q_{k\ell}}{a}\right)^{1/2} \exp\left(-\frac{\pi}{a}Q(x)\right). \qquad (2.20)$$

Remark. The volume element $D_a x$ has been chosen so as to be without physical dimension. In Feynman's dissertation, the volume element $\mathcal{D}x$ is the limit for $D = \infty$ of the discretized expression

$$\mathcal{D}x = \prod_i \mathrm{d}x(t_i) A^{-1}(\delta t_i). \tag{2.21}$$

The normalization factor was determined by requiring that the wave function for a free particle of mass m moving in one dimension be continuous. It was found to be

$$A(\delta t_k) = (2\pi i \hbar\, \delta t_k / m)^{1/2}. \tag{2.22}$$

A general expression for the absolute value of the normalization factor was determined by requiring that the short-time propagators be unitary [1]. For a system with action function $\mathcal{S}$, and paths taking their values in an n-dimensional configuration space,

$$|A(\delta t_k)| = \left| \det_{k,\ell} (2\pi\hbar)^{-1} \frac{\partial^2 \mathcal{S}(x^\mu(t_{k+1}), x^\nu(t_k))}{\partial x^\mu(t_{k+1}) \partial x^\nu(t_k)} \right|^{-1/2}. \tag{2.23}$$

The "intractable" product of the infinite number of normalization factors was found to be a Jacobian [1] later encountered by integrating out momenta from phase-space path integrals. Equation (2.5) suggests equation (2.17) and equation (2.19) in which $D_{a,Q}(x)$ provides a volume element that can be generalized to infinite-dimensional spaces without working through an infinite product of short-time propagators.

2.3 Gaussians on a Banach space

In infinite dimensions, the reduction of a quadratic form to a sum of squares of linear forms (see formula (2.11)) is very often inconvenient, and shall be bypassed. Instead, we take formulas (2.17) and (2.19) as our starting point.

The set-up

We denote by $\mathbb{X}$ a Banach space, which may consist of paths

$$x : \mathbb{T} \to \mathbb{R}^D, \tag{2.24}$$

where $\mathbb{T} = [t_a, t_b]$ is a time interval, and $\mathbb{R}^D$ is the configuration space in quantum mechanics. In quantum field theory, $\mathbb{X}$ may consist of fields, that is functions

$$\phi : \mathbb{M}^D \to \mathbb{C} \tag{2.25}$$

for scalar fields, or

$$\phi : \mathbb{M}^D \to \mathbb{C}^\nu \tag{2.26}$$

2.3 Gaussians on a Banach space

for spinor or tensor fields, where $\mathbb{M}^D$ is the physical space (or spacetime). To specify $\mathbb{X}$, we take into account suitable smoothness and/or boundary conditions on x, or on ϕ.

We denote by $\mathbb{X}'$ the dual of $\mathbb{X}$, that is the Banach space consisting of the continuous linear forms $x' : \mathbb{X} \to \mathbb{R}$; by $\langle x', x \rangle$ we denote the value taken by x' in $\mathbb{X}'$ on the vector x in $\mathbb{X}$.

Our formulas require the existence of two quadratic forms, $Q(x)$ for x in $\mathbb{X}$ and $W(x')$ for x' in $\mathbb{X}'$. By generalizing the fact that the matrices (Q_{kl}) in (2.11) and (W^{kl}) in (2.18) are inverse to each other, we require that the quadratic forms Q and W be inverse of each other in the following sense.

There exist two continuous linear maps

$$D : \mathbb{X} \to \mathbb{X}', \qquad G : \mathbb{X}' \to \mathbb{X}$$

with the following properties.

- They are the inverses of each other:

$$DG = 1 \quad (\text{on } \mathbb{X}'), \qquad GD = 1 \quad (\text{on } \mathbb{X}). \qquad (2.27)$$

- They are symmetric:

$$\langle Dx, y \rangle = \langle Dy, x \rangle \qquad \text{for } x, y \text{ in } \mathbb{X},$$
$$\langle x', Gy' \rangle = \langle y', Gx' \rangle \qquad \text{for } x', y' \text{ in } \mathbb{X}'.$$

- The quadratic forms are given by

$$Q(x) = \langle Dx, x \rangle, \qquad W(x') = \langle x', Gx' \rangle. \qquad (2.28)$$

We set also $W(x', y') := \langle x', Gy' \rangle$ for x', y' in $\mathbb{X}'$.

Definition

A gaussian volume element on the Banach space $\mathbb{X}$ is defined by its Fourier transform $\mathcal{F}\Gamma_{a,Q}$, namely

$$(\mathcal{F}\Gamma_{a,Q})(x') := \int_{\mathbb{X}} d\Gamma_{a,Q}(x) \exp(-2\pi i \langle x', x \rangle) = \exp(-a\pi W(x')) \quad (2.29)$$

for x' arbitrary in $\mathbb{X}'$. We also define formally a volume element $\mathcal{D}_{a,Q}x$ in $\mathbb{X}$ by[1]

$$d\Gamma_{a,Q}(x) \stackrel{\int}{=} \mathcal{D}_{a,Q}(x) \exp\left(-\frac{\pi}{a} Q(x)\right). \qquad (2.30)$$

---

[1] We use the qualified equality symbol $\stackrel{\int}{=}$ for terms that are equal after integration.

So far we have been working with $d\Gamma_{a,Q}$, where a was a positive number. **As long as gaussians are defined by their Fourier transforms we can replace a by $s \in \{1, i\}$. Hence we rewrite (2.29) and (2.30):**

$$(\mathcal{F}\Gamma_{s,Q})(x') := \int_{\mathbb{X}} d\Gamma_{s,Q}(x)\exp(-2\pi i \langle x', x\rangle) = \exp(-s\pi W(x')), \quad (2.29)_s$$

$$d\Gamma_{s,Q}(x) \stackrel{f}{=} \mathcal{D}_{s,Q}(x)\exp\left(-\frac{\pi}{s}Q(x)\right). \quad (2.30)_s$$

Important remark. Because of the presence of i in the exponent of the Feynman integral, it was (and occasionally still is) thought that the integral could not be made rigorous. The gaussian definition $(2.29)_s$ is rigorous for $Q(x) \geq 0$ when $s = 1$, and for $Q(x)$ real when $s = i$.

Physical interpretation

In our definitions, the case $s = 1$ (or $a > 0$) corresponds to problems in statistical mechanics, whereas the case $s = i$ corresponds to quantum physics via the occurrence of the phase factor $\exp[(i/\hbar)S(\psi)]$.

The volume-element definition corresponding to $(2.29)_s$ and $(2.30)_s$ can be written

$$\int \mathcal{D}\psi \exp\left(\frac{i}{\hbar}S(\psi) - i\langle J, \psi\rangle\right) = \exp\left(\frac{i}{\hbar}W(J)\right) = Z(J), \quad (2.31)$$

where ψ is either a self-interacting field, or a collection of interacting fields. But the generating functional $Z(J)$ is difficult to ascertain a priori for the following reason. Let $\Gamma(\bar\psi)$ be the Legendre transform of $W(J)$. For given $\bar\psi$, $J(\bar\psi)$ is the solution of the equation $\hbar\bar\psi = \delta W(J)/\delta J$ and

$$\Gamma(\bar\psi) := W(J(\bar\psi)) - \hbar\langle J(\bar\psi), \bar\psi\rangle. \quad (2.32)$$

Then $\Gamma(\bar\psi)$ is the inverse of $W(J)$ in the same sense as Q and W are the inverses of each other, (2.27) and (2.28), but $\Gamma(\bar\psi)$ is the *effective action* which is used for computing observables. If $S(\psi)$ is quadratic, the *bare action* $S(\psi)$ and the effective action $\Gamma(\psi)$ are identical, and the fields do not interact. In the case of interacting fields, the exact relation between the bare and effective actions is the main problem of quantum field theory (see Chapters 15–18).

Examples

In this chapter we define a volume element on a space Φ of fields ϕ on $\mathbb{M}^D$ by the equation

$$\int_\Phi \mathcal{D}_{s,Q}\phi \cdot \exp\left(-\frac{\pi}{s}Q(\phi)\right)\exp(-2\pi i\langle J, \phi\rangle) := \exp(-\pi s W(J)) \quad (2.33)$$

2.3 Gaussians on a Banach space

for given Q and W that are inverses of each other. For convenience, we will define instead the volume element $\mathrm{d}\mu_G$ by[2]

$$\int_\Phi \mathrm{d}\mu_G(\phi)\exp(-2\pi\mathrm{i}\langle J,\phi\rangle) := \exp(-\pi s W(J)). \qquad (2.34)$$

As before, W is defined by the covariance[3] G:

$$W(J) = \langle J, GJ\rangle; \qquad (2.35)$$

G is the inverse of the operator D defined by

$$Q(\phi) = \langle D\phi, \phi\rangle. \qquad (2.36)$$

It is also the two-point function

$$\frac{s}{2\pi}G(x,y) = \int_\Phi \mathrm{d}\mu_G(\phi)\phi(x)\phi(y). \qquad (2.37)$$

We shall construct covariances in quantum mechanics and quantum field theory in two simple examples.

In quantum mechanics: let $D = -\mathrm{d}^2/\mathrm{d}t^2$; its inverse on the space $\mathbb{X}_{ab}$ of paths with two fixed end points is an integral operator with kernel given by[4]

$$G(t,s) = \theta(s-t)(t-t_a)(t_b-t_a)^{-1}(t_b-s)$$
$$-\theta(t-s)(t-t_b)(t_a-t_b)^{-1}(t_a-s). \qquad (2.38)$$

In quantum field theory: let D be the operator $-\Sigma(\mathrm{d}/\mathrm{d}x j)^2$ on $\mathbb{R}^D$ (for $D \geq 3$); then the kernel of the integral operator G is given by

$$G(x,y) = \frac{C_D}{|x-y|^{D-2}}, \qquad (2.39)$$

with a constant C_D equal to

$$\Gamma\left(\frac{D}{2}-1\right)\Big/\left(4\pi^{D/2}\right). \qquad (2.40)$$

Notice that $G(t,s)$ is a continuous function. The function $G(x,y)$ is singular at the origin for euclidean fields and on the lightcone for minkowskian

---

[2] Hence $\mathrm{d}\mu_G$ is the same as $\mathrm{d}\Gamma_{s,Q}$, but with the emphasis now placed on the covariance G. Notice that the formulas (2.34), (2.35) and (2.37) retain their meaning for a symmetrical linear map $G : \mathbb{X}' \to \mathbb{X}$, which is *not necessarily invertible*.
[3] Even if the map G is not invertible. See e.g. (2.79).
[4] We denote by $\theta(u)$ the Heaviside function

$$\theta(u) = \begin{cases} 1 & \text{for } u > 0, \\ 0 & \text{for } u < 0, \\ \text{undefined} & \text{for } u = 0. \end{cases}$$

fields. However, we note that *the quantity of interest is not the covariance G, but the variance W*:

$$W(J) = \langle J, GJ \rangle,$$

which is singular only if J is a point-like source,

$$\langle J, \phi \rangle = c \cdot \phi(x_0),$$

where c is a constant and x_0 is a fixed point.

2.4 Variances and covariances

The quadratic form W on $\mathbb{X}'$ that characterizes the Fourier transform $\mathcal{F}\Gamma_{s,Q}$ of the gaussian which in turn characterizes the gaussian $\Gamma_{s,Q}$ is known in probability theory as the variance. The kernel G in (2.28) is known as the covariance of the gaussian distribution. In quantum theory G is the propagator of the system. It is also the "two-point function" since (2.47) gives as the particular case $n = 1$

$$\int_{\mathbb{X}} d\Gamma_{s,Q}(x)\langle x_1', x\rangle\langle x_2', x\rangle = \frac{s}{2\pi}W(x_1', x_2'). \qquad (2.41)$$

The exponential $\exp(-s\pi W(x'))$ is a generating functional, which yields the moments (2.43) and (2.44) and the polarization (2.47). It has been used extensively by Schwinger, who considers the term $\langle x', x \rangle$ in $(2.29)_s$ as a source.

In this section, we work only with the variance W. In Chapter 3 we work with gaussian volume elements, i.e. with the quadratic form Q on $\mathbb{X}$. In other words we move from the algebraic theory of gaussians (Chapter 2) to their differential theory (Chapter 3), which is commonly used in physics.

Moments

The integral of polynomials with respect to a gaussian volume element follows readily from the definition $(2.29)_s$ after replacing x' by $cx'/(2\pi i)$, i.e.

$$\int_{\mathbb{X}} d\Gamma_{s,Q}(x)\exp(-c\langle x', x\rangle) = \exp[c^2 sW(x')/(4\pi)]. \qquad (2.42)$$

Expanding both sides in powers of c yields

$$\int_{\mathbb{X}} d\Gamma_{s,Q}(x)\langle x', x\rangle^{2n+1} = 0 \qquad (2.43)$$

2.4 Variances and covariances

and

$$\int_X d\Gamma_{s,Q}(x)\langle x', x\rangle^{2n} = \frac{2n!}{n!}\left(\frac{sW(x')}{4\pi}\right)^n$$
$$= \frac{2n!}{2^n n!}\left(\frac{s}{2\pi}\right)^n W(x')^n. \tag{2.44}$$

Hint. $W(x')$ is an abbreviation of $W(x', x')$; therefore, nth-order terms in expanding the right-hand side are equal to $(2n)$th-order terms of the left-hand side.

Polarization 1

The integral of a multilinear expression,

$$\int_X d\Gamma_{s,Q}(x)\langle x'_1, x\rangle \cdots \langle x'_{2n}, x\rangle, \tag{2.45}$$

can readily be computed. Replacing x' in the definition $(2.29)_s$ by the linear combination $c_1 x'_1 + \cdots + c_{2n} x'_{2n}$ and equating the $c_1 c_2 \cdots c_{2n}$ terms on the two sides of the equation yields

$$\int_X d\Gamma_{s,Q}(x)\langle x'_1, x\rangle \cdots \langle x'_{2n}, x\rangle$$
$$= \frac{1}{2^n n!}\left(\frac{s}{2\pi}\right)^n \sum W(x'_{i_1}, x'_{i_2}) \cdots W(x'_{i_{2n-1}}, x'_{i_{2n}}), \tag{2.46}$$

where the sum is performed over all possible distributions of the arguments. However each term occurs $2^n n!$ times in this sum since $W(x'_{i_j}, x'_{i_k}) = W(x'_{i_k}, x'_{i_j})$ and since the product order is irrelevant. Finally[5]

$$\int_X d\Gamma_{s,Q}(x)\langle x'_1, x\rangle \cdots \langle x'_{2n}, x\rangle$$
$$= \left(\frac{s}{2\pi}\right)^n {\sum}' W(x'_{i_1}, x'_{i_2}) \cdots W(x'_{i_{2n-1}}, x'_{i_{2n}}), \tag{2.47}$$

where $\sum'$ is a sum without repetition of identical terms.[6]

Example. If $2n = 4$, the sum consists of three terms, which can be recorded by three diagrams as follows. Let $1, 2, 3, 4$ designate x'_1, x'_2, x'_3, x'_4, respectively, and a line from i_1 to i_2 records $W(x'_{i_1}, x'_{i_2})$. Then the sum in (2.47) is recorded by the three diagrams in figure 2.1.

---

[5] Corrected by Leila Javindpoor.
[6] For instance, we can assume the inequalities

$$i_1 < i_2, i_3 < i_4, \ldots, i_{2n-1} < i_{2n} \quad \text{and} \quad i_1 < i_3 < i_5 < \ldots < i_{2n-1}$$

in the summation.

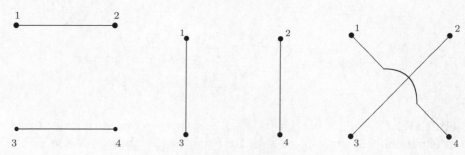

Fig. 2.1 Diagrams

Polarization 2

The following proof of the polarization formula (2.47) belongs also to several other chapters:

- Chapter 11, where the integration by parts (2.48) is justified;
- Chapter 3, where we introduce (in Section 3.3) the quadratic form Q on $\mathbb{X}$ which is the inverse of the quadratic form W on $\mathbb{X}'$ in the sense of equations (2.27) and (2.28).

Given the qualified equality $(2.30)_s$,

$$d\Gamma_{s,Q}(x) \stackrel{s}{=} \mathcal{D}_{s,Q}(x)\exp\left(-\frac{\pi}{s}Q(x)\right),$$

the gaussian defined in terms of W by $(2.29)_s$ is then expressed in terms of Q by $(2.30)_s$.

We consider the case in which $\mathbb{X}$ consists of paths $x = (x(t))_{t_a \leq t \leq t_b}$ in a one-dimensional space. Furthermore, $D = D_t$ is a differential operator.

The basic integration-by-parts formula

$$\int_{\mathbb{X}} \mathcal{D}_{s,Q}(x)\exp\left(-\frac{\pi}{s}Q(x)\right)\frac{\delta F(x)}{\delta x(t)}$$
$$:= -\int_{\mathbb{X}} \mathcal{D}_{s,Q}(x)\exp\left(-\frac{\pi}{s}Q(x)\right)F(x)\frac{\delta}{\delta x(t)}\left(-\frac{\pi}{s}Q(x)\right) \quad (2.48)$$

yields the polarization formula (2.47) when

$$Q(x) = \int_{t_a}^{t_b} dr\, Dx(r) \cdot x(r). \quad (2.49)$$

Indeed,

$$-\frac{\delta}{\delta x(t)}\frac{\pi}{s}Q(x) = -2\frac{\pi}{s}\int_{t_a}^{t_b} dr\, Dx(r)\delta(r-t) = -\frac{2\pi}{s}D_t x(t). \quad (2.50)$$

2.4 Variances and covariances

When
$$F(x) = x(t_1)\ldots x(t_n), \tag{2.51}$$
then
$$\frac{\delta F(x)}{\delta x(t)} = \sum_{i=1}^{n} \delta(t - t_i) x(t_1) \ldots \hat{x}(t_i) \ldots x(t_n), \tag{2.52}$$

where a $\hat{}$ over a term means that the term is deleted. The n-point function with respect to the quadratic action $S = \frac{1}{2}Q$ is by definition

$$G_n(t_1,\ldots,t_n) := \left(\frac{2\pi}{s}\right)^{n/2} \int \mathcal{D}_{s,Q}(x) \exp\left(-\frac{\pi}{s} Q(x)\right) x(t_1)\ldots x(t_n). \tag{2.53}$$

Therefore the left-hand side of the integration-by-parts formula (2.48) is

$$\int_{\mathbb{X}} \mathcal{D}_{s,Q}(x) \exp\left(-\frac{\pi}{s} Q(x)\right) \frac{\delta F(x)}{x(t)}$$
$$= \left(\frac{s}{2\pi}\right)^{(n-1)/2} \sum_{i=1}^{n} \delta(t - t_i) G_{n-1}(t_1,\ldots,\hat{t}_i,\ldots,t_n). \tag{2.54}$$

Given (2.50), the right-hand side of (2.48) is

$$-\int_{\mathbb{X}} \mathcal{D}_{s,Q}(x) \exp\left(-\frac{\pi}{s} Q(x)\right) x(t_1)\ldots x(t_n) \left(-\frac{2\pi}{s} D_t x(t)\right)$$
$$= \frac{2\pi}{s} \left(\frac{s}{2\pi}\right)^{(n+1)/2} D_t G_{n+1}(t, t_1,\ldots,t_n). \tag{2.55}$$

The integration-by-parts formula (2.48) yields a recurrence formula for the n-point functions, G_n, namely

$$D_t G_{n+1}(t, t_1,\ldots,t_n) = \sum_{i=1}^{n} \delta(t - t_i) G_{n-1}(t_1,\ldots,\hat{t}_i,\ldots,t_n); \tag{2.56}$$

equivalently (replace n by $n-1$)

$$D_{t_1} G_n(t_1,\ldots,t_n) = \sum_{i=2}^{n} \delta(t_1 - t_i) G_{n-2}(\hat{t}_1,\ldots,\hat{t}_i,\ldots,t_n). \tag{2.57}$$

The solution is given by the following rules:
- the two-point function G_2 is a solution of the differential equation
$$D_{t_1} G_2(t_1, t_2) = \delta(t_1 - t_2);$$
- the m-point function is 0 for m odd; and

- for $m = 2n$ even, the m-point function is given by

$$G_{2n}(t_1,\ldots,t_{2n}) = \sum G_2(t_{i_1}, t_{i_2})\ldots G_2(t_{i_{2n-1}}, t_{i_{2n}})$$

with the same restrictions on the sum as in (2.47).

Linear maps

Let $\mathbb{X}$ and $\mathbb{Y}$ be two Banach spaces, possibly two copies of the same space. Let L be a linear continuous map $L : \mathbb{X} \to \mathbb{Y}$ by $x \mapsto y$ and $\tilde{L} : \mathbb{Y}' \to \mathbb{X}'$ by $y' \mapsto x'$ defined by

$$\langle \tilde{L}y', x \rangle = \langle y', Lx \rangle. \tag{2.58}$$

If L maps $\mathbb{X}$ onto $\mathbb{Y}$, then we can associate with a gaussian $\Gamma_\mathbb{X}$ on $\mathbb{X}$ another gaussian $\Gamma_\mathbb{Y}$ on $\mathbb{Y}$ such that the Fourier transforms $\mathcal{F}\Gamma_\mathbb{X}$ and $\mathcal{F}\Gamma_\mathbb{Y}$ on $\mathbb{X}$ and $\mathbb{Y}$, respectively, satisfy the equation

$$\mathcal{F}\Gamma_\mathbb{Y} = \mathcal{F}\Gamma_\mathbb{X} \circ \tilde{L}, \tag{2.59}$$

i.e. for the variances

$$W_{\mathbb{Y}'} = W_{\mathbb{X}'} \circ \tilde{L}. \tag{2.60}$$

If the map L is invertible, then we have similarly $Q_\mathbb{X} = Q_\mathbb{Y} \circ L$, but there exists no simple relation between $Q_\mathbb{X}$ and $Q_\mathbb{Y}$ when L is not invertible. The diagram in figure 2.2 will be used extensively.

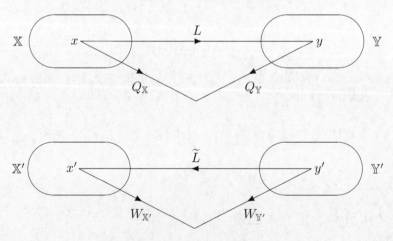

Fig. 2.2 Linear maps

2.5 Scaling and coarse-graining

In this section, we exploit the scaling properties of gaussian volume elements on spaces $\boldsymbol{\Phi}$ of fields ϕ on $\mathbb{M}^D$. These properties are valid for vector

2.5 Scaling and coarse-graining

space $\mathbb{M}^D$ with either euclidean or minkowskian signature. These properties are applied to the $\lambda\phi^4$ system in Section 16.2.

The gaussian volume element μ_G is defined according to the conventions described in formulas (2.34) and (2.35). The covariance G is the two-point function (2.37). Objects defined by the covariance G include (see Wick calculus in Appendix D) the following:

- convolution with volume element μ_G

$$(\mu_G * F)(\phi) := \int_\mathbb{X} \mathrm{d}\mu_G(\psi) F(\phi + \psi), \tag{2.61}$$

which yields

$$\mu_G * F = \exp\left(\frac{s}{4\pi}\Delta_G\right) F, \quad s \in \{1, \mathrm{i}\}, \tag{2.62}$$

where the functional Laplacian

$$\Delta_G := \int_{\mathbb{M}^D} \mathrm{d}^D x \int_{\mathbb{M}^D} \mathrm{d}^D y\, G(x,y) \frac{\delta^2}{\delta\phi(x)\delta\phi(y)}; \tag{2.63}$$

- the Bargmann–Segal transform defined by

$$B_G := \mu_G * = \exp\left(\frac{s}{4\pi}\Delta_G\right); \tag{2.64}$$

- and the Wick transform

$$:\ :_G\ := \exp\left(-\frac{s}{4\pi}\Delta_G\right). \tag{2.65}$$

Scaling

The scaling properties of covariances can be used for investigating the transformation (or the invariance) of some quantum systems under a change of scale.

The definition of a gaussian volume element μ_G (2.34) in quantum field theory reads

$$\int_\Phi \mathrm{d}\mu_G(\phi)\exp(-2\pi\mathrm{i}\langle J, \phi\rangle) := \exp(-\pi\mathrm{i}W(J)), \tag{2.66}$$

where ϕ is a field on spacetime (Minkowski or euclidean). The gaussian μ_G of covariance G can be decomposed into the convolution of any number of gaussians. For example, if

$$W = W_1 + W_2 \tag{2.67}$$

then

$$G = G_1 + G_2 \tag{2.68}$$

and
$$\mu_G = \mu_{G_1} * \mu_{G_2}. \qquad (2.69)$$

The convolution (2.69) can be defined as follows:

$$\int_\Phi \mathrm{d}\mu_G(\phi)\exp(-2\pi\mathrm{i}\langle J,\phi\rangle)$$
$$= \int_\Phi \mathrm{d}\mu_{G_2}(\phi_2) \int_\Phi \mathrm{d}\mu_{G_1}(\phi_1)\exp(-2\pi\mathrm{i}\langle J,\phi_1+\phi_2\rangle). \qquad (2.70)$$

Formally, this amounts to a decomposition

$$\phi = \phi_1 + \phi_2, \qquad (2.71)$$

where ϕ_1 and ϕ_2 are stochastically independent.

The additive property (2.68) makes it possible to express a covariance G as an integral over an independent scale variable. Let $\lambda \in [0,\infty[$ be an independent scale variable.[7] A scale variable has no physical dimension,

$$[\lambda] = 0. \qquad (2.72)$$

The scaling operator S_λ acting on a function f of physical length dimension $[f]$ is by definition

$$S_\lambda f(x) := \lambda^{[f]} f\left(\frac{x}{\lambda}\right), \qquad x \in \mathbb{R} \quad \text{or } x \in \mathbb{M}^D. \qquad (2.73)$$

A physical dimension is often given in powers of mass, length, and time. Here we set $\hbar = 1$ and $c = 1$, and the physical dimensions are physical *length* dimensions. We choose the length dimension rather than the more conventional mass dimension because we define fields on coordinate space, not on momentum space. The subscript of the scaling operator has no dimension.

The scaling of an interval $[a,b[$ is given by

$$S_\lambda[a,b[= \left\{\frac{s}{\lambda} \Big| s \in [a,b[\right\}, \qquad \text{i.e. } S_\lambda[a,b[= \left[\frac{a}{\lambda},\frac{b}{\lambda}\right[. \qquad (2.74)$$

By definition the (dimensional) scaling of a functional F is

$$(S_\lambda F)(\phi) = F(S_\lambda \phi). \qquad (2.75)$$

We use for lengths multiplicative differentials that are scale invariant:

$$\mathrm{d}^\times l = \mathrm{d}l/l, \qquad (2.76)$$
$$\partial^\times/\partial l = l\,\partial/\partial l. \qquad (2.77)$$

---

[7] Brydges *et al.* use $\lambda \in [1,\infty[$ and $\lambda^{-1} \in [0,1[$.

2.5 Scaling and coarse-graining

Scaled covariances

In order to control infrared problems at large distances, and ultraviolet divergences at short distances, in euclidean[8] field theory, of the covariance

$$G(x,y) = C_D/|x-y|^{D-2}, \qquad x,y \in \mathbb{R}^D, \tag{2.78}$$

one introduces a scaled (truncated) covariance

$$G_{[l_0,l[}(x,y) := \int_{l_0}^{l} d^\times s \cdot s^{2[\phi]} u\left(\frac{|x-y|}{s}\right), \tag{2.79}$$

where the length dimensions of the various symbols are

$$[l] = 1, \quad [l_0] = 1, \quad [s] = 1, \quad [u] = 0, \quad [G] = 2 - D. \tag{2.80}$$

In agreement with (2.37),

$$[G] = 2[\phi]. \tag{2.81}$$

The function u is chosen so that

$$\lim_{l_0=0, l=\infty} G_{[l_0,l[}(x,y) = G(x,y). \tag{2.82}$$

For $G(x,y)$ given by (2.78) the only requirement on u is

$$\int_0^\infty d^\times r \cdot r^{-2[\phi]} u(r) = C_D. \tag{2.83}$$

Example. For the ordinary laplacian

$$\Delta = -\sum_j (\partial/\partial x^j)^2, \tag{2.84}$$

the covariance G satisfies the equation

$$\Delta G = G\Delta = \mathbf{1} \tag{2.85}$$

if and only if the constant C_D is given by

$$C_D = \Gamma(D/2 - 1)/(4\pi^{D/2}). \tag{2.86}$$

The decomposition of the covariance G into scale-dependent contributions (2.79) is also written

$$G = \sum_{j=-\infty}^{+\infty} G_{[2^j l_0, 2^{j+1} l_0[}. \tag{2.87}$$

The contributions are self-similar in the following sense. According to (2.79) and using the definition $\xi = |x-y|$,

$$G_{[a,b[}(\xi) = \int_{[a,b[} d^\times s \cdot s^{2[\phi]} u(\xi/s); \tag{2.88}$$

---

[8] For the Minkowski case see Chapter 16 and [2].

given a scale parameter λ, replace the integration variable s by λs, hence

$$G_{[a,b[}(\xi) = \int_{a/\lambda}^{b/\lambda} d^\times s \, \lambda^{2[\phi]} s^{2[\phi]} u(\xi/\lambda s).$$

Hence, by (2.88),

$$G_{[a,b[}(\xi) = \lambda^{[2\phi]} G_{[a/\lambda, b/\lambda[}(\xi/\lambda). \qquad (2.89)$$

Henceforth the suffix G in the objects defined by covariances such as μ_G, Δ_G, B_G, : :$_G$ is replaced by the interval defining the scale-dependent covariance.

Example. Convolution

$$\mu_{[l_0,\infty[} * F = \mu_{[l_0,l[} * \left(\mu_{[l,\infty[} * F \right) \qquad (2.90)$$

for any functional F of the fields. That is, in terms of the Bargmann–Segal transform (2.64),

$$B_G = B_{G_1} B_{G_2}, \qquad (2.91)$$

where G, G_1, and G_2 correspond, respectively, to the intervals $[l_0, \infty[$, $[l_0, l[$, and $[l, \infty[$.

To the covariance decomposition (2.87) corresponds, according to (2.71), the field decomposition

$$\phi = \sum_{j=-\infty}^{+\infty} \phi_{[2^j l_0, 2^{j+1} l_0[}. \qquad (2.92)$$

We also write

$$\phi(x) = \sum_{j=-\infty}^{+\infty} \phi_j(l_0, x), \qquad (2.93)$$

where the component fields $\phi_j(l_0, x)$ are stochastically independent.

Brydges' coarse-graining operator P_l

D. C. Brydges, J. Dimock, and T. R. Hurd [3] introduced and developed the properties of a coarse-graining operator P_l, which rescales the Bargmann–Segal transform so that all integrals are performed with a scale-independent gaussian:

$$P_l F := S_{l/l_0} B_{[l_0,l[} F := S_{l/l_0} \left(\mu_{[l_0,l[} * F \right). \qquad (2.94)$$

Here l_0 is a fixed length and l runs over $[l_0, \infty[$.

2.5 Scaling and coarse-graining

The six following properties of the coarse-graining operator are frequently used in Chapter 16:

(i) P_l obeys a multiplicative semigroup property. Indeed,

$$P_{l_2} P_{l_1} = P_{l_2 l_1 / l_0} \qquad (2.95)$$

whenever $l_1 \geq l_0$ and $l_2 \geq l_0$.

Proof of multiplicative property (2.95)[9]:

$$P_{l_2} P_{l_1} F = S_{l_2/l_0}(\mu_{[l_0, l_2[} * (S_{l_1/l_0}(\mu_{[l_0, l_1[} * F)))$$
$$= S_{l_2/l_0} S_{l_1/l_0} (\mu_{[l_0 l_1 / l_0, l_2 l_1 / l_0]} * (\mu_{[l_0, l_1[} * F))$$
$$= S_{\frac{l_2 l_1}{l_0} \frac{1}{l_0}} \left(\mu_{[l_0, \frac{l_2 l_1}{l_0}[} * F \right).$$

$\square$

(ii) P_l does not define a group because convolution does not have an inverse. Information is lost by convolution. Physically, information is lost by integrating over some degrees of freedom.

(iii) Wick ordered monomials defined by (2.65) and in Appendix D are (pseudo-)eigenfunctions of the coarse-graining operator:

$$P_l \int_{\mathbb{M}^D} d^D x : \phi^n(x) :_{[l_0, \infty[}$$
$$= \left(\frac{l}{l_0}\right)^{n[\phi]+D} \int_{\mathbb{M}^D} d^D x : \phi^n(x) :_{[l_0, \infty[}. \qquad (2.96)$$

If the integral is over a finite volume, the volume is scaled down by S_{l/l_0}. Hence we use the expression "pseudo-eigenfunction" rather than "eigenfunction."

Proof of eigenfunction equation (2.96):

$$P_l : \phi^n(x) :_{[l_0, \infty[} = S_{l/l_0} \left(\mu_{[l_0, l[} * \left(\exp\left(-\frac{s}{4\pi} \Delta_{[l_0, \infty[}\right) \phi^n(x) \right) \right)$$
$$= S_{l/l_0} \left(\exp\left(\frac{s}{4\pi} \Delta_{[l_0, l[} - \frac{s}{4\pi} \Delta_{[l_0, \infty[}\right) \phi^n(x) \right)$$
$$= S_{l/l_0} \left(\exp\left(-\frac{s}{4\pi} \Delta_{[l, \infty[}\right) \phi^n(x) \right)$$

---

[9] In the proof we use the identity $S_\lambda(\mu_{[a,b[}) = \mu_{[a/\lambda, b/\lambda[}$, which follows easily from (2.89).

$$= \exp\left(-\frac{s}{4\pi}\Delta_{[l_0,\infty[}\right) S_{l/l_0} \phi^n(x)$$

$$= \exp\left(-\frac{s}{4\pi}\Delta_{[l_0,\infty[}\right) \left(\frac{l}{l_0}\right)^{n[\phi]} \phi^n\left(\frac{l_0}{l}x\right)$$

$$= \left(\frac{l}{l_0}\right)^{n[\phi]} : \phi^n\left(\frac{l_0}{l}x\right) :_{[l_0,\infty[}. \tag{2.97}$$

Note that P_l preserves the scale range. Integrating both sides of (2.97) over x gives, after a change of variable $(l_0/l)x \mapsto x'$, equation (2.96) and the scaling down of integration when the domain is finite. □

(iv) The coarse-graining operator satisfies a parabolic evolution equation, which is valid for $l \geq l_0$ with initial condition $P_{l_0} F(\phi) = F(\phi)$:

$$\left(\frac{\partial^\times}{\partial l} - \dot{S} - \frac{s}{4\pi}\dot{\Delta}\right) P_l F(\phi) = 0, \tag{2.98}$$

where

$$\dot{S} := \left.\frac{\partial^\times}{\partial l}\right|_{l=l_0} S_{l/l_0} \quad \text{and} \quad \dot{\Delta} := \left.\frac{\partial^\times}{\partial l}\right|_{l=l_0} \Delta_{[l_0,l[}. \tag{2.99}$$

Explicitly,

$$\dot{\Delta} F(\phi) = \int_{M^D} d^D x \int_{M^D} d^D y \left.\frac{\partial^\times}{\partial l}\right|_{l=l_0} G_{[l_0,l[}(|x-y|) \cdot \frac{\delta^2 F(\phi)}{\delta\phi(x)\delta\phi(y)}$$

with

$$\left.\frac{\partial^\times}{\partial l}\right|_{l=l_0} G_{[l_0,l[}(\xi) = \left.\frac{\partial^\times}{\partial l}\right|_{l=l_0} \int_{l_0}^{l} d^\times s\, s^{2[\phi]} u(\xi/s)$$

$$= l_0^{2[\phi]} u(\xi/l_0). \tag{2.100}$$

The final formula reads

$$\dot{\Delta} F(\phi) = \int_{M^D} d^D x \int_{M^D} d^D y\, l_0^{2[\phi]} u\left(\frac{|x-y|}{l_0}\right) \frac{\delta^2 F(\phi)}{\delta\phi(x)\delta\phi(y)}. \tag{2.101}$$

Remark. Frequently u is labeled $\dot{G}$, an abbreviation of (2.100) that is meaningful to the cognoscenti.

Proof of evolution equation (2.98). One computes $(\partial^\times/\partial l) P_l$ at $l = l_0$, then uses the semigroup property (2.95) to prove the validity of the evolution equation (2.98) for all l. Starting from the definition of P_l (2.94), one computes the convolution (2.61):

$$(\mu_{[l_0,l[} * F)(\phi) = \int_\Phi d\mu_{[l_0,l[}(\psi) F(\phi + \psi). \tag{2.102}$$

2.5 Scaling and coarse-graining

The functional Taylor expansion of $F(\phi + \psi)$ up to second order is sufficient for deriving (2.98):

$$(\mu_{[l_0,l[} * F)(\phi) = \int_\Phi d\mu_{[l_0,l[}(\psi) \left(F(\phi) + \frac{1}{2} F''(\phi) \cdot \psi\psi + \cdots \right) \quad (2.103)$$

$$= F(\phi) \int_\Phi d\mu_{[l_0,l[}(\psi) + \frac{s}{4\pi} \Delta_{[l_0,l[} F(\phi) \quad (2.104)$$

to second order only,

where $\Delta_{[l_0,l[}$ is the functional laplacian (2.63)

$$\Delta_{[l_0,l[} = \int_{M^D} d^D x \int_{M^D} d^D y\, G_{[l_0,l[}(x,y) \frac{\delta^2}{\delta\phi(x)\delta\phi(y)} \quad (2.105)$$

obtained by the ψ integration in (2.103) and the two-point function property (2.37):

$$\int_\Phi d\mu_{[l_0,l[}(\psi) \psi(x)\psi(y) = \frac{s}{2\pi} G_{[l_0,l[}(x,y). \quad (2.106)$$

Finally

$$\left. \frac{\partial^\times}{\partial l} \right|_{l=l_0} S_{l/l_0}(\mu_{[l_0,l[} * F)(\phi) = \left(\dot{S} + \frac{s}{4\pi} \dot{\Delta} \right) F(\phi). \quad (2.107)$$

$\square$

(v) The generator H of the coarse-graining operator is defined by

$$H := \left. \frac{\partial^\times}{\partial l} P_l \right|_{l=l_0}, \quad (2.108)$$

or equivalently

$$P_l = \exp\left(\frac{l}{l_0} H \right). \quad (2.109)$$

The evolution operator in (2.98) can therefore be written

$$\frac{\partial^\times}{\partial l} - \dot{S} - \frac{s}{4\pi}\dot{\Delta} = \frac{\partial^\times}{\partial l} - H. \quad (2.110)$$

The generator H operates on Wick monomials as follows (2.97):

$$H : \phi^n(x) :_{[l_0,\infty[} = \left. \frac{\partial^\times}{\partial l} P_l : \phi^n(x) : \right|_{l=l_0}$$

$$= \left. \frac{\partial^\times}{\partial l} \left(\frac{l}{l_0} \right)^{n[\phi]} : \phi^n\left(\frac{l_0}{l} x \right) :_{[l_0,\infty[} \right|_{l=l_0};$$

hence
$$H : \phi^n(x) :_{[l_0,\infty[} = n : [\phi]\phi^n(x) - \phi^{n-1}(x)E\phi(x) :_{[l_0,\infty[}, \quad (2.111)$$
where E is the Euler operator $\sum_{i=1}^{D} x^i \partial/\partial x^i$.

The second-order operator H, consisting of scaling and convolution, operates on Wick monomials as a first-order operator.

(vi) *Coarse-grained integrands in gaussian integrals.* The following equation is used in Section 16.1 for deriving the scale evolution of the effective action:
$$\langle \mu_{[l_0,\infty[}, A \rangle = \langle \mu_{[l_0,\infty[}, P_l A \rangle. \quad (2.112)$$

Proof of equation (2.112):
$$\langle \mu_{[l_0,\infty[}, A \rangle = \langle \mu_{[l,\infty[}, \mu_{[l_0,l]} * A \rangle$$
$$= \langle \mu_{[l_0,\infty[}, S_{l/l_0} \cdot \mu_{[l_0,l]} * A \rangle$$
$$= \langle \mu_{[l_0,\infty[}, P_l A \rangle.$$

The important step in this proof is the second one,
$$\mu_{[l,\infty[} = \mu_{[l_0,\infty[} S_{l/l_0}. \quad (2.113)$$

We check on an example[10] that $\mu_{[l_0,\infty[} S_{l/l_0} = \mu_{[l,\infty[}$:
$$\int d\mu_{[l_0,\infty[}(\phi) S_{l/l_0}[\phi(x)\phi(y)] = \int d\mu_{[l_0,\infty[}(\phi) \left(\frac{l}{l_0}\right)^{2[\phi]} \phi\left(\frac{l_0}{l}x\right)\phi\left(\frac{l_0}{l}y\right)$$
$$= \frac{s}{2\pi}\left(\frac{l}{l_0}\right)^{2[\phi]} G_{[l_0,\infty[}\left(\frac{l_0}{l}|x-y|\right)$$
$$= \frac{s}{2\pi} G_{[l,\infty[}(|x-y|)$$
$$= \int d\mu_{[l,\infty[}(\phi)\phi(x)\phi(y),$$
where we have used (2.89), then (2.37). □

From this example we learn the fundamental concepts involved in (2.113): the scaling operator S_{l/l_0} with $l/l_0 > 1$ shrinks the domain of the fields (first line), whereas a change of range from $[l_0, \infty[$ to $[l, \infty[$ (third line) restores the original domain of the fields. These steps, shrinking, scaling, and restoring, are at the heart of renormalization in condensed-matter physics. The second and fourth lines, which are convenient for having an explicit presentation of the process, relate

---

[10] The general case can be obtained similarly by integrating $\exp(-2\pi i \langle J, \phi \rangle)$ according to (2.66).

the gaussian μ_G to its covariance G. The self-similarity of covariances in different ranges makes this renormalization process possible.

References

The first four sections summarize results scattered in too many publications to be listed here. The following publication contains many of these results and their references: P. Cartier and C. DeWitt-Morette (1995). "A new perspective on functional integration," *J. Math. Phys.* **36**, 2237–2312.

[1] C. Morette (1951). "On the definition and approximation of Feynman's path integral," *Phys. Rev.* **81**, 848–852.
[2] A. Wurm (2002). Renormalization group applications in area-preserving nontwist maps and relativistic quantum field theory, unpublished Ph.D. Thesis, University of Texas at Austin.
[3] D. C. Brydges, J. Dimock, and T. R. Hurd (1998). "Estimates on renormalization group transformations," *Can. J. Math.* **50**, 756–793 and references therein.
D. C. Brydges, J. Dimock, and T. R. Hurd (1998). "A non-gaussian fixed point for ϕ^4 in $4 - \epsilon$ dimensions," *Commun. Math. Phys.* **198**, 111–156.

3
Selected examples

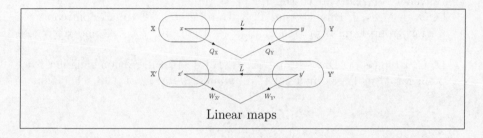
Linear maps

In this chapter, we apply properties of linear maps given in Section 2.4, and summarized in figure 2.2 reproduced above. In Sections 3.1 and 3.2, the key equation is

$$W_{\mathbb{Y}'} = W_{\mathbb{X}'} \circ \tilde{L}.$$

In Section 3.1, $\mathbb{X}$ is a space of continuous paths, and

$$\langle x', x \rangle_{\mathbb{X}} := \int_{\mathbb{T}} \mathrm{d}x'(t) x(t).$$

In Section 3.2, when $\mathbb{X}$ is an L^2-space,

$$\langle x', x \rangle_{\mathbb{X}} := \int \mathrm{d}t \, x'(t) x(t);$$

and when $\mathbb{X}$ is an $L^{2,1}$-space,

$$\langle x', x \rangle_{\mathbb{X}} := \int \mathrm{d}t \, \dot{x}'(t) \dot{x}(t).$$

In Section 3.3, the key equation is

$$Q_{\mathbb{X}} = Q_{\mathbb{Y}} \circ L.$$

3.1 The Wiener measure and brownian paths (discretizing a path integral)

Let $\mathbb{X}$ be the space of continuous paths x over a time interval $\mathbb{T} = [t_a, t_b]$:

$$x : \mathbb{T} \to \mathbb{R} \quad \text{by } t \mapsto x(t). \tag{3.1}$$

The dual $\mathbb{X}'$ of $\mathbb{X}$ is the space of bounded measures x' on $\mathbb{T}$ and the duality is given by

$$\langle x'|x\rangle_{\mathbb{X}} = \int_{\mathbb{T}} \mathrm{d}x'(t) x(t); \tag{3.2}$$

here x' is of bounded variation $\int_{\mathbb{T}} |\mathrm{d}x'(t)| < \infty$ and the integral can be interpreted as a Stieltjes integral. Let us discretize the time interval $\mathbb{T}$:

$$t_a = t_0 < t_1 < \cdots < t_n \leq t_b. \tag{3.3}$$

Let $\mathbb{Y}$ be the Wiener differential space consisting of the differences of two consecutive values of x on the discretized time interval ($0 \leq j < n$):

$$y^j = x(t_{j+1}) - x(t_j) = \langle \delta_{t_{j+1}} - \delta_{t_j}, x\rangle. \tag{3.4}$$

Therefore the discretizing map $L : \mathbb{X} \to \mathbb{Y}$ is a projection from the infinite-dimensional space $\mathbb{X}$ onto the n-dimensional space $\mathbb{Y}$.

Let $\Gamma_{\mathbb{X}}$ (or $\Gamma_{s,\mathbb{X}}$) be the Wiener gaussian defined by its Fourier transforms $\mathcal{F}\Gamma_{\mathbb{X}}$, i.e. by the variance

$$W_{\mathbb{X}'}(x') = \int_{\mathbb{T}} \mathrm{d}x'(t) \int_{\mathbb{T}} \mathrm{d}x'(s) \inf(t - t_a, s - t_a), \tag{3.5}$$

where $\inf(u, v)$ is the smaller of u and v:

$$\inf(t - t_a, s - t_a) = \theta(t - s)(s - t_a) + \theta(s - t)(t - t_a); \tag{3.6}$$

the step function θ is equal to 1 for positive arguments, equal to 0 for negative arguments, and discontinuous at the origin; here we must assign the value $1/2$ to $\theta(0)$. The transpose $\tilde{L}$ of L is defined by (2.58), i.e.[1]

$$\langle \tilde{L}y', x\rangle_{\mathbb{X}} = \langle y', Lx\rangle_{\mathbb{Y}} = \sum_j y'_j y^j$$

$$= \sum_j y'_j \langle \delta_{t_{j+1}} - \delta_{t_j}, x\rangle$$

$$= \sum_j \langle y'_j (\delta_{t_{j+1}} - \delta_{t_j}), x\rangle;$$

---

[1] The index j of summation runs over $0, 1, \ldots, n-1$.

hence
$$\tilde{L}y' = \sum_j y'_j(\delta_{t_{j+1}} - \delta_{t_j}) \tag{3.7}$$

and
$$W_{\mathbb{Y}'}(y') = (W_{\mathbb{X}'} \circ \tilde{L})(y') = W_{\mathbb{X}'}(\tilde{L}y')$$
$$= \sum_{j,k} y'_j y'_k (\inf(t_{j+1}, t_{k+1}) - \inf(t_{j+1}, t_k) - \inf(t_j, t_{k+1}) + \inf(t_j, t_k)).$$

The terms $j \neq k$ do not contribute to $W_{\mathbb{Y}'}(y')$. Hence
$$W_{\mathbb{Y}'}(y') = \sum_j (y'_j)^2 (t_{j+1} - t_j). \tag{3.8}$$

Choosing $\mathbb{Y}$ to be a Wiener differential space rather than a naively discretized space defined by $\{x(t_j)\}$ diagonalizes the variance $W_{\mathbb{Y}'}(y')$. The gaussian $\Gamma_{\mathbb{Y}}$ (or $\Gamma_{s,\mathbb{Y}}$) on $\mathbb{Y}$ defined by the Fourier transform
$$(\mathcal{F}\Gamma_{\mathbb{Y}})(y') = \exp\left(-\pi s \sum_{i,j} \delta^{ij} y'_i y'_j (t_{j+1} - t_j)\right) \tag{3.9}$$
is
$$d\Gamma_{\mathbb{Y}}(y) \stackrel{f}{=} dy^0 \cdots dy^{n-1} \frac{1}{\prod_{j=0}^{n-1}(s(t_{j+1} - t_j))^{1/2}} \exp\left(-\frac{\pi}{s} \sum_{i,j} \frac{\delta_{ij} y^i y^j}{t_{j+1} - t_j}\right). \tag{3.10}$$

Set
$$\Delta t_j := t_{j+1} - t_j \quad (0 \leq j \leq n-1),$$
$$\Delta x^j := (\Delta x)^j := x(t_{j+1}) - x(t_j) = y^j \quad (0 \leq j \leq n-1),$$
$$x^j := x(t_j) \quad (1 \leq j \leq n);$$

then
$$d\Gamma_{\mathbb{Y}}(\Delta x) \stackrel{f}{=} dx^1 \cdots dx^n \frac{1}{\prod_{j=0}^{n-1}(s\,\Delta t_j)^{1/2}} \exp\left(-\frac{\pi}{s} \sum_{j=0}^{n-1} \frac{(\Delta x^j)^2}{\Delta t_j}\right). \tag{3.11}$$

When $s = 1$, *we recover the marginals of the Wiener measure* (see eq. (1.41)) with $D = 1/(2\pi)$; hence the gaussian $\Gamma_{\mathbb{X}}$ of covariance $\inf(t - t_a, s - t_a)$ *is the Wiener measure*.

Deriving the distribution of brownian paths from the Wiener measure is a particular case of a general formula. Let $F : \mathbb{X} \to \mathbb{R}$ be a functional on X that can be decomposed into two maps $F = f \circ L$, where

$$L: \mathbb{X} \to \mathbb{Y} \text{ linearly,}$$
$$f: \mathbb{Y} \to \mathbb{R} \text{ integrable with respect to } \Gamma_\mathbb{Y}.$$

Then

$$\int_\mathbb{X} d\Gamma_\mathbb{X}(x) F(x) = \int_\mathbb{Y} d\Gamma_\mathbb{Y}(y) f(y), \qquad (3.12)$$

where the gaussians $\Gamma_\mathbb{X}$ and $\Gamma_\mathbb{Y}$ are characterized by the quadratic forms $W_{\mathbb{X}'}$ and $W_{\mathbb{Y}'}$, such that

$$W_{\mathbb{Y}'} = W_{\mathbb{X}'} \circ \tilde{L}. \qquad (3.13)$$

3.2 Canonical gaussians in L^2 and $L^{2,1}$

Wiener gaussians on spaces of continuous paths serve probabilists very well, but physicists prefer paths with a finite action, or, which amounts to the same thing, finite kinetic energy. The corresponding space of paths is denoted[2] by $L^{2,1}$; it is known in probability theory as the *Cameron–Martin space*.

Abstract framework

Let $\mathcal{H}$ be a real Hilbert space. Hence, on the vector space $\mathcal{H}$ there is given a scalar product $(x|y)$, which is bilinear and symmetric, and satisfies $(x|x) > 0$ for $x \neq 0$. We then define the norm $||x|| = (x|x)^{1/2}$ and assume that $\mathcal{H}$ is complete and separable for this norm. Well-known examples are the following:

the space ℓ^2 of sequences of real numbers

$$x = (x_1, x_2, \ldots)$$

with $\sum_{n \geq 1} x_n^2 < +\infty$, and the scalar product $(x|y) = \sum_{n \geq 1} x_n y_n$;
and
the space $L^2(\mathbb{T})$ of square-integrable real-valued functions on $\mathbb{T}$, with $(x|y) = \int_\mathbb{T} dt\, x(t) y(t)$.

We denote by $\mathcal{H}'$ the dual of the Banach space $\mathcal{H}$, and by $\langle x', x \rangle$ the duality between $\mathcal{H}$ and $\mathcal{H}'$. By a well-known result (F. Riesz), there exists an isomorphism $D: \mathcal{H} \to \mathcal{H}'$ characterized by

$$\langle Dx, y \rangle = (x|y) \qquad (3.14)$$

---

[2] Roughly "square-integrable first derivative." See Appendix G.

for x, y in $\mathcal{H}$. Henceforth, we shall identify $\mathcal{H}$ and $\mathcal{H}'$ under D. Then the quadratic form $Q(x) = (x|x)$ is equal to its inverse $W(x') = (x'|x')$, and, in the general setup of Section 2.3, we have to put $\mathbb{X} = \mathbb{X}' = \mathcal{H}$ and $G = D = \mathbf{1}_{\mathcal{H}}$. The *canonical gaussian* $\Gamma_{s,\mathcal{H}}$ on $\mathcal{H}$ is defined by its Fourier transform

$$\int_{\mathcal{H}} d\Gamma_{s,\mathcal{H}}(x) e^{-2\pi i (x|y)} = e^{-\pi s(x|x)}. \quad (3.15)$$

Suppose that $\mathcal{H}$ is of finite dimension N; then

$$d\Gamma_{s,\mathcal{H}}(x) = \begin{cases} e^{-\pi ||x||^2} d^N x & \text{for } s = 1, \\ e^{-\pi i N/4} e^{\pi i ||x||^2} d^N x & \text{for } s = i, \end{cases} \quad (3.16)$$

where we have explicitly

$$||x||^2 = \sum_{i=1}^{N} (x^i)^2, \qquad d^N x = dx^1 \ldots dx^N \quad (3.17)$$

in any orthonormal system of coordinates $(x^1, \ldots, x^N)$. If $\mathcal{H}_1$ and $\mathcal{H}_2$ are Hilbert spaces and L is an isometry of $\mathcal{H}_1$ onto $\mathcal{H}_2$, then obviously L transforms $\Gamma_{s,\mathcal{H}_1}$ into $\Gamma_{s,\mathcal{H}_2}$. In particular $\Gamma_{s,\mathcal{H}}$ and $\mathcal{D}_{s,\mathcal{H}} x$ *are invariant under the rotations in* $\mathcal{H}$.

We introduce now a Banach space $\mathbb{X}$ and a continuous linear map $P : \mathcal{H} \to \mathbb{X}$. Identifying $\mathcal{H}$ and $\mathcal{H}'$ as before, the transposed map $\tilde{P} : \mathbb{X}' \to \mathcal{H}$ is defined by

$$(\tilde{P}(x')|h) = \langle x', P(h) \rangle \quad (3.18)$$

for h in $\mathcal{H}$ and x' in $\mathbb{X}'$. The situation is summarized by a triple,

$$\mathbb{X}' \xrightarrow{\tilde{P}} \mathcal{H} \xrightarrow{P} \mathbb{X},$$

which is analogous to the *Gelfand triple of the white-noise theory*. In standard situations P is injective with a dense image, hence $\tilde{P}$ shares these properties. The linear map $G = P \circ \tilde{P}$ is symmetric (2.28) and the corresponding quadratic form W on $\mathbb{X}'$ satisfies

$$W(x') = \langle x', Gx' \rangle = ||\tilde{P}(x')||^2. \quad (3.19)$$

Therefore, *the corresponding gaussian $\Gamma_{s,W}$ on $\mathbb{X}$ is the image by the map $P : \mathcal{H} \to \mathbb{X}$ of the canonical gaussian $\Gamma_{s,\mathcal{H}}$ on the Hilbert space $\mathcal{H}$.*

Paths beginning at $a = (t_a, 0)$

Let us go back to the Wiener gaussian. We are considering again a time interval $\mathbb{T} = [t_a, t_b]$ and the space $\mathbb{X}_a$ of continuous paths $x : \mathbb{T} \to \mathbb{R}^D$,

3.2 Canonical gaussians in L^2 and $L^{2,1}$

with components x^j for $1 \leq j \leq D$, such that $x(t_a) = 0$. On the configuration space $\mathbb{R}^D$, we suppose given a positive quadratic form $g_{ij}x^i x^j$, with inverse $g^{ij}x_i x_j$ on the dual space $\mathbb{R}_D$. The Hilbert space $\mathcal{H}$ is defined as $L^2(\mathbb{T}; \mathbb{R}^D)$, the space of square-integrable functions $h : \mathbb{T} \to \mathbb{R}^D$ with the scalar product

$$(h|h') = \sum_{i,j} g_{ij} \int_\mathbb{T} dt\, h^i(t) h'^j(t). \tag{3.20}$$

The map P_- from $\mathcal{H}$ to $\mathbb{X}_a$ is defined by

$$(P_- h)^j(t) = \int_\mathbb{T} ds\, \theta(t-s) h^j(s)$$
$$= \int_{t_a}^{t} ds\, h^j(s). \tag{3.21}$$

Otherwise stated, $x = P_- h$ is characterized by the properties

$$x(t_a) = 0, \qquad \dot{x}(t) = h(t).$$

We claim that *the Wiener gaussian is the image under P_- of the canonical gaussian $\Gamma_{1,\mathcal{H}}$ on the Hilbert space $\mathcal{H}$*. The elements of the dual $\mathbb{X}'_a$ of $\mathbb{X}_a$ are described by

$$\langle x', x \rangle = \sum_{j=1}^{D} \int_\mathbb{T} dx'_j(t) x^j(t), \tag{3.22}$$

where the components x'_j are of bounded variation, and the integral is a Stieltjes integral. From the formulas (3.18)–(3.22), one derives easily

$$\tilde{P}_-(x')^j(s) = \int_\mathbb{T} g^{jk}\, dx'_k(t') \theta(t'-s), \tag{3.23$_a$}$$

$$G_-(x')^j(t) = \int_\mathbb{T} ds\, \theta(t-s) \tilde{P}_-(x')^j(s)$$
$$= \int_\mathbb{T} g^{jk}\, dx'_k(t') G_-(t,t'), \tag{3.23$_b$}$$

$$W_-(x') = \int_\mathbb{T} dx'_j(t) \int_\mathbb{T} dx'_k(t') g^{jk} G_-(t,t'), \tag{3.23$_c$}$$

with the covariance

$$G_-(t,t') = \int_\mathbb{T} ds\, \theta(t-s)\theta(t'-s) = \inf(t-t_a, t'-t_a). \tag{3.24}$$

On comparing these with (3.5), we see that, in the case $D = 1$, the variance (3.23$_c$) is equal to $W_{\mathbb{X}'_a}$; hence our claim is established. In the multidimensional case, we can transform to the case $g^{jk} = \delta^{jk}$ by a linear change of coordinates, and then the image of $\Gamma_{s,\mathcal{H}}$ by P_- is equal to $d\Gamma_{s,\mathbb{X}_a}(x^1)\ldots d\Gamma_{s,\mathbb{X}_a}(x^D)$ for $x=(x^1,\ldots,x^D)$. In probabilistic terms (for $s=1$), *the components $x^1,\ldots,x^D$ are stochastically independent Wiener processes.*

Remark. The map $P_-: L^2(\mathbb{T};\mathbb{R}^D) \to \mathbb{X}_a$ is injective, and its image is the space $L_-^{2,1}(\mathbb{T};\mathbb{R}^D)$ (abbreviated $L_-^{2,1}$) of absolutely continuous functions $x(t)$ vanishing at $t=t_a$, whose derivative $\dot{x}(t)$ is square integrable. The scalar product is given by

$$(x|y) = \int_{\mathbb{T}} dt\, g_{jk}\dot{x}^j(t)\dot{y}^k(t). \tag{3.25}$$

We refer to Appendix G for the properties of the space $L^{2,1}$. Since P_- defines an isometry P'_- of L^2 to $L_-^{2,1}$, it transforms the canonical gaussian on[3] L^2 into the canonical gaussian on $L_-^{2,1}$. Moreover, P_- is the composition

$$L^2 \xrightarrow{P'_-} L_-^{2,1} \xrightarrow{i} \mathbb{X}_a,$$

where i is the injection of $L_-^{2,1}$ into $\mathbb{X}_a$. In conclusion, *i maps the canonical gaussian $\Gamma_{s,L_-^{2,1}}$ into the Wiener gaussian $\Gamma_{s,\mathbb{X}_a}$ on $\mathbb{X}_a$.* The Hilbert subspace $L_-^{2,1}$ embedded in $\mathbb{X}_a$ is the so-called Cameron–Martin subspace.

Remark. Injections are not trivial: they do change the environment. For example, x^2 for x a positive number has a unique inverse, but x^2 for x a real number does not have a unique inverse.

Paths ending at $b=(t_b,0)$

We consider the set $\mathbb{X}_b$ of continuous paths $x:\mathbb{T}\to\mathbb{R}^D$ such that $x(t_b)=0$. The map $P_+: L^2 \to \mathbb{X}_b$ is defined by

$$\begin{aligned}(P_+ h)^j(t) &= \int_{\mathbb{T}} ds\, \theta(s-t) h^j(s) \\ &= \int_t^{t_b} ds\, h^j(s).\end{aligned} \tag{3.26}$$

Its image is the subspace $L_+^{2,1}$ of $L^{2,1}$ consisting of paths x such that $x(t_b)=0$.

The final result can be stated as follows. *The map P_+ transforms the canonical gaussian Γ_{s,L^2} into the gaussian Γ_{s,W_+} on $\mathbb{X}_b$ corresponding to*

---

[3] We abbreviate $L^2(\mathbb{T};\mathbb{R}^D)$ as L^2.

the variance W_+ on $\mathbb{X}'_b$ given by

$$W_+ = \int_\mathbb{T} dx'_j(t) \int_\mathbb{T} dx'_k(t') g^{jk} G_+(t,t'), \tag{3.23_+}$$

$$G_+(t,t') = \int_\mathbb{T} ds\, \theta(s-t)\theta(s-t') = \inf(t_b - t, t_b - t'). \tag{3.24_+}$$

The mappings $P_\pm$ and i defined in figure 3.1 have been generalized by A. Maheshwari [1] and were used in [2] for various versions of the Cameron–Martin formula and for Fredholm determinants.

3.3 The forced harmonic oscillator

Up to this point we have exploited properties of gaussian $\Gamma_\mathbb{X}$ on $\mathbb{X}$ defined by quadratic forms W on the dual $\mathbb{X}'$ of $\mathbb{X}$, but we have not used the quadratic form Q on $\mathbb{X}$ inverse of W in the formulas (2.27) and (2.28), namely

$$Q(x) =: \langle Dx, x \rangle,$$
$$W(x') =: \langle x', Gx' \rangle, \tag{3.27}$$

with symmetric maps $G: \mathbb{X}' \to \mathbb{X}$ and $D: \mathbb{X} \to \mathbb{X}'$ satisfying

$$DG = \mathbf{1}_{\mathbb{X}'}, \qquad GD = \mathbf{1}_\mathbb{X}. \tag{3.28}$$

The gaussian volume element $d\Gamma_{s,Q}$ can be expressed in terms of the quadratic form Q:

$$\mathcal{D}_{s,Q}(x) \cdot \exp\left(-\frac{\pi}{s} Q(x)\right) \stackrel{\int}{=} d\Gamma_{s,Q}(x); \tag{3.29}$$

this is a qualified equality, meaning that both expressions are defined by the same integral equation (2.29). In this section, we use gaussians defined by Q with $s = \mathrm{i}$ and $\mathcal{D}_{s,Q}$ is abbreviated to $\mathcal{D}_Q$.

The first path integral proposed by Feynman was

$$\langle B, t_b | A, t_a \rangle = \int_{\mathbb{X}_{a,b}} \mathcal{D}x\, \exp(\mathrm{i}S(x)/\hbar) \tag{3.30}$$

with

$$S(x) = \int_{t_a}^{t_b} dt \left(\frac{m}{2}(\dot{x}(t))^2 - V(x(t))\right), \tag{3.31}$$

where $\langle B, t_b | A, t_a \rangle$ is the probability amplitude that a particle in the state A at time t_a be found in the state B at time t_b.

In the following the domain of integration $\mathbb{X}_{a,b}$ is the space of paths $x: [t_a, t_b] \to \mathbb{R}$ such that $A = x(t_a) \equiv x_a$ and $B = x(t_b) \equiv x_b$.

The domain of integration $\mathbb{X}_{a,b}$ and the normalization of its volume element

$\mathbb{X}_{a,b}$ is not a vector space[4] unless $x_a = 0$ and $x_b = 0$. Satisfying *only one* of these two vanishing requirements is easy and beneficial.

It is easy; indeed, choose the origin of the coordinates of $\mathbb{R}$ to be either x_a or x_b. The condition $x(t_a) = 0$ is convenient for problems in diffusion; the condition $x(t_b) = 0$ is convenient for problems in quantum mechanics.

It is beneficial; a space of pointed paths is contractible (see Section 7.1) and can be mapped into a space of paths on $\mathbb{R}^D$.

Therefore we rewrite (3.30) as an integral over $\mathbb{X}_b$, the space of paths vanishing at t_b; the other requirement is achieved by introducing $\delta(x(t_a) - x_a)$:

$$\langle x_b, t_b | x_a, t_a \rangle = \int_{\mathbb{X}_b} \mathcal{D}x \, \exp(\mathrm{i}S(x)/\hbar) \delta(x(t_a) - x_a). \tag{3.32}$$

This is a particular case of

$$\langle x_b, t_b | \phi, t_a \rangle = \int_{\mathbb{X}_b} \mathcal{D}x \, \exp(\mathrm{i}S(x)/\hbar) \phi(x(t_a)), \tag{3.33}$$

which is useful for solving Schrödinger equations, given an initial wave function ϕ.

An affine transformation from $\mathbb{X}_{a,b}$ onto $\mathbb{X}_{0,0}$ expresses "the first path integral," (3.30) and (3.32), as an integral over a Banach space. This affine transformation is known as the *background method*. We present it here in the simplest case of paths with values in $\mathbb{R}^D$. For more general cases see Chapter 4 on semiclassical expansions; and for the general case of a map from a space of pointed paths on a riemannian manifold $\mathbb{M}^D$ onto a space of pointed paths on $\mathbb{R}^D$ see Chapter 7.

Let $x \in \mathbb{X}_{0,0}$ and y be a fixed arbitrary path in $\mathbb{X}_{a,b}$, possibly a classical path defined by the action functional, but not necessarily so. Then

$$y + x \in \mathbb{X}_{a,b}, \qquad x \in \mathbb{X}_{0,0}. \tag{3.34}$$

Let $\Gamma_{s,Q}$, abbreviated to Γ, be the gaussian defined by

$$\int_{\mathbb{X}_b} \mathrm{d}\Gamma(x) \exp(-2\pi\mathrm{i}\langle x', x \rangle) := \exp(-\pi s W_b(x')), \tag{3.35}$$

where $\mathbb{X}_b$ *is the space of paths vanishing at* t_b,

$$x(t_b) = 0.$$

---

[4] Let $x \in \mathbb{X}_{a,b}$ and $y \in \mathbb{X}_{a,b}$; then $x + y$ lies in $\mathbb{X}_{a,b}$ only if $x(t_a) + y(t_a) = x(t_a)$ and $x(t_b) + y(t_b) = x(t_b)$.

3.3 The forced harmonic oscillator

The gaussian volume defined by (3.35) is normalized to unity:[5]

$$\Gamma(\mathbb{X}_b) := \int_{\mathbb{X}_b} \mathrm{d}\Gamma(x) = 1. \tag{3.36}$$

The gaussian volume element on $\mathbb{X}_{0,0}$ is readily computed by means of the linear map

$$L : \mathbb{X}_{b=0} \longrightarrow \mathbb{R}^D \quad \text{by } x^i \longrightarrow u^i := \langle \delta_{t_a}, x^i \rangle$$

whose transposed map $\tilde{L} : \mathbb{R}_D \to \mathbb{X}'_{b=0}$ is readily identified by

$$(\tilde{L}u')_i = u'_i \delta_{t_a} \tag{3.37}$$

for $u' = (u'_i)$ in $\mathbb{R}_D$. Hence, we get

$$\begin{aligned}\Gamma_{0,0}(\mathbb{X}_{0,0}) &:= \int_{\mathbb{X}_{b=0}} \mathrm{d}\Gamma(x) \delta(x^1(t_a)) \ldots \delta(x^D(t_a)) \\ &= \int_{\mathbb{R}} \mathrm{d}\Gamma^L(u) \delta(u^1) \ldots \delta(u^D), \end{aligned} \tag{3.38}$$

where the gaussian Γ^L on $\mathbb{R}^D$ is defined by the variance

$$W^L = W_b \circ \tilde{L}.$$

If we express the variance W_b by the two-point function $G(t,s)$, namely

$$W_b(x') = \int_\mathbb{T} \int_\mathbb{T} \mathrm{d}x'_i(s) \mathrm{d}x'_j(t) G^{ij}(s,t),$$

we obtain from (3.37)

$$W^L(u') = u'_i u'_j G^{ij}(t_a, t_a)$$

and

$$\mathrm{d}\Gamma^L(u) \stackrel{i}{=} \mathrm{d}^D u (\det G_{ij}(t_a, t_a)/s)^{-1/2} \exp\left(-\frac{\pi}{s} u^i u^j G_{ij}(t_a, t_a)\right),$$

where $G^{ij} G_{jk} = \delta^i_k$.
Finally,

$$\Gamma_{0,0}(\mathbb{X}_{0,0}) = e^{\pi i D/4} (\det G_{ij}(t_a, t_a)/s)^{-1/2}. \tag{3.39}$$

An affine transformation preserves gaussian normalization; therefore,

$$\Gamma_{a,b}(\mathbb{X}_{a,b}) = \Gamma_{0,0}(\mathbb{X}_{0,0}). \tag{3.40}$$

An affine transformation "shifts" a gaussian, and multiplies its Fourier transform by a phase. This property is most simply seen in one dimension:

---

[5] Hint: set $x' = 0$ in equation (3.35).

it follows from
$$\int_{\mathbb{R}} \frac{\mathrm{d}x}{\sqrt{a}} \exp\left(-\pi \frac{x^2}{a} - 2\pi\mathrm{i}\langle x', x\rangle\right) = \exp(-\pi a x'^2)$$
that
$$\int_{\mathbb{R}} \frac{\mathrm{d}x}{\sqrt{a}} \exp\left(-\pi \frac{(x+l)^2}{a} - 2\pi\mathrm{i}\langle x', x\rangle\right) = \exp(2\pi\mathrm{i}\langle x', l\rangle)\exp(-\pi a x'^2). \quad (3.41)$$
Under the affine transformation $x \in \mathbb{X}_{0,0} \mapsto y + x \in \mathbb{X}_{a,b}$, the gaussian $\Gamma_{0,0}$ goes into $\Gamma_{a,b}$; their respective Fourier transforms differ only by a phase, since their gaussian volumes are equal. Since
$$\int_{\mathbb{X}_{0,0}} \mathrm{d}\Gamma_{0,0}(x)\exp(-2\pi\mathrm{i}\langle x', x\rangle) = \Gamma_{0,0}(\mathbb{X}_{0,0})\exp(-\pi\mathrm{i}W_{0,0}(x')), \quad (3.42)$$
we obtain
$$\int_{\mathbb{X}_{a,b}} \mathrm{d}\Gamma_{a,b}(x)\exp(-2\pi\mathrm{i}\langle x', x\rangle)$$
$$= e^{\pi\mathrm{i}D/4}(\det G_{ij}(t_a, t_a)/s)^{-1/2} \exp(2\pi\mathrm{i}\langle x', y\rangle)\exp(-\pi\mathrm{i}W_{0,0}(x')). \quad (3.43)$$

Normalization dictated by quantum mechanics[6]

The first path integral (3.30),
$$\left\langle b \left| \exp\left(-\frac{\mathrm{i}}{\hbar}H(t_b - t_a)\right) \right| a \right\rangle = \int_{\mathbb{X}_{a,b}} \mathcal{D}x \cdot \exp(\mathrm{i}S(x)/\hbar), \quad (3.44)$$
implies a relationship between the normalization of volume elements in path integrals and the normalization of matrix elements in quantum mechanics, itself dictated by the physical meaning of such matrix elements. Two examples are a free particle and a simple harmonic oscillator. The most common normalization in quantum mechanics is
$$\langle x''|x'\rangle = \delta(x'' - x'), \qquad \langle p''|p'\rangle = \delta(p'' - p'); \quad (3.45)$$
it implies [3] the normalizations
$$\langle x'|p'\rangle = \frac{1}{\sqrt{h}} \exp(2\pi\mathrm{i}\langle p', x'\rangle/h) \quad (3.46)$$
and
$$|p\rangle = \int_{\mathbb{R}} \mathrm{d}x \frac{1}{\sqrt{h}} \exp(2\pi\mathrm{i}\langle p, x\rangle/h)|x\rangle, \quad (3.47)$$
i.e.
$$|p_a = 0\rangle = \frac{1}{\sqrt{h}} \int_{\mathbb{R}} \mathrm{d}x |x\rangle. \quad (3.48)$$

---
[6] Contributed by Ryoichi Miyamoto.

3.3 The forced harmonic oscillator

The hamiltonian operator H_0 of a free particle of mass m is

$$H_0 := p^2/(2m) \qquad (3.49)$$

and the matrix element

$$\left\langle x_b = 0 \left| \exp\left(-\frac{2\pi i}{h} H_0(t_b - t_a)\right) \right| p_a = 0 \right\rangle = \frac{1}{\sqrt{h}}. \qquad (3.50)$$

The normalization of this matrix element corresponds to the normalization of the gaussian $\Gamma_{s,Q_0/h}$ on $\mathbb{X}_b$,

$$\int_{\mathbb{X}_b} d\Gamma_{s,Q_0/h}(x) = 1, \qquad (3.51)$$

which follows on putting $x' = 0$ in the definition (3.35) of $\Gamma_{Q_0/h}$, which is valid for $s = 1$, and $s = i$.

Proof. Set $s = i$, then

$$Q_0 := \int_{t_a}^{t_b} dt \, m\dot{x}^2(t), \qquad (3.52)$$

and

$$\int_{\mathbb{X}_b} d\Gamma_{i,Q_0/h}(x)\delta(x(t_a) - x_a) = \left\langle x_b = 0 \left| \exp\left(-\frac{2\pi i}{h} H_0(t_b - t_a)\right) \right| x_a \right\rangle; \qquad (3.53)$$

equivalently, with $S_0 := \frac{1}{2}Q_0$,

$$\int_{\mathbb{X}_b} \mathcal{D}_{i,Q_0/h}(x)\exp\left(\frac{2\pi i}{h} S_0(x)\right)\delta(x(t_a) - x_a)$$
$$= \left\langle x_b = 0 \left| \exp\left(-\frac{2\pi i}{h} H_0(t_b - t_a)\right) \right| x_a \right\rangle. \qquad (3.54)$$

In order to compare the operator and path-integral normalizations, we integrate both sides of (3.53) with respect to x_a:

$$\int_{\mathbb{X}_b} d\Gamma_{i,Q_0/h}(x) = \int dx_a \left\langle x_b = 0 \left| \exp\left(-\frac{2\pi i}{h} H_0(t_b - t_a)\right) \right| x_a \right\rangle$$
$$= \sqrt{h} \left\langle x_b = 0 \left| \exp\left(-\frac{2\pi i}{h} H_0(t_b - t_a)\right) \right| p_a = 0 \right\rangle, \text{ by (3.48)}$$
$$= 1, \text{ by (3.50).} \qquad (3.55)$$

$\square$

The quantum-mechanical normalization (3.50) implies the functional path-integral normalization (3.51), with Q_0 given by equation (3.52).

The harmonic oscillator

The hamiltonian operator $H_0 + H_1$ for a simple harmonic oscillator is

$$H_0 + H_1 = \frac{p^2}{2m} + \frac{1}{2}m\omega^2 x^2. \tag{3.56}$$

There is a choice of quadratic form for defining the gaussian volume element, namely Q_0 or $Q_0 - Q_1$, with

$$\frac{1}{2}(Q_0 - Q_1) \text{ corresponding to } S_0 + S_1 \text{ and to } H_0 + H_1, \text{ respectively.} \tag{3.57}$$

We use first $\Gamma_{i,Q_0/h}$; the starting point (3.53) or (3.54) now reads

$$\int_{\mathbb{X}_b} d\Gamma_{i,Q_0/h}(x) \exp\left(-\frac{\pi i}{h} Q_1(x)\right) \delta(x(t_a) - x_a)$$

$$\equiv \int_{\mathbb{X}_b} \mathcal{D}_{i,Q_0/h}(x) \exp\left(\frac{2\pi i}{h}(S_0 + S_1)(x)\right) \delta(x(t_a) - x_a)$$

$$= \left\langle x_b = 0 \left| \exp\left(-\frac{2\pi i}{h}(H_0 + H_1)(t_b - t_a)\right) \right| x_a \right\rangle. \tag{3.58}$$

This matrix element can be found, for instance, in [3, (2.5.18)]; it is equal to

$$A(x_a) = \left(\frac{m\omega}{ih\sin(\omega(t_b - t_a))}\right)^{1/2} \cdot \exp\left(\frac{\pi i m\omega}{h\sin(\omega(t_b - t_a))} x_a^2 \cos(\omega(t_b - t_a))\right).$$

By integration over x_a, we derive

$$\int_{\mathbb{X}_b} d\Gamma_{i,Q_0/h}(x) \exp\left(-\frac{\pi i}{h} Q_1(x)\right) = \int_{\mathbb{R}} dx_a \cdot A(x_a) = (\cos(\omega(t_b - t_a)))^{-1/2}. \tag{3.59}$$

On the other hand, the left-hand side of (3.59) is computed in Section 4.2 and found equal, in (4.21), to the following ratio of determinants:

$$\int_{\mathbb{X}_b} d\Gamma_{i,Q_0/h}(x) \exp\left(-\frac{\pi i}{h} Q_1(x)\right) \equiv \int_{\mathbb{X}_b} \mathcal{D}_{i,Q_0/h}(x) \exp\left(\frac{\pi i}{h}(Q_0(x) - Q_1(x))\right)$$

$$= \left|\frac{\text{Det } Q_0}{\text{Det}(Q_0 - Q_1)}\right|^{1/2} i^{-\text{Ind}((Q_0 - Q_1)/Q_0)}. \tag{3.60}$$

The ratio[7] of these infinite-dimensional determinants is equal to a

---

[7] This ratio should be more vigorously denoted $\text{Det}((Q_0 - Q_1)/Q_0)^{-1}$ (see Section 4.2). The same remark applies to formulas (3.64), (3.66), and (3.67).

3.3 The forced harmonic oscillator

finite-dimensional determinant (Appendix E):

$$\int_{\mathbb{X}_b} d\Gamma_{i,Q_0/h}(x) \exp\left(-\frac{\pi i}{h} Q_1(x)\right) = (\cos(\omega(t_b - t_a)))^{-1/2}. \quad (3.61)$$

In conclusion, the path integral (3.61) is indeed a representation of the matrix element on the right-hand side of (3.59). This result confirms the normalizations checked in the simpler case of the free particle.

Remark. Equation (3.61) also shows that one would be mistaken in assuming that the gaussian $d\Gamma_{i,(Q_0-Q_1)/h}(x)$ is equal to $d\Gamma_{i,Q_0/h}(x) \cdot \exp(-(\pi i/h)Q_1(x))$. The reason is that

$$d\Gamma_{i,Q_0/h}(x) \stackrel{f}{=} \mathcal{D}_{i,Q_0/h} x \cdot \exp(\pi i Q_0(x)/h), \quad (3.62)$$

$$d\Gamma_{i,(Q_0-Q_1)/h}(x) \stackrel{f}{=} \mathcal{D}_{i,(Q_0-Q_1)/h} x \cdot \exp(\pi i (Q_0 - Q_1)(x)/h), \quad (3.63)$$

and

$$\mathcal{D}_{i,Q_0/h} x / \mathcal{D}_{i,(Q_0-Q_1)/h} x \stackrel{f}{=} |\text{Det } Q_0/\text{Det}(Q_0 - Q_1)|^{1/2}. \quad (3.64)$$

□

In the case of the forced harmonic oscillator (in the next subsection) it is simpler to work with $\Gamma_{(Q_0-Q_1)/h}$ rather than $\Gamma_{Q_0/h}$. Given the action

$$S(x) = \frac{1}{2}(Q_0(x) - Q_1(x)) - \lambda \int_{\mathbb{T}} dt\, f(t) x(t), \quad (3.65)$$

we shall express the integral with respect to $\Gamma_{Q_0/h}$ as an integral with respect to $\Gamma_{(Q_0-Q_1)/h}$,

$$\int_{\mathbb{X}_b} d\Gamma_{i,Q_0/h}(x) \exp\left(-\frac{\pi i}{h} Q_1(x)\right) \exp\left(-\frac{2\pi i \lambda}{h} \int_{\mathbb{T}} dt\, f(t) x(t)\right)$$

$$= \left|\frac{\text{Det } Q_0}{\text{Det}(Q_0 - Q_1)}\right|^{1/2} \int_{\mathbb{X}_b} d\Gamma_{i,(Q_0-Q_1)/h}(x) \exp\left(-\frac{2\pi i \lambda}{h} \int_{\mathbb{T}} dt\, f(t) x(t)\right),$$

$$(3.66)$$

and use Appendix E for the explicit value of the ratio of these infinite-dimensional determinants, namely

$$|\text{Det } Q_0/\text{Det}(Q_0 - Q_1)|^{1/2} = (\cos(\omega(t_b - t_a)))^{-1/2}. \quad (3.67)$$

Remark. As noted in Section 1.1, Wiener showed the key role played by differential spaces. Here the kinetic energy Q_0 can be defined on a differential space, whereas $Q_0 - Q_1$ cannot. Therefore, one often needs to begin with Q_0 before introducing more general gaussians.

Selected examples

Choosing a quadratic form Q on $\mathbb{X}_b$

The integrand, $\exp(iS(x)/\hbar)$, suggests three choices for Q:

Q can be the kinetic energy;
Q can be the kinetic energy plus any existing quadratic terms in V; and
Q can be the second term in the functional Taylor expansion of S.

The third option is, in general, the best one because the corresponding gaussian term contains the most information. See for instance, in Section 4.3, the anharmonic oscillator. The expansion of S around its value for the classical path (the solution of the Euler–Lagrange equation) is called the semiclassical expansion of the action. It will be exploited in Chapter 4.

The forced harmonic oscillator

The action of the forced harmonic oscillator is

$$S(x) = \int_\mathbb{T} dt \left(\frac{m}{2} \dot{x}^2(t) - \frac{m}{2} \omega^2 x^2(t) - \lambda f(t) x(t) \right). \tag{3.68}$$

The forcing term $f(t)$ is assumed to be without physical dimension. Therefore the physical dimension of $\lambda/\hbar$ is $L^{-1}T^{-1}$ when $x(t)$ is of dimension L.

We choose the quadratic form $Q(x)$ to be

$$Q(x) := Q_0(x) - Q_1(x) = \frac{m}{h} \int_\mathbb{T} dt (\dot{x}^2(t) - \omega^2 x^2(t)), \qquad h = 2\pi\hbar, \tag{3.69}$$

and the potential contribution to be

$$\int_\mathbb{T} dt\, V(x(t), t) = 2\pi \frac{\lambda}{h} \langle f, x \rangle. \tag{3.70}$$

Remark. In the context of this application $Q(x)$ in (3.69) differs from $Q(x)$ in (3.52) by a factor h. The reason for not "hiding" h in the volume element in (3.69) and in (3.82) is the expansion (3.83) in powers of λ/h.

The quantum-mechanical transition amplitudes are given by path integrals of the type (3.66). To begin with, we compute

$$I := \int_{\mathbb{X}_{0,0}} d\Gamma_{0,0}(x) \exp\left(-2\pi i \frac{\lambda}{h} \langle f, x \rangle \right), \tag{3.71}$$

where

$$d\Gamma_{0,0}(x) \stackrel{f}{=} \mathcal{D}_Q(x) \cdot \exp(\pi i Q(x)) \tag{3.72}$$

3.3 The forced harmonic oscillator

is defined by equation (3.42), namely

$$\int_{\mathbb{X}_{0,0}} \mathrm{d}\Gamma_{0,0}(x)\exp(-2\pi\mathrm{i}\langle x', x\rangle) = \Gamma_{0,0}(\mathbb{X}_{0,0})\exp(-\mathrm{i}\pi W_{0,0}(x')). \qquad (3.73)$$

The kernel $G_{0,0}$ of $W_{0,0}$ is the unique Green function of the differential operator D defined by $Q(x)$ on $\mathbb{X}_{0,0}$ (see Appendix E, equation (E.30c)):

$$Q(x) = \langle Dx, x\rangle,$$

$$D = -\frac{m}{h}\left(\frac{\mathrm{d}^2}{\mathrm{d}t^2} + \omega^2\right), \qquad (3.74)$$

$$G_{0,0}(r,s) = \frac{h}{m}\frac{1}{\omega}\theta(s-r)\sin(\omega(r-t_a))\sin(\omega(t_b-t_a))^{-1}\sin(\omega(t_b-s))$$
$$- \frac{h}{m}\frac{1}{\omega}\theta(r-s)\sin(\omega(r-t_b))\sin(\omega(t_a-t_b))^{-1}\sin(\omega(t_a-s)), \qquad (3.75)$$

where θ is the Heaviside step function, which is equal to unity for positive arguments and zero otherwise.

The gaussian volume of $\mathbb{X}_{0,0} \subset \mathbb{X}_{b=0}$ is given by (3.39) in terms of the kernel of W_b defining the gaussian Γ on $\mathbb{X}_{b=0}$ by (3.35), i.e. in terms of the Green function (see Appendix E, equation (E.30a))

$$G_b(r,s) = \frac{h}{m}\frac{1}{\omega}\theta(s-r)\cos(\omega(r-t_a))\frac{1}{\cos(\omega(t_a-t_b))}\sin(\omega(t_b-s))$$
$$- \frac{h}{m}\frac{1}{\omega}\theta(r-s)\sin(\omega(r-t_b))\frac{1}{\cos(\omega(t_b-t_a))}\cos(\omega(t_a-s)), \qquad (3.76)$$

$$G_b(t_a, t_a) = \frac{h}{m}\frac{1}{\omega}\sin(\omega(t_b-t_a))\frac{1}{\cos(\omega(t_b-t_a))}, \qquad (3.77)$$

$$\Gamma_{0,0}(\mathbb{X}_{0,0}) = \left(\frac{\mathrm{i}m\omega}{h}\frac{\cos(\omega(t_b-t_a))}{\sin(\omega(t_b-t_a))}\right)^{1/2}. \qquad (3.78)$$

A quick calculation of (3.71)

The integrand being the exponential of a linear functional, we can use the Fourier transform (3.42) of the volume element with $x' = (\lambda/h)f$,

$$I = \Gamma_{0,0}(\mathbb{X}_{0,0})\exp\left(-\mathrm{i}\pi W_{0,0}\left(\frac{\lambda}{h}f\right)\right), \qquad (3.79)$$

where $\Gamma_{0,0}(\mathbb{X}_{0,0})$ is given by (3.78) and

$$W_{0,0}\left(\frac{\lambda}{h}f\right) = \left(\frac{\lambda}{h}\right)^2 \int_{\mathbb{T}} \mathrm{d}r \int_{\mathbb{T}} \mathrm{d}s\, f(r)f(s)G_{0,0}(r,s) \qquad (3.80)$$

is given explicitly by (3.75). Finally, on bringing together these equations with (3.66) and (3.67), one obtains

$$\langle 0, t_b | 0, t_a \rangle = \left(\frac{im\omega}{h \sin(\omega(t_b - t_a))} \right)^{1/2}$$
$$\times \exp\left(-i\pi \left(\frac{\lambda}{h}\right)^2 \int_{\mathbb{T}} dr \int_{\mathbb{T}} ds \, f(r)f(s)G_{0,0}(r,s) \right). \tag{3.81}$$

This amplitude is identical with the amplitude computed by L. S. Schulman [4] when $x(t_a) = 0$ and $x(t_b) = 0$.

A general technique valid for time-dependent potential

The quick calculation of (3.71) does not display the power of linear maps. We now compute the more general expression

$$I := \int_{\mathbb{X}_{0,0}} d\Gamma_{0,0}(x) \exp\left(-2\pi i \frac{\lambda}{h} \int_{\mathbb{T}} dt \, V(x(t)) \right), \tag{3.82}$$

where $V(x(t), t)$ is abbreviated to $V(x(t))$. Following the traditional method:

(i) expand the exponential,

$$I =: \sum I_n \tag{3.83}$$

$$I_n = \frac{1}{n!} \left(\frac{\lambda}{i\hbar}\right)^n \int_{\mathbb{X}_{0,0}} d\Gamma_{0,0}(x) \left(\int_{\mathbb{T}} dt \, V(x(t)) \right)^n;$$

(ii) exchange the order of integrations,

$$I_n = \frac{1}{n!} \left(\frac{\lambda}{i\hbar}\right)^n \int_{\mathbb{T}} dt_1 \ldots \int_{\mathbb{T}} dt_n \int_{\mathbb{X}_{0,0}} d\Gamma_{0,0}(x) V(x(t_1)) \ldots V(x(t_n)). \tag{3.84}$$

If $V(x(t))$ is a polynomial in $x(t)$, use a straightforward generalization of the polarization formula (2.47) for computing gaussian integrals of multilinear polynomials.

If $V(x(t)) = f(t)x(t)$, then

$$I_n = \frac{1}{n!} \left(\frac{\lambda}{i\hbar}\right)^n \int_{\mathbb{T}} dt_1 \, f(t_1) \ldots \int_{\mathbb{T}} dt_n \, f(t_n)$$
$$\times \int_{\mathbb{X}_{0,0}} d\Gamma_{0,0}(x) \langle \delta_{t_1}, x \rangle \ldots \langle \delta_{t_n}, x \rangle. \tag{3.85}$$

Each individual integral I_n can be represented by diagrams as shown after (2.47). A line $G(r, s)$ is now attached to $f(r)$ and $f(s)$, which encode the potential. Each term $f(r)$ is attached to a vertex.

The "quick calculation" bypassed the expansion of the exponential and the (tricky) combinatorics of the polarization formula.

Remarks

- A quadratic action functional S_0 defines a gaussian volume element by (2.29) and (2.30), and hence a covariance G by (2.28). On the other hand, a covariance is also a two-point function (2.37) – therefore it is the Feynman propagator for S_0.
- On the space $\mathbb{X}_{0,0}$, the paths are loops, and the expansion $I = \sum_n I_n$ is often called *loop expansion*. An expansion terminating at I_n is said to be of n-loop order. The physical-dimension analysis of the integral over $\mathbb{X}_{0,0}$ shows that the loop expansion is an expansion in powers of h. Indeed, a line representing G is of order h, whereas a vertex is of order h^{-1}. See (3.75) and (3.70). Let L be the number of lines, V the number of vertices, and K the number of independent closed loops; then

$$L - V = K - 1.$$

Therefore every diagram with K independent closed loops has a value proportional to $h^{K-1/2}$.

3.4 Phase-space path integrals

The example in Section 3.3 begins with the action functional (3.68) of the system; the paths take their values in a configuration space. In this section we construct gaussian path integrals over paths taking their values in phase space. As before, the domain of integration is not the limit $n = \infty$ of $\mathbb{R}^{2n}$, but a function space. The method of discretizing path integrals presented in Section 3.1 provides a comparison for earlier heuristic results obtained by replacing a path by a finite number of its values [6–8].

Notation

$(\mathbb{M}^D, g)$: the configuration space of the system, metric g
$T\mathbb{M}^D$: the tangent bundle over $\mathbb{M}$
$T^*\mathbb{M}^D$: the cotangent bundle over $\mathbb{M}$, i.e. the phase space
$L: T\mathbb{M}^D \times \mathbb{T} \longrightarrow \mathbb{R}$: the lagrangian
$H: T^*\mathbb{M}^D \times \mathbb{T} \longrightarrow \mathbb{R}$: the hamiltonian
(q, p): a classical path in phase space

(x, y): an arbitrary path in phase space; for instance, in the flat space $\mathbb{R}^D$,

$$x : \mathbb{T} \longrightarrow \mathbb{R}^D, \qquad y : \mathbb{T} \longrightarrow \mathbb{R}_D, \qquad \mathbb{T} = [t_a, t_b].$$

A path (x, y) satisfies D initial vanishing boundary conditions and D final vanishing boundary conditions.

$\theta := \langle p, \mathrm{d}q \rangle - H \, \mathrm{d}t$, the canonical 1-form (a relative integral invariant of the hamiltonian Pfaff system)

$F := \mathrm{d}\theta$, the canonical 2-form, a symplectic form on $\mathbb{T}^*\mathrm{M}$.

The phase-space action functional is

$$S(x, y) := \int \langle y(t), \mathrm{d}x(t) \rangle - H(y(t), x(t), t) \mathrm{d}t. \tag{3.86}$$

The action functions, solutions of the Hamilton–Jacobi equation for the various boundary conditions,

$$q(t_a) = x_a, \qquad q(t_b) = x_b, \qquad p(t_a) = p_a, \qquad p(t_b) = p_b,$$

are

$$\begin{align}
S(x_b, x_a) &= S(q, p), \tag{3.87}\\
S(x_b, p_a) &= S(q, p) + \langle p_a, q(t_a) \rangle, \tag{3.88}\\
S(p_b, x_a) &= S(q, p) - \langle p_b, q(t_b) \rangle, \tag{3.89}\\
S(p_b, p_a) &= S(q, p) - \langle p_b, q(t_b) \rangle + \langle p_a, q(t_a) \rangle, \tag{3.90}
\end{align}$$

where (q, p) is the classical path satisfying the said boundary conditions.

The Jacobi operator ([2] and [6])

The Jacobi operator in configuration space is obtained by varying a one-parameter family of paths in the action functional (Appendix E). The Jacobi operator in phase space is obtained by varying a two-parameter family of paths in the phase-space action functional. Let

$$u, v \in [0, 1]$$

and let $\bar{\gamma}(u, v) : \mathbb{T} \longrightarrow T^*\mathrm{M}$ be given in local coordinates by $(\bar{\alpha}^i(u), \bar{\beta}_i(u, v))$, where

$$\begin{align}
\bar{\alpha}^i(0) &= q^i, & \bar{\alpha}^i(1) &= x^i, \\
\bar{\beta}_i(0, 0) &= p_i, & \bar{\beta}_i(1, 1) &= y_i,
\end{align} \tag{3.91}$$

and set

$$\bar{\alpha}^i(u)(t) = \alpha^i(u, t), \qquad \bar{\beta}_i(u, v)(t) = \beta_i(u, v, t).$$

For $\mathrm{M}^D = \mathbb{R}^D$, the family $\bar{\beta}$ depends only on v.

3.4 Phase-space path integrals

Let ζ be the $2D$-dimensional vector (D contravariant and D covariant components):

$$\zeta := \begin{pmatrix} \xi \\ \eta \end{pmatrix},$$

where

$$\xi^i := \left.\frac{d\bar{\alpha}^i(u)}{du}\right|_{u=0}, \qquad \eta_i := \left.\frac{\partial\bar{\beta}_i(u,v)}{\partial v}\right|_{v=0}. \tag{3.92}$$

The expansion of the action functional $S(x,y)$ around $S(q,p)$ is

$$(S \circ \bar{\gamma})(1,1) = \sum_{n=0}^{\infty} \frac{1}{n!}(S \circ \bar{\gamma})^{(n)}(0,0).$$

The first variation vanishes for paths satisfying the Hamilton set of equations. The second variation defines the Jacobi operator.

Example. The phase-space Jacobi equation in $\mathbb{R}^D \times \mathbb{R}_D$ with coordinates q^i, p_i. The expansion of the action functional (3.86) gives the following Jacobi equation:

$$\begin{pmatrix} -\partial^2 H/\partial q^i\,\partial q^j & -\delta_i{}^j\partial/\partial t - \partial^2 H/\partial q^i\,\partial p_j \\ \delta^i{}_j\partial/\partial t - \partial^2 H/\partial p_i\,\partial q^j & -\partial^2 H/\partial p_i\,\partial p_j \end{pmatrix} \begin{pmatrix} \xi^j \\ \eta_j \end{pmatrix} = 0. \tag{3.93}$$

Example. A free particle in the riemannian manifold $\mathbb{M}^D$ with metric g. Varying u results in changing the fiber $T^*_{\alpha(u,t)}\mathbb{M}^D$; the momentum $\beta(u,v,t)$ is conveniently chosen to be the momentum parallel transported along a geodesic beginning at $\alpha(0,t)$; the momentum is *then* uniquely defined by

$$\nabla_u \beta = 0,$$

the vanishing of the covariant derivative of β along the path $\alpha(\cdot,t) : u \mapsto \alpha(u,t)$.

The action functional is

$$S(p,q) = \int_\mathbb{T} dt \left(\langle p(t), \dot{q}(t) \rangle - \frac{1}{2m}(p(t)|p(t)) \right). \tag{3.94}$$

The bracket $\langle\,,\,\rangle$ is the duality pairing of $T\mathbb{M}$ and $T^*\mathbb{M}$ and the parenthesis $(\,|\,)$ is the scalar product defined by the inverse metric g^{-1}.

The Jacobi operator is

$$\mathcal{J}(q,p) = \begin{pmatrix} -(1/m)R^l_{ikj}g^{km}p_m p_l & -\delta^j_i \nabla_t \\ \delta^i_j \nabla_t & -(1/m)g^{ij} \end{pmatrix}. \tag{3.95}$$

Covariances

The second variation of the action functional in phase space provides a quadratic form Q on the space parametrized by (ξ, η) with

$$\xi \in T_x \mathbb{M}^D, \qquad \eta \in T_y^* \mathbb{M}^D,$$
$$Q(\xi, \eta) = \langle \mathcal{J}(q,p) \cdot (\xi, \eta), (\xi, \eta) \rangle, \qquad (3.96)$$

where $\mathcal{J}(q,p)$ is the Jacobi operator defined by a classical path (q,p). There exists an inverse quadratic form W, corresponding to the quadratic form Q; it is defined by the Green functions of the Jacobi operator,

$$W(\xi', \eta') = (\xi'_\alpha, \eta'^\alpha) \begin{pmatrix} G_1^{\alpha\beta} & G_{2\beta}^{\alpha} \\ G_\alpha^{3\beta} & G_{\alpha\beta}^4 \end{pmatrix} \begin{pmatrix} \xi'_\beta \\ \eta'^\beta \end{pmatrix}, \qquad (3.97)$$

where

$$\mathcal{J}_r(q,p) G(r,s) = \mathbf{1} \delta_s(r); \qquad (3.98)$$

$\mathcal{J}_r$ acts on the r-argument of the $2D \times 2D$ matrix G made of the four blocks (G_1, G_2, G^3, G^4).

Once the variance W and the covariance G have been identified, path integrals over phase space are constructed as in the previous examples. For explicit expressions that include normalization, correspondences with configuration-space path integrals, discretization of phase-space integrals, physical interpretations of the covariances in phase space, and infinite-dimensional Liouville volume elements, see, for instance, John LaChapelle's Ph.D. dissertation "Functional integration on symplectic manifolds" [9].

References

[1] A. Maheshwari (1976). "The generalized Wiener–Feynman path integrals," *J. Math. Phys.* **17**, 33–36.

[2] C. DeWitt-Morette, A. Maheshwari, and B. Nelson (1979). "Path integration in non relativistic quantum mechanics," *Phys. Rep.* **50**, 266–372.

[3] J. J. Sakurai (1985). *Modern Quantum Mechanics* (Menlo Park, CA, Benjamin/Cummings).

[4] L. S. Schulman (1981). *Techniques and Applications of Path Integration* (New York, John Wiley).

[5] C. DeWitt-Morette (1974). "Feynman path integrals; I. Linear and affine techniques; II. The Feynman–Green function," *Commun. Math. Phys.* **37**, 63–81.

[6] C. DeWitt-Morette, A. Maheshwari, and B. Nelson (1977). "Path integration in phase space," *General Relativity and Gravitation* **8**, 581–593.

[7] M. M. Mizrahi (1978). "Phase space path integrals, without limiting procedure," *J. Math. Phys.* **19**, 298–307.

[8] C. DeWitt-Morette and T.-R. Zhang (1983). "A Feynman–Kac formula in phase space with application to coherent state transitions," *Phys. Rev.* **D28**, 2517–2525.

[9] J. LaChapelle (1995). Functional integration on symplectic manifolds, unpublished Ph.D. Dissertation, University of Texas at Austin.

4
Semiclassical expansion; WKB

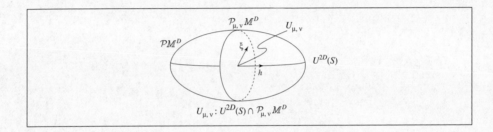

4.1 Introduction

Classical mechanics is a natural limit of quantum mechanics; therefore the expansion of the action functional S around its value at (or near) a classical system leads to interesting results[1] in quantum physics.

The expansion, written as

$$S(x) = S(q) + S'(q) \cdot \xi + \frac{1}{2!} S''(q) \cdot \xi\xi + \frac{1}{3!} S'''(q) \cdot \xi\xi\xi + \cdots, \qquad (4.1)$$

is constructed in Appendix E.

The role of classical solutions in quantum physics is comparable to the role of equilibrium points in classical dynamical systems; there the time evolution of the system is governed by the dynamical equation

$$\mathrm{d}x(t) = f(x(t))\mathrm{d}t, \qquad (4.2)$$

---

[1] The semiclassical expansion in path integrals was introduced [1] in 1951 for defining the absolute value of the normalization left undefined by Feynman, and for computing approximate values of path integrals. By now there is a very rich literature on semiclassical expansion. For details and proofs of several results presented in this chapter see [2].

4.1 Introduction

and an equilibrium point $\bar{x}$ satisfies $f(\bar{x}) = 0$. The nature of the equilibrium point is found from the long-time behavior of the nearby motions $x(t)$, which in turn is determined, to a large extent, by the derivative $f'(\bar{x})$ evaluated at the equilibrium point. For example, if the dynamical equation (4.2) reads

$$\frac{dq(t)}{dt} = v(t),$$
$$\frac{dv(t)}{dt} = -\nabla V(q(t)), \tag{4.3}$$

then the equilibrium point $(\bar{q}, \bar{v})$ is a critical point $(\nabla V(\bar{q}) = 0)$ of the potential $V(q)$; its nature is determined by the hessian of V at $\bar{q}$, namely

$$-\frac{\partial^2 V(q)}{\partial q^\alpha \, \partial q^\beta}\bigg|_{q=\bar{q}}. \tag{4.4}$$

Given an action functional S on a space $\mathbb{X}$ of functions x, its critical point q, $S'(q) = 0$, is a classical solution; its hessian at q is

$$\mathrm{Hess}(q; \xi, \eta) := S''(q) \cdot \xi \eta.$$

The hessian of the action functional determines the nature of the semiclassical expansion of functional integrals, such as

$$I = \int_\mathbb{X} \mathcal{D}x \, \exp\left(\frac{i}{\hbar} S(x)\right) \cdot \phi(x(t_a)). \tag{4.5}$$

Remark. The hessian also provides a quadratic form for expressing an action-functional integral as a gaussian integral over the variable $\xi \in T_q \mathbb{X}$.

The arena for semiclassical expansion consists of the intersection of two spaces: the space U (or U^{2D}, or $U^{2D}(S)$) of classical solutions (that is, the critical points) of the action functional S of the system, and the domain of integration $\mathcal{P}_{\mu,\nu} \mathbb{M}^D$ of the functional integral.

Consider a space $\mathcal{P}\mathbb{M}^D$ of paths $x : \mathbb{T} \to \mathbb{M}^D$ on a D-dimensional manifold $\mathbb{M}^D$ (or simply $\mathbb{M}$). The action functional S is a function

$$S : \mathcal{P}\mathbb{M}^D \to \mathbb{R}, \tag{4.6}$$

defined by a Lagrangian L,

$$S(x) = \int_\mathbb{T} dt \, L(x(t), \dot{x}(t), t). \tag{4.7}$$

Two subspaces of $\mathcal{P}\mathbb{M}^D$ and their intersection dominate the semiclassical expansion:

- $U \subset \mathcal{P}\mathbb{M}^D$ (or U^{2D}, or $U^{2D}(S)$), the space of critical points of S,

$$q \in U \Leftrightarrow S'(q) = 0,$$

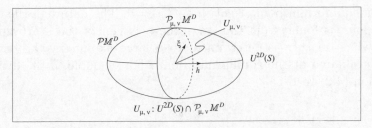

Fig. 4.1 $U_{\mu,\nu} := U^{2D}(S) \cap \mathcal{P}_{\mu,\nu}\mathbb{M}^D$

i.e. q is a solution of the Euler–Lagrange equations; it follows that U is $2D$-dimensional.
- $\mathcal{P}_{\mu,\nu}\mathbb{M}^D \subset \mathcal{P}\mathbb{M}^D$, the space of paths satisfying D initial conditions (μ) and D final conditions (ν).
- Let $U_{\mu,\nu}$ be their intersection,

$$U_{\mu,\nu} := U \cap \mathcal{P}_{\mu,\nu}\mathbb{M}^D.$$

In Section 4.2 the intersection $U_{\mu,\nu}$ consists of only one point q, or several isolated points $\{q_i\}$. In Section 5.1 the intersection $U_{\mu,\nu}$ is of dimension $\ell > 0$. In Section 5.2 the intersection $U_{\mu,\nu}$ is a multiple root of $S'(q) \cdot \xi = 0$. In Section 5.3 the intersection $U_{\mu,\nu}$ is an empty set.

One can approach $U_{\mu,\nu}$ either from the tangent space $T_q U^{2D}(S)$ to the space U of classical solutions of S, or from the tangent space $T_q \mathcal{P}_{\mu,\nu}\mathbb{M}^D$ to the domain of integration $\mathcal{P}_{\mu,\nu}\mathbb{M}^D$.

One can also introduce a variation $S + \delta S$ of the action S and approach $U^{2D}(S)$ from $U^{2D}(S + \delta S)$.

4.2 The WKB approximation

Let the intersection $U_{\mu,\nu}$ of the space U of classical solutions q and the space $\mathcal{P}_{\mu,\nu}\mathbb{M}^D$ of paths x on $\mathbb{M}^D$ satisfying D initial conditions (μ) and D final conditions (ν) consist of one simple root q of the Euler–Lagrange equation. The case of $U_{\mu,\nu}$ consisting of a collection of *isolated* points $\{q_i\}$ (e.g. the anharmonic oscillator) is similar. Expand the action functional around its value at q,

$$S(x) = S(q) + \frac{1}{2}S''(q) \cdot \xi\xi + \Sigma(q;x) \qquad \text{for } \xi \in T_q\mathcal{P}_{\mu,\nu}\mathbb{M}^D. \qquad (4.8)$$

The WKB approximation consists of dropping $\Sigma(q;x)$ and integrating over the space $T_q\mathcal{P}_{\mu,\nu}\mathbb{M}^D$ of vector fields ξ:

$$I_{\text{WKB}} := \int_{\mathbb{X}_b} \mathcal{D}\xi \, \exp\left(\frac{i}{\hbar}\left(S(q) + \frac{1}{2}S''(q) \cdot \xi\xi\right)\right) \cdot \phi(x(t_a)). \qquad (4.9)$$

4.2 The WKB approximation

The domain of integration $\mathbb{X}_b$ stands for the tangent space $T_q\mathcal{P}_{\mu,\nu}\mathbb{M}^D$ at q of $\mathcal{P}_{\mu,\nu}\mathbb{M}^D$ with D vanishing final conditions (ν),

$$x(t_b) \text{ a fixed point} \Rightarrow \xi(t_b) = 0, \qquad (4.10)$$

and D arbitrary initial conditions (μ), to be specified by the choice of the initial "wave function" ϕ. For position-to-position transition, choose

$$\phi(x(t_a)) = \delta(x(t_a) - x_a). \qquad (4.11)$$

For momentum-to-position transitions, if $\mathbb{M}^D \neq \mathbb{R}^D$, choose

$$\phi(x(t_a)) = \exp\left(\frac{i}{\hbar}S_0(x(t_a))\right) \cdot \mathcal{T}(x(t_a)), \qquad (4.12)$$

where $\mathcal{T}$ is a smooth function on $\mathbb{M}^D$ of compact support and S_0 is an arbitrary, but well-behaved, function on $\mathbb{M}^D$. Equation (4.12) generalizes plane waves on $\mathbb{R}^D$; it is not a momentum eigenstate, but it is adequate[2] for computing WKB approximations of momentum-to-position transitions (see the next subsection).

The space of paths is such that $S''(q) \cdot \xi\xi$ is finite, i.e. an $L^{2,1}$ space (with square-integrable first derivatives) and $\mathcal{D}\xi$ is the canonical gaussian on $L^{2,1}$ defined by the norm on $L^{2,1}$ (see Section 3.2),

$$\|\xi\|_{L^{2,1}} = \left(\int_\mathbb{T} dt\ g_{ij}\dot{\xi}^i(t)\dot{\xi}^j(t)\right)^{1/2}. \qquad (4.13)$$

When is it justified to replace an integral (4.5) over x by the integral (4.9) over ξ?

If $x(t) \in \mathbb{R}^D$, then in $\mathcal{P}_{\mu,\nu}\mathbb{R}^D$ we can write $x = q + \xi$ and hence $\mathcal{D}x = \mathcal{D}\xi$. If

$$Q(x) = \frac{1}{2}\int dt(\dot{x}(t))^2,$$

then

$$Q(q+\xi) = \frac{1}{2}\int dt(\dot{q}(t))^2 + \int dt\ \dot{q}(t)\dot{\xi}(t) + \frac{1}{2}\int dt(\dot{\xi}(t))^2. \qquad (4.14)$$

The first term contributes to $S(q)$, the second contributes to $S'(q) \cdot \xi = 0$, the third term is equal to $Q(\xi)$, and there is no problem in replacing the integral over x by an integral over ξ. The WKB approximation is strictly equivalent to dropping $\Sigma(q;x)$ in the expansion of the action functional S. The neglected contributions can easily be computed.

When $\mathbb{M}^D \neq \mathbb{R}^D$, the contractibility of $\mathbb{X}_b$ allows one to map it onto the tangent space of $\mathcal{P}_{\mu,\nu}\mathbb{M}^D$ at q. This technique plays a major role

---

[2] See the justification in [3].

in Chapter 7 and will be developed there *ab initio*. At this point, we simply compute I_{WKB} given by (4.9) for ϕ a plane wave of momentum p_a smoothly limited to a neighborhood of $x(t_a)$

$$\phi(x(t_a)) = \exp\left(\frac{i}{\hbar}\langle p_a, x(t_a)\rangle\right) \mathcal{T}(x(t_a)). \tag{4.15}$$

Momentum-to-position transitions

The WKB approximation (4.9) is then given explicitly, under the assumption (4.15), by the well-known formula

$$I_{\text{WKB}}(x_b, t_b; p_a, t_a) = \exp\left(\frac{i}{\hbar}\mathcal{S}(x_{\text{cl}}(t_b), p_{\text{cl}}(t_a))\right)$$
$$\times \left(\det\left(\frac{\partial^2 \mathcal{S}}{\partial x^i_{\text{cl}}(t_b)\partial p_{\text{cl}j}(t_a)}\right)\right)^{1/2}. \tag{4.16}$$

In this formula, we denote by $(x_{\text{cl}}(t), p_{\text{cl}}(t))$ the classical motion in phase space under the boundary conditions $p_{\text{cl}}(t_a) = p_a$ and $x_{\text{cl}}(t_b) = x_b$. We give an outline of the proof.

The terms independent of ξ, namely

$$S(q) + \langle p_a, x(t_a)\rangle = \mathcal{S}(x_{\text{cl}}(t_b), p_{\text{cl}}(t_a)), \tag{4.17}$$

combine to give the action function. The integral over ξ yields the determinant of the hessian of the action function. First we compute

$$\int_{\mathbb{X}} \mathcal{D}\xi \exp\left(\frac{i}{2\hbar}(Q_0(\xi) + Q(\xi))\right), \tag{4.18}$$

where $Q_0(\xi) + Q(\xi) = S''(q) \cdot \xi\xi$ and $Q_0(\xi)$ is the kinetic-energy contribution to the second variation. Moreover, $\mathcal{D}\xi$ is what was denoted more explicitly by $\mathcal{D}_{i,Q_0/\hbar}\xi$ in Chapter 2, equation $(2.30)_s$. Integrals similar to (4.18) occur frequently, and we proceed to calculate them.

More generally, consider a space Ξ endowed with two quadratic forms Q and Q_1. According to our general theory, we have two volume elements $\mathcal{D}\xi$ and $\mathcal{D}_1\xi$ characterized by

$$\int_{\Xi} \mathcal{D}\xi \cdot e^{-\pi Q(\xi)/s} = \int_{\Xi} \mathcal{D}_1\xi \cdot e^{-\pi Q_1(\xi)/s} = 1, \tag{4.19}$$

where s is equal to 1 or i.

Proposition

(a) *If $s = 1$, and therefore Q and Q_1 are positive definite, then*

$$\int_{\Xi} \mathcal{D}\xi \cdot e^{-\pi Q_1(\xi)} = \text{Det}(Q_1/Q)^{-1/2}. \tag{4.20}$$

4.2 The WKB approximation

(b) If $s = i$, i.e. for oscillatory integrals, and Q is positive-definite, then

$$\int_\Xi \mathcal{D}\xi \cdot e^{\pi i Q_1(\xi)} = |\mathrm{Det}(Q_1/Q)|^{-1/2} i^{-\mathrm{Ind}(Q_1/Q)}, \qquad (4.21)$$

where $\mathrm{Ind}(Q_1/Q)$ counts the number of negative eigenvalues of Q_1 with respect to Q.

To calculate (4.18), we need to specialize $Q = Q_0/h$ and $Q_1(\xi) = S''(q) \cdot \xi\xi/h$. The factor h disappears when we take the determinant of Q_1/Q as explained below.

Let us give the proof in a finite-dimensional space Ξ. Write

$$Q(\xi) = q_{ij}\xi^i \xi^j, \qquad Q_1(\xi) = q^1_{ij}\xi^i \xi^j,$$

for a vector ξ with N components ξ^i. Assuming first that Q and Q_1 are positive definite, and writing $\mathrm{d}^N\xi$ for $\mathrm{d}\xi^1 \ldots \mathrm{d}\xi^N$, we know that

$$\mathcal{D}\xi = \mathrm{d}^N\xi \cdot (\det q_{ij})^{1/2}, \qquad (4.22)$$

$$\mathcal{D}_1\xi = \mathrm{d}^N\xi \cdot (\det q^1_{ij})^{1/2}; \qquad (4.23)$$

hence

$$\mathcal{D}\xi = \mathcal{D}_1\xi \cdot \left(\det(q^1_{ij})/\det(q_{ij})\right)^{-1/2} \qquad (4.24)$$

and the quotient of determinants is the determinant of the matrix $(q_{ij})^{-1} \cdot (q^1_{ij}) = (u^i{}_j)$. The meaning of $u = (u^i{}_j)$ is the following: introduce two invertible linear maps D and D_1 mapping from Ξ to its dual Ξ' such that

$$\begin{aligned} Q(\xi) &= \langle D\xi, \xi \rangle, \qquad Q_1(\xi) = \langle D_1\xi, \xi \rangle, \\ \langle D\xi, \eta \rangle &= \langle D\eta, \xi \rangle, \qquad \langle D_1\xi, \eta \rangle = \langle D_1\eta, \xi \rangle. \end{aligned} \qquad (4.25)$$

Then u is the operator $\Xi \to \Xi$ such that

$$D_1 = Du. \qquad (4.26)$$

Once we know that $\mathcal{D}\xi = c\mathcal{D}_1\xi$ with a constant $c \neq 0$, then

$$\int_\Xi \mathcal{D}\xi \cdot e^{-\pi Q_1(\xi)} = c \int_\Xi \mathcal{D}_1\xi \cdot e^{-\pi Q_1(\xi)} = c \qquad (4.27)$$

by (4.19). Hence equation (4.20) follows.

Consider now the case $s = i$. Here we use the value of the Fresnel integral

$$\int \mathrm{d}u \exp(\pm \pi i a u^2) = a^{-1/2} e^{\pm i\pi/4} \qquad (4.28)$$

for $a > 0$. Hence, if Q is an indefinite quadratic form, put it into the diagonal form $Q(\xi) = \sum_{i=1}^{N} a_i \xi_i^2$, then

$$\mathcal{D}\xi = |a_1 \ldots a_N|^{1/2} \, \mathrm{d}^N \xi \, \mathrm{e}^{-\mathrm{i}\pi N/4} \mathrm{i}^\sigma \tag{4.29}$$

when σ is the number of negative numbers in the sequence $a_1, \ldots, a_N$. We have a similar formula for $\mathcal{D}_1 \xi$, namely

$$\mathcal{D}_1 \xi = |a_1^1 \ldots a_N^1|^{1/2} \, \mathrm{d}^N \xi \, \mathrm{e}^{-\mathrm{i}\pi N/4} \mathrm{i}^{\sigma_1} \tag{4.30}$$

and it follows that $\mathcal{D}\xi = c\,\mathcal{D}_1 \xi$, with $c = \Pi_{i=1}^{N} |a_i/a_i^1|^{1/2} \mathrm{i}^{\sigma - \sigma_1}$. The absolute value is $|\det u|^{-1/2}$, that is $|\mathrm{Det}(Q_1/Q)|^{-1/2}$, and since Q is assumed to be positive-definite, one has $\sigma = 0$ and σ_1 is the number of negative eigenvalues of u, that is the index $\mathrm{Ind}(Q_1/Q)$. Once we know that $\mathcal{D}\xi = c\,\mathcal{D}_1 \xi$ with $c = |\mathrm{Det}(Q_1/Q)|^{-1/2} \mathrm{i}^{-\mathrm{Ind}(Q_1/Q)}$, equation (4.21) follows immediately from (4.19).

Let us comment on the changes needed when Ξ is infinite-dimensional. We can introduce invertible maps D and D_1 mapping from Ξ to Ξ' satisfying (4.25) and then define $u : \Xi \to \Xi$ by (4.26). Assuming that $u - 1$ is trace-class ("nuclear" in Grothendieck's terminology), one can define the determinant of u, and we put $\mathrm{Det}(Q_1/Q) := \det u$ by definition. In the indefinite case, we have to assume that Ξ can be decomposed into a direct sum $\Xi_1 \oplus \Xi_2$ such that Q and Q_1 are positive-definite on Ξ_1 and that Ξ_2 is finite-dimensional, and Ξ_1 and Ξ_2 are orthogonal w.r.t. both Q and Q_1. This immediately reduces the proof to the cases in which Q and Q_1 are both positive-definite, or the finite-dimensional case treated before. $\square$

The determinants in (4.20) and (4.21) are infinite-dimensional.[3] It remains to prove that their ratio is a finite Van Vleck–Morette determinant [5] – the finite determinant in (4.16) for momentum-to-position transitions, or the finite determinant (4.45) for position-to-position transitions. There are many lengthy proofs of this remarkably useful fact. Each author favors the proof using his/her previously established results. Following Kac's advice [6, p. 128] that "one should try to formulate even familiar things in as many different ways as possible," we shall give the gist of six proofs with references and apologies for not quoting all the proofs.

Finite-dimensional determinants

(i) In the discretized version of path integrals [1] the infinite limit of the product of the short-time normalizations is shown to be the square root of a jacobian, which appears if one changes the description from

---

[3] See [3], pp. 2295–2301, the section on infinite-dimensional determinants. See also Section 11.2 and Appendix E.

momenta (at discretized times) to points (at discretized times). In modern parlance, this says that the position-to-position determinant (4.45) can be obtained from a phase-space path integral [7] that has a short-time propagator normalized to unity.

(ii) The prodistribution definition of path integrals (see [4, 8]) generalizes the definition of oscillatory integrals over $\mathbb{R}^n$ to integrals over spaces that are not necessarily locally compact. Prodistributions can be used on topological vector spaces $\mathbb{X}$ that are Hausdorff and locally convex. In brief, one works with the projective system of $\mathbb{X}$ – that is a family of finite-dimensional quotient spaces of $\mathbb{X}$ indexed by closed subspaces of finite codimension. The word "prodistribution" stands for "projective family of distributions," like "promeasure" stands for "projective family of measures." In spite of their sophisticated names, prodistributions are a very practical tool – to quote only one application, they are useful in obtaining a closed form glory-scattering cross section [9].

Showing that the infinite-dimensional determinants (4.20) and (4.21) on $\mathbb{X}$ are equal to finite Van Vleck–Morette determinants can be done as follows. Choose the projective system on $\mathbb{X}$ indexed by closed subspaces V, W, etc., where

$$V \text{ is the set of functions } f \text{ vanishing on a partition } \theta_v = \{t_1, \ldots, t_v\}. \tag{4.31}$$

The quotient $\mathbb{X}/V$ is the v-dimensional space of functions f defined by $\{f(t_1), \ldots, f(t_v)\}$.

The reader may wonder whether it is worthwhile to introduce projective systems when they lead to spaces such as $\mathbb{X}/V$ that discretization introduces right away. The answer is that discretization deals with meaningless limits of integrals over $\mathbb{R}^n$ when n tends to infinity, whereas projective systems deal with families of meaningful projections of $\mathbb{X}$ on $\mathbb{R}^v$, $\mathbb{R}^w$, etc. There are many choices of projective systems of a space $\mathbb{X}$. We choose one that shows how prodistributions relate to discretization; it is also a practical choice for the computation of ratio of infinite-dimensional determinants. The work consists of a certain amount of algebra on finite-dimensional spaces.

(iii) A generalized Cameron–Martin formula [4, 10].[4] Determinants are objects defined by linear mappings; therefore, the Cameron–Martin formula, which deals with gaussian integrals under a linear change

---

[4] There are generalizations and several versions of the original Cameron–Martin formula that are useful in a variety of applications. A number of them can be found in the above references.

of variable of integration, provides a tool for computing. Let (see figure 2.2)

$$L : \mathbb{X} \to \mathbb{Y} \text{ by } \xi \mapsto \eta \quad \text{such that } Q_{\mathbb{X}} = Q_{\mathbb{Y}} \circ L,$$
$$\tilde{L} : \mathbb{Y}' \to \mathbb{X}' \text{ by } \eta' \mapsto \xi' \quad \text{such that } W_{\mathbb{Y}'} = W_{\mathbb{X}'} \circ \tilde{L}, \quad (4.32)$$
$$W_{\mathbb{X}'}(x') = \langle x', G_{\mathbb{X}'} x' \rangle \text{ and similarly for } W_{\mathbb{Y}'}.$$

Then

$$\text{Det}(Q_{\mathbb{X}}/Q_{\mathbb{Y}})^{1/2} = \text{Det} L = \text{Det} \tilde{L} = \text{Det}(W_{\mathbb{Y}'}/W_{\mathbb{X}'})^{1/2}. \quad (4.33)$$

For a momentum-to-position transition (Appendix E, equation (E.24a))

$$G_{\mathbb{X}}(r,s) = \theta(s-r)K(r,t_a)N(t_a,t_b)J(t_b,s)$$
$$- \theta(r-s)J(r,t_b)\tilde{N}(t_b,t_a)\tilde{K}(t_a,s). \quad (4.34)$$

In (4.33) the quadratic form $Q_{\mathbb{Y}}$ is the kinetic energy. For paths $\xi : \mathbb{T} \to \mathbb{R}$ and $m = 1$, $Q_{\mathbb{Y}}(\xi) = \frac{1}{2} \int_{\mathbb{T}} dt \, \dot{x}^2(t)$, and

$$G_{\mathbb{Y}}(r,s) = \theta(s-r)(t_b - s) - \theta(r-s)(r-t_b). \quad (4.35)$$

The linear map $\tilde{L} : \mathbb{Y}' \to \mathbb{X}'$ by $\eta' \mapsto \xi'$ is explicitly given by

$$\xi'(t) = \eta'(t) + \int_{\mathbb{T}} dr \, \theta(r-t) \frac{dK}{dr}(r,t_a)N(t_a,r)\eta'(r). \quad (4.36)$$

With an explicit expression for $\tilde{L}$, we use the fundamental relation between determinant and trace

$$d \ln \text{Det} \tilde{L} = \text{Tr}(\tilde{L}^{-1} d\tilde{L}). \quad (4.37)$$

After some calculations[5]

$$\text{Det} \tilde{L} = (\det K(t_b, t_a)/\det K(t_a, t_a))^{1/2} \quad (4.38)$$
$$= (\det K(t_b, t_a))^{1/2}.$$

The inverse of the Jacobi matrix K is the hessian of the classical action function $\mathcal{S}(x_{\text{cl}}(t_b), p_{\text{cl}}(t_a))$, hence

$$\text{Det} \tilde{L} = \left(\det \left(\frac{\partial^2 \mathcal{S}}{\partial x^i_{\text{cl}}(t_b) \partial p_{\text{cl}\, j}(t_a)} \right) \right)^{-1/2}. \quad (4.39)$$

Equation (4.16) follows from (4.21) and (4.33). □

---

[5] The calculations are explicitly given in [10] and [4]; they can be simplified by using the qualified equality $d\Gamma_{\mathbb{X}}(x) \stackrel{\int}{=} \mathcal{D}_{\mathbb{X}} x \cdot \exp(-(\pi/s)Q_{\mathbb{X}}(x))$.

(iv) Varying the integration [3]. Equation (4.16) is fully worked out in [3, pp. 2254–2259]. The strategy for computing $\mathrm{Det}(Q_{\mathrm{X}}/Q_{\mathrm{Y}})$, where

$$Q_{\mathrm{Y}}(\xi) = Q_0(\xi), \qquad Q_{\mathrm{X}}(\xi) = Q_0(\xi) + Q(\xi) = S''(q) \cdot \xi\xi,$$

consists of introducing a one-parameter action functional $S_\nu(x)$ such that

$$S''_\nu(q) \cdot \xi\xi = Q_0(\xi) + \nu Q(\xi) =: Q_\nu(\xi). \tag{4.40}$$

Physical arguments as well as previous calculations suggest that $\mathrm{Det}(Q_0/Q_1)$ is equal to $\det K^i_j(t_b, t_a)$, where the Jacobi matrix K is defined (Appendix E) by

$$K^i_j(t_b, t_a) = \partial x^i_{\mathrm{cl}}(t_b)/\partial x^j_{\mathrm{cl}}(t_a). \tag{4.41}$$

From the one-parameter family of action functionals $S_\nu(x) \equiv S(\nu; x)$, one gets a one-parameter family of Jacobi matrices

$$K^i_j(\nu; t_b, t_a) = \partial x^i_{\mathrm{cl}}(\nu; t_b)/\partial x^j_{\mathrm{cl}}(\nu; t_a). \tag{4.42}$$

Set

$$c(\nu) := \det K(\nu; t_b, t_a)/K(0; t_b, t_a). \tag{4.43}$$

A nontrivial calculation [3] proves that $c(\nu)$ satisfies the same differential equation and the same boundary condition as $\mathrm{Det}(Q_0/Q_\nu)$.

(v) Varying the end points of a classical path (Appendix E, equations (E.33)–(E.44)). This method gives the Van Vleck–Morette determinant for the position-to-position transitions (4.45). Let $u = (x_a, x_b)$ characterize the end points of the classical path $x_{\mathrm{cl}}(u)$. To the action functional $S(x_{\mathrm{cl}}(u))$ corresponds an action function $\mathcal{S}$:

$$S(x_{\mathrm{cl}}(u)) = \mathcal{S}(x_{\mathrm{cl}}(t_a; u), x_{\mathrm{cl}}(t_b; u)). \tag{4.44}$$

By taking the derivatives with respect to x_a and x_b (i.e. w.r.t. the D components of x_a and the D components of x_b) of both sides of (4.44), one obtains an equality between the ratio of two infinite-dimensional determinants (left-hand side) and a finite-dimensional determinant (right-hand side). The calculation is spelled out in Appendix E.

(vi) Varying the action functional [11]. The method is worked out explicitly for the case of position-to-position transitions in (4.45) [11, pp. 239–241]. The variation of the action functional induces a variation of its hessian more general than the variation of the hessian in (4.40). Consider two Green functions of the same Jacobi operator, but with different boundary conditions such that their difference is expressed in terms of a Jacobi matrix. The variation of the ratio of the determinants of these Green functions can be expressed in terms of the trace of their differences (4.37), i.e. in terms of the trace of a Jacobi matrix. Using again the trace determinant equation (4.37), one arrives at an equation between the ratio of infinite-dimensional determinants and a finite-dimensional determinant.

Position-to-position transitions

A derivation similar to the derivation of (4.16) yields (omitting phase factors)

$$I_{\text{WKB}}(x_b, t_b; x_a, t_a) = \exp\left(\frac{i}{\hbar}\mathcal{S}(x_{\text{cl}}(t_b), x_{\text{cl}}(t_a))\right) \quad (4.45)$$

$$\times \left(\det\left(\frac{\partial^2 \mathcal{S}/\hbar}{\partial x_{\text{cl}}^i(t_b)\partial x_{\text{cl}}^j(t_a)}\right)\right)^{1/2},$$

where $\mathcal{S}$ is the classical action function evaluated for the classical path defined by $x_{\text{cl}}(t_b) = x_b$, $x_{\text{cl}}(t_a) = x_a$.

4.3 An example: the anharmonic oscillator

The one-dimensional quartic anharmonic oscillator is a particle of mass m in a potential

$$V(x) = \frac{1}{2}m\omega^2 x^2 + \frac{1}{4}\lambda x^4. \quad (4.46)$$

Maurice Mizrahi [12] used path integrals, as developed here,[6] for computing the semiclassical expansion of the anharmonic oscillator propagator:

$$\langle x_b, t_b | x_a, t_a \rangle = \int_{\mathbb{X}_{a,b}} \mathcal{D}x \, \exp\left(\frac{i}{\hbar}\int_{\mathbb{T}} dt\left(\frac{m}{2}\dot{x}^2(t) - V(x(t))\right)\right). \quad (4.47)$$

In this section we present only the highlights of Mizrahi's work; we refer to his publications [12] for proofs and explicit formulas.

The potential

The shape of the potential changes drastically[7] as λ changes sign, and the semiclassical expansion is not analytic in λ. In some neighborhood

---

[6] In early works equation $(2.29)_s$ reads

$$\int d\omega(x) \exp(-i\langle x', x\rangle) = \exp(iW(x')),$$

where $d\omega(x)$ has not been decomposed into a translation-invariant term $\mathcal{D}_Q x$ and the exponential of a quadratic term $Q(x)$, the "inverse" of $W(x')$. The "inverse" of $W(x')$ was labeled $W^{-1}(x)$.

[7] We are not concerned here with the change of the shape of the potential when $m\omega^2$ changes sign, i.e. we do not consider the double-well potential.

of the origin, $x = 0$, there exists a potential well and a harmonic motion for $\lambda > 0$ and $\lambda < 0$; for $\lambda > 0$ there is a stable ground state, but for $\lambda < 0$ the ground state is unstable since there is a finite probability of the particle "tunneling" through the well. For large x, the x^4-term dominates the x^2-term, regardless of the magnitude of λ. Consequently perturbation expansions in powers of λ are doomed to failure. Non-analyticity at $\lambda = 0$ does not mean singularity at $\lambda = 0$.

Perturbation expansions in powers of the coupling constant λ have led to the conclusion that there is a singularity at $\lambda = 0$, but the conclusion of Mizrahi's nonperturbative calculation reads as follows:

> The folkloric singularity at $\lambda = 0$ does not appear at any stage, whether in the classical paths, the classical action, the Van Vleck–Morette function, the Jacobi fields, or the Green's functions, in spite of reports to the contrary.

We present Mizrahi's proof because it is simple and because it clarifies the use of boundary conditions in ordinary differential equations. Recall that boundary conditions are necessary for defining the domain of integration of a path integral.

The classical system

The dynamical equation satisfied by the classical solution q is

$$\ddot{q}(t) + \omega^2 q(t) + \frac{\lambda}{m} q^3(t) = 0. \tag{4.48}$$

If we set $m = 1$ (as we shall do), then the mass can be restored by replacing λ by λ/m. The dynamical equation can be solved in terms of biperiodic elliptic functions [13]. The solution may be expressed in terms of any elliptic function. For example, assuming $\lambda > 0$, we choose

$$q(t) = q_m \operatorname{cn}(\Omega(t - t_0), k) \tag{4.49}$$

and insert $q(t)$ into (4.48) for determining Ω and k

$$\Omega^2 = \omega^2 + \lambda q_m^2,$$

$$k^2 = \frac{1}{2}\lambda\left(\frac{q_m}{\Omega}\right)^2; \quad \text{note that } 0 \leq 2k^2 \leq 1.$$

Therefore,

$$q_m = \left(\frac{2k^2\omega^2}{\lambda(1 - 2k^2)}\right)^{1/2} \tag{4.50}$$

$$q(t) = q_m \operatorname{cn}\left(\frac{\omega(t - t_0)}{(1 - 2k^2)^{1/2}}, k\right).$$

For arbitrary values of the constants of integration (k^2, t_0) the classical solution has a $\lambda^{-1/2}$ singularity at $\lambda = 0$. If we instead choose "physical" boundary conditions such as
$$q(t_a) = q_a \quad \text{and} \quad \dot{q}(t_a) = \sqrt{a},$$
then, for $\lambda = 0$, the classical solution
$$q(t) = A^{1/2} \cos(\omega(t - t_a) - \arccos(q_a A^{-1/2})),$$
$$A = \left(\frac{\sqrt{a}}{\omega}\right)^2 + q_a^2 \tag{4.51}$$
has no singularity; it approaches harmonic motion as λ tends to 0.

There is a countably infinite number of possible values for k^2.

For the quantum system (4.47) one needs to define the classical paths by end-point boundary conditions
$$q(t_a) = x_a, \qquad q(t_b) = x_b. \tag{4.52}$$
The end-point boundary conditions (4.52) are satisfied only when k^2/λ is a constant (possibly depending on ω). There is only one solution[8] for k that satisfies (4.50) and (4.52) *and* goes to zero when λ goes to zero. This solution corresponds to the lowest-energy system; this is the classical solution we choose for expanding the action functional. This solution is not singular at $\lambda = 0$; indeed, it coincides with the (generally) unique harmonic oscillator path between the two fixed end points.

Remark. The use of "physical" boundary conditions versus the use of arbitrary boundary conditions, possibly chosen for mathematical simplicity, can be demonstrated on a simpler example: consider the fall of a particle, slowed by friction at a rate proportional to k; the solution of the dynamical equation appears to be singular when $k \longrightarrow 0$ if expressed in terms of arbitrary constants of integration; but it tends to free fall when expressed in terms of physical constants of integration.

The quantum system

The WKB approximation of a point-to-point transition is given by (4.45). The quantities appearing in (4.45) can all be constructed from the solutions of the Jacobi operator equation (see Appendix E)
$$\left(-\frac{d^2}{dt^2} - \omega^2 - 3\lambda q^2(t)\right)\xi(t) = 0. \tag{4.53}$$

---

[8] The infinite number of classical paths for the harmonic oscillator that exist when $\omega(t_b - t_a) = n\pi$ is a different phenomenon, which is analyzed in Section 5.3 on caustics.

4.3 An example: the anharmonic oscillator

The action function $\mathcal{S}(x_a, x_b)$ and the determinant of its hessian have been computed explicitly in terms of elliptic functions by Mizrahi [12] for the anharmonic oscillator.

The expansion of the action functional (4.1) provides a quadratic form,

$$Q(\xi) := S''(q) \cdot \xi\xi, \qquad (4.54)$$

which is useful for defining a gaussian volume element

$$\mathcal{D}_Q\xi \exp\left(\frac{i}{\hbar}Q(\xi)\right) =: \mathrm{d}\gamma_Q(\xi). \qquad (4.55)$$

The full propagator (4.47) is given by a sum of gaussian integrals of polynomials in ξ, with the gaussian volume element $\mathrm{d}\gamma_Q(\xi)$. The full propagator is an asymptotic expansion,

$$K = I_{\mathrm{WKB}}(1 + \hbar K_1 + \hbar^2 K_2 + \cdots). \qquad (4.56)$$

The term of order n consists of an n-tuple ordinary integral of elliptic functions; see [12] for explicit expressions of I_{WKB} and K_1. The expansion is asymptotic because the elliptic functions depend implicitly on λ.

A prototype for the $\lambda\phi^4$ model

To what extent is the anharmonic oscillator a prototype for the $\lambda\phi^4$ model in quantum field theory?

Both systems are nonlinear systems that admit a restricted superposition principle. Indeed, the dynamical equation for the $\lambda\phi^4$ self-interaction, namely $(\Box + m^2)\phi + \lambda\phi^3 = 0$, can readily be reduced to the dynamical equation for the one-dimensional anharmonic oscillator, namely

$$\bar{\phi}'' + m^2\bar{\phi}/K^2 + \lambda\bar{\phi}^3/K^2 = 0, \qquad \text{where } \phi(x_1, x_2, x_3, x_4) \equiv \bar{\phi}(K \cdot x), \qquad (4.57)$$

where K is an arbitrary 4-vector (plane-wave solution). The elliptic functions which are solutions of this equation are periodic, and admit a restricted superposition principle, which is rare for nonlinear equations: if an elliptic cosine (cn) with a certain modulus k_1 is a solution, and if an elliptic sine (sn) with another modulus k_2 is also a solution, then the linear combination $(\mathrm{cn} + \mathrm{i}\,\mathrm{sn})$ is also a solution [14], but the common modulus k_3 is different from k_1 and k_2.

The gaussian technology developed for path integrals can be used to a large extent for functional integrals in quantum field theory. In the case of the anharmonic oscillator, the gaussian volume elements for free particles and for harmonic oscillators lead to a "folkloric" singularity. Mizrahi's work has shown that, for coherent results, one must use the gaussian volume element defined by the second variation of the action functional.

To the best of our knowledge, such a gaussian volume element has not been used in the $\lambda\phi^4$ model (see the footnote in Section 16.2).

A major difference between systems in quantum mechanics and quantum field theory comes from the behavior of the two-point functions, which are the preferred covariance for gaussian volume elements. Consider, for example, covariances defined by $-\mathrm{d}^2/\mathrm{d}t^2$,

$$-\frac{\mathrm{d}^2}{\mathrm{d}t^2}G(t,s) = \delta(t,s), \qquad (4.58)$$

and by a laplacian in D dimensions,

$$\Delta_x G(x,y) = \delta(x,y). \qquad (4.59)$$

The covariance $G(t,s)$ is continuous; the covariance $G(x,y)$ is singular at $x = y$. The situation $G(x,y) \simeq |x-y|^{2-D}$ is discussed in Section 2.5. Renormalization schemes are needed in quantum field theory and are introduced in Chapters 15–17.

4.4 Incompatibility with analytic continuation

A word of caution for semiclassical physics: the laplacian, or its generalizations, is the highest-order derivative term and determines the nature of the equation, but the laplacian contribution vanishes when $\hbar = 0$. Therefore the solutions of the Schrödinger equation have an essential singularity at the origin of a complexified $\hbar$-plane. Care must be taken when using analytic continuation together with expansions in powers of $\hbar$. Michael Berry [15] gives an illuminating example from the study of scent weakly diffusing in random winds. Berry works out the probability that the pheromones emitted by a female moth reach a male moth, given a realistic diffusion constant D. Diffusion alone cannot account for males picking up scents from hundreds of meters away. In fact, females emit pheromones only when there is a wind so that diffusion is aided by convection. The concentration of pheromones looks like a Schrödinger equation if one sets $D = \mathrm{i}\hbar$. However, the limit as $D \longrightarrow 0$ (convection-dominated transport) is utterly different from the limit $\hbar \longrightarrow 0$. This reflects the fact that the solutions of the concentration equation have an essential singularity at the origin of the complexified D-plane, and behave differently depending on whether $D = 0$ is approached along the real axis (convection–diffusion) or imaginary axis (quantum mechanics). The creative aspects of singular limits, in particular the quantum–classical limit, are praised by Berry in an article in *Physics Today* (May 2002), whereas in 1828 Niels Henrik Abel wrote that divergent series "are an invention of the devil."

4.5 Physical interpretation of the WKB approximation

One often hears that classical mechanics is the limit of quantum mechanics when Planck's constant, $h = 2\pi\hbar$, tends to zero. First of all "h tends to zero" is meaningless since h is not a dimensionless quantity; its value depends on the choice of units. The statement cannot imply that h is negligible in classical physics, because many macroscopic classical phenomena would simply not exist in the limit $h = 0$. The WKB approximation of a wave function gives a mathematical expression of the idea of "classical physics as a limit of quantum physics." Let us consider a classical path q defined by its initial momentum p_a and its final position b.

$$p(t_a) = p_a, \qquad q(t_b) = b, \qquad q(t) \in \mathbb{M}. \tag{4.60}$$

Let us consider a flow of classical paths of momentum p_a in a neighborhood $\mathbb{N} \subset \mathbb{M}$ of $q(t_a)$. This flow generates a group of transformations on $\mathbb{N}$:

$$\phi_t : \mathbb{N} \to \mathbb{M} \qquad \text{by } a \mapsto q(t; a, p_a). \tag{4.61}$$

Under the transformation ϕ_t a volume element $d\omega(q(t_a))$ becomes a volume element

$$d\omega_t := \det(\partial q^\alpha(t)/\partial q^\beta(t_a)) d\omega(q(t_a)). \tag{4.62}$$

This equation says that $\det(\partial q^\alpha(t)/\partial q^\beta(t_a))$ gives the rate at which the flow $\{q(t; a, p_a); a \in \mathbb{N}\}$ diverges or converges around $q(t; q(t_a), p_a)$. In classical physics a system in $\mathbb{N}$ at time t_a will be found in $\phi_t(\mathbb{N})$ at time t with probability unity.

In quantum physics, the localization of a system on its configuration space $\mathbb{M}$ is the support of its wave function. Consider an initial wave function ϕ representing a particle localized in $\mathbb{N}$ and with momentum p_a. Let

$$\phi(a) := \exp(i\mathcal{S}_0(a)/\hbar) \mathcal{T}(a), \tag{4.63}$$

where $\mathcal{T}$ is an arbitrary well-behaved function on $\mathbb{M}$ of support in $\mathbb{N}$, and $\mathcal{S}_0$ is the initial value of the solution of the Hamilton–Jacobi equation for the given system. The initial probability density is

$$\rho(a) = \phi^*(a)\phi(a) = |\mathcal{T}(a)|^2. \tag{4.64}$$

The initial current density is

$$j = \frac{1}{2}\frac{\hbar}{im}(\phi^* \nabla \phi - (\nabla\phi)^* \phi); \tag{4.65}$$

in the limit $\hbar = 0$

$$j(a) = \rho(a) \nabla \mathcal{S}_0(a)/m. \tag{4.66}$$

The initial wave function can be said to represent a particle of momentum

$$p_a = \nabla \mathcal{S}_0(a). \tag{4.67}$$

In the limit $\hbar = 0$, the wave function (4.63) is an eigenstate of the momentum operator $-i\hbar \nabla$ with eigenvalue $\nabla \mathcal{S}_0(a)$,

$$-i\hbar \nabla \phi(a) = \nabla \mathcal{S}_0(a)\phi(a). \tag{4.68}$$

The WKB approximation of the wave function Ψ with initial wave function ϕ is given by (4.15) or equivalently (4.38) by

$$\Psi^{\mathrm{WKB}}(t_b, b) = (\det K(t_a, t_a)/\det K(t_b, t_a))^{1/2} \exp\left(\frac{i}{\hbar}\mathcal{S}(t_b, b)\right) \mathcal{T}(q(t_a)),$$

where K is the Jacobi matrix (see Appendix E)

$$K^{\alpha}{}_{\beta}(t, t_a) = \partial q^{\alpha}(t)/\partial q^{\beta}(t_a).$$

Therefore, on any domain Ω in the range of $\phi_t(\mathbb{N})$,

$$\int_{\Omega} |\Psi^{\mathrm{WKB}}(t, b)|^2 \, d\omega_t(b) = \int_{\phi_t^{-1}\Omega} |\phi(a)|^2 \, d\omega(a).$$

In the WKB approximation, the probability of finding in Ω at time t a system known to be in $\phi_t^{-1}\Omega$ at time t_a is unity.

This discussion clarifies the phrase "limit of a quantum system when $\hbar$ tends to zero." It does not imply that all quantum systems have a classical limit. It does not imply that all classical systems are the limit of a quantum system.

References

[1] C. Morette (1951). "On the definition and approximation of Feynman's path integral," *Phys. Rev.* **81**, 848–852.

[2] C. DeWitt-Morette (1976). "The semi-classical expansion," *Ann. Phys.* **97**, 367–399 and **101**, 682–683.
P. Cartier and C. DeWitt-Morette (1997). "Physics on and near caustics," in *NATO-ASI Proceedings, Functional Integration: Basics and Applications* (Cargèse, 1996) eds. C. De Witt-Morette, P. Cartier, and A. Folacci (New York, Plenum Publishing Co.), pp. 51–66.
P. Cartier and C. DeWitt-Morette (1999). "Physics on and near caustics. A simpler version," in *Mathematical Methods of Quantum Physics*, eds. C. C. Bernido, M. V. Carpio-Bernido, K. Nakamura, and K. Watanabe (New York, Gordon and Breach), pp. 131–143.

[3] P. Cartier and C. DeWitt-Morette (1995). "A new perspective on functional integration," *J. Math. Phys.* **36**, 2237–2312. (Located on the World Wide Web at http://godel.ph.utexas.edu/Center/Papers.html and http://babbage.sissa.it/list/funct-an/9602.)

[4] C. DeWitt-Morette, A. Maheshwari, and B. Nelson (1979). "Path integration in non relativistic quantum mechanics," *Phys. Rep.* **50**, 266–372.

[5] Ph. Choquard and F. Steiner (1996). "The story of Van Vleck's and Morette–Van Hove's determinants," *Helv. Phys. Acta* **69**, 636–654.

[6] M. Kac (1956). "Some stochastic problems in physics and mathematics," Magnolia Petroleum Co. Colloquium Lectures, October 1956, partially reproduced in 1974 in *Rocky Mountain J. Math.* **4**, 497–509.

[7] L. D. Faddeev and A. A. Slavnov (1980). *Gauge Fields, Introduction to Quantum Theory* (Reading, MA, Benjamin Cummings).

[8] C. DeWitt-Morette (1974). "Feynman path integrals; I. Linear and affine techniques; II. The Feynman–Green function," *Commun. Math. Phys.* **37**, 63–81.
C. DeWitt-Morette (1976). "The semi-classical expansion," *Ann. Phys.* **97**, 367–399 and **101**, 682–683.

[9] C. DeWitt-Morette (1984). "Feynman path integrals. From the prodistribution definition to the calculation of glory scattering," in *Stochastic Methods and Computer Techniques in Quantum Dynamics*, eds. H. Mitter and L. Pittner, *Acta Phys. Austriaca Suppl.* **26**, 101–170. Reviewed in *Zentralblatt für Mathematik* 1985.
C. DeWitt-Morette and B. Nelson (1984). "Glories – and other degenerate critical points of the action," *Phys. Rev.* **D29**, 1663–1668.
C. DeWitt-Morette and T.-R. Zhang (1984). "WKB cross section for polarized glories," *Phys. Rev. Lett.* **52**, 2313–2316.

[10] A. Maheshwari (1976). "The generalized Wiener–Feynman path integrals," *J. Math. Phys.* **17**, 33–36.

[11] B. DeWitt (2003). *The Global Approach to Quantum Field Theory* (Oxford, Oxford University Press; with corrections 2004).

[12] M. M. Mizrahi (1975). An investigation of the Feynman path integral formulation of quantum mechanics, unpublished Ph.D. Dissertation (University of Texas at Austin).
M. M. Mizrahi (1979). "The semiclassical expansion of the anharmonic oscillator propagator," *J. Math. Phys.* **20**, 844–855.

[13] P. F. Byrd and M. D. Friedman (1954). *Handbook of Elliptic Integrals for Engineers and Physicists* (Berlin, Springer).

[14] G. Petiau (1960). "Les généralisations non-linéaires des équations d'ondes de la mécanique ondulatoire," *Cahiers de Phys.* **14**, 5–24.
D. F. Kurdgelaidzé (1961). "Sur la solution des équations non linéaires de la théorie des champs physiques," *Cahiers de Phys.* **15**, 149–157.

[15] M. Berry. "Scaling and nongaussian fluctuations in the catastophe theory of waves":
(in Italian) M. V. Berry (1985). *Prometheus* **1**, 41–79 (A Unesco publication, eds. P. Bisigno and A. Forti).
(in English) M. V. Berry (1986), in *Wave Propogation and Scattering*, ed. B. J. Uscinski (Oxford, Clarendon Press), pp. 11–35.

5
Semiclassical expansion; beyond WKB

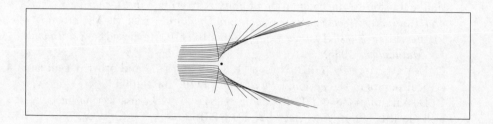

5.1 Introduction

When "WKB breaks down..." then rainbows and glory scattering may appear. In other words, in the notation of Chapter 4, when the hessian $S''(q)$ of the functional S is degenerate for some $q \in \mathcal{P}_{\mu,\nu}\mathbb{M}^D$, there exists some $h \neq 0$ in $T_q\mathcal{P}_{\mu,\nu}\mathbb{M}^D$ such that

$$S''(q) \cdot h\xi = 0 \qquad (5.1)$$

for all ξ in $T_q\mathcal{P}_{\mu,\nu}\mathbb{M}^D$. Stated otherwise, there is at least one nonzero Jacobi field h along q,

$$S''(q)h = 0, \qquad h \in T_q U^{2D}, \qquad (5.2)$$

with D vanishing initial conditions $\{\mu\}$ and D vanishing final conditions $\{\nu\}$.

Notation.

$\mathcal{P}\mathbb{M}^D$: the space of paths $x : \mathbb{T} \to \mathbb{M}^D$, $\mathbb{T} = [t_a, t_b]$;
$\mathcal{P}_{\mu,\nu}\mathbb{M}^D$: the space of paths with D boundary conditions $\{\mu\}$ at t_a and D boundary conditions $\{\nu\}$ at t_b;
U^{2D} : the space of critical points of S (also known as solutions of the Euler–Lagrange equation $S'(q) = 0$).

5.1 Introduction

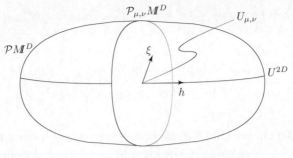

Fig. 5.1

The properties of the intersection

$$U_{\mu,\nu} := U^{2D} \cap \mathcal{P}_{\mu,\nu}\mathbb{M}^D \tag{5.3}$$

dominate this chapter. In the previous chapter, the intersection $U_{\mu,\nu}$ consists of only one point q, or several isolated points. In Section 5.2, the action functional is invariant under automorphisms of the intersection $U_{\mu,\nu}$. The intersection is of dimension $\ell > 0$. In Section 5.3, the classical flow has an envelope, and the intersection $U_{\mu,\nu}$ is a multiple root of $S'(q) = 0$. Glory scattering (Section 5.4) is an example in which both kinds of degeneracies, conservation law and caustics, occur together. In Section 5.5, the intersection $U_{\mu,\nu}$ is the empty set.

We reproduce here as figure 5.1 the picture from Chapter 4 which is the framework for semiclassical expansions.

Degeneracy in finite dimensions

The finite-dimensional case can be used as an example for classifying the various types of degeneracy of $S''(q) \cdot \xi\xi$. Let

$$S: \mathbf{R}^2 \to \mathbf{R}, \qquad x = (x^1, x^2) \in \mathbb{R}^2.$$

Example 1. $S(x) = x^1 + (x^2)^2$ with *relative critical points* (see Section 5.2). The first derivatives are $\partial S/\partial x^1 = 1$ and $\partial S/\partial x^2 = 2x^2$; there is no critical point x_0 such that $S'(x_0) = 0$. However, on any subspace $x^1 = $ constant, $x^2 = 0$ is a critical point. We say that $x^2 = 0$ is a relative critical point (relative to the projection $(x^1, x^2) \to x^1$ onto the horizontal):

$$S''(x) = \begin{pmatrix} 0 & 0 \\ 0 & 2 \end{pmatrix}, \quad \begin{pmatrix} 0 & 0 \\ 0 & 2 \end{pmatrix}\begin{pmatrix} a^1 \\ a^2 \end{pmatrix} = \begin{pmatrix} 0 \\ 2a^2 \end{pmatrix} \tag{5.4}$$

$S''(x)_{ij}a^i b^j = 2a^2 b^2$ is 0 for all b when $a = \begin{pmatrix} a^1 \\ 0 \end{pmatrix} \neq 0$.

Hence $S''(x)$ is degenerate.

Example 2. $S(x) = (x^2)^2 + (x^1)^3$ with double roots of $S'(x_0) = 0$ (see Section 5.3):

$$S'(x_0) = 0 \quad \Rightarrow \quad \left(x_0^1\right)^2 = 0, \; x_0^2 = 0, \; \text{that is } x_0 = (0,0),$$

$$S''(x) = \begin{pmatrix} 6x^1 & 0 \\ 0 & 2 \end{pmatrix}, \quad S''(x_0) = \begin{pmatrix} 0 & 0 \\ 0 & 2 \end{pmatrix}. \tag{5.5}$$

$S''(x)$ is not degenerate in general, but $S''(x_0)$ is degenerate at the critical point. In this case we can say the critical point x_0 is degenerate.

The tangent spaces at the intersection

The intersection $U_{\mu,\nu}$ is analyzed in terms of the tangent spaces at q to $\mathcal{P}_{\mu,\nu}\mathbb{M}^D$ and U^{2D}.

(i) *A basis for* $T_q\mathcal{P}_{\mu,\nu}\mathbb{M}^D$, $q \in U^{2D}$, $S : \mathcal{P}_{\mu,\nu}\mathbb{M}^D \to \mathbf{R}$. Consider the restriction of the action functional S to $\mathcal{P}_{\mu,\nu}\mathbb{M}^D$. Expand $S(x)$ around $S(q)$:

$$S(x) = S(q) + S'(q) \cdot \xi + \frac{1}{2!}S''(q) \cdot \xi\xi + \frac{1}{3!}S'''(q) \cdot \xi\xi\xi + \cdots. \tag{5.6}$$

The vector field $\xi \in T_q\mathcal{P}_{\mu,\nu}\mathbb{M}^D$ has D vanishing boundary conditions at t_a and D vanishing boundary conditions at t_b dictated by the (μ) and (ν) conditions satisfied by all $x \in \mathcal{P}_{\mu,\nu}\mathbb{M}^D$. The second variation is the hessian

$$S''(q) \cdot \xi\xi = \langle \mathcal{J}(q) \cdot \xi, \xi \rangle, \tag{5.7}$$

where $\mathcal{J}(q)$ is the Jacobi operator on $T_q\mathcal{P}_{\mu,\nu}\mathbb{M}^D$. A good basis for $T_q\mathcal{P}_{\mu,\nu}\mathbb{M}^D$ diagonalizes the hessian: in its diagonal form the hessian may consist of a finite number of zeros, say ℓ, and a nondegenerate quadratic form of codimension ℓ. The hessian is diagonalized by the eigenvectors of the Jacobi operator. (See Appendix E, equations (E.45)–(E.48).) Let $\{\Psi_k\}_k$ be a complete set of orthonormal eigenvectors of $\mathcal{J}(q)$:

$$\mathcal{J}(q) \cdot \Psi_k = \alpha_k \Psi_k, \quad k \in \{0, 1, \ldots\}. \tag{5.8}$$

Let $\{u^k\}$ be the coordinates of ξ in the $\{\Psi_k\}$ basis

$$\xi^a(t) = \sum_{k=0}^{\infty} u^k \Psi_k^\alpha(t), \tag{5.9}$$

$$u^k = \int_{t_a}^{t_b} dt (\xi(t)|\Psi_k(t)). \tag{5.10}$$

5.1 Introduction

Therefore

$$S''(q) \cdot \xi\xi = \sum_{k=0}^{\infty} \alpha_k (u^k)^2 \qquad (5.11)$$

and the space of the $\{u^k\}$ is the hessian space l^2 of points u such that $\sum \alpha_k (u^k)^2$ is finite.

(ii) *A basis for $T_q U^{2D}$: a complete set of linearly independent Jacobi fields.* Let h be a Jacobi field, i.e. let h satisfy the Jacobi equation

$$\mathcal{J}(q) \cdot h = 0. \qquad (5.12)$$

The Jacobi fields can be obtained without solving this equation. The critical point q is a function of t and of the two sets of constants of integration. The derivatives of q with respect to the constants of integration are Jacobi fields.

If one eigenvalue of the Jacobi operator corresponding to the field Ψ_0 vanishes, then Ψ_0 is a Jacobi field with vanishing boundary conditions. Let α_0 be this vanishing eigenvalue. Then

$$\mathcal{J}(q) \cdot \Psi_0 = 0, \qquad \Psi_0 \in T_q \mathcal{P}_{\mu,\nu} \mathbb{M}^D. \qquad (5.13)$$

The $2D$ Jacobi fields are not linearly independent; $S''(q) \cdot \xi\xi$ is degenerate. When there exists a Jacobi field Ψ_0 together with vanishing boundary conditions, the end points $q(t_a)$ and $q(t_b)$ are said to be conjugates of each other. The above construction of Jacobi fields needs to be modified slightly.

(iii) *Construction of Jacobi fields with vanishing boundary conditions.*
The Jacobi matrices, J, K, $\tilde{K}$, and L, defined in Appendix E in (E.18), can be used for constructing Jacobi fields such as Ψ_0 defined by (5.13).

If $\Psi_0(t_a) = 0$,	then $\Psi_0(t) = J(t, t_a)\dot{\Psi}_0(t_a).$	(5.14)
If $\dot{\Psi}_0(t_a) = 0$,	then $\Psi_0(t) = K(t, t_a)\Psi_0(t_a).$	(5.15)
If $\dot{\Psi}_0(t_b) = 0$,	then $\Psi_0(t) = \Psi_0(t_b)\tilde{K}(t_b, t).$	(5.16)

To prove (5.14)–(5.16) note that both sides satisfy the same differential equation and the same boundary conditions at t_a or t_b.

For constructing a Jacobi field such that $\Psi_0(t_a) = 0$ and $\Psi_0(t_b) = 0$, it suffices to choose $\dot{\Psi}_0(t_a)$ in the kernel of $J(t_b, t_a)$. A similar procedure applies to the two other cases. The case of derivatives vanishing both at t_a and t_b is best treated with phase-space path integrals [1, 2].

In conclusion, a zero-eigenvalue Jacobi eigenvector is a Jacobi field with vanishing boundary conditions. The determinant of the corresponding Jacobi matrix (E.18) vanishes.

5.2 Constants of the motion

The intersection $U_{\mu,\nu} := U^{2D} \cap \mathcal{P}_{\mu,\nu}\mathbb{M}^D$ can be said to exist at the crossroads of calculus of variation and functional integration. On approaching the crossroads from functional integration one sees classical physics as a limit of quantum physics. Conservation laws, in particular, emerge at the classical limit.

The Euler–Lagrange equation $S'(q) = 0$, which determines the critical points $q \in U^{2D}_{\mu,\nu}$, consists of D coupled equations

$$S'_\alpha(q) := \delta S(x)/\delta x^\alpha(t)|_{x=q}. \tag{5.17}$$

It can happen, possibly after a change of variable in the space of paths $\mathcal{P}_{\mu,\nu}\mathbb{M}^D$, that this system of equations splits into two sets:

$$S'_a(q) = g_a \quad \text{for } a \in \{1,\ldots,\ell\}, \tag{5.18}$$
$$S'_A(q) = 0 \quad \text{for } A \in \{\ell+1,\ldots,D\}. \tag{5.19}$$

The ℓ equations (5.18) are constraints; the $D - \ell$ equations (5.19) determine $D - \ell$ coordinates q^A of q. The space of critical points $U_{\mu,\nu}$ is of dimension ℓ. The second set (5.19) defines a relative critical point under the constraints $S'_a(q) = g_a$. The expansion (5.6) of the action functional now reads

$$S(x) = S(q) + g_a\xi^a + \frac{1}{2}S''_{AB}(q)\xi^A\xi^B + \mathcal{O}(|\xi|^3). \tag{5.20}$$

The variables $\{\xi^a\}$ play the role of Lagrange multipliers of the system S.

In a functional integral, symbolically written

$$\int_\Xi \{\mathcal{D}\xi^a\}\{\mathcal{D}\xi^A\}\exp(2\pi i S(x)/h), \tag{5.21}$$

integration with respect to $\{\mathcal{D}\xi^a\}$ gives δ-functions in $\{g_a\}$.

The decomposition of $\mathcal{D}\xi$ into $\{\mathcal{D}\xi^a\}\{\mathcal{D}\xi^A\}$ is conveniently achieved by the linear change of variable of integration (5.9) and (5.10),

$$L: T_q\mathcal{P}_{\mu,\nu}\mathbb{M}^D \to \mathbb{X} \text{ by } \xi \mapsto u. \tag{5.22}$$

The fact that ℓ eigenvectors $\boldsymbol{\Psi}_k$, $k \in \{0,\ldots,\ell-1\}$ have zero eigenvalues is not a problem. For simplicity, let us assume $\ell = 1$. The domain of integration $\mathbb{X}$ that is spanned by the complete set of eigenvectors can be decomposed into a one-dimensional space $\mathbb{X}^1$ that is spanned by $\boldsymbol{\Psi}_0$ and an infinite-dimensional space $\mathbb{X}^\infty$ of codimension 1. Under the change of variables (5.22), the expansion of the action functional reads

$$S(x) = S(q) + c_0 u^0 + \frac{1}{2}\sum_{k=1}^\infty \alpha_k(u^k)^2 + \mathcal{O}(|u|^3) \tag{5.23}$$

with

$$c_0 = \int_\mathbb{T} dt \, \frac{\delta S}{\delta q^j(t)} \Psi_0^j(t). \tag{5.24}$$

Remark. For finite dimension, according to the Morse lemma, there is a nonlinear change of variable that can be used to remove the terms of order greater than 2 in a Taylor expansion. For infinite dimension, there is no general prescription for removing $\mathcal{O}(|u|^3)$.

The integral over u^0 contributes a δ-function $\delta(c_0/h)$ to the propagator $\delta(c_0/h)$. The propagator vanishes unless the conservation law

$$\frac{1}{h} \int_\mathbb{T} dt \, \frac{\delta S}{\delta q^j(t)} \Psi_0^j(t) = 0 \tag{5.25}$$

is satisfied.

In conclusion, conservation laws appear in the classical limit of quantum physics. It is not an anomaly for a quantum system to have less symmetry than its classical limit.

5.3 Caustics

In Section 5.2 we approached the crossroads $U_{\mu,\nu}$ from $T_q \mathcal{P}_{\mu,\nu} \mathbb{M}^D$. Now we approach the same crossroads from $T_q U^{2D}$. As we approach quantum physics from classical physics, we see that a caustic (the envelope of a family of classical paths) is "softened up" by quantum physics.

Caustics abound in physics. We quote only four examples corresponding to the boundary conditions (initial or final position or momentum) we discuss in Appendix E.

(i) *The soap-bubble problem* [3]. The "paths" are the curves defining (by rotation around an axis) the surface of a soap bubble held by two rings. The "classical flow" is a family of catenaries with one fixed point. The caustic is the envelope of the catenaries.

(ii) *The scattering of particles by a repulsive Coulomb potential.* The flow is a family of Coulomb paths with fixed initial momentum. Its envelope is a parabola.

The two following examples are not readily identified as caustic problems because the flows do not have envelopes in physical space. The vanishing of boundary conditions of the Jacobi field at the caustic corresponds to the vanishing of first derivatives. In phase space the projection of the flow onto the momentum space has an envelope.

(iii) *Rainbow scattering from a point source.*
(iv) *Rainbow scattering from a source at infinity.*

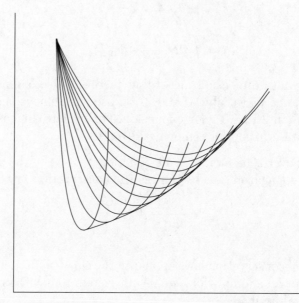

Fig. 5.2   For a point on the "dark" side of the caustic there is no classical path; for a point on the "bright" side there are two classical paths, which coalesce into a single path as the intersection of the two paths approaches the caustic. Note that the paths do not arrive at an intersection at the same time; the paths do not intersect in a space–time diagram.

The relevant features can be analyzed on a specific example, e.g. the scattering of particles by a repulsive Coulomb potential (figure 5.3). For other examples see [4].

Let q and q^Δ be two solutions of the same Euler–Lagrange equation with slightly different boundary conditions at t_b, i.e. $q \in \mathcal{P}_{\mu,\nu}\mathbb{M}^D$ and $q^\Delta \in \mathcal{P}_{\mu,\nu^\Delta}\mathbb{M}^D$:

$$p(t_a) = p_a, \qquad q(t_b) = x_b,$$
$$p^\Delta(t_a) = p_a, \qquad q^\Delta(t_b) = x_b^\Delta.$$

Assume that p_a and x_b are conjugates along q with multiplicity 1; i.e. the Jacobi fields h along q satisfying

$$\dot{h}(t_a) = 0, \qquad h(t_b) = 0$$

form a one-dimensional space. Assume that p_a and x_b^Δ are not conjugate along q^Δ.

We shall compute the probability amplitude $K(x_b^\Delta, t_b; p_a, t_a)$ when x_b^Δ is close to the caustic on the "bright" side and on the "dark" side. We shall *not* compute K by expanding S around q^Δ for the following reasons.

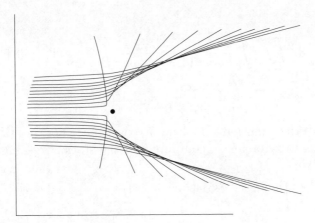

Fig. 5.3 A flow on configuration space of charged particles in a repulsive Coulomb potential

- If $x_b^\triangle$ is on the dark side, then $q^\triangle$ does not exist.
- If $x_b^\triangle$ is on the bright side, then one could consider K to be the limit of the sum of two contributions corresponding to the two paths q and $q^\triangle$ that intersect at $b^\triangle$,

$$K(x_b, t_b; p_a, t_a) = \lim_{\Delta = 0} \left[K_q(x_b^\triangle, t_b + \Delta t; p_a, t_a) + K_{q^\triangle}(x_b^\triangle, t_b; p_a, t_a) \right].$$

At $x_b^\triangle$, q has touched the caustic and "picked up" an additional phase equal to $-\pi/2$; both limits are infinite and their sum is not defined.

We compute $K(x_b^\triangle, t_b; p_a, t_a)$ by expanding S around q, using (5.6) and possibly higher derivatives. The calculation requires some care (see [5] for details) because $q(t_b) \neq b^\triangle$. In other words, q is not a critical point of the action restricted to the space of paths such that $x(t_b) = x_b^\triangle$. We approach the intersection in a direction other than tangential to $\mathcal{U}^{2D}$.

As before, we make the change of variable $\xi \mapsto u$ of (5.9) and (5.10) that diagonalizes $S''(q) \cdot \xi\xi$. Again we decompose the domain of integration in the u-variables

$$\mathbb{X} = \mathbb{X}^1 \times \mathbb{X}^\infty.$$

The second variation restricted to $\mathbb{X}^\infty$ is once again nonsingular, and the calculation of the integral over $\mathbb{X}^\infty$ proceeds as usual for the strict WKB approximation. Denote by Ψ_0 the Jacobi field with vanishing boundary conditions for the path q. The integral over $\mathbb{X}^1$ is

$$I(\nu, c) = \int_\mathbb{R} du^0 \exp\left(i\left(cu^0 - \frac{\nu}{3}(u^0)^3\right)\right) = 2\pi\nu^{-1/3} \operatorname{Ai}(-\nu^{-1/3}c), \quad (5.26)$$

where

$$\nu = \frac{\pi}{h} \int_{\mathbf{T}} dr \int_{\mathbf{T}} ds \int_{\mathbf{T}} dt \, \frac{\delta^3 S}{\delta q^\alpha(r) \delta q^\beta(s) \delta q^\gamma(t)} \, \Psi_0^\alpha(r) \Psi_0^\beta(s) \Psi_0^\gamma(t), \quad (5.27)$$

$$c = -\frac{2\pi}{h} \int_{\mathbf{T}} dt \, \frac{\delta S}{\delta q^\alpha(t)} \cdot \Psi_0^\alpha(t) \big(|x_b^\Delta - x_b|\big), \quad (5.28)$$

and Ai is the Airy function. The leading contribution of the Airy function when h tends to zero can be computed by the stationary-phase method. At $v^2 = \nu^{-1/3} c$

$$2\pi \mathrm{Ai}(-\nu^2) \simeq \begin{cases} 2\sqrt{\pi} v^{-1/2} \cos\big(\tfrac{2}{3} v^3 - \pi/4\big) & \text{for } v^2 > 0, \\ \sqrt{\pi} (-v^2)^{-1/4} \exp\big(-\tfrac{2}{3}(-v^2)^{3/2}\big) & \text{for } v^2 < 0; \end{cases} \quad (5.29)$$

v is the critical point of the phase in the integrand of the Airy function and is of order $h^{-1/3}$. For $v^2 > 0$, x_b^Δ exists in the illuminated region, and the probability amplitude oscillates rapidly as h tends to zero. For $v^2 < 0$, x_b^Δ exists in the shadow region, and the probability amplitude decays exponentially.

The probability amplitude $K(x_b^\Delta, t_b; a, t_a)$ does not become infinite when x_b^Δ tends to x_b. Quantum mechanics softens up the caustics.

Remark. The normalization and the argument of the Airy function can be expressed solely in terms of the Jacobi fields.

Remark. Other cases, such as position-to-momentum, position-to-position, momentum-to-momentum, and angular-momentum-to-angular-momentum transitions have been treated explicitly in [4, 5].

5.4 Glory scattering

The backward scattering of light along a direction very close to the direction of the incoming rays has a long and interesting history (see for instance [6] and references therein). This type of scattering creates a bright halo around a shadow, and is usually called glory scattering. Early derivations of glory scattering were cumbersome, and used several approximations. It has been computed from first principles by functional integration using only the expansion in powers of the square root of Planck's constant [5, 7].

The classical cross section for the scattering of a beam of particles in a solid angle $d\Omega = 2\pi \sin\theta \, d\theta$ by an axisymmetric potential is

$$d\sigma_{c\ell}(\Omega) = 2\pi B(\theta) dB(\theta). \quad (5.30)$$

The deflection function $\Theta(B)$ that gives the scattering angle θ as a function of the impact parameter B is assumed to have a unique inverse $B(\Theta)$.

5.4 Glory scattering

We can write

$$d\sigma_{c\ell}(\Omega) = B(\Theta)\frac{dB(\Theta)}{d\Theta}\bigg|_{\Theta=\theta}\frac{d\Omega}{\sin\theta} \qquad (5.31)$$

abbreviated henceforth

$$d\sigma_{c\ell}(\Omega) = B(\theta)\frac{dB(\theta)}{d\theta}\frac{d\Omega}{\sin\theta}. \qquad (5.32)$$

It can happen that, for a certain value of B, say B_g (g for glory), the deflection function vanishes,

$$\theta = \Theta(B_g) \quad \text{is 0 or } \pi, \qquad (5.33)$$

which implies that $\sin\theta = 0$, and renders (5.32) useless.

The classical glory-scattering cross section is infinite on two accounts.

(i) There exists a conservation law: the final momentum $p_b = -p_a$ the initial momentum.
(ii) Near glory paths, particles with impact parameters $B_g + \delta B$ and $-B_g + \delta B$ exit at approximately the same angle, namely π + terms of order $(\delta B)^3$.

The glory cross section [5, 7] can be computed using the methods presented in Sections 5.2 and 5.3. The result is

$$d\sigma(\Omega) = 4\pi^2 h^{-1}|p_a|B^2(\theta)\frac{dB(\theta)}{d\theta}J_0(2\pi h^{-1}|p_a|B(\theta)\sin\theta)^2\,d\Omega, \qquad (5.34)$$

where J_0 is the Bessel function of order 0.

A similar calculation [5, 6, 8] gives the WKB cross section for polarized glories of massless waves in curved spacetimes,

$$d\sigma(\Omega) = 4\pi^2\lambda^{-1}B_g^2\frac{dB}{d\theta}J_{2s}(2\pi\lambda^{-1}B_g\sin\theta)^2\,d\Omega, \qquad (5.35)$$

where

$s = 0$ for scalar waves; when $\sin\theta = 0$, the Bessel function $J_0(0) \neq 0$;
$s = 1$ for electromagnetic waves; when $\sin\theta = 0$, the Bessel function $J_2(0) = 0$;
$s = 2$ for gravitational waves; when $\sin\theta = 0$, the Bessel function $J_4(0) = 0$; and
λ is the wave length of the incoming wave.

Equation (5.35) agrees with the numerical calculations [6] of R. Matzner based on the partial wave-decomposition method.

Semiclassical expansion; beyond WKB

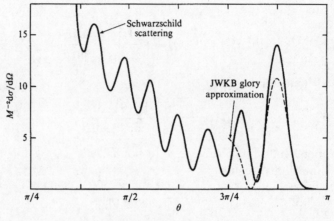

Fig. 5.4

$$\frac{d\sigma}{d\Omega} = 2\pi\omega B_g^2 \left.\frac{dB}{d\theta}\right|_{\theta=\pi} J_{2s}(\omega B_g \sin\theta)^2$$

$B_g = B(\pi)$ glory impact parameter

$$\omega = 2\pi\lambda^{-1}$$

$s = 2$ for gravitational wave

analytic cross section: dashed line

numerical cross section: solid line

5.5 Tunneling

A quantum transition between two points **a** and **b** that are not connected by a classical path is called tunneling [9]. Quantum particles can go through potential barriers, e.g. in nuclear α-decay, α-particles leave a nucleus although they are inhibited by a potential barrier. To prepare the study of semiclassical tunneling we recall the finite-dimensional case: the stationary-phase approximation[1] in which the critical point of the phase lies outside the domain of the integration.

Introduction

In its simplest form, the stationary-phase approximation is an asymptotic approximation for large λ of integrals of the form

$$F(\lambda) = \int_{\mathbb{X}} d\mu(x) h(x) \exp\left(i\lambda f(x)\right), \tag{5.36}$$

---

[1] See for instance [10], Vol. I, p. 593 and [11].

5.5 Tunneling

where h is a real-valued smooth function of *compact* support on the D-dimensional Riemannian manifold $\mathbb{X}$ with volume element $d\mu(x)$. The critical points of f (i.e. the solutions of $f'(y) = 0$, $y \in \mathbb{X}$) are assumed to be nondegenerate: the determinant of the hessian $\partial^2 f / \partial x^i \, \partial x^j$ does not vanish at the critical point $x = y$. In this simple case

$$F(\lambda) \approx \mathcal{O}(\lambda^{-N}), \qquad \text{for any } N \text{ if } f \text{ has no critical point on the support of } h; \tag{5.37}$$

$$F(\lambda) \approx \mathcal{O}(\lambda^{-D/2}), \qquad \text{if } f \text{ has a finite number of non-degenerate critical points on the support of } h. \tag{5.38}$$

These results have been generalized for the case in which the critical point $y \in \text{supp } h$ is degenerate.[2] We recall the following case which paved the way for semiclassical tunneling.

Stationary-phase approximation with non-vanishing integrand on the boundary $\partial \mathbb{X}$ of the domain of integration $\mathbb{X}$

The simple results (5.37) and (5.38) are obtained by integrating (5.36) by parts under the assumption that the boundary terms vanish. Here, the boundary terms do not vanish. To illustrate,[3] let $\mathbb{X} = [a, b] \subset \mathbb{R}$, then

$$F(\lambda) = \frac{1}{i\lambda} \frac{h(x)}{f'(x)} \exp(i\lambda f(x)) \bigg|_a^b - \frac{1}{i\lambda} \int_a^b \left(\frac{h(x)}{f'(x)} \right)' \exp(i\lambda f(x)) dx. \tag{5.39}$$

After N integrations by parts, the boundary terms are expressed up to phase factors $e^{i\lambda f(a)}$ or $e^{i\lambda f(b)}$, as a polynomial in $(i\lambda)^{-1}$ of order N. As in the simple case, the remaining integral is of order λ^{-N} for any N if f has no critical point in the domain of integration and of order $\lambda^{-D/2}$ otherwise. Therefore, for large λ, the leading term is the boundary term in (5.39).

The leading term of $F(\lambda)$ in equation (5.38) is, after integration by parts, the boundary term. It can be rewritten[4] as an integral over the $(D-1)$-dimensional boundary $\partial \mathbb{X}$,

$$F(\lambda) \approx \frac{1}{i\lambda} \int_{\partial \mathbb{X}} h(x) \exp(i\lambda f(x)) \boldsymbol{\omega} \cdot d\boldsymbol{\sigma}(x), \tag{5.40}$$

---

[2] See Sections 5.3 and 5.4 for the infinite-dimensional counterpart of this case.
[3] For the integration by parts of the general case (5.36), see [11]. The example is sufficient for computing the powers of λ.
[4] Rewriting the boundary terms as an integral over the boundary is a classic exercise in integration over riemannian spaces; see details in [11], pp. 328–329.

where the boundary $\partial \mathbb{X}$ is regular and compact. $\boldsymbol{\omega}$ is the derivation at x defined by

$$\boldsymbol{\omega}(x) := \frac{\mathbf{v}(x)}{|\mathbf{v}(x)|^2}, \quad \text{with } \mathbf{v}(x) = g^{ij}(x) \frac{\partial f}{\partial x^i} \frac{\partial}{\partial x^j}, \tag{5.41}$$

and $d\boldsymbol{\sigma}$ is the $(D-1)$-dimensional surface element on $\partial \mathbb{X}$. The asymptotic expression (5.40) can then be obtained by the stationary-phase method. The integral is of order $\lambda^{-1}\lambda^{-N}$ for any N if $f|_{\partial \mathbb{X}}$ has no critical point and of order $\lambda^{-1}\lambda^{-(D-1)/2}$ otherwise.

The phase f, which is restricted to $\partial \mathbb{X}$, obtains its extrema when the components of $\operatorname{grad} f$ in $T_y \partial \mathbb{X}$ (with $y \in \partial \mathbb{X}$) vanish, i.e. when $\operatorname{grad} f(y)$ is normal to $\partial \mathbb{X}$.

The wedge problem has a long and distinguished history.[5] In 1896, Sommerfeld, then a *Privatdozent*, computed the diffraction of light from a perfectly reflecting knife edge, i.e. a wedge with external angle $\theta = 2\pi$. For this he employed an *Ansatz* suggested by Riemann's computation of the potential of a point charge outside a conducting wedge of external angle $\theta = \mu\pi/\nu$ with μ/ν rational.

In 1938, on the occasion of Sommerfeld's seventieth birthday, Pauli dedicated his paper on wedges to his "old teacher."

The wedge is a boundary-value problem whose solution exploits the method of images, Riemann surfaces for multivalued functions, properties of special functions, etc., a combination of powerful analytic tools. It is therefore amazing that L. S. Schulman [13] found a simple exact path integral representing the probability amplitude $K(\mathbf{b}, t; \mathbf{a})$ that a particle that is at $\mathbf{a}$ at time $t = 0$ will be found at $\mathbf{b}$ at time t, when a knife edge precludes the existence of a classical path from $\mathbf{a}$ to $\mathbf{b}$ (figure 5.5).

The mathematical similarity[6] of the three techniques used by Riemann, Sommerfeld, and Schulman is often mentioned when two-slit interference is invoked for introducing path integration. This similarity is rarely addressed in the context of diffraction.

Schulman's result is challenging but, upon reflection, not unexpected. Functional integration is ideally suited for solving boundary-value problems. Indeed, it incorporates the boundary conditions in the choice of its domain of integration, whereas boundary conditions in differential calculus are *additional* requirements satisfied by a solution of a partial differential equation (PDE) whose compatibility must be verified. Moreover, a PDE states only local relationships between a function and its derivatives,

---

[5] Some historical references can be found in [1] (see the bibliography) dedicated to John Archibald Wheeler on the occasion of his seventieth birthday.
[6] The similarity is spelled out in [11].

5.5 Tunneling

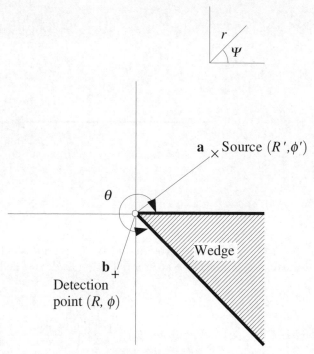

Fig. 5.5 The source is at **a** with polar coordinates (R', ϕ'). The detector is at **b** with coordinates (R, ϕ). The wedge angle θ is the external angle. When $\theta = 2\pi$, the wedge is a half-plane barrier called a knife edge.

whereas the domain of integration of a path integral consists of paths that probe the global properties of their ranges.

Schulman's computation of the knife-edge tunneling

Consider a free particle of unit mass, which cannot pass through a thin barrier along the positive axis.

The knife-edge problem can be solved in $\mathbb{R}^2$ without the loss of its essential characteristics. The probability amplitude $K(\mathbf{b}, t; \mathbf{a})$ that a particle that is at **a** at time $t = 0$ will be found at **b** at time t can always be expressed as an integral on $\mathbb{R}^2$,

$$K(\mathbf{b}, t; \mathbf{a}) = \int_{\mathbb{R}^2} d^2\mathbf{c}\ K(\mathbf{b}, t - t_c; \mathbf{c}) K(\mathbf{c}, t_c; \mathbf{a}) \qquad (5.42)$$

for an arbitrary intermediate t_c. It is convenient to choose t_c equal to the time a free particle going from **a** to **0**, and then from **0** to **b** in time t, takes to reach the edge **0** of the knife. When **c** is visible both from **a** and from **b** (shown in figure 5.6), there are direct contributions to the

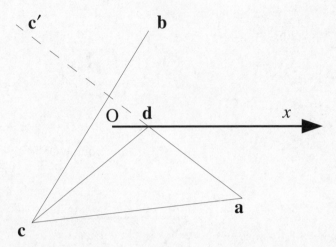

Fig. 5.6 The point **c** is visible both from **a** and **b**. A path **ac** is called "direct," whereas a path **adc**, which reaches **c** after a reflection at **d**, is called "reflected".

probability amplitude[7]

$$K_D(\mathbf{b}, t - t_c; \mathbf{c}) = (2\pi i\hbar(t - t_c))^{-1} \exp\left(i\frac{|\mathbf{bc}|^2}{2\hbar(t - t_c)}\right) \qquad (5.43)$$

and

$$K_D(\mathbf{c}, t_c; \mathbf{a}) = (2\pi i\hbar t_c)^{-1} \exp\left(i\frac{|\mathbf{ca}|^2}{2\hbar t_c}\right), \qquad (5.44)$$

and the following reflected contribution:

$$K_R(\mathbf{c}, t_c; \mathbf{a}) = (2\pi i\hbar t_c)^{-1} \exp\left(i\frac{|\mathbf{c'a}|^2}{2\hbar t_c}\right), \qquad (5.45)$$

where $\mathbf{c}'$ is symmetric to $\mathbf{c}$ with respect to the barrier plane. Let $K_{DD}(\mathbf{b}, t; \mathbf{a})$ be the probability amplitude obtained by inserting (5.43) and (5.44) into (5.42), and $K_{DR}(\mathbf{b}, t; \mathbf{a})$ be the probability amplitude obtained by inserting (5.43) and (5.45) into (5.42). It is understood that each term appears only in its classically allowed region. The total probability amplitude $K(\mathbf{b}, t; \mathbf{a})$ is a linear combination of K_{DD} and K_{DR}. Schulman considers two possible linear combinations,

$$K^{\mp}(\mathbf{b}, t; \mathbf{a}) = K_{DD}(\mathbf{b}, t; \mathbf{a}) \mp K_{RD}(\mathbf{b}, t; \mathbf{a}). \qquad (5.46)$$

After some calculations [12], he obtains

$$K^{\mp}(\mathbf{b}, t; \mathbf{a}) = K_0\big(e^{-im^2} h(-m) \mp e^{-in^2} h(-n)\big), \qquad (5.47)$$

---

[7] In his calculation Schulman assumes the mass of the particle to be 1/2. The mass dimension disappears from his formulas.

where

$$K_0 = \frac{1}{2\pi i \hbar t} \exp\left(\frac{i}{2\hbar t}(|\mathbf{a0}|^2 + |\mathbf{0b}|^2)\right), \qquad (5.48)$$

$$m = (2/\hbar\gamma)^{1/2} v \sin\omega_2, \qquad n = (2/\hbar\gamma)\frac{1}{2}v \sin\omega_1, \qquad (5.49)$$

$$\omega_2 = \frac{1}{2}(\phi' - \phi) - \frac{\pi}{2}, \qquad \omega_1 = \frac{1}{2}(\phi' + \phi), \qquad (5.50)$$

$$\gamma = t_c^{-1} + (t - t_c)^{-1}, \qquad v = |\mathbf{a0}|/t_c = |\mathbf{0b}|/(t - t_c), \qquad (5.51)$$

and

$$h(m) = \pi^{-1/2} \exp(-i\pi/4) \int_{-\infty}^{m} \exp(it^2) dt.$$

The propagators $K^{\mp}$ are identical to the propagators obtained by Carslaw. In 1899, these propagators had been used by Sommerfeld to continue his investigation of the knife-edge problem. The propagator K^- consists of two contributions that interfere destructively with one another on the far side of the knife. This propagator can be said to vanish on the knife, i.e. it satisfies the Dirichlet boundary condition. The propagator K^+ consists of the sum of two contributions whose normal derivatives (derivatives along the direction perpendicular to the knife) interfere destructively on the far side of the knife. This propagator can be said to satisfy Neumann boundary conditions.

Note that, for $\phi - \phi' < \pi$, there is a straight line in the space of allowed paths, and the stationary-phase approximation of $K(\mathbf{b}, t; \mathbf{a})$ is the free propagator K_0 in $\mathbb{R}^2$ proportional to $\hbar^{-1}$. If $\phi - \phi' > \pi$ the stationary-phase approximation is dominated by the boundary terms, and $K(\mathbf{b}, t; \mathbf{a})$ is of order $\hbar^{1/2}$.

Independently of Schulman, F. W. Wiegel and J. Boersma [13] and Amar Shiekh [11] have solved the knife-edge problem using other novel path-integral techniques. Wiegel and Boersma base their derivation on results obtained in the solution of the polymer problem, another case in which the boundary of the domain of integration of a path integral determines the solution.

Shiekh [11] bases his derivation on the properties of free propagators on a two-sheeted Riemann surface, which are constructed as a linear combination of Aharonov–Bohm propagators for different fluxes.

References

Chapter 5 is based on two articles on "Physics on and near caustics" by the authors of this book, which have appeared in the following proceedings:

- *Functional Integration: Basics and Applications*. eds. C. DeWitt-Morette, P. Cartier, and A. Folacci (New York, Plenum Press, 1997); and
- RCTP Proceedings of the second Jagna Workshop (January 1998), *Mathematical Methods of Quantum Physics*, eds. C. C. Bernido, M. V. Carpio-Bernido, K. Nakamura, and K. Watanabe (New York, Gordon and Breach, 1999).

[1] C. DeWitt-Morette and T.-R. Zhang (1983). "A Feynman–Kac formula in phase space with application to coherent state transitions," *Phys. Rev.* **D28**, 2517–2525.

[2] C. DeWitt-Morette (1984). "Feynman path integrals. From the prodistribution definition to the calculation of glory scattering," in *Stochastic Methods and Computer Techniques in Quantum Dynamics*, eds. H. Mitter and L. Pittner, *Acta Phys. Austriaca Suppl.* **26**, 101–170.

[3] C. DeWitt-Morette (1976). "Catastrophes in Lagrangian systems," and C. DeWitt-Morette and P. Tshumi (1976). "Catastrophes in Lagrangian systems. An example," both in *Long Time Prediction in Dynamics*, eds. V. Szebehely and B. D. Tapley (Dordrecht, D. Reidel), pp. 57–69. Also Y. Choquet-Bruhat and C. DeWitt-Morette (1982). *Analysis, Manifolds and Physics* (Amsterdam, North Holland, 1982–1996), Vol. I, pp. 105–109 and 277–281.

[4] C. DeWitt-Morette, B. Nelson, and T.-R. Zhang (1983). "Caustic problems in quantum mechanics with applications to scattering theory," *Phys. Rev.* **D28**, 2526–2546.

[5] C. DeWitt-Morette (1984). Same as [2].

[6] R. A. Matzner, C. DeWitt-Morette, B. Nelson, and T.-R. Zhang (1985). "Glory scattering by black holes," *Phys. Rev.* **D31**, 1869–1878.

[7] C. DeWitt-Morette and B. Nelson (1984). "Glories – and other degenerate points of the action," *Phys. Rev.* **D29**, 1663–1668.

[8] T.-R. Zhang and C. DeWitt-Morette (1984). "WKB cross-section for polarized glories of massless waves in curved space-times," *Phys. Rev. Lett.* **52**, 2313–2316.

[9] See for instance S. Gasiorowicz (1974). *Quantum Physics* (New York, John Wiley and Sons Inc.).
For tunneling effects in fields other than physics as well as in physics, see for instance "Tunneling from alpha particle decay to biology" (APS Meeting, Indianapolis, 18 March 2002) in *History of Physics Newsletter*, Vol. VIII no. 5 (Fall 2002), pp. 2–3. Reports by I. Giaever, J. Onuchic, E. Merzbacher, and N. Makri.

[10] Y. Choquet-Bruhat and C. DeWitt-Morette (1996 and 2000). *Analysis, Manifolds, and Physics*, Part I with M. Dillard, revised edn. and Part II, revised and enlarged edn. (Amsterdam, North Holland).

[11] C. DeWitt-Morette, S. G. Low, L. S. Schulman, and A. Y. Shiekh (1986). "Wedges I," *Foundations of Physics* **16**, 311–349.

[12] L. S. Schulman (1982). "Exact time-dependent Green's function for the half plane barrier," *Phys. Rev. Lett.* **49**, 559–601.
L. S. Schulman (1984). "Ray optics for diffraction: a useful paradox in a path-integral context," in *The Wave Particle Dualism* (Conference in honor of Louis de Broglie) (Dordrecht, Reidel).

[13] F.W. Wiegel and J. Boersma (1983). "The Green function for the half plane barrier: derivation from polymer entanglement probabilities," *Physica A* **122**, 325–333.

Note. Our references on glory scattering fail to include an earlier paper by Alex B. Gaina:

A. B. Gaina (1980). "Scattering and absorption of scalar particles and fermions in Reissner–Nordstrom field," VINITI (All-Union Institute for Scientific and Technical Information) 1970–80 Dep. 20 pp.

6

Quantum dynamics: path integrals and operator formalism

$$\langle b|\hat{O}|a\rangle = \int_{\mathcal{P}_{a,b}} O(\gamma)\exp(\mathrm{i}S(\gamma)/\hbar)\mu(\gamma)\mathcal{D}\gamma$$

We treat the Schrödinger equation together with an initial wave function as our first example of the rich and complex relationship of functional integration with the operator formalism. We treat successively free particles, particles in a scalar potential V, and particles in a vector potential A. In Chapters 12 and 13 we construct path-integral solutions of PDEs other than parabolic.

6.1 Physical dimensions and expansions

The Schrödinger equation reads

$$\mathrm{i}\hbar\frac{\mathrm{d}\psi}{\mathrm{d}t} = \hat{H}\psi \quad \text{or} \quad \psi(t,x) = \mathrm{e}^{-\mathrm{i}\hat{H}t/\hbar}\psi(0,x). \tag{6.1}$$

There are different expansions of the Schrödinger operator. For a particle in a scalar potential V, the dimensionless operator $\hat{H}t/\hbar$ can be written[1]

$$\hat{H}t/\hbar = -\frac{1}{2}A\Delta + BV, \tag{6.2}$$

where Δ has the physical dimension of a laplacian, namely $[\Delta] = L^{-2}$, and the potential V has the physical dimension of energy, $[V] = ML^2T^{-2}$. Therefore $[A] = L^2$ and $[B] = M^{-1}L^{-2}T^2$. Set $A = \hbar t/m$ and $B = \lambda t/\hbar$,

---

[1] S. A. Fulling, private communication.

where λ is a dimensionless coupling constant. The following expansions are adapted to various problems.

Expansion in powers of the coupling λ (i.e. in powers of B) for perturbations.
Expansion in powers of m^{-1} (i.e. in A).
Expansion in powers of t (i.e. in $(AB)^{1/2}$) for short-time propagators.
Expansion in powers of $\hbar$ (i.e. in $(A/B)^{1/2}$) for semiclassical physics.

In particular, the short-time propagator is not necessarily equal to the semiclassical propagator.

There are other possible expansions. For instance, a brownian path defined by its distribution (3.11) is such that $(\Delta z)^2/\Delta t$ is dimensionless. Hence

$$[z]^2 = T. \tag{6.3}$$

A semiclassical expansion obtained by expanding a path x around a classical path x_{cl} of dimension L is often written

$$x = x_{\text{cl}} + (\hbar/m)^{1/2} z. \tag{6.4}$$

Equation (6.4) provides an expansion in powers of $(\hbar/m)^{1/2}$. In elementary cases the odd powers vanish because the gaussian integral of odd polynomials vanishes.

6.2 A free particle

The classical hamiltonian of a free particle is

$$H_0 = \frac{1}{2m}|p|^2. \tag{6.5}$$

Here the position vector is labeled $x = (x^1, \ldots, x^D)$ in $\mathbb{R}^D$ and the momentum $p = (p_1, \ldots, p_D)$ in the dual space $\mathbb{R}_D$. We use the scalar product

$$\langle p, x \rangle = p_j x^j \tag{6.6}$$

and the quadratic form

$$|p|^2 = \delta^{ij} p_i p_j = \sum_{j=1}^{D} (p_j)^2. \tag{6.7}$$

According to the quantization rule

$$\hat{p}_j = -i\hbar\, \partial_j \quad \text{with } \partial_j = \partial/\partial x^j, \tag{6.8}$$

116   *Quantum dynamics: path integrals and operator formalism*

the quantum hamiltonian of a free particle is[2]

$$\hat{H}_0 = -\frac{\hbar^2}{2m}\Delta_x \quad \text{with } \Delta_x := \sum_j (\partial_j)^2. \tag{6.9}$$

With these definitions, the Schrödinger equation reads as

$$i\hbar\partial_t \psi(t,x) = -\frac{\hbar^2}{2m}\Delta_x \psi(t,x), \tag{6.10}$$
$$\psi(0,x) = \phi(x).$$

The path-integral solution of this equation can be read off the general path-integral solution of a parabolic PDE constructed in Section 7.2. We outline a different method based on the Fourier transform and the path integral of a translation operator [1, pp. 289–290].

(i) *The Fourier transform.*
We define the Fourier transform Φ of ψ by

$$\Phi(t,p) = \int_{\mathbb{R}^D} d^D x\, \psi(t,x) \cdot \exp\left(-\frac{i}{\hbar}\langle p, x\rangle\right). \tag{6.11}$$

In momentum space, the Schrödinger equation is expressed as

$$i\hbar\,\partial_t \Phi(t,p) = \frac{1}{2m}|p|^2 \Phi(t,p), \tag{6.12}$$

with the solution

$$\Phi(T,p) = \exp\left(-\frac{i}{\hbar}\frac{|p|^2 T}{2m}\right)\cdot \Phi(0,p). \tag{6.13}$$

(ii) *The functional integral.*
We want to express the factor

$$\exp\left(-\frac{i}{\hbar}\frac{|p|^2 T}{2m}\right)$$

as a path integral, and, for this purpose, we use the methods of Section 3.2. We introduce the space $\mathbb{X}_b = L_+^{2,1}$ of absolutely continuous maps $\xi: [0,T] \to \mathbb{R}^D$ with square integrable derivative such that $\xi(T) = 0$, and the quadratic form

$$Q(\xi) = \frac{m}{\hbar}\int_0^T dt\,|\dot\xi(t)|^2 \tag{6.14}$$

---

[2] A preferable but less common definition of the laplacian is $\Delta_x = -\sum_j (\partial_j)^2$.

on $\mathbb{X}_b$. We identify the dual $\mathbb{X}'_b$ of $\mathbb{X}_b$ with the space L^2 of square integrable maps $\kappa : [0, T] \to \mathbb{R}_D$. The duality is given by

$$\langle \kappa, \xi \rangle = \frac{1}{\hbar} \int_0^T dt\, \kappa_j(t) \dot{\xi}^j(t) \tag{6.15}$$

and the inverse quadratic form W on $\mathbb{X}'_b = L^2$ is given by

$$W(\kappa) = \frac{1}{m\hbar} \int_0^T dt\, |\kappa(t)|^2. \tag{6.16}$$

When κ is constant, namely $\kappa(t) = p$ for $0 \leq t \leq T$, we obtain

$$W(p) = \frac{|p|^2 T}{m\hbar}. \tag{6.17}$$

We use now the fundamental equation $(2.29)_s$ for $s = \mathrm{i}$ and $x' = p$ to get

$$\int \mathcal{D}\xi \cdot \exp\left[\frac{\mathrm{i}}{\hbar} \int_0^T dt\left(\frac{m}{2}|\dot{\xi}(t)|^2 - p_j \dot{\xi}^j(t)\right)\right] = \exp\left(-\frac{\mathrm{i}}{\hbar}\frac{|p|^2 T}{2m}\right). \tag{6.18}$$

We write simply $\mathcal{D}\xi$ for $\mathcal{D}_{\mathrm{i},Q}\xi$.

(iii) *The operator formula.*
The identity (6.18) is valid for arbitrary real numbers $p_1, \ldots, p_D$. Quantizing it means replacing $p_1, \ldots, p_D$ by pairwise commuting self-adjoint operators $\hat{p}_1, \ldots, \hat{p}_D$, according to (6.8). We obtain a *path-integral representation* for the propagator

$$\exp\left(-\frac{\mathrm{i}}{\hbar}\hat{H}_0 T\right) = \int \mathcal{D}\xi \cdot \exp\left[\frac{\mathrm{i}}{\hbar} \int_0^T dt\left(\frac{m}{2}|\dot{\xi}(t)|^2 - \hat{p}_j \dot{\xi}^j(t)\right)\right]. \tag{6.19}$$

Since any ξ in $\mathbb{X}$ is such that $\xi(T) = 0$, we obtain

$$-\int_0^T dt\, \hat{p}_j \dot{\xi}^j(t) = \hat{p}_j \xi^j(0). \tag{6.20}$$

Moreover, the operator $\exp((\mathrm{i}/\hbar)\hat{p}_j \xi^j(0)) = \exp(\xi^j(0)\partial_j)$ acts on the space of functions $\phi(x)$ as a translation operator, transforming it into $\phi(x + \xi(0))$.

(iv) *Conclusion.*
By applying the operator equation (6.19) to a function $\phi(x)$ and taking into account the previous result, we can express the solution

to the Schrödinger equation (6.10) as

$$\psi(T,x) = \int \mathcal{D}\xi \cdot \exp\left(\frac{\mathrm{i}}{\hbar}\int_0^T \mathrm{d}t\, \frac{m}{2}|\dot{\xi}(t)|^2\right)\phi(x+\xi(0)). \qquad (6.21)$$

Changing the notation slightly, let us introduce a time interval $\mathbb{T} = [t_a, t_b]$, and the space $\mathbb{X}_b$ of paths $\xi : \mathbb{T} \to \mathbb{R}^D$ with $\xi(t_b) = x_b$ and a finite (free) action

$$S_0(\xi) = \frac{m}{2}\int_{\mathbb{T}} \mathrm{d}t\,|\dot{\xi}(t)|^2. \qquad (6.22)$$

The time evolution of the wave function is then given by

$$\psi(t_b, x_b) = \int_{\mathbb{X}_b} \mathcal{D}\xi \cdot \exp\left(\frac{\mathrm{i}}{\hbar}S_0(\xi)\right)\psi(t_a, \xi(t_a)). \qquad (6.23)$$

Remark. On introducing two spacetime points $a = (t_a, x_a)$ and $b = (t_b, x_b)$, the previous formula can be written as a path-integral representation of the propagator $\langle b|a\rangle = \langle t_b, x_b|t_a, x_a\rangle$, namely (see Section 6.5)

$$\langle b|a\rangle = \int_{\mathcal{P}_{a,b}} \mathcal{D}\xi \cdot \exp\left(\frac{\mathrm{i}}{\hbar}S_0(\xi)\right), \qquad (6.23_{\mathrm{bis}})$$

where $\mathcal{P}_{a,b}$ denotes the space of paths ξ from a to b, that is $\xi(t_a) = x_a$, $\xi(t_b) = x_b$.

6.3 Particles in a scalar potential V

Introduction

The solution of the Schrödinger equation

$$\mathrm{i}\hbar\,\partial_t\psi(t,x) = (\hat{H}_0 + e\hat{V})\psi(t,x) \quad \text{with } \hat{H}_0 = -\frac{\hbar^2}{2m}\Delta_x \qquad (6.24)$$
$$\psi(0,x) = \phi(x)$$

cannot be written

$$\psi(t,x) \text{``=''} \exp\left(-\frac{\mathrm{i}}{\hbar}(\hat{H}_0 + e\hat{V})t\right)\phi(x), \qquad (6.25)$$

because the operators $\hat{H}_0$ and $\hat{V}$ do not commute and the exponential is meaningless as written. There are several methods for defining exponentials of sums of noncommuting operators, using

- product integrals (the Feynman–Kac formula);
- a multistep method (the Trotter–Kato–Nelson formula);
- a time-ordered exponential (Dyson's series); and
- a variational method (Schwinger's variational principle).

6.3 Particles in a scalar potential V

We recall briefly here and in Appendix B the use of product integrals and the use of the Trotter–Kato–Nelson formula for solving the Schrödinger equation (6.24). We present in greater detail the time-ordered exponentials and the variational method because they serve as prototypes for functional integrals in quantum field theory.

- *Product integrals; the Feynman–Kac formula* (see Appendix B)

Together with the representation (6.19) of $e^{-i\hat{H}_0 t/\hbar}$ as an averaged translation operator, product integrals have been used for the solution of (6.24) and for the construction of the Møller wave operators. Not unexpectedly, the solution of (6.24) generalizes the solution (6.23) of the free particle. It is the functional integral, known as the Feynman–Kac formula for the wave function,

$$\psi(t_b, x) = \int_{\mathbb{X}_b} d\Gamma^W(\xi) \exp\left(-\frac{ie}{\hbar}\int_{\mathbb{T}} dt\, V(x+\xi(t))\right) \cdot \phi(x+\xi(t_a)), \tag{6.26}$$

where the variable of integration is

$$\xi : \mathbb{T} \to \mathbb{R}^D \quad \text{such that } \xi(t_b) = 0.$$

The measure $d\Gamma^W(\xi) \doteq e^{\pi i Q(\xi)} \mathcal{D}\xi$ is the gaussian constructed in the previous paragraph for solving the free-particle case (see formulas (6.14)–(6.16)).

If the factor m/h is not incorporated in the quadratic form (6.14) defining Γ^W, then $\xi(t)$ must be replaced by $\sqrt{h/m}\,\xi(t)$ in the integrand.

- *The Trotter–Kato–Nelson formula: a multistep method*

In the Schrödinger equation, the hamiltonian $\hat{H}$ is a sum of two operators, $\hat{H} = \hat{H}_0 + e\hat{V}$. Solving the free-particle case gave us an expression for the operator:

$$P^t = e^{-i\hat{H}_0 t/\hbar}. \tag{6.27}$$

Since $\hat{V}$ is simply a multiplication operator, transforming a wave function $\phi(x)$ into $V(x)\phi(x)$, we obtain immediately an expression for the operator

$$Q^t = e^{-ie\hat{V}t/\hbar} \tag{6.28}$$

as the one transforming $\phi(x)$ into $e^{-ieV(x)t/\hbar}\phi(x)$. The Trotter–Kato–Nelson formula is given by

$$U^t = \lim_{n=\infty} \left(P^{t/n} Q^{t/n}\right)^n \tag{6.29}$$

Time-ordered exponentials; Dyson series

Time-ordered exponentials appeared in Feynman operator calculus, then colloquially called the "disentangling of non-commuting operators" [2].[3] Feynman remarked that the order of two operators, say A and B, need not be specified by their positions, say AB, but can be given by an ordering label $A(s_2)B(s_1)$. In this notation, the operators can be manipulated nearly as if they commuted, the proper order being restored when needed. In his landmark paper [3] that convinced reluctant physicists of the worth of Feynman's formulation of quantum electrodynamics, Dyson introduced time-ordered exponentials,[4] which can be expanded into "Dyson's series."

Time-ordered exponentials can be introduced as the solution of the ordinary differential equation

$$\frac{\mathrm{d}x(t)}{\mathrm{d}t} = A(t)x(t) \quad \text{with } x(t) \in \mathbb{R}^D, \, A(t) \in \mathbb{R}^{D \times D}, \tag{6.30}$$

$$x(t_0) = x_0.$$

Indeed, this equation can be written

$$x(t) = x(t_0) + \int_{t_0}^{t} \mathrm{d}s \, A(s)x(s), \tag{6.31}$$

and solved by the following iteration in the pointed space of functions on $[t_0, t]$ with values in $\mathbb{R}^D$,

$$x_{n+1}(t) = x_n(t_0) + \int_{t_0}^{t} \mathrm{d}s \, A(s)x_n(s), \tag{6.32}$$

$$x_n(t_0) = x_0.$$

The solution of (6.30) can be written

$$x(t) = U(t, t_0)x(t_0), \tag{6.33}$$

$$U(t, t_0) := \sum_{n=0}^{\infty} \int_{\Delta_n} A(s_n) \ldots A(s_1) \mathrm{d}s_1 \ldots \mathrm{d}s_n, \tag{6.34}$$

where

$$\Delta_n := \{t_0 \leq s_1 \leq s_2 \leq \ldots \leq s_n \leq t\} \tag{6.35}$$

---

[3] This paper was presented orally several years before its publication in 1951.
[4] Time ordering is labeled P in Dyson's paper [3].

6.3 Particles in a scalar potential V

and
$$U(t_0, t_0) = 1.$$

Note the composition property
$$U(t_2, t_1)U(t_1, t_0) = U(t_2, t_0), \tag{6.36}$$

which is easily proved by showing that both sides of equation (6.36) satisfy the same differential equation in t_2, whose solution, together with the initial condition
$$U(t_0, t_0) = 1, \tag{6.37}$$

is unique.

Dyson remarked that $U(t, t_0)$ is a time-ordered exponential,
$$U(t, t_0) = T \exp\left(\int_{t_0}^{t} A(s)\,\mathrm{d}s\right) \tag{6.38}$$

$$= \sum_{n=0}^{\infty} \frac{1}{n!} T \int_{\square_n} A(s_n)\ldots A(s_1)\,\mathrm{d}s_1 \ldots \mathrm{d}s_n \tag{6.39}$$

$$= \sum_{n=0}^{\infty} \frac{1}{n!} \int_{\square_n} T[A(s_n)\ldots A(s_1)]\,\mathrm{d}s_1 \ldots \mathrm{d}s_n, \tag{6.40}$$

where $\square_n$ is the n-cube of side length $t - t_0$. Here the T operator commutes with summation and integration and restores a product $A(s_n)\ldots A(s_1)$ in the order such that the labels $s_n \ldots s_1$ increase from right to left.

Consider, for instance, the second term in Dyson's series (6.40),

$$\int_{t_0}^{t} A(s_2) \left(\int_{t_0}^{s_2} A(s_1)\,\mathrm{d}s_1\right) \mathrm{d}s_2 = \int_{t_0}^{t} A(s_1) \left(\int_{t_0}^{s_1} A(s_2)\,\mathrm{d}s_2\right) \mathrm{d}s_1$$

$$= \frac{1}{2!} \int_{\square_2} T[A(s_2)A(s_1)]\,\mathrm{d}s_1\,\mathrm{d}s_2.$$

The domains of integration of these three integrals are, respectively, the shaded areas in figures 6.1(a)–(c).

The same analysis can be used to construct the higher terms in (6.40).

The Schrödinger equation

The process demonstrated in equations (6.30)–(6.34) applies also to the Schrödinger equation. Because the wave function $\psi(t, x)$ is the value of the state vector $|\psi_t\rangle$ in the x-representation of the system

$$\psi(t, x) = \langle x|\psi_t\rangle, \tag{6.41}$$

122   Quantum dynamics: path integrals and operator formalism

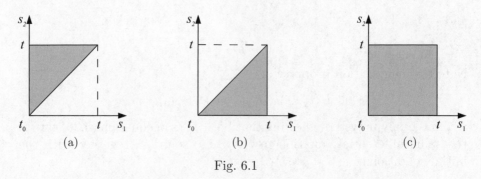

Fig. 6.1

one can replace the Schrödinger equation for the wave function ψ,

$$\begin{aligned}
i\partial_t \psi(t,x) &= \hbar^{-1}\hat{H}\psi(t,x) \\
&= \left[-\frac{1}{2}\left(\frac{\hbar}{m}\right)\Delta_x + \hbar^{-1}eV(x)\right]\psi(t,x) \\
&= \hbar^{-1}(\hat{H}_0 + e\hat{V})\psi(t,x),
\end{aligned} \quad (6.42)$$

$$\psi(t_0,x) = \phi(x),$$

by the Schrödinger equation for the time-evolution operator $U(t,t_0)$ acting on the state vector $|\psi_{t_0}\rangle$,

$$\begin{aligned}
i\partial_t U(t,t_0) &= \hbar^{-1}\hat{H}U(t,t_0), \\
U(t_0,t_0) &= \mathbf{1}.
\end{aligned} \quad (6.43)$$

Can one apply equation (6.34) to our case? The answer is yes, but not readily. Proceeding like Dyson, we first construct the solution of (6.43) in the interaction representation. This allows us to compare the time evolution of an interacting system with the time evolution of the corresponding noninteracting system.

Let $U_0(t,t_0)$ be the (unitary) time-evolution operator for the free hamiltonian H_0, given as the solution of the differential equation

$$\begin{aligned}
i\partial_t U_0(t,t_0) &= \hbar^{-1}\hat{H}_0 U_0(t,t_0), \\
U_0(t_0,t_0) &= \mathbf{1}.
\end{aligned} \quad (6.44)$$

We introduce two new operator families:

$$R(t) = U_0(t,t_0)^\dagger U(t,t_0), \quad (6.45)$$

$$A(t) = U_0(t,t_0)^\dagger \hat{V} U(t,t_0). \quad (6.46)$$

Since $\hat{H} = \hat{H}_0 + e\hat{V}$, a simple calculation gives

$$i\partial_t R(t) = e\hbar^{-1} A(t) R(t). \quad (6.47)$$

6.3 Particles in a scalar potential V

On repeating the derivation of equation (6.34), we obtain

$$R(t) = \sum_{n=0}^{\infty} \left(-\frac{ie}{\hbar}\right)^n \int_{\Delta_n} A(s_n)\ldots A(s_1) ds_1 \ldots ds_n. \quad (6.48)$$

Since $U(t, t_0)$ is equal to $U_0(t, t_0)R(t)$, we obtain the final expression

$$U(t, t_0) = \sum_{n=0}^{\infty} \left(-\frac{ie}{\hbar}\right)^n \int_{\Delta_n} B(s_n, \ldots, s_1) ds_1 \ldots ds_n \quad (6.49)$$

with

$$B(s_n, \ldots, s_1) = U_0(t, s_n)\hat{V}U_0(s_n, s_{n-1})\hat{V} \ldots U_0(s_2, s_1)\hat{V}U_0(s_1, t_0). \quad (6.50)$$

Path-integral representation

The action functional corresponding to the hamiltonian operator

$$\hat{H} = -\frac{\hbar^2}{2m}\Delta + eV$$

is of the form

$$S(\xi) = S_0(\xi) + eS_1(\xi), \quad (6.51)$$

where $\xi : \mathbb{T} \to \mathbb{R}^D$ is a path and

$$S_0(\xi) = \frac{m}{2}\int_{\mathbb{T}} dt |\dot{\xi}(t)|^2, \qquad S_1(\xi) = -\int_{\mathbb{T}} dt\, V(\xi(t)). \quad (6.52)$$

On developing an exponential and taking into account the *commutativity* of the factors $V(\xi(s))$, we get

$$\exp\left(\frac{ie}{\hbar}S_1(\xi)\right) = \exp\left(-\frac{ie}{\hbar}\int_{\mathbb{T}} dt\, V(\xi(t))\right)$$
$$= \sum_{n=0}^{\infty}\left(-\frac{ie}{\hbar}\right)^n \int_{\Delta_n} V(\xi(s_n))\ldots V(\xi(s_1))ds_1\ldots ds_n. \quad (6.53)$$

An important property of the path integral is the "markovian principle": let us break a path $\xi : \mathbb{T} \to \mathbb{R}^D$ by introducing intermediate positions $x_1 = \xi(s_1), \ldots, x_n = \xi(s_n)$, and the $n+1$ partial paths $\xi_0, \xi_1, \ldots, \xi_n$, where ξ_i runs from time s_i to s_{i+1} (with the convention $\mathbb{T} = [s_0, s_{n+1}]$). Then $\mathcal{D}\xi$ can be replaced by $\mathcal{D}\xi_0\, d^D x_1\, \mathcal{D}\xi_1\, d^D x_2 \ldots d^D x_{n-1}\, \mathcal{D}\xi_{n-1}\, d^D x_n\, \mathcal{D}\xi_n$. For the markovian principle to hold, it is crucial to use the normalization of $\mathcal{D}$ given by the formulas (3.39), (3.40), (3.72) and (3.73).

We take now the final step. We start from the representation (6.23) and (6.23$_{\text{bis}}$) of the free propagator U_0 as a path integral, plug this expression into the perturbation expansion (6.49) and (6.50) for the propagator U, and perform the integration of $B(s_n, \ldots, s_1)$ over Δ_n using the markovian principle. We obtain the propagator $\langle b|a\rangle = \langle t_b, x_b|t_a, x_a\rangle$ as an integral $\int_{\mathcal{P}_{a,b}} \mathrm{d}\Gamma^W(\xi) e^{\mathrm{i} S_1(\xi)/\hbar}$. Since the gaussian volume element $\mathrm{d}\Gamma^W(\xi)$ is nothing other than $\mathcal{D}\xi \cdot e^{\mathrm{i} S_0(\xi)/\hbar}$, we take the product of the two exponentials and, using the splitting (6.51) of the action, we obtain finally

$$\langle b|a\rangle = \int_{\mathcal{P}_{a,b}} \mathcal{D}\xi \cdot e^{\mathrm{i} S(\xi)/\hbar}. \tag{6.54}$$

This formula is obviously equivalent to the Feynman–Kac formula (6.26).

Remark. In our derivation, we assumed that the scalar potential $V(x)$ is time-independent. It requires a few modifications only to accommodate a time-varying potential $V(x,t)$: simply replace $S_1(\xi)$ by $-\int_{\mathbb{T}} \mathrm{d}t\, V(\xi(t), t)$.

Dyson proved the equivalence of the Tomonaga–Schwinger theory for quantum electrodynamics and its Feynman functional-integral formulation by comparing two series, one of which is obtained from the operator formalism and the other from the functional-integral formalism.

Schwinger's variational method

Let $U := U(b; a|V)$ be the propagator for a particle going from $a = (t_a, x_a)$ to $b = (t_b, x_b)$ in the potential $V = V(x,t)$. We shall compute the functional derivative

$$\frac{\delta U}{\delta V(x,t)}$$

for two cases:

- $U = U_S$ is the propagator of Schrödinger's equation

$$\mathrm{i}\hbar\, \partial_t \psi = \left(-\frac{\hbar^2}{2m}\Delta_x + eV(x,t)\right)\psi = (\hat{H}_0 + e\hat{V})\psi = \hat{H}\psi, \tag{6.55}$$

$$\mathrm{i}\hbar\, \partial_t U_S(t, t_a) = \hat{H}(t) U_S(t, t_a); \tag{6.56}$$

- $U = U_F$ is the propagator given by the Feynman integral,

$$U_F = \int_{\mathcal{P}_{a,b}} \mathrm{d}\Gamma^W(\xi) \exp\left(-\frac{\mathrm{i}e}{\hbar}\int_{\mathbb{T}} \mathrm{d}t\, V(\xi(t), t)\right), \tag{6.57}$$

$$\mathrm{d}\Gamma^W(\xi) \stackrel{!}{=} \mathcal{D}\xi\, \exp\left(\frac{\mathrm{i}}{\hbar} S_0(\xi)\right). \tag{6.58}$$

The Schrödinger equation

Introduce a one-parameter family of potentials V_λ, which reduces to V when $\lambda = 0$. Set

$$\delta V(x,t) := \frac{\mathrm{d}}{\mathrm{d}\lambda} V_\lambda(x,t)\bigg|_{\lambda=0}. \tag{6.59}$$

The propagator $U_{S,\lambda}$ is the solution of

$$i\hbar\, \partial_t U_{S,\lambda}(t,t_a) = \hat{H}_\lambda(t) U_{S,\lambda}(t,t_a),$$
$$U_{S,\lambda}(t_a,t_a) = \mathbf{1}. \tag{6.60}$$

Take the ordinary derivative of this equation with respect to λ and set $\lambda = 0$. The result is the Poincaré equation

$$i\hbar\, \partial_t \delta U_S(t,t_a) = \delta\hat{H}(t) U_S(t,t_a) + \hat{H}(t)\delta U_S(t,t_a), \tag{6.61}$$

whose operator solution is

$$i\hbar\, \delta U_S(t_b,t_a) = e \int_{t_a}^{t_b} \mathrm{d}t\, U_S(t_b,t) \cdot \delta V(\cdot,t) \cdot U_S(t,t_a), \tag{6.62}$$

or, using explicitly the kernels,

$$i\hbar\, \delta U_S(b;a) = e \int_{t_a}^{t_b} \mathrm{d}t\, \mathrm{d}^D x\, U_S(b;x,t) \cdot \delta V(x,t) \cdot U_S(x,t;a). \tag{6.63}$$

In conclusion, it follows from (6.63) that

$$i\hbar\, \frac{\delta U_S(b;a)}{\delta V(x,t)} = e U_S(b;x,t) \cdot U_S(x,t;a). \tag{6.64}$$

Path-integral representation

From (6.57) one finds that

$$i\hbar\, \delta U_F = e \int_{\mathcal{P}_{a,b}} \mathrm{d}\Gamma^W(\xi) \exp\left(-\frac{ie}{\hbar} \int_{\mathbb{T}} \mathrm{d}t\, V(\xi(t),t)\right) \cdot \int_{\mathbb{T}} \mathrm{d}t\, \delta V(\xi(t),t) \tag{6.65}$$

$$= e \int_{t_a}^{t_b} \mathrm{d}t \int_{\mathcal{P}_{a,b}} \mathcal{D}\xi\, \exp\left(\frac{i}{\hbar} S(\xi)\right) \delta V(\xi(t),t). \tag{6.66}$$

By breaking the time interval $[t_a,t_b]$ into $[t_a,t[$ and $]t,t_b]$, one displays the markovian property of the path integral (6.66). Let ξ'_t be a path from a to $\xi(t) = x$, and let ξ''_t be a path from $\xi(t) = x$ to b (figure 6.2). Then

$$\mathcal{D}\xi = \mathcal{D}\xi'_t \cdot \mathrm{d}^D x \cdot \mathcal{D}\xi''_t. \tag{6.67}$$

126   *Quantum dynamics: path integrals and operator formalism*

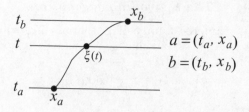

Fig. 6.2

The integral (6.66) is an integral over $\xi'_t \in \mathcal{P}_{a,(x,t)}$, $\xi''_t \in \mathcal{P}_{(x,t),b}$, and (x,t). In conclusion, it follows from (6.66) and (6.67) that

$$i\hbar \frac{\delta U_\mathrm{F}}{\delta V(x,t)} = eU_\mathrm{F}(b;x,t) \cdot U_\mathrm{F}(x,t;a). \qquad (6.68)$$

In Section 6.2, we proved that, when $V = 0$, $U_\mathrm{S} = U_\mathrm{F}$. Therefore, it follows from (6.64) and (6.68) that $U_\mathrm{F} = U_\mathrm{S}$, by virtue of the uniqueness of solutions of variational differential equations. □

The reader will notice the strong analogy between Dyson's and Schwinger's methods. Both are based on perturbation theory, but whereas Dyson performs explicitly the expansion to all orders, Schwinger writes a first-order variational principle.

6.4 Particles in a vector potential $\vec{A}$

The quantum hamiltonian for a particle in a vector potential $\vec{A}$ and a scalar potential V is

$$\begin{aligned}\hat{H} &= \frac{1}{2m}|-i\hbar\vec{\nabla} - e\vec{A}|^2 + eV \qquad (6.69) \\ &= \hat{H}_0 + e\hat{H}_1 + e^2\hat{H}_2,\end{aligned}$$

where

$$\hat{H}_0 = -\frac{\hbar^2}{2m}\Delta, \qquad \hat{H}_1 = -\frac{1}{2m}(\hat{p}\cdot\vec{A} + \vec{A}\cdot\hat{p}) + V, \qquad \hat{H}_2 = \frac{1}{2m}|\vec{A}|^2, \qquad (6.70)$$

$\hat{p}$ is the operator $-i\hbar\nabla$, and the factor ordering is such that $\hat{H}_1$ is self-adjoint.

The corresponding action functional is

$$S(\xi) = S_0(\xi) + eS_1(\xi), \qquad (6.71)$$

where

$$S_0(\xi) = \frac{m}{2}\int_\xi \frac{|d\vec{x}|^2}{dt}, \quad S_1(\xi) = \int_\xi (\vec{A}\cdot d\vec{x} - V\,dt) =: \int_\xi \mathcal{A}. \quad (6.72)$$

Note that, in contrast to the hamiltonian, the action functional contains no term in e^2. One must therefore compute the path integral to second-order perturbation (two-loop order) when identifying it as the solution of a Schrödinger equation.

The new integral, $\int_\xi \vec{A}\cdot d\vec{x}$, is an integral over t along ξ. It can be defined either as an Ito integral or as a Stratonovich integral [4] as follows. Divide the time interval $[t_a, t_b]$ into increments Δt_i such that

$$\begin{aligned}\Delta t_i &:= t_{i+1} - t_i, \\ \overrightarrow{\Delta x_i} &= x(t_{i+1}) - x(t_i) =: x_{i+1} - x_i.\end{aligned} \quad (6.73)$$

Let

$$\vec{A}_i := \vec{A}(x_i), \quad \vec{A}_{i,\alpha\beta} = \partial_\alpha \vec{A}_\beta(x_i).$$

An Ito integral for $x(t) \in \mathbb{R}^D$ is, by definition,

$$\int_\xi^I \vec{A}\cdot d\vec{x} \simeq \sum_i \vec{A}_i \cdot \overrightarrow{\Delta x_i}. \quad (6.74)$$

A Stratonovich integral is, by definition,

$$\int_\xi^S \vec{A}\cdot d\vec{x} \simeq \sum_i \frac{1}{2}(\vec{A}_i + \vec{A}_{i+1})\cdot \overrightarrow{\Delta x_i}. \quad (6.75)$$

Feynman referred to the definition of the Stratonovich integral as the *midpoint rule*. Whereas the Ito integral is nonanticipating and therefore serves probability theory well, the Stratonovich integral serves quantum physics well for the following reasons.

- It is invariant under time reversal; it codifies the principle of detailed balance.
- It corresponds to the factor ordering $\frac{1}{2}(\hat{p}\cdot\vec{A} + \vec{A}\cdot\hat{p})$ chosen for the hamiltonian (6.70), which is necessary in order for $\hat{H}$ to be self-adjoint.
- It is coherent with the functional space $L^{2,1}$, i.e. the space of functions which are continuous and whose derivatives (in the sense of distributions) are square integrable.

Spaces of $L^{2,1}$ paths were chosen as domains of integration in order for the kinetic energy to be finite. We shall now show that this choice requires the use of Stratonovich integrals. Let δ be the difference between a Stratonovich integral and an Ito integral for the same integrand.

$$\delta := \int_\xi^S \vec{A} \cdot \mathrm{d}\vec{x} - \int_\xi^I \vec{A} \cdot \mathrm{d}\vec{x} \tag{6.76}$$

$$\delta \simeq \frac{1}{2} \sum (\vec{A}_{i+1} - \vec{A}_i) \cdot \overrightarrow{\Delta x_i}$$

$$\simeq \frac{1}{2} \sum A_{i,\alpha\beta} \Delta x_i^\alpha \Delta x_i^\beta$$

$$\simeq \frac{1}{2} \sum A_{i,\alpha\beta} \delta^{\alpha\beta} \Delta t_i.$$

Hence

$$\delta = \frac{1}{2} \int \mathrm{d}t \, \vec{\nabla} \cdot \vec{A}. \tag{6.77}$$

The non-negligible term δ does not belong to spaces of continuous paths, but is well defined in $L^{2,1}$ spaces since $|\Delta x_i|^2$ is of the order of Δt_i.

Remark. There is an extensive literature on the correspondence between factor ordering in the operator formalism and the definition of functional integrals. M. Mizrahi [5] has given a summary of these correspondences.

First-order perturbation, $A \neq 0$, $V = 0$

To first order in e, the probability amplitude for the transition from a to b of a particle in a vector potential $\vec{A}$ is

$$K(b;a) = \frac{ie}{\hbar} \int_{\mathcal{P}_{a,b}} \mathrm{d}\Gamma^W(\xi) \int_\xi \vec{A} \cdot \mathrm{d}\vec{x}, \tag{6.78}$$

where $\int_\xi \vec{A} \cdot \mathrm{d}\vec{x}$ is a Stratonovich integral defined by the midpoint rule. The strategy for computing (6.78) is similar to the strategy used for computing first-order perturbation by a scalar potential, namely, interchanging the time integral along the path ξ and the gaussian integral over the path ξ. Divide the time interval into two time intervals (see figure 6.2), $[t_a, t[$ and $]t, t_b]$. As before (6.67) is

$$\mathcal{D}\xi = \mathcal{D}\xi' \cdot \mathrm{d}^D x \cdot \mathcal{D}\xi'', \tag{6.79}$$

where ξ' is a path from x_a to $x(t)$, and ξ'' is a path from $x(t)$ to x_b.

To compute the probability amplitude to second order in e, one divides the time interval into three time intervals, $[t_a, t_1[, [t_1, t_2[$, and $[t_2, t_b[$.

6.5 Matrix elements and kernels

The formula at the beginning of this chapter captures the essence of the relation between the path integral and the operator formalisms of quantum mechanics:

$$\langle b|\hat{O}|a\rangle = \int_{\mathcal{P}_{a,b}} O(\gamma)\exp(iS(\gamma)/\hbar)\mu(\gamma)\mathcal{D}\gamma. \tag{6.80}$$

It has been derived for many systems by a variety of methods (see for instance Section 1.4). Equation (6.80) is used by Bryce DeWitt [6] as the definition of functional integrals in quantum field theory. Some of the "remarkable properties of $\mu(\phi)$" presented in Chapter 18 are derived from this definition of functional integrals. The particular case $O(\gamma) \equiv 1$ gives a path-integral representation for the *propagator* $\langle b|a\rangle$ since $\hat{O} = \mathbf{1}$.

On the right-hand side of (6.80), $O(\gamma)$ is a functional of the path γ in the space $\mathcal{P}_{a,b}$ of paths from $a = (t_a, x_a)$ to $b = (t_b, x_b)$. Colloquially one says that $\hat{O}$ is the "time-ordered" operator corresponding to the functional O. Some explanation is needed in order to make sense of this expression: $\hat{O}$ is an operator on a Hilbert space $\mathcal{H}$. "Time-ordering" of operators on Hilbert spaces is not defined outside equations such as (6.80). The time-ordering comes from the representation of the left-hand side of (6.80) by a path integral. A path γ is parametrized by time, hence its values can be time-ordered. Recall, in Section 6.3, the remark introducing time-ordered exponentials: "the order of two operators, say A and B, need not be specified by their positions, say AB, but can be given by an ordering label $A(s_2)B(s_1)$; [then] the operators can be manipulated nearly as if they commuted, the proper order being restored when needed." The path-integral representation of the left-hand side of (6.80) does just that.

We shall spell out the formalism reflecting the idea of attaching ordering labels to operators.

Let $\mathbb{T}$ be the time interval $[t_a, t_b]$, and let $\mathbb{M}^D$ be the configuration space of a system S. Set

$$\mathbb{E} = \mathbb{M}^D \times \mathbb{T} \quad \text{with the projection } \Pi : \mathbb{E} \to \mathbb{T}.$$

Let $\mathcal{P}$ be the space of paths

$$\gamma : \mathbb{T} \to \mathbb{E},$$
$$\Pi(\gamma(t)) = t.$$

Let $\mathcal{P}_{a,b}$ be the space of paths from a to b:

$$\gamma(t_a) = x_a,$$
$$\gamma(t_b) = x_b.$$

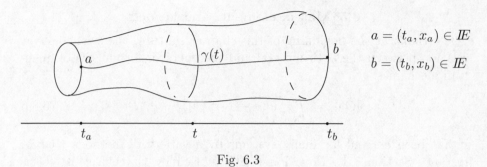

Fig. 6.3

Let $O(\gamma) := x(t)x(t')$ be an observable. Provided that there is a quantum rule associating an observable $x(t)$ with a quantum operator $\hat{x}(t)$, the functional $O(\gamma)$ defines the time-ordered product

$$\hat{O} = T(\hat{x}(t)\hat{x}(t'))$$

by providing an ordering label to the operators.

Matrix elements and kernels

Let $\hat{X}$ be an operator mapping a Hilbert space $\mathcal{H}_{t_a}$ into a Hilbert space $\mathcal{H}_{t_b}$,

$$\hat{X}: \mathcal{H}_{t_a} \to \mathcal{H}_{t_b},$$

and let

$$\psi_a \in \mathcal{H}_{t_a} \quad \text{and} \quad \psi_b \in \mathcal{H}_{t_b}.$$

The matrix element $\langle \psi_b | \hat{X} | \psi_a \rangle$ is given by integrating the kernel $\langle b | X | a \rangle$:

$$\langle \psi_b | \hat{X} | \psi_a \rangle = \iint d^D x_a \, d^D x_b \, \psi_b(t_b, x_b)^* \langle t_b, x_b | \hat{X} | t_a, x_a \rangle \psi_a(t_a, x_a). \tag{6.81}$$

Now apply this equation to the kernel $\langle b | \hat{O} | a \rangle$ given by (6.80) in order to give a representation of the matrix element in terms of a functional integral over the space $\mathcal{P}$ of paths $\gamma: \mathbb{T} \to \mathbb{E}$ with free end points:

$$\langle \psi_b | \hat{O} | \psi_a \rangle = \int_{\mathcal{P}} \psi_b(t_b, \gamma(t_b))^* O(\gamma) \psi_a(t_a, \gamma(t_a)) \exp(iS(\gamma)/\hbar) \mu(\gamma) \mathcal{D}\gamma. \tag{6.82}$$

References

[1] C. DeWitt-Morette, A. Maheshwari, and B. Nelson (1979). "Path integration in non-relativisitic quantum mechanics," *Phys. Rep.* **50**, 266–372.

[2] R. P. Feynman (1951). "An operator calculus having applications in quantum electrodynamics," *Phys. Rev.* **84**, 108–128.
[3] F. J. Dyson (1949). "The radiation theories of Tomonaga, Schwinger, and Feynman," *Phys. Rev.* **75**, 486–502.

In the book edited by Christopher Sykes, *No Ordinary Genius: The Illustrated Richard Feynman* (New York, W. W. Norton & Co., 1994), Dyson recalls his work and his discussions with Feynman that led to his publication of the radiation theories.

[4] C. DeWitt-Morette and K. D. Elworthy (1981). "A stepping stone to stochastic analysis," *Phys. Rep.* **77**, 125–167.
[5] M. M. Mizrahi (1981). "Correspondence rules and path integrals," *Il Nuovo Cimento* **61B**, 81–98.
[6] B. DeWitt (2003). *The Global Approach to Quantum Field Theory* (Oxford, Oxford University Press; with corrections, 2004).

Part III
Methods from differential geometry

7
Symmetries

> Symmetry is one idea by which man has tried to comprehend and create order, beauty, and perfection.
>
> H. Weyl [1]

7.1 Groups of transformations. Dynamical vector fields

In the Louis Clark Vaneuxem Lectures he gave[1] in 1951, Hermann Weyl introduced symmetry in these words: "Symmetry, as wide or as narrow as you may define its meaning, is one idea by which man through the ages has tried to comprehend and create order, beauty, and perfection." In his last lecture, Weyl extracted the following guiding principle from his survey of symmetry: "Whenever you have to do with a structure-endowed entity Σ try to determine its group of automorphisms[2]... You can expect to gain a deep insight into the constitution of Σ in this way."

In Chapters 2–6, classical actions served as building elements of functional integrals in physics. In Chapter 7, we begin with groups of transformations of the configuration spaces of physical systems.

Let $\mathbb{N}^D$ be a manifold, possibly multiply connected. Let $\{\sigma(r)\}$ be a one-parameter group of transformations on $\mathbb{N}^D$ and let X be its generator. Let $\mathbf{x}, \mathbf{x}_0$ be points in $\mathbb{N}^D$. By definition,

$$\frac{\mathrm{d}}{\mathrm{d}r}(\mathbf{x}_0 \cdot \sigma(r)) = X(\mathbf{x}_0 \cdot \sigma(r)) \tag{7.1}$$

---

[1] His "swan-song on the eve of [his] retirement from the Institute for Advanced Study" [1].
[2] This means "the group of those element-wise transformations which leave all structural relations undisturbed" [1].

with
$$X(\mathbf{x}) = \frac{d}{dr}(\mathbf{x} \cdot \sigma(r))|_{r=0}. \quad (7.2)$$

In the general case, we consider N one-parameter groups of transformations $\sigma_A(r)$, $A \in \{1, \ldots, N\}$. In the examples treated in Sections 7.2 and 7.3, we consider $N = D$ one-parameter groups of transformations on $\mathbb{N}^D$; we consider a set of one-parameter groups $\sigma_{(A)}(r)$ generated by $X_{(A)}$, where $A \in \{1, \ldots, D\}$. Equation (7.1) is replaced by

$$d(\mathbf{x}_0 \cdot \sigma_{(A)}(r)) = X_{(A)}(\mathbf{x}_0 \cdot \sigma_{(A)}(r))dr. \quad (7.3)$$

In general, the vector fields $X_{(A)}$ do not commute:

$$[X_{(A)}, X_{(B)}] \neq 0. \quad (7.4)$$

The quantization of the system S with configuration space $\mathbb{N}^D$ endowed with $\{\sigma_{(A)}(r)\}_A$ groups of transformations is achieved by substituting the following equation into (7.3):

$$dx(t,z) = \sum_A X_{(A)}(x(t,z))dz^A(t). \quad (7.5)$$

The pointed paths $x \in \mathcal{P}_0 \mathbb{N}^D \equiv \mathbb{X}$,

$$x : \mathbb{T} \longrightarrow \mathbb{N}^D, \qquad x = \{x^\alpha\}, \\ x(t_0) = \mathbf{x}_0, \quad (7.6)$$

were parametrized by $r \in \mathbb{R}$ in (7.3); in (7.5) they are parametrized by pointed paths $z \in \mathcal{P}_0 \mathbb{R}^D \equiv \mathbb{Z}$:

$$z : \mathbb{T} \longrightarrow \mathbb{R}^D, \qquad z = \{z^A\}, \\ z(t_0) = 0. \quad (7.7)$$

If z is a brownian path then (7.5) is a stochastic differential equation. Here we consider $L^{2,1}$ paths, i.e. paths such that, for a given constant metric h_{AB},

$$\int_\mathbb{T} dt\, h_{AB} \dot{z}^A(t) \dot{z}^B(t) =: Q_0(z) < \infty. \quad (7.8)$$

The quadratic form Q_0 is used for defining the gaussian (see Chapter 2) $\mathcal{D}_{s,Q_0}(z) \exp(-(\pi/s) Q_0(z))$ normalized to unity.

A space of pointed paths is contractible, therefore (7.5) leads to a map P between the contractible spaces $\mathbb{Z}$ and $\mathbb{X}$.

$$P : \mathbb{Z} \longrightarrow \mathbb{X} \qquad \text{by } z \mapsto x. \quad (7.9)$$

In general, the path x parametrized by z is not a function of t and $z(t)$ but rather a function of t and a functional of z. One can rarely solve (7.5), but one can prove [2] that the solution has the form

$$x(t, z) = \mathbf{x}_0 \cdot \Sigma(t, z), \qquad (7.10)$$

where $\Sigma(t, z)$ is an element of a group of transformations on $\mathbb{N}^D$:

$$\mathbf{x}_0 \cdot \Sigma(t + t', z \times z') = \mathbf{x}_0 \cdot \Sigma(t, z) \cdot \Sigma(t', z'). \qquad (7.11)$$

The path z defined on $]t', t + t']$ is followed by the path z' on $]t, t']$.

The group $\{\Sigma(t, z)\}$ of transformations on $\mathbb{N}^D$ is defined by the generators $\{X_{(A)}\}$. This group contains the building elements of functional integrals in physics and has a variety of applications. The generators $\{X_{(A)}\}$ are dynamical vector fields.

In Chapters 2–6, the classical path became a point in a space of paths; the action functional on the space of paths contains the necessary information for the quantum world formalism.

In Chapter 7, the groups of transformations on $\mathbb{N}^D$, parametrized by $r \in \mathbb{R}$, are replaced by groups of transformations on $\mathbb{N}^D$ parametrized by a path $z \in \mathcal{Z}$. The generators $\{X_{(A)}\}$ of the classical groups of transformations provide the necessary information for developing the quantum world formalism.

In both cases, action and symmetry formalisms, the role previously played by classical paths is now played by *spaces of paths*. Locality in the configuration space has lost its importance: the quantum picture is richer than the sharp picture of classical physics.

7.2 A basic theorem

It has been proved [2] that the functional integral in the notation of the previous section,

$$\Psi(t, \mathbf{x}_0) := \int_{\mathcal{P}_0 \mathbb{R}^D} \mathcal{D}_{s, Q_0}(z) \cdot \exp\left(-\frac{\pi}{s} Q_0(z)\right) \phi(\mathbf{x}_0 \cdot \Sigma(t, z)), \qquad (7.12)$$

is the solution of the parabolic equation

$$\frac{\partial \Psi}{\partial t} = \frac{s}{4\pi} h^{AB} \mathcal{L}_{X_{(A)}} \mathcal{L}_{X_{(B)}} \Psi, \qquad (7.13)$$

$$\Psi(t_0, \mathbf{x}) = \phi(\mathbf{x}) \qquad \text{for any } \mathbf{x} \in \mathbb{N},$$

where $\mathcal{L}_X$ is the Lie derivative with respect to the generator X.

So far we have given the simplest form of the basic theorem for systems without an external potential. If there is an external potential, then the basic equations (7.5) and (7.13) read

$$dx(t,z) = X_{(A)}(x(t,z))dz^A(t) + Y(x(t,z))dt \qquad (7.14)$$

$$\frac{\partial \Psi}{\partial t} = \frac{s}{4\pi} h^{AB} \mathcal{L}_{X_{(A)}} \mathcal{L}_{X_{(B)}} \Psi + \mathcal{L}_Y \Psi, \qquad (7.15)$$

$$\Psi(t_0, \mathbf{x}) = \phi(\mathbf{x}).$$

The solution of (7.14) can again be expressed in terms of a group of transformations $\{\Sigma(t,z)\}$. The functional integral representing the solution of (7.15) is the same expression as (7.12) in terms of the new group of transformations Σ.

The functional integral (7.12) is clearly a generalization of the integral constructed in Chapter 6 for a free particle.

Example: paths in non-cartesian coordinates

We begin with a simple, almost trivial, example, which has been for many years a bone of contention in the action formalism of path integrals. Consider the case in which paths $x(t) \in \mathbb{N}^D$ are expressed in non-cartesian coordinates. Equation (7.5) can be used for expressing non-cartesian differentials $\{dx^\alpha(t)\}$ in terms of cartesian differentials $\{dz^A(t)\}$. In this case, x is not a functional of z, but simply a function of $z(t)$. Polar coordinates in $\mathbb{N}^2 \equiv \mathbb{R}^+ \times [0, 2\pi[$ are sufficient for displaying the general construction of the relevant dynamical vector fields $\{X_{(1)}, X_{(2)}\}$.

Let us abbreviate $z^A(t)$ to z^A, $x^1(t)$ to r, and $x^2(t)$ to θ; it follows from

$$z^1 = r \cdot \cos\theta, \qquad z^2 = r \cdot \sin\theta,$$

that equation (7.5) reads

$$dr \equiv dx^1 = \cos\theta \cdot dz^1 + \sin\theta \cdot dz^2 =: X^1_{(1)} dz^1 + X^1_{(2)} dz^2,$$

$$d\theta \equiv dx^2 = -\frac{\sin\theta}{r} dz^1 + \frac{\cos\theta}{r} dz^2 =: X^2_{(1)} dz^1 + X^2_{(2)} dz^2. \qquad (7.16)$$

The dynamical vector fields are therefore

$$X_{(1)} = \cos\theta \frac{\partial}{\partial r} - \frac{\sin\theta}{r} \frac{\partial}{\partial \theta},$$

$$X_{(2)} = \sin\theta \frac{\partial}{\partial r} + \frac{\cos\theta}{r} \frac{\partial}{\partial \theta}. \qquad (7.17)$$

Let the quadratic form Q_0, which defines the gaussian volume element in the functional integral (7.12), be given by (7.8) with $h_{AB} = \delta_{AB}$, the metric in cartesian coordinates. According to equation (7.13) the

Schrödinger equation in polar coordinates for a free particle is

$$\begin{aligned}\frac{\partial \Psi}{\partial t} &= \frac{s}{4\pi}\delta^{AB}\mathcal{L}_{X_{(A)}}\mathcal{L}_{X_{(B)}}\Psi \\ &= \frac{s}{4\pi}\delta^{\alpha\beta}X_{(\alpha)}X_{(\beta)}\Psi \\ &= \frac{s}{4\pi}\left(\frac{\partial^2}{\partial r^2} + \frac{1}{r^2}\frac{\partial^2}{\partial \theta^2} + \frac{1}{r}\frac{\partial}{\partial r}\right)\Psi. \end{aligned} \qquad (7.18)$$

In Section 7.3 we use the symmetry formalism for constructing the functional integral over paths taking their values in a riemannian manifold.

7.3 The group of transformations on a frame bundle

On a principal bundle, the dynamical vector fields $\{X_{(A)}\}$, which are necessary for the construction (7.12) of the solution of the parabolic equation (7.13), are readily obtained from connections. On a riemannian manifold, there exists a unique metric connection such that the torsion vanishes.[3] Whether unique or not, a connection defines the horizontal lift of a vector on the base space. A point $\rho(t)$ in the frame bundle is a pair: $x(t)$ is a point on the base space $\mathbb{N}^D$, and $u(t)$ is a frame in the tangent space at $x(t)$ to $\mathbb{N}^D$, $T_{x(t)}\mathbb{N}^D$,

$$\rho(t) = (x(t), u(t)). \qquad (7.19)$$

The connection σ defines the horizontal lift of a vector $\dot{x}(t)$, namely

$$\dot{\rho}(t) = \sigma(\rho(t)) \cdot \dot{x}(t). \qquad (7.20)$$

In order to express this equation in the desired form,

$$\dot{\rho}(t) = X_{(A)}(\rho(t))\dot{z}^A(t), \qquad (7.21)$$

we exploit the linearity of the map $u(t)$:

$$u(t): \mathbb{R}^D \to T_{x(t)}\mathbb{N}^D. \qquad (7.22)$$

Let

$$\dot{z}(t) := u(t)^{-1}\dot{x}(t). \qquad (7.23)$$

Choose a basis $\{e_{(A)}\}$ in $\mathbb{R}^D$ and a basis $\{e_{(\alpha)}\}$ in $T_{x(t)}\mathbb{N}^D$ such that

$$\dot{z}(t) = \dot{z}^A(t)e_{(A)} = u(t)^{-1}(\dot{x}^\alpha(t)e_{(\alpha)}). \qquad (7.24)$$

---

[3] This unique connection is the usual riemannian (Levi-Civita) connection, which is characterized by the vanishing of the covariant derivative of the metric tensor. For the definition of a metric connection and its equivalence with the riemannian (Levi-Civita) connection see for instance [3, Vol. I, p. 381].

Then
$$\dot{\rho}(t) = \sigma(\rho(t) \circ u(t) \circ u(t)^{-1} \cdot \dot{x}(t)) \qquad (7.25)$$
can be written in the form (7.21), where
$$X_{(A)}(\rho(t)) = (\sigma(\rho(t)) \circ u(t)) \cdot e_{(A)}. \qquad (7.26)$$

The construction (7.5)–(7.13) gives a parabolic equation on the bundle. If the connection in (7.20) is the metric connection, then the parabolic equation on the bundle gives, by projection onto the base space, the parabolic equation with the Laplace–Beltrami operator. Explicitly, equation (7.13) for Ψ a scalar function on the frame bundle is

$$\frac{\partial}{\partial t_b}\Psi(t_b, \rho_b) = \frac{s}{4\pi}h^{AB}X_{(A)}X_{(B)}\Psi(t_b, \rho_b), \qquad (7.27)$$
$$\Psi(t_a, \cdot) = \Phi(\cdot),$$

where $X_{(A)} = X_{(A)}^\lambda(\rho_b)(\partial/\partial\rho_b^\lambda)$. Let Π be the projection of the bundle onto its base space $\mathbf{N}^D$,

$$\Pi : \rho_b \longmapsto x_b. \qquad (7.28)$$

Let ψ and ϕ be defined on $\mathbf{N}^D$, with Ψ and Φ defined on the frame bundle by

$$\Psi =: \psi \circ \Pi, \qquad \Phi =: \phi \circ \Pi; \qquad (7.29)$$

then by projection

$$\frac{\partial}{\partial t_b}\psi(t_b, x_b) = \frac{s}{4\pi}g^{ij}D_iD_j\psi(t_b, x_b), \qquad (7.30)$$

where D_i is the covariant derivative defined by the riemannian connection σ. That is

$$\frac{\partial}{\partial t_b}\psi(t_b, x_b) = \frac{s}{4\pi}\Delta\psi(t_b, x_b), \qquad (7.31)$$
$$\psi(t_a, x) = \phi(x),$$

where Δ is the Laplace–Beltrami operator on $\mathbf{N}^D$.

In summary, starting with the dynamical vector fields $\{X_{(A)}\}$ given by (7.26) and the group of transformations (7.10) defined by (7.5), we construct the path integral (7.12) which is the solution of the parabolic equation (7.13), or the particular case (7.27) for Ψ a scalar function on the frame bundle. The projection of (7.27) onto the base space yields (7.31) and the projection of (7.12) onto the base space is

$$\psi(t_b, x_b) = \int_{\mathcal{P}_b\mathbb{R}^D} \mathcal{D}_{s,Q_0}(z)\exp\left(-\frac{\pi}{s}Q_0(z)\right)\phi((\text{Dev } z)(t_a)). \qquad (7.32)$$

Dev is the Cartan development map; it is a bijection from the space of pointed paths z on $T_b\mathbb{N}^D$ (identified to $\mathbb{R}^D$ via the frame u_b) into a space of pointed paths x on $\mathbb{N}^D$ (paths such that $x(t_b) = b$):

$$\text{Dev}: \mathcal{P}_0 T_b \mathbb{N}^D \longrightarrow \mathcal{P}_b \mathbb{N}^D \qquad \text{by } z \longmapsto x. \tag{7.33}$$

Explicitly

$$(\Pi \circ \rho)(t) =: (\text{Dev } z)(t). \tag{7.34}$$

The path x is said to be the development of z if $\dot{x}(t)$ parallel transported along x from $x(t)$ to x_b is equal to $\dot{z}(t)$ trivially transported to the origin of $T_{x_b}\mathbb{N}^D$, for every $t \in \{t_a, t_b\}$.

Remark. It has been shown in [3, Part II, p. 496] that the covariant laplacian Δ at point x_b of $\mathbb{N}^D$ can be lifted to a sum of products of Lie derivatives $h^{AB}X_{(A)}X_{(B)}$ at the frame ρ_b; the integral curves of the set of vector fields $\{X_{(A)}\}$ starting from ρ_b at time t_b are the horizontal lifts of a set of geodesics at x_b, tangent to the basis $\{e_A\}$ of $T_{x_b}\mathbb{N}^D$ corresponding to the frame ρ_b.

7.4 Symplectic manifolds[4]

The steps leading from dynamical vector fields $\{X_{(A)}, Y\}$ to a functional integral representation (7.12) of the solution Ψ of a parabolic equation (7.15) have been taken in the previous section on configuration spaces. In this section we outline the same steps on phase spaces, or more generally on symplectic manifolds.

A symplectic manifold $(\mathbb{M}^{2N}, \Omega)$, $2N = D$ is an even-dimensional manifold together with a closed nondegenerate 2-form Ω of rank D. The symplectic form Ω can to some extent be thought of as an antisymmetric metric (see Section 11.1). Two generic examples of symplectic manifolds are cotangent bundles of configuration space (e.g. phase space) and Kähler manifolds. A Kähler manifold is a real manifold of dimension $2N$ with a symplectic structure Ω and a complex structure J compatible with Ω.

Groups of transformations

A *symplectomorphism* (canonical transformation) from $(\mathbb{M}_1, \Omega_1)$ onto $(\mathbb{M}_2, \Omega_2)$ is a diffeomorphism $f: \mathbb{M}_1 \to \mathbb{M}_2$ such that

$$f^*\Omega_2 = \Omega_1. \tag{7.35}$$

---

[4] This section is based on the work of John LaChapelle in [2, p. 2274] and [4]. For basic properties of symplectic manifolds see [6], [5], and also [3].

The set of symplectomorphisms of $(\mathbb{M}, \Omega)$ onto itself is a group labeled $\mathrm{SP}(\mathbb{M}, \Omega)$.

Let $\mathbb{L} \subset \mathbb{M}$. The *symplectic complement* $(T_m\mathbb{L})^\perp$ of the tangent space to $\mathbb{L}$ at m is the set of vectors $Y \in T_m\mathbb{M}$ such that $\Omega(X, Y) = 0$ for all X in $T_m\mathbb{L}$. The use of the superscript $\perp$ may be misleading because it may suggest metric orthogonality, but it is commonplace.

A submanifold $\mathbb{L} \subset \mathbb{M}$ is called *lagrangian* if, and only if, for each point $m \in \mathbb{L}$, the tangent space $T_m\mathbb{L} \subset T_m\mathbb{M}$ is its own symplectic complement,

$$T_m\mathbb{L} = (T_m\mathbb{L})^\perp. \tag{7.36}$$

Example. Let $T^*\mathbb{M}$ be a cotangent bundle over $\mathbb{M}$. The base space $\mathbb{M}$ (embedded into $T^*\mathbb{M}$ by the zero cross section) is a lagrangian submanifold of $T^*\mathbb{M}$. Indeed, in canonical coordinates

$$\Omega = \mathrm{d}p_\alpha \wedge \mathrm{d}q^\alpha. \tag{7.37}$$

Vector fields Y on the bundle and X on the base space are given in the canonical dual basis by, respectively,

$$Y = Y^\alpha \frac{\partial}{\partial q^\alpha} + Y_\alpha \frac{\partial}{\partial p_\alpha}, \qquad X = X^\alpha \frac{\partial}{\partial q^\alpha}. \tag{7.38}$$

Contract Ω with Y and X:

$$i_Y(\mathrm{d}p_\alpha \wedge \mathrm{d}q^\alpha) = -Y^\alpha \mathrm{d}p_\alpha + Y_\alpha \mathrm{d}q^\alpha,$$
$$i_X(-Y^\alpha \mathrm{d}p_\alpha + Y_\alpha \mathrm{d}q^\alpha) = X^\alpha Y_\alpha.$$

We have $\Omega(X, Y) = 0$ for all X iff $Y_\alpha = 0$ for all α, hence $Y = Y^\alpha(\partial/\partial q^\alpha)$. This proves our claim that $\mathbb{M}$ is lagrangian.

Example. Fibers in cotangent bundles are submanifolds of constant q. An argument similar to the argument in the previous example shows that they are lagrangian submanifolds.

A real *polarization* is a foliation of a symplectic manifold whose leaves are lagrangian submanifolds, necessarily of dimension N.

Example. The fibration of a cotangent bundle is a polarization. Here we shall consider the subgroup of symplectomorphisms that preserve polarization. We refer the reader to John LaChapelle's work [4] for transition amplitudes when the initial and final states are given in different polarizations; his work includes applications of "twisted" polarizations and applications of Kähler polarizations to coherent states, a strategy suggested by the work of Berezin [7] and Bar-Moshe and Marinov [8].

Let us assume for simplicity that the symplectic manifold $\mathbb{M}$ is globally polarized. For instance, if $\mathbb{M}$ is a cotangent bundle, we assume

$$\mathbb{M} = (\text{base space}) \times (\text{typical fiber}). \tag{7.39}$$

7.4 Symplectic manifolds

We shall not write $\mathbb{M} = \mathbb{Q} \times \mathbb{P}$ because this may give the impression that a symplectic manifold is only a glorified name of phase space. Set

$$\mathbb{M} = \hat{\mathbb{X}} \times \check{\mathbb{X}}. \tag{7.40}$$

Both spaces are of dimension N. The paths x in $\mathbb{M}$ satisfy N initial and N final boundary conditions. Let $\mathcal{P}_{a,b}\mathbb{M}$ be the space of such paths, where a summarizes the initial conditions and b the final ones. The space $\mathcal{P}_{a,b}\mathbb{M}$ can be parametrized by the space $\mathcal{P}_{0,0}\mathbb{Z}$ of paths z in $\mathbb{R}^N \times \mathbb{R}_N$. For example, let x_{cl} be the classical path in $\mathbb{M}$ defined by an action functional $\mathcal{S}$, then the equation

$$x = x_{\text{cl}} + \alpha z \tag{7.41}$$

for some constant α parametrizes $\mathcal{P}_{a,b}\mathbb{M}$ by $\mathcal{P}_{0,0}\mathbb{Z}$. The general case can be treated by the method developed in Section 7.1 culminating in the basic theorem (7.14) and (7.15). Equation (7.41) is generalized to

$$\mathrm{d}x(t) = X_{(A)}(x(t))\mathrm{d}z^A(t) + Y(x(t))\mathrm{d}t \tag{7.42}$$

with $2N$ boundary conditions. Generically (7.42) cannot be solved explicitly but the solution can be written in terms of a group of transformations on $\mathbb{M}$. On a polarized symplectic manifold $\mathbb{M}$, let the path $x = (\hat{x}, \check{x})$ have, for instance, the following boundary conditions:

$$\hat{x}(t_a) = \hat{\mathbf{x}}_a, \qquad \check{x}(t_b) = \check{\mathbf{x}}_b. \tag{7.43}$$

Then, the solution of (7.42) can be written

$$\hat{x}(t) = \hat{\mathbf{x}}_a \cdot \Sigma(t, z), \tag{7.44}$$
$$\check{x}(t) = \check{\mathbf{x}}_b \cdot \Sigma(t, z). \tag{7.45}$$

For constructing a path integral on $\mathbb{M}$, we choose a gaussian volume element defined by

$$\int_{\mathcal{P}_{0,0}\mathbb{Z}} \mathcal{D}_{s,Q_0}(\check{z}, \hat{z}) \exp\left(-\frac{\pi}{s} Q_0(\check{z}, \hat{z}) - 2\pi i \langle (\hat{z}', \hat{z}'), (\check{z}, \hat{z}) \rangle \right) \tag{7.46}$$
$$= \exp(-\pi s W(\check{z}', \hat{z}')),$$

where $\mathcal{P}_{0,0}\mathbb{Z} = \mathcal{P}_0\hat{\mathbb{Z}} \times \mathcal{P}_0\check{\mathbb{Z}}$, $\hat{z} \in \mathcal{P}_0\hat{\mathbb{Z}}$, $\check{z} \in \mathcal{P}_0\check{\mathbb{Z}}$, and as usual s is either 1 or i. Choosing the volume element consists of choosing the quadratic form Q_0. Here Q_0 has a unique inverse[5] W dictated by the boundary conditions imposed on the paths $x \in \mathcal{P}_{a,b}\mathbb{M}$. For a system with action functional $\mathcal{S}$, the quadratic form Q_0 can be either the kinetic energy of

---

[5] An "inverse" in the sense of (2.26) and (2.27).

the corresponding free particle or the hessian of $\mathcal{S}$. For example

$$Q_0(z) = \int_{t_a}^{t_b} dt(\check{z}^i, \hat{z}^i) \begin{pmatrix} -\dfrac{\partial^2 H}{\partial \check{z}^i \, \partial \check{z}^j} & \delta_{ij}\dfrac{\partial}{\partial t} - \dfrac{\partial^2 H}{\partial \check{z}^i \, \partial \hat{z}^j} \\ -\delta_{ij}\dfrac{\partial}{\partial t} - \dfrac{\partial^2 H}{\partial \hat{z}^i \, \partial \check{z}^j} & -\dfrac{\partial^2 H}{\partial \hat{z}^i \, \partial \hat{z}^j} \end{pmatrix}\begin{pmatrix} \check{z}^j \\ \hat{z}^j \end{pmatrix} \tag{7.47}$$

for H a scalar functional on $\mathcal{P}_{0,0}\mathbb{Z}$. See explicit examples in [9].

The basic theorem (7.14) and (7.15) can be applied to symplectic manifolds before polarization. However, in quantum physics one cannot assign precise values to position and momentum at the same time. Consequently the path integral over the whole symplectic manifold does not represent a quantum state wave function. However, upon restriction to a lagrangian submanifold, the path integral will correspond to a quantum state wave function.

A polarized manifold makes it possible to use the basic theorem by projection onto a lagrangian submanifold. For example, if we assume that the group of transformations $\Sigma(t, z)$ preserves the polarization, so that $\{X_{(A)}\} = \{X_{(a)}, X_{(a')}\}$, where $X_{(a)} \in T_{\check{\mathbf{x}}}\check{\mathbb{X}}$ and $X_{(a')} \in T_{\hat{\mathbf{x}}}\hat{\mathbb{X}}$, then the path integral

$$\psi(t_b, \check{\mathbf{x}}_b) := \int_{\mathcal{P}_{0,0}\mathbb{Z}} \mathcal{D}_{s,Q_0}(z) \exp\left(-\frac{\pi}{s}Q_0(z)\right)\phi(\check{\mathbf{x}}_b \cdot \Sigma(t_b, z)) \tag{7.48}$$

is the solution of

$$\frac{\partial \psi}{\partial t_b} = \left(\frac{s}{4\pi}G^{ab}\mathcal{L}_{X_{(a)}}\mathcal{L}_{X_{(b)}} + \mathcal{L}_Y\right)\psi, \tag{7.49}$$

$$\psi(t_a, \check{\mathbf{x}}) = \phi(\check{\mathbf{x}}),$$

where G is the Green function defined by Q_0 as in Chapter 2; namely, given Q_0 and W in the definition (7.46), they define a differential operator D and a Green function G:

$$Q_0(z) = \langle Dz, z \rangle \quad \text{and} \quad W(z') = \langle z', Gz' \rangle,$$
$$DG = 1.$$

The vector fields $X_{(a)}$ and Y are the projections onto the chosen polarization of the vector fields $X_{(A)}$ and Y. For explicit expressions of G^{ab} in a number of transitions, including coherent state transitions, see [9].

References

[1] H. Weyl (1952). *Symmetry* (Princeton, NJ, Princeton University Press).
[2] P. Cartier and C. DeWitt-Morette (1995). "A new perspective on functional integration," *J. Math. Phys.* **36**, 2237–2312. (Located on the

World Wide Web at http://godel.ph.utexas.edu/Center/Papers.html and http://babbage.sissa.it/list/funct-an/9602.)
[3] Y. Choquet-Bruhat and C. DeWitt-Morette (1996 and 2000). *Analysis, Manifolds, and Physics*, Part I: Basics (with M. Dillard-Bleick), revised edn., and Part II, revised and enlarged edn. (Amsterdam, North Holland).
[4] J. LaChapelle (1995). Functional integration on symplectic manifolds, unpublished Ph.D. Dissertation, University of Texas at Austin.
[5] P. Libermann and C.-M. Marle (1987). *Symplectic Geometry and Analytic Mechanics* (Dordrecht, D. Reidel).
[6] N. M. J. Woodhouse (1992). *Geometric Quantization*, 2nd edn. (Oxford, Clarendon Press).
[7] F. A. Berezin (1974). "Quantization," *Izv. Math. USSR* **8**, 1109–1165.
[8] D. Bar-Moshe and M. S. Marinov (1996). "Berezin quantization and unitary representations of Lie groups," in *Topics in Statistical and Theoretical Physics: F. A. Berezin Memorial Volume*, eds. R. L. Dobrushin, A. A. Minlos, M. A. Shubin and A. M. Vershik (Providence, RI, American Mathematical Society), pp. 1–21. also hep-th/9407093.

D. Bar-Moshe and M. S. Marinov (1994). "Realization of compact Lie algebras in Kähler manifolds," *J. Phys. A* **27**, 6287–6298.
[9] C. DeWitt-Morette, A. Maheshwari, and B. Nelson (1979). "Path integration in non-relativistic quantum mechanics," *Phys. Rep.* **50**, 266–372.

8
Homotopy

$$\left|K(b,t_b;a,t_a)\right| = \left|\sum_\alpha \chi(\alpha) K^\alpha(b,t_b;a,t_a)\right|$$

8.1 An example: quantizing a spinning top

An example worked out by Schulman [1] displays a novel feature of path integrals that arises when the paths take their values in multiply connected spaces. Schulman computes propagators of spin-one particles whose configuration space[1] $\mathbb{N}^6$ is identical to that of a spinning top:

$$\mathbb{N}^6 := \mathbb{R}^3 \times \mathrm{SO}(3). \tag{8.1}$$

For a free particle, the probability amplitude is the product of two amplitudes, one on $\mathbb{R}^3$ and one on SO(3). The SO(3)-manifold is the projective space P^3, a 3-sphere with identified antipodes. Schulman notes the benefit of working with path integrals rather than multivalued functions defined on group manifolds: in path-integral theory, we work directly with the paths. Distinct homotopy classes of paths enter the sum over paths with arbitrary relative phase factors. The selection of these phase factors gives rise to the various multivalued representations. There exist two classes of paths between given end points in SO(3). Depending on the

---

[1] Characterizing a spin particle by an element of $\mathbb{N}^6$ may be justified in one of two ways: as in the work of F. Bopp and R. Haag, "Über die Möglichkeit von Spinmodellen," *Z. Naturforsch.* **5**a, 644 (1950), or by treating the spin as an odd 2-current on $\mathbb{R}^3$ as in L. Schwartz *Théorie des distributions* (Paris, Hermann, 1966), p. 355.

relative phase with which these are added, one obtains the propagator for a top of integral or half-integral spin.

A calculation of the SO(3) propagators similar to Schulman's is presented explicitly in Section 8.2. Schulman concludes his lengthy and meticulous calculation modestly: "It is unlikely (to say the least) that the techniques described here can compete in utility with the customary Pauli spin formalism." Little did he know. His work was the first step leading to the challenging and productive connection between functional integration and topology. It motivated the formulation of the basic homotopy theorem for path integration [2, 3] given in Section 8.3. The theorem is applied to systems of indistinguishable particles in Section 8.4, and to the Aharanov–Bohm effect in Section 8.5.

8.2 Propagators on SO(3) and SU(2) [4]

The group SO(3) is the group of real orthogonal 3×3 matrices of determinant equal to 1:

$$M \in \mathrm{SO}(3) \Leftrightarrow M \text{ real}, \qquad MM^{\mathrm{T}} = \mathbf{1}, \qquad \det M = 1. \tag{8.2}$$

The group SO(3) is a Lie group. The group manifold is a 3-sphere whose antipodal points are identified, namely, it is the projective space P^3. The covering space of the "half 3-sphere" SO(3) is the 3-sphere SU(2). See for instance [5, pp. 181–190] for many explicit properties of SO(3) and SU(2). Here we use the coordinate system defined by the Euler angles (θ, ϕ, ψ). Hence an element of SU(2) is parametrized as $e^{i\sigma_3\phi/2}e^{i\sigma_2\theta/2}e^{i\sigma_3\psi/2}$, where σ_1, σ_2, and σ_3 are the Pauli matrices.

On SO(3),

$$0 \leq \theta < \pi, \qquad 0 \leq \psi < 2\pi, \qquad 0 \leq \phi < 2\pi. \tag{8.3}$$

On SU(2),

$$0 \leq \theta < \pi, \qquad 0 \leq \psi < 2\pi, \qquad 0 \leq \phi < 4\pi. \tag{8.4}$$

In these coordinates, the line element is

$$ds^2 = d\theta^2 + d\psi^2 + d\phi^2 + 2\cos\theta \, d\phi \, d\psi, \tag{8.5}$$

and the lagrangian of a free particle on SO(3) or SU(2) is

$$L = \frac{1}{2}I(\dot{\theta}^2 + \dot{\psi}^2 + \dot{\phi}^2 + 2\cos\theta \, \dot{\psi} \, \dot{\phi}), \tag{8.6}$$

where I has the physical dimension of a moment of inertia. The radius of curvature is

$$R = 2. \tag{8.7}$$

Homotopy

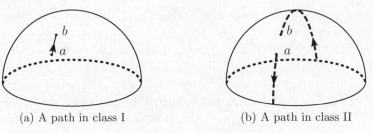

(a) A path in class I   (b) A path in class II

Fig. 8.1

Geodesics on SO(3) and SU(2)

Paths on SO(3) from a to b belong to two different classes of homotopy, the contractible and the noncontractible paths labeled henceforth class I and class II, respectively (figure 8.1).

Let Γ_0 be the length of the shortest geodesic from a to b in class I (in the unit provided by the line element (8.5)). The length of a geodesic in class I is

$$\Gamma_n^I = \Gamma_0 + 2n\pi, \qquad n \text{ even}; \tag{8.8}$$

the length of a geodesic in class II is

$$\Gamma_n^{II} = \Gamma_0 + 2n\pi, \qquad n \text{ odd}. \tag{8.9}$$

The double covering SU(2) of SO(3) is simply connected; all paths on SU(2) are in the same homotopy class, the trivial (the zero) homotopy class. The length of a geodesic on SU(2) is

$$\Gamma_n = \Gamma_0 + 4n\pi, \qquad n \text{ an integer.} \tag{8.10}$$

Example. For a and b antipodal in SU(2), and hence identified in SO(3), the lengths of the shortest geodesics are

$$\Gamma_0^I = 0, \qquad \Gamma_0^{II} = 2\pi, \qquad \Gamma_0 = 2\pi. \tag{8.11}$$

Propagators

Since all the paths in SU(2) are in the trivial homotopy class, the short-time propagator computed by standard methods is

$$K_{SU(2)}(b, t_a + \epsilon; a, t_a) = \left(\frac{I}{2\pi i \hbar \epsilon}\right)^{3/2} e^{i\hbar\epsilon/(8I)} \cdot$$

$$\cdot \sum_{n=-\infty}^{+\infty} \frac{\Gamma_0 + 4n\pi}{2\sin(\Gamma_0/2)} \exp\left(\frac{iI(\Gamma_0 + 4n\pi)^2}{2\hbar\epsilon}\right). \tag{8.12}$$

8.2 Propagators on SO(3) and SU(2)

Convolutions of this short-time propagator are self-reproducing, and equation (8.12) is valid for $\epsilon = T$ finite.

Schulman has computed the propagator (8.12) in terms of the eigenfunctions Φ^j_{mk} and eigenvalues E_{jmk} of the hamiltonian H of a free particle on SO(3) and SU(2):

$$H = -\frac{\hbar^2}{2I}\Delta, \qquad (8.13)$$

where the laplacian on SO(3) and SU(2) is

$$\Delta = \frac{\partial^2}{\partial\theta^2} + \cot\theta\frac{\partial}{\partial\theta} + \frac{1}{\sin^2\theta}\left(\frac{\partial^2}{\partial\psi^2} + \frac{\partial^2}{\partial\phi^2} - 2\cos\theta\frac{\partial^2}{\partial\psi\,\partial\phi}\right). \qquad (8.14)$$

The normalized eigenfunctions of the laplacian Δ are

$$\Phi^j_{mk}\text{ (on SU(2))} = \left(\frac{2j+1}{16\pi^2}\right)^{1/2} D^j_{mk}{}^*(\theta,\psi,\phi) \qquad (8.15)$$

$$\text{for } j = 0, \frac{1}{2}, 1, \frac{3}{2}, \ldots$$

$$\Phi^j_{mk}\text{ (on SO(3))} = \left(\frac{2j+1}{8\pi^2}\right)^{1/2} D^j_{mk}{}^*(\theta,\psi,\phi) \qquad (8.16)$$

$$\text{for } j = 0, 1, 2, 3, \ldots,$$

where j is the total spin eigenvalue.

The Ds form a matrix representation of SU(2). The eigenvalues of Δ are

$$E_{jmk} = \frac{\hbar^2}{2I}j(j+1) \qquad (8.17)$$

and the SU(2) propagator is

$$K_{SU(2)}(b, t_b; a, t_a) = \sum_{j,m,k} \Phi^j_{mk}{}^*(b)\exp\left(-\frac{i\hbar(t_b-t_a)}{2I}j(j+1)\right)\Phi^j_{mk}(a). \qquad (8.18)$$

Schulman was able to reexpress the propagator (8.18) as the sum of two terms,

$$K_{SU(2)} = K_{SU(2)}(\text{integer } j) + K_{SU(2)}(\text{half-integer } j), \qquad (8.19)$$

and to establish the following relations:

$$2K_{SU(2)}(\text{integer } j) = K^I_{SO(3)} - K^{II}_{SO(3)}, \qquad (8.20)$$

$$2K_{SU(2)}(\text{half-integer } j) = K^I_{SO(3)} + K^{II}_{SO(3)}. \qquad (8.21)$$

Schulman's calculation was a *tour de force*. The result of the calculation is strikingly simple. By adding the contributions of the paths in both homotopy classes of SO(3), one obtains a half-integer spin propagator; by subtracting one of the contributions from the other, one obtains an integer spin propagator.

Path integrals reflect the global properties of the manifold where the paths lie. Therefore, motivated by Schulman's result, Cécile DeWitt and her graduate student Michael Laidlaw worked out the properties of path integrals where the paths take their values in multiply connected spaces. They applied their homotopy theorem to systems of indistinguishable particles. They showed, in particular, that in $\mathbb{R}^2$ indistinguishable particles cannot be labeled "bosons" or "fermions."

8.3 The homotopy theorem for path integration

Theorem. The probability amplitude for a given transition is, up to a phase factor, a linear combination of partial probability amplitudes; each of these amplitudes is obtained by integrating over paths in the same homotopy class. The coefficients in the linear combination form a unitary representation of the fundamental group of the range of the paths. The tensorial nature of the representation is determined by the tensorial nature of the propagator.

Proof. The theorem can be proved in terms of covering spaces [3, 6]. We give the original proof, which requires a minimum of prior results.

The principle of superposition of quantum states requires the probability amplitude to be a linear combination of partial probability amplitudes. The construction of the path integral is defined only for paths in the same homotopy class. It follows that the absolute value of the probability amplitude for a transition from the state a at t_a to the state b at t_b will have the form

$$|K(b, t_b; a, t_a)| = \left| \sum_\alpha \chi(\alpha) \mathcal{K}^\alpha(b, t_b; a, t_a) \right|, \qquad (8.22)$$

where $\mathcal{K}^\alpha$ is the integral over paths in the same homotopy class, and $\chi(\alpha)$ is a phase factor that must be deduced from first principles. The homotopy theorem states that $\{\chi(\alpha)\}$ form a representation of the fundamental group.[2] In order to assign the label α both to a group element and to a homotopy class of paths from a to b, one needs to select a homotopy "mesh" based at an arbitrary point c and assign homotopy classes

---

[2] For making a group, one needs to consider loops, that is paths beginning and ending at the same point.

8.4 Systems of indistinguishable particles. Anyons

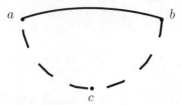

Fig. 8.2 Given a point c and paths ca and cb, one can assign the same label to the group element $cabc$ and to the homotopy class of the path ab

of paths from c to a and from c to b. A homotopy class of paths from a to b can then be assigned the label of the group element $cabc$ (figure 8.2). The key element of the proof is the requirement that the total probability amplitude be independent of the chosen homotopy mesh. For a full presentation of the proof see [2].

In conclusion, if the configuration space of a given system is multiply connected, then there exist as many propagators as there are representations of the propagator tensoriality. □

8.4 Systems of indistinguishable particles. Anyons

A configuration space is useful if it is in one-to-one correspondence with the space of states of the system. It follows that the configuration space of a system of n indistinguishable particles in $\mathbb{R}^3$ is not $\mathbb{R}^{3\otimes n}$ but

$$\mathrm{N}^{3n} = \mathbb{R}^{3\otimes n}/\mathrm{S}_n, \tag{8.23}$$

where S_n is the symmetric group, and coincidence points are excluded from N^{3n}. Note that

- for n indistinguishable particles in $\mathbb{R}$, $\mathbb{R}^{1\otimes n}$ is not connected;
- for n indistinguishable particles in $\mathbb{R}^2$, $\mathbb{R}^{2\otimes n}$ is multiply connected; and
- for n indistinguishable particles in $\mathbb{R}^m$, for $m \geq 3$, $\mathbb{R}^{m\otimes n}$ is simply connected.

Because the coincidence points have been excluded, S_n acts effectively on $\mathbb{R}^{3\otimes n}$. Because $\mathbb{R}^{3\otimes n}$ is simply connected, and S_n acts effectively on $\mathbb{R}^{3\otimes n}$, the fundamental group of N^{3n} is isomorphic to S_n. There are only two scalar unitary representations of the symmetric group:

$$\chi^B : \alpha \in \mathrm{S}_n \mapsto 1 \quad \text{for all permutations } \alpha, \tag{8.24}$$

$$\chi^F : \alpha \in \mathrm{S}_n \mapsto \begin{cases} 1 & \text{for even permutations,} \\ -1 & \text{for odd permutations.} \end{cases} \tag{8.25}$$

Therefore, in $\mathbb{R}^3$, there are two different propagators of indistinguishable particles:

$$K^{\text{Bose}} = \sum_\alpha \chi^{\text{B}}(\alpha)\mathcal{K}^\alpha \qquad \text{is a symmetric propagator;} \qquad (8.26)$$

$$K^{\text{Fermi}} = \sum_\alpha \chi^{\text{F}}(\alpha)\mathcal{K}^\alpha \qquad \text{is an antisymmetric propagator.} \qquad (8.27)$$

The argument leading to the existence of bosons and fermions in $\mathbb{R}^3$ fails in $\mathbb{R}^2$. Statistics cannot be assigned to particles in $\mathbb{R}^2$; particles without statistics are called *anyons*.

8.5 A simple model of the Aharanov–Bohm effect

The system consists of a wire perpendicular to a plane $\mathbb{R}^2$ through its origin, carrying a magnetic flux F. The probability amplitude for a particle of mass m, and charge e, to go from $a = (t_a, x_a)$ to $b = (t_b, x_b)$ can be obtained from a path integral over the space $\mathcal{P}_{a,b}$ of paths from a to b. The homotopy classes of paths in $\mathcal{P}_{a,b}$ can be labeled by the number of path-loops around the origin.

The setup

The basic manifold $\mathbb{N} = \mathbb{R}^2 \setminus \{0\}$, $x \in \mathbb{N}$, cartesian coordinates (x^1, x^2).
The universal covering of $\mathbb{N}$ is the set $\widetilde{\mathbb{N}}$ of pairs of real numbers r, θ, with $r > 0$.
The projection $\Pi : \widetilde{\mathbb{N}} \to \mathbb{N}$ by $(r, \theta) \mapsto x^1 = r \cos\theta$, $x^2 = r \sin\theta$.
The magnetic potential A with components

$$A_1 = -\frac{F}{2\pi} \frac{x^2}{|x|^2}, \qquad A_2 = \frac{F}{2\pi} \frac{x^1}{|x|^2}, \qquad (8.28)$$

where $|x|^2 = (x^1)^2 + (x^2)^2$.
The magnetic field has one component B perpendicular to the plane, vanishing outside the origin:

$$B = \frac{\partial A_1}{\partial x^2} - \frac{\partial A_2}{\partial x^1}. \qquad (8.29)$$

The loop integral $\oint A_1 \, dx^1 + A_2 \, dx^2$ around the origin is equal to F:

$$B = F\delta(x). \qquad (8.30)$$

The action functional

$$S(x) = S_0(x) + S_{\text{M}}(x), \qquad (8.31)$$

8.5 A simple model of the Aharanov–Bohm effect

where the kinetic action is

$$S_0(x) = \frac{m}{2} \int_{\mathbb{T}} \frac{|dx|^2}{dt} \qquad (8.32)$$

and the magnetic action is

$$S_M(x) = \frac{eF}{2\pi} \int_{\mathbb{T}} \frac{x^1\, dx^2 - x^2\, dx^1}{(x^1)^2 + (x^2)^2}. \qquad (8.33)$$

Set the dimensionless quantity $eF/(2\pi\hbar)$ equal to a constant c:

$$c := eF/(2\pi\hbar). \qquad (8.34)$$

Let $\mathbb{X}_{a,b}$ be the space of paths on $\mathbb{N}$:

$$x : \mathbb{T} \to \mathbb{N} \quad \text{such that } x(t_a) = x_a,\ x(t_b) = x_b,$$

$$x_a = (r_a \cos\theta_a, r_a \sin\theta_a), \quad x_b = (r_b \cos\theta_b, r_b \sin\theta_b). \qquad (8.35)$$

Let $\widetilde{\mathbb{X}}_b$ be the space of paths on $\widetilde{\mathbb{N}}$:

$$\tilde{x} : \mathbb{T} \to \widetilde{\mathbb{N}} \quad \text{such that } \tilde{x}(t_b) = (r_b, \theta_b). \qquad (8.36)$$

Two paths in $\mathbb{X}_{a,b}$ are in the same homotopy class if, and only if, the lifted paths in $\widetilde{\mathbb{N}}$ correspond to the same determination of the angular coordinate θ_a of x_a. An equivalent condition is that they have the same magnetic action

$$S_M(x)/\hbar = c \cdot (\theta_b - \theta_a). \qquad (8.37)$$

Transition amplitudes

The fundamental group is the group $\mathbb{Z}$ of integers n acting on $\widetilde{\mathbb{N}}$ by

$$(r, \theta) \cdot n = (r, \theta + 2\pi n). \qquad (8.38)$$

In the absence of the magnetic field the probability amplitude is

$$\langle t_b, x_b | t_a, x_a \rangle_0 = \frac{1}{r_a} \sum_{n \in \mathbb{Z}} \langle t_b, r_b, \theta_b | t_a, r_a, \theta_a + 2\pi n \rangle. \qquad (8.39)$$

The detailed calculation can be found in [7, pp. 2264–2265]. In brief, it uses the construction developed in Chapter 7 when the dynamical vector fields are given by (7.17) and the wave function ϕ in (7.12) is

$$\delta(r\cos\theta - r_a\cos\theta_a)\delta(r\sin\theta - r_a\sin\theta_a) = \frac{1}{r_a} \sum_{n \in \mathbb{Z}} \delta(r - r_a)\delta(\theta - \theta_a - 2n\pi).$$

$$(8.40)$$

Equation (8.39) is nothing other than a point-to-point transition amplitude in cartesian coordinates. To obtain the transition amplitude in polar coordinates we shall invert the transition amplitude $\langle t_b, x_b | t_a, x_a \rangle_F$ when there is a magnetic field $F = 2\pi\hbar c/e$:

$$\langle t_b, x_b | t_a, x_a \rangle_F = \frac{1}{r_a} \sum_{n \in \mathbb{Z}} \exp(ic(\theta_b - \theta_a - 2n\pi)) \langle t_b, r_b, \theta_b | t_a, r_a, \theta_a + 2n\pi \rangle. \tag{8.41}$$

This equation follows readily from the fact that the magnetic action (8.37) is independent of the path x in a given homotopy class, and can be taken outside the functional integral. Equation (8.41) can be inverted to give the value of the transition amplitude in polar coordinates:

$$\langle t_b, r_b, \theta_b | t_a, r_a, \theta_a \rangle = r_a \int_0^1 dc \, \exp(-ic(\theta_b - \theta_a)) \langle t_b, x_b | t_a, x_a \rangle_F. \tag{8.42}$$

As a function of F, the absolute value of the amplitude (8.41) admits a period $2\pi\hbar/e$. $\square$

N.B. The exact rule is as follows: adding $2\pi\hbar/e$ to F multiplies $\langle t_b, x_b | t_a, x_a \rangle_F$ by the phase factor $e^{i(\theta_b - \theta_a)}$ depending on x_a and x_b only.

Gauge transformations on $\mathbb{N}$ and $\widetilde{\mathbb{N}}$

The following combines the theorems established in Chapter 7 using dynamical vector fields and the theorems established in Chapter 8 for paths in multiply connected spaces. In Chapter 7 we applied the sequence dynamical vector fields, functional integral, Schrödinger equation to two examples: paths in polar coordinates and paths on a frame bundle, i.e. an SO(n)-principal bundle. In this section we consider a U(1)-principal bundle on $\mathbb{N} = \mathbb{R}^2 \setminus \{0\}$ and its covering space $\widetilde{\mathbb{N}}$. The choice of the group U(1) is dictated by the fact that there is a linear representation of the fundamental group $\mathbb{Z}$ of $\mathbb{N}$ into U(1):

$$\text{reps: } \mathbb{Z} \to U(1) \qquad \text{by } n \mapsto \exp(2\pi i n c). \tag{8.43}$$

Let

$$\mathbb{P} = \mathbb{N} \times U(1) \qquad \text{coordinates } x^1, x^2, \Theta, \tag{8.44}$$

$$\widetilde{\mathbb{P}} = \widetilde{\mathbb{N}} \times U(1) \qquad \text{coordinates } r, \theta, \Theta. \tag{8.45}$$

The dynamical vector fields on $\mathbb{P}$ are [7, equation IV 83]

$$X_{(\alpha)} = \frac{\partial}{\partial x^\alpha} - i\frac{e}{\hbar} A_\alpha(x) \frac{\partial}{\partial \Theta}, \qquad \alpha \in \{1, 2\}, \tag{8.46}$$

8.5 A simple model of the Aharanov–Bohm effect

and the Schrödinger equation for the Aharanov–Bohm system is

$$\frac{\partial \psi_F}{\partial t} = \frac{i\hbar}{2m}\delta^{\alpha\beta}X_{(\alpha)}X_{(\beta)}\psi_F =: \frac{i\hbar}{2m}L\psi_F, \tag{8.47}$$

where

$$L = \left(\frac{\partial}{\partial x^1}\right)^2 + \left(\frac{\partial}{\partial x^2}\right)^2 - \frac{2ci}{|x|^2}\left(x^1\frac{\partial}{\partial x^2} - x^2\frac{\partial}{\partial x^1}\right) - \frac{c^2}{|x|^2}. \tag{8.48}$$

The Schrödinger equation lifted on the covering $\widetilde{\mathbb{N}}$ of $\mathbb{N}$ is

$$\frac{\partial \widetilde{\psi}_F}{\partial t} = \frac{i\hbar}{2m}\widetilde{L}\widetilde{\psi}_F, \tag{8.49}$$

where

$$\widetilde{L} = \frac{\partial^2}{\partial r^2} + \frac{1}{r^2}\frac{\partial^2}{\partial \theta^2} + \frac{1}{r}\frac{\partial}{\partial r} - \frac{2ci}{r^2}\frac{\partial}{\partial \theta} - \frac{c^2}{r^2}. \tag{8.50}$$

We note that, under the gauge transformation

$$\widetilde{\psi}_F(t,r,\theta) \mapsto \widetilde{\psi}_0(t,r,\theta) = \exp(-ci\theta)\widetilde{\psi}_F(t,r,\theta), \tag{8.51}$$

the new function $\widetilde{\psi}_0$ satisfies the Schrödinger equation for a free particle in polar coordinates in the absence of a magnetic field:

$$\frac{\partial \widetilde{\psi}_0}{\partial t} = \frac{i\hbar}{2m}\Delta\widetilde{\psi}_0 \tag{8.52}$$

with

$$\Delta = \frac{\partial^2}{\partial r^2} + \frac{1}{r^2}\frac{\partial^2}{\partial \theta^2} + \frac{1}{r}\frac{\partial}{\partial r}. \tag{8.53}$$

For a quick proof of (8.52), note that $\widetilde{L} = e^{ci\theta}\Delta e^{-ci\theta}$. The gauge transformation (8.51) apparently removes the magnetic potential, but $\widetilde{\psi}_F$ and $\widetilde{\psi}_0$ do not have the same periodicity condition:

$$\widetilde{\psi}_F(t,r,\theta + 2\pi) = \widetilde{\psi}_F(t,r,\theta), \tag{8.54}$$

whereas

$$\widetilde{\psi}_0(t,r,\theta + 2\pi) = e^{-2\pi ic}\widetilde{\psi}_0(t,r,\theta). \tag{8.55}$$

$\widetilde{\psi}_0$ is a section of a line bundle associated with the principal bundle $\mathbb{P}$ on $\mathbb{N}$.

For more properties of the dynamical vector fields and the wave functions on $\mathbb{N}$ and $\widetilde{\mathbb{N}}$ we refer the reader to [7, pp. 2276–2281].

References

[1] L. S. Schulman (1968). "A path integral for spin," *Phys. Rev.* **176**, 1558–1569; (1971) "Approximate topologies," *J. Math Phys.* **12**, 304–308. From the examples worked out in his 1971 paper, Schulman developed an intuition of the homotopy theorem independently of Laidlaw and DeWitt.

[2] M. G. G. Laidlaw and C. M. DeWitt (1971). "Feynman functional integrals for systems of indistinguishable particles," *Phys. Rev.* **D3**, 1375–1378.

[3] M. G. G. Laidlaw (1971). Quantum mechanics in multiply connected spaces, unpublished Ph.D. Thesis, University of North Carolina, Chapel Hill.

[4] C. M. DeWitt (1969). "L'intégrale fonctionnelle de Feynman. Une introduction," *Ann. Inst. Henri Poincaré* **XI**, 153–206.

[5] Y. Choquet-Bruhat and C. DeWitt-Morette (1996). *Analysis, Manifolds, and Physics*, Vol. I (with M. Dillard-Bleick), revised edition (Amsterdam, North Holland).

[6] J. S. Dowker (1972). "Quantum mechanics and field theory on multiply connected and on homogenous spaces," *J. Phys. A: Gen. Phys.* **5**, 936–943.

[7] P. Cartier and C. DeWitt-Morette (1995). "A new perspective in functional integration," *J. Math. Phys.* **36**, 2237–2312.

9
Grassmann analysis: basics

Parity

9.1 Introduction

Parity is ubiquitous, and Grassmann analysis is a tool well adapted for handling systematically parity and its implications in all branches of algebra, analysis, geometry, and topology. Parity describes the behavior of a product under exchange of its two factors. Koszul's so-called parity rule states that "*Whenever you interchange two factors of parity 1, you get a minus sign.*" Formally the rule defines graded commutative products

$$AB = (-1)^{\tilde{A}\tilde{B}} BA, \qquad (9.1)$$

where $\tilde{A} \in \{0, 1\}$ denotes the parity of A. Objects with parity zero are called *even*, and objects with parity one *odd*. The rule also defines graded anticommutative products. For instance,

$$A \wedge B = -(-1)^{\tilde{A}\tilde{B}} B \wedge A. \qquad (9.2)$$

- A *graded* commutator $[A, B]$ can be either a commutator $[A, B]_- = AB - BA$, or an anticommutator $[A, B]_+ = AB + BA$.
- A *graded* anticommutative product $\{A, B\}$ can be either an anticommutator $\{A, B\}_+$, or a commutator $\{A, B\}_-$.

Most often, the context makes it unnecessary to use the $+$ and $-$ signs.

There are no (anti)commutative rules for vectors and matrices. Parity is assigned to such objects in the following way.

- The parity of a vector is determined by its behavior under multiplication by a scalar z:

$$zX = (-1)^{\tilde{z}\tilde{X}} Xz. \qquad (9.3)$$

- A matrix is even if it preserves the parity of graded vectors. A matrix is odd if it inverts the parity.

Scalars, vectors and matrices do not necessarily have well-defined parity, but they can always be decomposed into a sum of even and odd parts.

The usefulness of Grassmann analysis in physics became apparent in the works of F. A. Berezin [1] and M. S. Marinov [2]. We refer the reader to [3–7] for references and recent developments. The next section summarizes the main formulas of Grassmann analysis.

As a rule of thumb, it is most often sufficient to insert the word "graded" into the corresponding ordinary situation. For example, an ordinary differential form is an antisymmetric covariant tensor. A Grassmann form is a graded antisymmetric covariant tensor: $\omega_{...\alpha\beta...} = -(-1)^{\tilde{\alpha}\tilde{\beta}} \omega_{...\beta\alpha...}$, where $\tilde{\alpha} \in \{0, 1\}$ is the grading of the index α. Therefore a Grassmann form is symmetric under the interchange of two Grassmann odd indices.

9.2 A compendium of Grassmann analysis
Contributed by Maria E. Bell[1]

This section is extracted from the Master's Thesis [8] of Maria E. Bell, "Introduction to supersymmetry." For convenience, we collect here formulas that are self-explanatory, as well as formulas whose meanings are given in the following sections.

Basic graded algebra

- $\tilde{A} :=$ parity of $A \in \{0, 1\}$.
- Parity of a product:

$$\widetilde{AB} = \tilde{A} + \tilde{B} \text{ mod } 2. \qquad (9.4)$$

- Graded commutator:

$$[A, B] := AB - (-1)^{\tilde{A}\tilde{B}} BA \quad \text{or} \quad [A, B]_{\mp} = AB \mp BA. \qquad (9.5)$$

- Graded anticommutator:

$$\{A, B\} := AB + (-1)^{\tilde{A}\tilde{B}} BA \quad \text{or} \quad \{A, B\}_{\pm} = AB \pm BA. \qquad (9.6)$$

---
[1] For an extended version see [8].

- Graded Leibniz rule for a differential operator:
$$D(A \cdot B) = DA \cdot B + (-1)^{\tilde{A}\tilde{D}}(A \cdot DB) \qquad (9.7)$$
(referred to as "anti-Leibniz" when $\tilde{D} = 1$).
- Graded symmetry: $A^{\cdots\alpha\beta\cdots}$ has graded symmetry if
$$A^{\cdots\alpha\beta\cdots} = (-1)^{\tilde{\alpha}\tilde{\beta}} A^{\cdots\beta\alpha\cdots}. \qquad (9.8)$$
- Graded antisymmetry: $A^{\cdots\alpha\beta\cdots}$ has graded antisymmetry if
$$A^{\cdots\alpha\beta\cdots} = -(-1)^{\tilde{\alpha}\tilde{\beta}} A^{\cdots\beta\alpha\cdots}. \qquad (9.9)$$
- Graded Lie derivative:
$$\mathcal{L}_X = [i_X, \mathrm{d}]_+ \text{ for } \tilde{X} = 0 \quad \text{and} \quad \mathcal{L}_\Xi = [i_\Xi, \mathrm{d}]_- \text{ for } \tilde{\Xi} = 1. \qquad (9.10)$$

Basic Grassmann algebra

- Grassmann generators $\{\xi^\mu\} \in \Lambda_\nu, \Lambda_\infty, \Lambda$, algebra generated respectively by ν generators, an infinite or an unspecified number:
$$\xi^\mu \xi^\sigma = -\xi^\sigma \xi^\mu; \qquad \Lambda = \Lambda^{\text{even}} \oplus \Lambda^{\text{odd}}. \qquad (9.11)$$
- Supernumber (real) $z = u + v$, where u is even (has $\tilde{u} = 0$) and v is odd (has $\tilde{v} = 1$). Odd supernumbers anticommute among themselves; they are called a-numbers. Even supernumbers commute with everything; they are called c-numbers. The set $\mathbb{R}_c$ of all c-numbers is a commutative subalgebra of Λ. The set $\mathbb{R}_c$ of all a-numbers is not a subalgebra:
$$z = z_B + z_S, \qquad z_B \in \mathbb{R} \text{ is the body, } z_S \text{ is the soul.} \qquad (9.12)$$
A similar definition applies for complex supernumbers, and $\mathbb{C}_c, \mathbb{C}_a$.
- Complex conjugation of a complex supernumber:
$$(zz')^* = z^* z'^*. \qquad (9.13)_a$$
- Hermitian conjugation of supernumbers:
$$(zz')^\dagger = z'^\dagger z^\dagger. \qquad (9.13)_b$$

Note that in [4] complex conjugation satisfies $(zz')^* = z'^* z^*$, that is the hermitian conjugation rule. It follows that in [4] the product of two real supernumbers is purely imaginary.

A supernumber
$$\psi = c_0 + c_i \xi^i + \frac{1}{2!} c_{ij} \xi^i \xi^j + \cdots \qquad (9.14)$$
is *real* if all its coefficients $c_{i_1 \ldots i_p}$ are real numbers. Let
$$\psi = \rho + i\sigma,$$

where both ρ and σ have real coefficients. Define complex conjugation by

$$(\rho + i\sigma)^* = \rho - i\sigma. \tag{9.15}$$

Then the generators $\{\xi^i\}$ are real, and the sum and product of two real supernumbers are real. Furthermore,

$$\psi \text{ is real} \Leftrightarrow \psi^* = \psi. \tag{9.16}$$

- Superpoints. Real coordinates $x, y \in \mathbb{R}^n$, $x = (x^1, \ldots, x^n)$. Superspace coordinates

$$(x^1, \ldots, x^n, \xi^1, \ldots, \xi^\nu) \in \mathbb{R}^{n|\nu} \tag{9.17}$$

are also written in condensed notation $x^A = (x^a, \xi^\alpha)$:

$$(u^1, \ldots, u^n, v^1, \ldots, v^\nu) \in \mathbb{R}_c^n \times \mathbb{R}_a^\nu. \tag{9.18}$$

- Supervector space, i.e. a graded module over the ring of supernumbers:

$$X = U + V, \quad \text{where } U \text{ is even and } V \text{ is odd};$$
$$X = e_{(A)}{}^A X;$$
$$X^A = (-1)^{\tilde{X}\tilde{A}}{}^A X.$$

The even elements of the basis $(e_{(A)})_A$ are listed first. A supervector is even if each of its coordinates ${}^A X$ has the same parity as the corresponding basis element $e_{(A)}$. It is odd if the parity of each ${}^A X$ is opposite to the parity of $e_{(A)}$. Parity cannot be assigned in other cases.
- Graded matrices. Four different uses of graded matrices follow.

Given $V = e_{(A)}{}^A V = \bar{e}_{(B)}{}^B \bar{V}$ with $A = (a, \alpha)$ and $e_{(A)} = \bar{e}_{(B)}{}^B M_A$, then ${}^B \bar{V} = {}^B M_A {}^A V$.

Given $\langle \omega, V \rangle = \omega_A {}^A V = \bar{\omega}_B {}^B \bar{V}$, where $\omega = \omega_A {}^{(A)}\theta = \bar{\omega}_B {}^{(B)}\bar{\theta}$, then $\langle \omega, V \rangle = \omega_A \langle {}^{(A)}\theta, e_{(B)} \rangle {}^B V$ implies $\langle {}^{(A)}\theta, e_{(B)} \rangle = {}^A \delta_B$, $\omega_A = \bar{\omega}_B {}^B M_A$, and ${}^{(B)}\bar{\theta} = {}^B M_A {}^{(A)}\theta$.

- Matrix parity:

$$\tilde{M} = 0, \quad \text{if, for all } A \text{ and } B, \; \widetilde{{}^B M_A} + \widetilde{\text{row } B} + \widetilde{\text{column } A} = 0 \bmod 2; \tag{9.19}$$

$$\tilde{M} = 1, \quad \text{if, for all } A \text{ and } B, \; \widetilde{{}^B M_A} + \widetilde{\text{row } B} + \widetilde{\text{column } A} = 1 \bmod 2. \tag{9.20}$$

Parity cannot be assigned in other cases. Multiplication by an even matrix preserves the parity of the vector components; multiplication by an odd matrix inverts the parity of the vector components.

9.2 A compendium of Grassmann analysis

- **Supertranspose:** supertransposition, labeled "ST," is defined so that the following basic rules apply:
$$(M^{ST})^{ST} = M,$$
$$(MN)^{ST} = (-1)^{\tilde{M}\tilde{N}} N^{ST} M^{ST}. \tag{9.21}$$

- **Superhermitian conjugate:**
$$M^{SH} := (M^{ST})^* = (M^*)^{ST}, \tag{9.22}$$
$$(MN)^{SH} = (-1)^{\tilde{M}\tilde{N}} N^{SH} M^{SH}. \tag{9.23}$$

- **Graded operators on Hilbert spaces.** Let $|\Omega\rangle$ be a simultaneous eigenstate of Z and Z' with eigenvalues z and z':
$$ZZ'|\Omega\rangle = zz'|\Omega\rangle, \tag{9.24}$$
$$\langle\Omega|Z'^{SH}Z^{SH} = \langle\Omega|z'^* z^*. \tag{9.25}$$

- **Supertrace:**
$$\mathrm{Str}\, M = (-1)^{\tilde{A}\,A} M_A. \tag{9.26}$$

Example. A matrix of order (p, q). Assume the p even rows and columns written first:

$$M_0 = \begin{pmatrix} \boxed{} \end{pmatrix} = \begin{pmatrix} A_0 & C_1 \\ D_1 & B_0 \end{pmatrix}, \quad M_1 = \begin{pmatrix} \boxed{} \end{pmatrix} = \begin{pmatrix} A_1 & C_0 \\ D_0 & B_1 \end{pmatrix}.$$

These are two matrices of order $(1, 2)$. The shaded areas indicate even elements. The matrix on the left is even; the matrix on the right is odd. Given the definitions above,

$$\mathrm{Str}\, M_0 = \mathrm{tr}\, A_0 - \mathrm{tr}\, B_0; \quad \mathrm{Str}\, M_1 = \mathrm{tr}\, A_1 - \mathrm{tr}\, B_1. \tag{9.27}$$

In general $M = M_0 + M_1$.

- **Superdeterminant** (also known as berezinian). It is defined so that it satisfies the basic properties
$$\mathrm{Ber}(MN) = \mathrm{Ber}\, M\, \mathrm{Ber}\, N, \tag{9.28}$$
$$\delta \ln \mathrm{Ber}\, M = \mathrm{Str}(M^{-1} \delta M), \tag{9.29}$$
$$\mathrm{Ber}(\exp M) = \exp(\mathrm{Str}\, M), \tag{9.30}$$
$$\mathrm{Ber}\begin{pmatrix} A & C \\ D & B \end{pmatrix} := \det(A - CB^{-1}D)(\det B)^{-1}. \tag{9.31}$$

The determinants on the right-hand side are ordinary determinants defined only when the entries commute. It follows that the definition

(9.31) applies only to the berezinian of even matrices, i.e. A and B even, C and D odd.
- Parity assignments for differentials:

$$\tilde{d} = 1, \qquad \widetilde{(dx)} = \tilde{d} + \tilde{x} = 1, \qquad \widetilde{(d\xi)} = \tilde{d} + \tilde{\xi} = 0, \qquad (9.32)$$

where x is an ordinary variable, and ξ is a Grassmann variable,

$$\widetilde{(\partial/\partial x)} = \tilde{x} = 0, \qquad \widetilde{(\partial/\partial \xi)} = \tilde{\xi} = 1, \qquad (9.33)$$

$$\tilde{i} = 1, \qquad \widetilde{i_X} = \tilde{i} + \tilde{X} = 1, \qquad \widetilde{i_\Xi} = \tilde{i} + \tilde{\Xi} = 0. \qquad (9.34)$$

Here X and Ξ are vector fields, with X even and Ξ odd.
Parity of real p-forms: even for $p = 0$ mod 2, odd for $p = 1$ mod 2.
Parity of Grassmann p-forms: always even.
Graded exterior product: $\omega \wedge \eta = (-1)^{\tilde{\omega}\tilde{\eta}} \eta \wedge \omega$.

Forms and densities of weight 1

Forms and densities will be introduced in Section 9.4. We list first definitions and properties of objects defined on ordinary manifolds $\mathbb{M}^D$ without metric tensors, then those on riemannian manifolds $(\mathbb{M}^D, g)$.

$(\mathcal{A}^\bullet, d)$ Ascending complex of forms $d : \mathcal{A}^p \to \mathcal{A}^{p+1}$.
$(\mathcal{D}_\bullet, \nabla$ or $b)$ Descending complex of densities $\nabla : \mathcal{D}_p \to \mathcal{D}_{p-1}$.
$\mathcal{D}_p \equiv \mathcal{D}^{-p}$, used for ascending complex in negative degrees.

- Operators on $\mathcal{A}^\bullet(\mathbb{M}^D)$:
  $M(f) : \mathcal{A}^p \to \mathcal{A}^p$, multiplication by a scalar function $f : \mathbb{M}^D \to \mathbb{R}$;
  $e(f) : \mathcal{A}^p \to \mathcal{A}^{p+1}$ by $\omega \mapsto df \wedge \omega$;
  $i(X) : \mathcal{A}^p \to \mathcal{A}^{p-1}$ by contraction with the vector field X;
  $\mathcal{L}_X \equiv \mathcal{L}(X) = i(X)d + di(X)$ maps $\mathcal{A}^p \to \mathcal{A}^p$ by the Lie derivative with respect to X.
- Operators on $\mathcal{D}_\bullet(\mathbb{M}^D)$:
  $M(f) : \mathcal{D}_p \to \mathcal{D}_p$, multiplication by scalar function $f : \mathbb{M}^D \to \mathbb{R}$;
  $e(f) : \mathcal{D}_p \to \mathcal{D}_{p-1}$ by $\mathcal{F} \mapsto df \cdot \mathcal{F}$ (contraction with the form df);
  $i(X) : \mathcal{D}_p \to \mathcal{D}_{p+1}$ by multiplication and partial antisymmetrization;
  $\mathcal{L}_X \equiv \mathcal{L}(X) = i(X)\nabla + \nabla i(X)$ maps $\mathcal{D}_p \to \mathcal{D}_p$ by the Lie derivative with respect to X.
- Forms and densities of weight 1 on a riemannian manifold $(\mathbb{M}^D, g)$:
  $C_g : \mathcal{A}^p \to \mathcal{D}_p$ (see equation (9.59));
  $* : \mathcal{A}^p \to \mathcal{A}^{D-p}$ such that $\mathcal{T}(\omega|\eta) = \omega \wedge *\eta$ (see equation (9.61));
  $\delta : \mathcal{A}^{p+1} \to \mathcal{A}^p$ is the metric transpose defined by

$$[d\omega|\eta] =: [\omega|\delta\eta] \text{ s.t. } [\omega|\eta] := \int \mathcal{T}(\omega|\eta);$$

9.2 A compendium of Grassmann analysis

$\delta = C_g^{-1} \, \mathrm{b} C_g$ (see equation (9.66));
$\beta : \mathcal{D}_p \to \mathcal{D}_{p+1}$ is defined by $C_g \, \mathrm{d} C_g^{-1}$.

We now list definitions and properties of objects defined on Grassmann variables.

Grassmann calculus on $\xi^\lambda \in \Lambda_\nu, \Lambda_\infty, \Lambda$

dd $= 0$ remains true, therefore

$$\frac{\partial}{\partial \xi^\lambda} \frac{\partial}{\partial \xi^\mu} = -\frac{\partial}{\partial \xi^\mu} \frac{\partial}{\partial \xi^\lambda}, \tag{9.35}$$

$$\mathrm{d}\xi^\lambda \wedge \mathrm{d}\xi^\mu = \mathrm{d}\xi^\mu \wedge \mathrm{d}\xi^\lambda. \tag{9.36}$$

- Forms and densities of weight -1 on $\mathbb{R}^{0|\nu}$.
  Forms are graded *totally symmetric covariant* tensors. Densities are graded *totally symmetric contravariant* tensors of weight -1.
  $(\mathcal{A}^\bullet(\mathbb{R}^{0|\nu}), \mathrm{d})$, ascending complex of forms not limited above.
  $(\mathcal{D}_\bullet(\mathbb{R}^{0|\nu}), \nabla$ or $\mathrm{b})$, descending complex of densities not limited above.
- Operators on $\mathcal{A}^\bullet(\mathbb{R}^{0|\nu})$.
  $M(\varphi) : \mathcal{A}^p(\mathbb{R}^{0|\nu}) \to \mathcal{A}^p(\mathbb{R}^{0|\nu})$, multiplication by a scalar function φ;
  $e(\varphi) : \mathcal{A}^p(\mathbb{R}^{0|\nu}) \to \mathcal{A}^{p+1}(\mathbb{R}^{0|\nu})$ by $\omega \mapsto \mathrm{d}\varphi \wedge \omega$;
  $i(\Xi) : \mathcal{A}^p(\mathbb{R}^{0|\nu}) \to \mathcal{A}^{p-1}(\mathbb{R}^{0|\nu})$ by contraction with the vector field Ξ;
  $\mathcal{L}_\Xi \equiv \mathcal{L}(\Xi) := i(\Xi)\mathrm{d} - \mathrm{d}i(\Xi)$ maps $\mathcal{A}^p(\mathbb{R}^{0|\nu}) \to \mathcal{A}^p(\mathbb{R}^{0|\nu})$ by Lie derivative with respect to Ξ.
- Operators on $\mathcal{D}_\bullet(\mathbb{R}^{0|\nu})$.
  $M(\varphi) : \mathcal{D}_p(\mathbb{R}^{0|\nu}) \to \mathcal{D}_p(\mathbb{R}^{0|\nu})$, multiplication by scalar function φ;
  $e(\varphi) : \mathcal{D}_p(\mathbb{R}^{0|\nu}) \to \mathcal{D}_{p-1}(\mathbb{R}^{0|\nu})$ by $\mathcal{F} \mapsto \mathrm{d}\varphi \cdot \mathcal{F}$ (contraction with the form $\mathrm{d}\varphi$);
  $i(\Xi) : \mathcal{D}_p(\mathbb{R}^{0|\nu}) \to \mathcal{D}_{p+1}(\mathbb{R}^{0|\nu})$ by multiplication and partial symmetrization;
  $\mathcal{L}_\Xi \equiv \mathcal{L}(\Xi) = i(\Xi)\nabla - \nabla i(\Xi)$ maps $\mathcal{D}_p(\mathbb{R}^{0|\nu}) \to \mathcal{D}_p(\mathbb{R}^{0|\nu})$ by Lie derivative with respect to Ξ.

In Section 10.2 we will construct a supersymmetric Fock space. The operators e and i defined above can be used for representing the following:

fermionic creation operators: $e(x^m)$;
fermionic annihilation operators: $i(\partial/\partial x^m)$;
bosonic creation operators: $e(\xi^\mu)$;
bosonic annihilation operators: $i(\partial/\partial \xi^\mu)$.

We refer to [3–7] for the definitions of graded manifolds, supermanifolds, supervarieties, superspace, and sliced manifolds. Here we consider simply superfunctions F on $\mathbb{R}^{n|\nu}$: functions of n real variables $\{x^a\}$ and

ν Grassmann variables $\{\xi^\alpha\}$:

$$F(x,\xi) = \sum_{p=0}^{\nu} \frac{1}{p!} f_{\alpha_1...\alpha_p}(x)\xi^{\alpha_1}...\xi^{\alpha_p}, \qquad (9.37)$$

where the functions $f_{\alpha_1...\alpha_p}$ are smooth functions on $\mathbb{R}^n$ that are antisymmetric in the indices $\alpha_1,...,\alpha_p$.

9.3 Berezin integration[2]

A Berezin integral is a derivation

The fundamental requirement on a definite integral is expressed in terms of an integral operator I and a derivative operator D on a space of functions, and is

$$DI = ID = 0. \qquad (9.38)$$

The requirement $DI = 0$ for functions of real variables $f : \mathbb{R}^D \to \mathbb{R}$ states that the *definite integral* does not depend upon the variable of integration:

$$\frac{d}{dx} \int f(x)dx = 0, \qquad x \in \mathbb{R}. \qquad (9.39)$$

The requirement $ID = 0$ on the space of functions that vanish on their domain boundaries states $\int df = 0$, or explicitly

$$\int \frac{d}{dx} f(x)dx = 0. \qquad (9.40)$$

Equation (9.40) is the foundation of integration by parts,

$$0 = \int d(f(x)g(x)) = \int df(x) \cdot g(x) + \int f(x) \cdot dg(x), \qquad (9.41)$$

and of Stokes' theorem for a form ω of compact support

$$\int_M d\omega = \int_{\partial M} \omega = 0, \qquad (9.42)$$

since ω vanishes on ∂M. We shall use the requirement $ID = 0$ in Section 11.3 for imposing a condition on volume elements.

We now use the fundamental requirements on Berezin integrals defined on functions f of the Grassmann algebra Λ_ν. The condition $DI = 0$ states

$$\frac{\partial}{\partial \xi^\alpha} I(f) = 0 \qquad \text{for } \alpha \in \{1,...,\nu\}. \qquad (9.43)$$

---

[2] See Project 19.4.1, "Berezin functional integrals. Roepstorff's formulation."

9.3 Berezin integration

Any operator on Λ_ν can be set in normal ordering[3]

$$\sum C_K^J \xi^K \frac{\partial}{\partial \xi^J}, \qquad (9.44)$$

where J and K are ordered multi-indices, where $K = (\alpha_1 < \ldots < \alpha_q)$, $J = (\beta_1 < \ldots < \beta_p)$, $\xi^K = \xi^{\alpha_1} \ldots \xi^{\alpha_q}$, and $\partial/\partial \xi^J = \partial/\partial \xi^{\beta_1} \ldots \partial/\partial \xi^{\beta_p}$.
Therefore the condition $DI = 0$ implies that I is a polynomial in $\partial/\partial \xi^i$,

$$I = Q\left(\frac{\partial}{\partial \xi^1}, \ldots, \frac{\partial}{\partial \xi^\nu}\right). \qquad (9.45)$$

The condition $ID = 0$ states that

$$Q\left(\frac{\partial}{\partial \xi^1}, \ldots, \frac{\partial}{\partial \xi^\nu}\right) \frac{\partial}{\partial \xi^\mu} = 0 \quad \text{for every } \mu \in \{1, \ldots, \nu\}. \qquad (9.46)$$

Equation (9.46) implies

$$I = C^\nu \frac{\partial}{\partial \xi^\nu} \ldots \frac{\partial}{\partial \xi^1} \quad \text{with } C \text{ a constant.} \qquad (9.47)$$

A Berezin integral is a derivation. Nevertheless, we write

$$I(f) = \int \delta\xi \, f(\xi),$$

where the symbol $\delta\xi$ is different from the differential form $d\xi$ satisfying (9.36). In a Berezin integral, one does not integrate a differential form. Recall ((9.36) and parity assignment (9.32)) that $d\xi$ is even:

$$d\xi^\lambda \wedge d\xi^\mu = d\xi^\mu \wedge d\xi^\lambda.$$

On the other hand, it follows from the definition of the Berezin integral (9.47) that

$$\int \delta\eta \, \delta\xi \cdot F(\xi, \eta) = C^2 \frac{\partial}{\partial \eta} \cdot \frac{\partial}{\partial \xi} F(\xi, \eta).$$

Since the derivatives on the right-hand side are odd,

$$\delta\xi \, \delta\eta = -\delta\eta \, \delta\xi;$$

hence $\delta\xi$, like its counterpart dx in ordinary variables, is odd.

---

[3] This ordering is also the operator normal ordering, in which the creation operator is followed by the annihilation operator, since $e(\xi^\mu)$ and $i(\partial/\partial \xi^\mu)$ can be interpreted as creation and annihilation operators (see Section 10.2).

The Fourier transform and the normalization constant

The constant C is a normalization constant chosen for convenience in the given context. Notice that $C = \int \delta\xi \cdot \xi$. Typical choices include 1, $(2\pi i)^{1/2}$, and $(2\pi i)^{-1/2}$. We choose $C = (2\pi i)^{-1/2}$ for the following reason.

The constant C in (9.47) can be obtained from the Dirac δ-function defined by two equivalent conditions:

$$\langle \delta, f \rangle = \int \delta\xi \, \delta(\xi) f(\xi) = f(0), \qquad \mathcal{F}\delta = 1, \qquad (9.48)_a$$

where $\mathcal{F}\delta$ is the Fourier transform of δ.

The first condition implies

$$\delta(\xi) = C^{-1}\xi. \qquad (9.48)_b$$

Define the Fourier transform $\tilde{f}$ (or $\mathcal{F}f$) of a function f by

$$\tilde{f}(\kappa) := \int \delta\xi \, f(\xi) \exp(-2\pi i\kappa\xi). \qquad (9.49)$$

The inverse Fourier transform is

$$f(\xi) = \int \delta\kappa \, \tilde{f}(\kappa) \exp(2\pi i\kappa\xi).$$

Then

$$f(\xi) = \int \delta p \, \delta(\xi - p) f(p),$$

provided that

$$\delta(\xi - p) = \int \delta\kappa \, \exp(2\pi i\kappa(\xi - p)). \qquad (9.50)$$

Hence, according to the definition of Fourier transforms given above,

$$\mathcal{F}\delta = 1.$$

According to (9.50) with $p = 0$,

$$\delta(\xi) = 2\pi i \int \delta\kappa \, \kappa\xi = 2\pi i C\xi.$$

Together $(9.48)_b$ and (9.50) imply

$$C^{-1}\xi = 2\pi i C\xi$$

and therefore

$$C^2 = (2\pi i)^{-1}. \qquad (9.51)$$

9.3 Berezin integration

Exercise. Use the Fourier transforms to show that
$$1 = \iint \delta\kappa\,\delta\xi \exp(-2\pi i \kappa \xi).$$
□

Conclusion

Let
$$f(\xi) = \sum_{\alpha_1 < \ldots < \alpha_p} c_{\alpha_1 \ldots \alpha_p} \xi^{\alpha_1} \ldots \xi^{\alpha_p} \quad \text{and} \quad \delta^\nu \xi = \delta\xi^\nu \ldots \delta\xi^1;$$

then
$$\int \delta^\nu \xi\, f(\xi^1, \ldots, \xi^\nu) = (2\pi i)^{-\nu/2} c_{1 \ldots \nu}.$$

Remark. Berezin's integrals satisfy Fubini's theorem:
$$\iint \delta\eta\,\delta\xi\, f(\xi, \eta) = \int \delta\eta\, g(\eta),$$
where
$$g(\eta) = \int \delta\xi\, f(\xi, \eta).$$

Remark. Using the antisymmetry $\kappa\xi = -\xi\kappa$, we can rewrite the Fourier transformation in the perfectly symmetric way
$$f(\xi) = \int \delta\kappa\, \tilde{f}(\kappa) \exp(2\pi i \kappa \xi),$$
$$\tilde{f}(\kappa) = \int \delta\xi\, f(\xi) \exp(2\pi i \xi \kappa).$$

Change of variable of integration

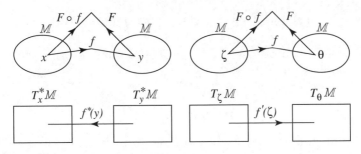

Fig. 9.1

Since integrating $f(\xi^1, \ldots, \xi^\nu)$ is equivalent to taking its derivatives with respect to $\xi^1, \ldots, \xi^\nu$, a change of variable of integration is most easily performed on the derivatives. Recall the induced transformations on the tangent and cotangent spaces given a change of coordinates f. Let $y = f(x)$ and $\theta = f(\zeta)$. For an ordinary integral

$$dy^1 \wedge \ldots \wedge dy^D = dx^1 \wedge \ldots \wedge dx^D \left(\det \frac{\partial y^i}{\partial x^j} \right) \tag{9.52}$$

and

$$\int dx^1 \wedge \ldots \wedge dx^D (F \circ f)(x) \left(\det \frac{\partial f^i}{\partial x^j} \right) = \int dy^1 \wedge \ldots \wedge dy^D F(y). \tag{9.53}$$

For an integral over Grassmann variables, the antisymmetry leading to a determinant is the antisymmetry of the product $\partial_1 \ldots \partial_\nu$, where $\partial_\alpha = \partial/\partial \zeta^\alpha$. Also

$$\left(\frac{\partial}{\partial \zeta^1} \cdots \frac{\partial}{\partial \zeta^\nu} \right)(F \circ f)(\zeta) = \left(\det \frac{\partial f^\lambda}{\partial \zeta^\mu} \right) \frac{\partial}{\partial \theta^1} \cdots \frac{\partial}{\partial \theta^\nu} F(\theta). \tag{9.54}$$

The determinant is now on the right-hand side; *it will become an inverse determinant* when brought to the left-hand side as in (9.53).

Exercise. A quick check of (9.54). Let $\theta = f(\zeta)$ be a *linear map*

$$\theta^\lambda = c^\lambda{}_\mu \zeta^\mu, \qquad \frac{\partial f^\lambda}{\partial \zeta^\mu} = \frac{\partial \theta^\lambda}{\partial \zeta^\mu} = c^\lambda{}_\mu, \qquad \frac{\partial}{\partial \zeta^\mu} = c^\lambda{}_\mu \frac{\partial}{\partial \theta^\lambda},$$

$$\frac{\partial}{\partial \zeta^\nu} \cdots \frac{\partial}{\partial \zeta^1} = \det\left(\frac{\partial f^\lambda}{\partial \zeta^\mu} \right) \frac{\partial}{\partial \theta^\nu} \cdots \frac{\partial}{\partial \theta^1}.$$

□

Exercise. Change of variable of integration in the Fourier transform of the Dirac δ

$$\delta(\xi) = \int \delta \kappa \, \exp(2\pi i \kappa \xi) = 2\pi \int \delta \alpha \, \exp(i\alpha \xi),$$

where $\alpha = 2\pi\kappa$ and $\delta\alpha = [1/(2\pi)]\delta\kappa$. Notice that, for differential forms, $\alpha = 2\pi\kappa$ implies $d\alpha = 2\pi \, d\kappa$.

9.4 Forms and densities

On an ordinary manifold M^D, a volume form is an exterior differential form of degree D. It is called a "top form" because there are no forms of degree higher than D on M^D; this is a consequence of the antisymmetry of forms. In Grassmann calculus, forms are symmetric. There are forms of arbitrary degrees on $\mathbb{R}^{0|\nu}$; therefore, there are *no* "top forms" on $\mathbb{R}^{0|\nu}$. We

9.4 Forms and densities

require another concept of volume element on M^D that can be generalized to $\mathbb{R}^{0|\nu}$.

In the 1930s [9], densities were used extensively in defining and computing integrals. Densities fell into disfavor, possibly because they do not form an algebra as forms do. On the other hand, complexes (ascending and descending) can be constructed with densities as well as with forms, both in ordinary and in Grassmann variables.

A form (an exterior differential form) is a *totally antisymmetric covariant tensor*. A density (a linear tensor density) is a *totally antisymmetric contravariant* tensor density of weight 1.[4]

Recall properties of forms and densities on ordinary D-dimensional manifolds M^D that can be established in the absence of a metric tensor. These properties can therefore be readily generalized to Grassmann calculus.

An ascending complex of forms on M^D

Let $\mathcal{A}^p$ be the space of p-forms on M^D, and let d be the exterior differentiation

$$\mathrm{d} : \mathcal{A}^p \to \mathcal{A}^{p+1}. \tag{9.55}$$

Explicitly,

$$\mathrm{d}\omega_{\alpha_1\ldots\alpha_{p+1}} = \sum_{j=1}^{p+1}(-1)^{j-1}\partial_{\alpha_j}\omega_{\alpha_1\ldots\alpha_{j-1}\alpha_{j+1}\ldots\alpha_p}.$$

Since $\mathrm{dd} = 0$, the graded algebra $\mathcal{A}^\bullet$ is an ascending complex with respect to the operator d:

$$\mathcal{A}^0 \xrightarrow{\mathrm{d}} \mathcal{A}^1 \xrightarrow{\mathrm{d}} \ldots \xrightarrow{\mathrm{d}} \mathcal{A}^D. \tag{9.56}$$

A descending complex of densities on M^D

Let $\mathcal{D}_p$ be the space of p-densities on M^D, and let ∇ be the divergence operator, also labeled b,

$$\nabla : \mathcal{D}_p \to \mathcal{D}_{p-1}, \qquad \nabla \equiv b. \tag{9.57}$$

Explicitly,

$$\nabla \mathcal{F}^{\alpha_1\ldots\alpha_{p-1}} = \partial_\alpha \mathcal{F}^{\alpha\alpha_1\ldots\alpha_{p-1}}$$

(ordinary derivative, not covariant derivative).

---

[4] See equation (9.75) for the precise meaning of weight 1.

Since bb = 0, $\mathcal{D}_\bullet$ (which is not a graded algebra) is a descending complex with respect to the divergence operator:

$$\mathcal{D}_0 \xleftarrow{b} \mathcal{D}_1 \xleftarrow{b} \ldots \xleftarrow{b} \mathcal{D}_D. \tag{9.58}$$

Metric-dependent and dimension-dependent transformations

The metric tensor g provides a correspondence C_g between a p-form and a p-density. Set

$$C_g : \mathcal{A}^p \to \mathcal{D}_p. \tag{9.59}$$

Example. The electromagnetic field F is a 2-form with components $F_{\alpha\beta}$ and

$$\mathcal{F}^{\alpha\beta} = \sqrt{\det g_{\mu\nu}} F_{\gamma\delta} g^{\alpha\gamma} g^{\beta\delta}$$

are the components of the 2-density $\mathcal{F}$. The metric g is used twice: (1) when raising indices; and (2) when introducing weight 1 by multiplication with $\sqrt{\det g}$. This correspondence does not depend on the dimension D.

On an orientable manifold, the dimension D can be used for transforming a p-density into a $(D-p)$-form. Set

$$\lambda^D : \mathcal{D}_p \to \mathcal{A}^{D-p}. \tag{9.60}$$

Example. Let $D = 4$ and $p = 1$, and define

$$t_{\alpha\beta\gamma} := \epsilon^{1234}_{\alpha\beta\gamma\delta} \mathcal{F}^\delta,$$

where the alternating symbol ϵ defines an orientation, and $t_{\alpha\beta\gamma}$ are the components of a 3-form.

The star operator (Hodge–de Rham operator, see [10], p. 295) is the composition of the dimension-dependent transformation λ^D with the metric-dependent one C_g. It transforms a p-form into a $(D-p)$-form:

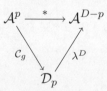

Let $\mathcal{T}$ be the volume element defined by $\mathcal{T} = *1$. It satisfies

$$\mathcal{T}(\omega|\eta) = \omega \wedge *\eta, \tag{9.61}$$

where the scalar product of two p-forms ω and η is

$$(\omega|\eta) = \frac{1}{p!} g^{i_1 j_1} \ldots g^{i_p j_p} \omega_{i_1 \ldots i_p} \eta_{j_1 \ldots j_p}. \tag{9.62}$$

9.4 Forms and densities

See (9.67) for the explicit expression.

We shall exploit the correspondence mentioned in the first paragraph,

$$C_g : \mathcal{A}^p \to \mathcal{D}_p, \tag{9.63}$$

for constructing a descending complex on $\mathcal{A}^\bullet$ with respect to the metric transpose δ of d ([10], p. 296),

$$\delta : \mathcal{A}^{p+1} \to \mathcal{A}^p, \tag{9.64}$$

and an ascending complex on $\mathcal{D}_\bullet$,

$$\beta : \mathcal{D}_p \to \mathcal{D}_{p+1}, \tag{9.65}$$

where β is defined by the following diagram:

$$\begin{array}{c} \mathcal{A}^p \xleftarrow[d]{\delta} \mathcal{A}^{p+1} \\ C_g \downarrow \quad \downarrow C_g \\ \mathcal{D}_p \xleftarrow[b]{\beta} \mathcal{D}_{p+1} \end{array} \iff \begin{cases} \delta = C_g^{-1} b C_g \\ \beta = C_g \, d\, C_g^{-1}. \end{cases} \tag{9.66}$$

Example 1. A volume element on an oriented D-dimensional riemannian manifold. By definition

$$\mathcal{T} = *1 = \lambda^D C_g 1.$$

Here 1 is a 0-form, $\mathcal{I} := C_g 1$ is a 0-density with component $\sqrt{\det g_{\mu\nu}}$ and λ^D transforms the 0-density with component $\mathcal{I}$ into the D-form

$$\mathcal{T} := dx^1 \wedge \ldots \wedge dx^D \mathcal{I} = dx^1 \wedge \ldots \wedge dx^D \sqrt{\det g}. \tag{9.67}$$

Under the change of coordinates $x'^j = A^j{}_i x^i$, the scalar density $\mathcal{I}$ transforms into $\mathcal{I}'$ such that

$$\mathcal{I}' = \mathcal{I} |\det A|^{-1} \tag{9.68}$$

and the D-form

$$dx'^1 \wedge \ldots \wedge dx'^D = |\det A| dx^1 \wedge \ldots \wedge dx^D. \tag{9.69}$$

Hence $\mathcal{I}$ remains invariant.

Example 2. The electric current. In the 1930s the use of densities was often justified by the fact that, in a number of useful examples, it reduces the number of indices. For example, by virtue of (9.60), the vector density $\mathcal{T}$ in $\mathbb{M}^4$ of components

$$\mathcal{T}^l = \epsilon^{ijkl}_{1234} T_{ijk} \tag{9.70}$$

can replace the 3-form T. An axial vector in $\mathbb{R}^3$ can replace a 2-form.

Grassmann forms

The two following properties of forms on real variables remain true for forms on Grassmann variables:

$$\mathrm{d}\mathrm{d}\omega = 0, \qquad (9.71)$$

$$\mathrm{d}(\omega \wedge \theta) = \mathrm{d}\omega \wedge \theta + (-1)^{\tilde{\omega}\tilde{d}} \omega \wedge \mathrm{d}\theta, \qquad (9.72)$$

where $\tilde{\omega}$ and $\tilde{d} = 1$ are the parities of ω and d, respectively. A form on Grassmann variables is a graded totally antisymmetric covariant tensor; this means that a Grassmann p-form is always even.

Since a Grassmann p-form is symmetric, the ascending complex $\mathcal{A}^*(\mathbb{R}^{0|\nu})$ does not terminate at ν-forms.

Grassmann densities

The two following properties of densities on real variables remain true for densities $\mathcal{F}$ on Grassmann variables:

$$\nabla\nabla\mathcal{F} = 0. \qquad (9.73)$$

Since a density is a tensor of weight 1, multiplication by a tensor of weight zero is the only possible product which maps a density into a density. Then

$$\nabla \cdot (XF) = (\nabla \cdot X) \cdot F + (-1)^{\tilde{X}\tilde{\nabla}} X \nabla \cdot F, \qquad (9.74)$$

where X is a vector field.

Since a density on Grassmann variables is a symmetric contravariant tensor, the descending complex $\mathcal{D}_\bullet(\mathbb{R}^{0|\nu})$ of Grassmann densities with respect to ∇ does not terminate at ν-densities.

Volume elements

The purpose of introducing densities was to arrive at a definition of volume elements suitable both for ordinary and for Grassmann variables. In example 1 (equation (9.67)) we showed how a scalar density enters a volume element on $\mathbb{M}^D$, and we gave the transformation (equation (9.68)) of a scalar density under a change of coordinates in $\mathbb{M}^D$. However, in order to generalize scalar densities to Grassmann volume elements, we start from Pauli's definition ([11], p. 32), which follows Weyl's terminology ([12], p. 109). "If $\int \mathcal{F} \, \mathrm{d}x$ is an invariant [under a change of coordinate system] then $\mathcal{F}$ is called a scalar density."

Under the change of variable $x' = f(x)$, the integrand $\mathcal{F}$ obeys the rule (9.68):

$$\mathcal{F} = |\det(\partial x'^j/\partial x^i)|\mathcal{F}'. \qquad (9.75)$$

If the Berezin integral

$$\int \delta\xi^\nu \ldots \delta\xi^1 \, f(\xi^1, \ldots, \xi^\nu) = \frac{\partial}{\partial \xi^\nu} \ldots \frac{\partial}{\partial \xi^1} \, f(\xi^1, \ldots, \xi^\nu)$$

is invariant under the change of coordinates $\theta(\xi)$, then f is a Grassmann scalar density. It follows from the formula for change of variable of integration (9.54) that a Grassmann scalar density obeys the rule

$$f = \det(\partial \theta^\lambda / \partial \xi^\kappa)^{-1} f' \qquad (9.76)$$

with the inverse of the determinant.

References

[1] F. A. Berezin (1965). *The Method of Second Quantization* (Moscow, Nauka) [in Russian] (English translation: Academic Press, New York, 1966).

[2] F. A. Berezin and M. S. Marinov (1975). "Classical spin and Grassmann algebra," *JETP Lett.* **21**, 320–321.
F. A. Berezin and M. S. Marinov (1977). "Particle spin dynamics as the Grassmann variant of classical mechanics," *Ann. Phys.* **104**, 336–362.
M. S. Marinov (1980). "Path integrals in quantum theory," *Phys. Rep.* **60**, 1–57.

[3] P. Cartier, C. DeWitt-Morette, M. Ihl, and Ch. Sämann, with an appendix by M. E. Bell. "Supermanifolds – applications to supersymmetry" (arXiv:math-ph/0202026 v1 19 Feb 2002). Michael Marinov Memorial Volume *Multiple Facets of Quantization and Supersymmetry*, eds. M. Olshanetsky and A. Vainshtein (River Edge, NJ, World Scientific, 2002), pp. 412–457.

[4] B. DeWitt (1992). *Supermanifolds*, 2nd edn. (Cambridge, Cambridge University Press). In this book complex conjugation is defined by $(zz')^* = z'^* z^*$.

[5] *Seminar on Supermanifolds*, known as SoS, over 2000 pages written by D. Leites and his colleagues, students, and collaborators from 1977 to 2000 (expected to be available electronically from arXiv).

[6] Y. Choquet-Bruhat (1989). *Graded Bundles and Supermanifolds* (Naples, Bibliopolis).

[7] T. Voronov (1992). "Geometric integration theory on supermanifolds," *Sov. Sci. Rev. C. Math. Phys.* **9**, 1–138.

[8] M. E. Bell (2002). Introduction to supersymmetry, unpublished Master's Thesis, University of Texas at Austin.

[9] L. Brillouin (1938). *Les tenseurs en mécanique et en élasticité* (Paris, Masson).

[10] Y. Choquet-Bruhat and C. DeWitt-Morette (1996 and 2000). *Analysis, Manifolds, and Physics*, Part I: Basics (with M. Dillard-Bleick) revised edn., and Part II, revised and enlarged edn. (Amsterdam, North Holland).

[11] W. Pauli (1958). *Theory of Relativity* (New York, Pergamon Press); translated with supplementary notes by the author from "Relativitätstheorie" in

Enzyclopädie der mathematischen Wissenschaften, Vol. 5, Part 2, pp. 539–775 (Leipzig, B. G. Teubner, 1921).

[12] H. Weyl (1921). *Raum–Zeit–Materie* (Berlin, Springer) [English translation: *Space–Time–Matter* (New York, Dover, 1950)]. The word "skew" is missing from the English translation of *schiefsymmetrisch*.

[13] G. Roepstorff (1994). *Path Integral Approach to Quantum Physics, An Introduction.* (Berlin, Springer).

10
Grassmann analysis: applications

$$\xi^1 \xi^2 = -\xi^2 \xi^1$$

10.1 The Euler–Poincaré characteristic

A characteristic class is a topological invariant defined on a bundle over a base manifold $\mathbb{X}$. Let $\mathbb{X}$ be a $2n$-dimensional oriented compact riemannian or pseudoriemannian manifold. Its *Euler number* $\chi(\mathbb{X})$ is the integral over $\mathbb{X}$ of the *Euler class* γ:

$$\chi(\mathbb{X}) = \int_{\mathbb{X}} \gamma. \tag{10.1}$$

The *Euler–Poincaré characteristic is equal to the Euler number* $\chi(\mathbb{X})$. The definition of the Euler–Poincaré characteristic can start from the definition of the Euler class, or from the definition of the *Betti numbers* b_p (i.e. the dimension of the p-homology group of $\mathbb{X}$). Chern [1] called the Euler characteristic "the source and common cause of a large number of geometrical disciplines." See e.g. [2, p. 321] for a diagram connecting the Euler–Poincaré characteristic to more than half a dozen topics in geometry, topology, and combinatorics. In this section we compute $\chi(\mathbb{X})$ by means of a supersymmetric path integral [3–5].

The supertrace of $\exp(-\Delta)$

We recall some classic properties of the Euler number $\chi(\mathbb{X})$, beginning with its definition as an alternating sum of Betti numbers,

$$\chi(\mathbb{X}) = \sum_{p=0}^{2n} (-1)^p b_p. \tag{10.2}$$

It follows from the Hodge theorem that the sum of the even Betti numbers is equal to the dimension d_e of the space of harmonic forms ω of even degrees, and similarly the sum of the odd Betti numbers is equal to the dimension d_o of the space of odd harmonic forms. Therefore

$$\chi(\mathbb{X}) = d_e - d_o. \tag{10.3}$$

By definition a form ω is said to be harmonic if

$$\Delta \omega = 0, \tag{10.4}$$

where Δ is the laplacian. On a compact manifold, Δ is a positive self-adjoint operator with discrete spectrum $\lambda_0 = 0, \lambda_1, \ldots, \lambda_n, \ldots$. Its trace in the Hilbert space spanned by its eigenvectors is

$$\operatorname{Tr} \exp(-\Delta) = \sum_{n=0}^{\infty} \nu_n \exp(-\lambda_n), \tag{10.5}$$

where ν_n is the (finite) dimension of the space spanned by the eigenvectors corresponding to λ_n. Let $\mathcal{H}^+$ and $\mathcal{H}^-$ be the Hilbert spaces of even and odd forms on $\mathbb{X}$, respectively. Let $\mathcal{H}_\lambda^\pm$ be the eigenspaces of Δ corresponding to the eigenvalues $\lambda \geq 0$. Then

$$\chi(\mathbb{X}) = \operatorname{Tr} \exp(-\Delta)|_{\mathcal{H}^+} - \operatorname{Tr} \exp(-\Delta)|_{\mathcal{H}^-} =: \operatorname{Str} \exp(-\Delta). \tag{10.6}$$

Proof. Let d be the exterior derivative and δ the metric transpose; then[1]

$$\Delta = (d + \delta)^2. \tag{10.7}$$

Let

$$Q = d + \delta. \tag{10.8}$$

Q is a self-adjoint operator such that

$$Q : \mathcal{H}^\pm \longrightarrow \mathcal{H}^\mp, \tag{10.9}$$
$$Q : \mathcal{H}_\lambda^\pm \longrightarrow \mathcal{H}_\lambda^\mp. \tag{10.10}$$

Equation (10.10) follows from $\Delta Q f = Q \Delta f = \lambda Q f$ together with (10.9),

$$Q^2|\mathcal{H}_\lambda^\pm = \lambda. \tag{10.11}$$

If $\lambda \neq 0$, then $\lambda^{-1/2} Q | \mathcal{H}_\lambda^\pm$ and $\lambda^{-1/2} Q | \mathcal{H}_\lambda^\mp$ are inverses of each other, and[2]

$$\dim \mathcal{H}_\lambda^+ = \dim \mathcal{H}_\lambda^-. \tag{10.12}$$

---

[1] Note that (10.7) defines a positive operator; that is a laplacian with sign opposite to the usual definition $g^{-1/2} \partial^i g^{1/2} \partial_i$.
[2] Each eigenvalue λ has finite multiplicity; hence the spaces $\mathcal{H}_\lambda^+$ and $\mathcal{H}_\lambda^-$ have a finite dimension.

10.1 The Euler–Poincaré characteristic

By definition
$$\operatorname{Str}\exp(-\Delta) = \sum_\lambda \exp(-\lambda)\left(\dim \mathcal{H}_\lambda^+ - \dim \mathcal{H}_\lambda^-\right) \quad (10.13)$$
$$= \dim \mathcal{H}_0^+ - \dim \mathcal{H}_0^-$$
$$= d_e - d_o$$

since a harmonic form is an eigenstate of Δ with eigenvalue 0.

The definition (10.2) of the Euler number belongs to a graded algebra. Expressing it as a supertrace (10.6) offers the possibility of computing it by a supersymmetric path integral.

Scale invariance

Since the sum (10.13) defining $\operatorname{Str}\exp(-\Delta)$ depends only on the term $\lambda = 0$,
$$\operatorname{Str}\exp(-\Delta) = \operatorname{Str}\exp(z\,\Delta) \quad (10.14)$$
for any $z \in \mathbb{C}$. In particular, the laplacian scales like the inverse metric tensor, but according to (10.14) $\operatorname{Str}\exp(-\Delta)$ is scale-invariant.

Supersymmetry

When Bose and Fermi systems are combined into a single system, new kinds of symmetries and conservation laws can occur. The simplest model consists of combining a Bose and a Fermi oscillator. The action functional $S(x,\xi)$ for this model is an integral of the lagrangian
$$L(x,\xi) = \frac{1}{2}(\dot{x}^2 - \omega^2 x^2) + \frac{i}{2}(\xi^T \dot{\xi} + \omega \xi^T M \xi)$$
$$= L_{\text{bos}}(x) + L_{\text{fer}}(\xi). \quad (10.15)$$

The *Fermi oscillator* is described by two real a-type dynamical variables:
$$\xi := \begin{pmatrix} \xi_1 \\ \xi_2 \end{pmatrix}, \quad \xi^T := (\xi_1, \xi_2), \quad \text{and} \quad M := \begin{pmatrix} 0 & 1 \\ -1 & 0 \end{pmatrix}. \quad (10.16)$$

Remark. M is an even antisymmetric matrix:
$$\xi^T M \eta = \eta^T M \xi = \xi_1 \eta_2 + \eta_1 \xi_2.$$

The dynamical equation for the Fermi trajectory is
$$\dot{\xi} + \omega M \xi = 0. \quad (10.17)$$

Since $M^2 = -\mathbf{1}_2$, it implies
$$\ddot{\xi} + \omega^2 \xi = 0. \quad (10.18)$$

The general solution of (10.17) is
$$\xi(t) = fu(t) + f^\dagger u^*(t), \tag{10.19}$$
where
$$u(t) := \begin{pmatrix} 1/\sqrt{2} \\ i/\sqrt{2} \end{pmatrix} e^{-i\omega t}, \tag{10.20}$$

f is an arbitrary complex a-number, and $f^\dagger$ is the hermitian conjugate (9.13)$_b$ of f. Upon quantization, the number f becomes an operator $\hat{f}$ satisfying the graded commutators
$$[\hat{f}, \hat{f}]_+ = 0, \qquad [\hat{f}, \hat{f}^\dagger]_+ = 1. \tag{10.21}$$

The Bose oscillator is described by one real c-type dynamical variable x. Its dynamical equation is
$$\ddot{x} + \omega^2 x = 0. \tag{10.22}$$

The general solution of (10.22) is
$$x(t) = \frac{1}{\sqrt{2\omega}}(be^{-i\omega t} + b^\dagger e^{i\omega t}), \tag{10.23}$$

where b is an arbitrary complex c-number; upon quantization, it becomes an operator $\hat{b}$ satisfying the graded commutators
$$[\hat{b}, \hat{b}]_- = 0, \qquad [\hat{b}, \hat{b}^\dagger]_- = 1. \tag{10.24}$$

Bosons and fermions have vanishing graded commutators:
$$[\hat{f}, \hat{b}]_- = 0, \qquad [\hat{f}, \hat{b}^\dagger]_- = 0. \tag{10.25}$$

The action functional $S(x, \xi)$ is invariant under the following infinitesimal changes of the dynamical variables generated by the real a-numbers $\delta\alpha = \begin{pmatrix} \delta\alpha_1 \\ \delta\alpha_2 \end{pmatrix}$:
$$\begin{aligned} \delta x &= i\xi^T \delta\alpha, \\ \delta\xi &= (\dot{x}\mathbf{1}_2 - \omega x M)\delta\alpha. \end{aligned} \tag{10.26}$$

The action functional $S(x, \xi)$ is called *supersymmetric* because it is invariant under the transformation (10.26) that defines δx by ξ and $\delta\xi$ by x. The supersymmetry occurs because the frequencies ω of the Bose and Fermi oscillators are equal.

The transformation (10.26) is an infinitesimal global supersymmetry because $\delta\alpha$ is assumed to be time-independent.

10.1 The Euler–Poincaré characteristic

Remark. Global finite supersymmetry is implemented by a unitary operator. Let us introduce new dynamical variables $b = (\omega x + i\dot{x})/\sqrt{2\omega}$ and $f = (\xi_1 - i\xi_2)/\sqrt{2}$ as well as their hermitian conjugates $b^\dagger$ and $f^\dagger$. The equation of motion reads as[3]

$$\dot{b} = -i\omega b, \qquad \dot{f} = -i\omega f,$$

and the hamiltonian is $H = \omega(b^\dagger b + f^\dagger f)$. Upon introducing the time-independent Grassmann parameter $\beta = \sqrt{\omega}(\alpha_2 + i\alpha_1)$ and its hermitian conjugate $\beta^\dagger = \sqrt{\omega}(\alpha_2 - i\alpha_1)$, the infinitesimal supersymmetric transformation is given now by

$$\delta b = -f\,\delta\beta^\dagger, \qquad \delta f = -b\,\delta\beta.$$

By taking hermitian conjugates, we get

$$\delta b^\dagger = f^\dagger\,\delta\beta, \qquad \delta f^\dagger = -b^\dagger\,\delta\beta^\dagger.$$

After quantization, the dynamical variables b and f correspond to operators $\hat{b}$ and $\hat{f}$ obeying the commutation rules (10.21), (10.24), and (10.25). The finite global supersymmetry is implemented by the unitary operator

$$T = \exp(\hat{b}^\dagger \hat{f}\beta^\dagger + \hat{f}^\dagger \hat{b}\beta),$$

where β and $\beta^\dagger$ are hermitian conjugate Grassmann parameters commuting with $\hat{b}$ and $\hat{b}^\dagger$, and anticommuting with $\hat{f}$ and $\hat{f}^\dagger$. Hence any quantum dynamical variable $\hat{A}$ is transformed into $T\hat{A}T^{-1}$.

Equation (10.26) is modified when x and ξ are arbitrary functions of time. For the modified supersymmetric transformation see [5, p. 292].

In quantum field theory supersymmetry requires bosons to have fermion partners of the same mass.

A supersymmetric path integral[4]

Consider the following superclassical system on an $(m, 2m)$ supermanifold where the ordinary dynamical variables $x(t) \in \mathbb{X}$, an m-dimensional riemannian manifold:

$$S(x, \xi) = \int_\mathbb{T} dt\left(\frac{1}{2}g_{ij}\dot{x}^i\dot{x}^j + \frac{1}{2}ig_{ij}\xi^i_\alpha \dot{\xi}^j_\alpha + \frac{1}{8}R_{ijkl}\xi^i_\alpha \xi^j_\alpha \xi^k_\beta \xi^l_\beta\right), \qquad (10.27)$$

---

[3] The old f and b (see equations (10.19) and (10.23)) were time-independent. The new f and b vary in time!

[4] This section has been extracted from [5, pp. 370–389], where the detailed calculations are carried out. For facilitating the use of [5], we label the dimension of the riemannian manifold m (rather than D). In [5] the symbol $\delta\xi$ introduced in Chapter 9 is written $d\xi$.

where $i, j \in \{1, \ldots, m\}$ and $\alpha \in \{1, 2\}$. Moreover, g is a positive metric tensor, the Riemann tensor is

$$R^i{}_{jkl} = -\Gamma^i{}_{jk,l} + \Gamma^i{}_{jl,k} + \Gamma^i{}_{mk}\Gamma^m{}_{jl} - \Gamma^i{}_{ml}\Gamma^m{}_{jk}, \tag{10.28}$$

and $\Gamma^a{}_{bc}$ are the components of the connection ∇. Introducing two sets $(\xi^1_\alpha, \ldots, \xi^m_\alpha)$ for $\alpha \in \{1, 2\}$ is necessary in order for the contribution of the curvature term in (10.27) to be nonvanishing. This is obvious when $m = 2$. It can be proved by calculation in the general case.

The action (10.27) is invariant under the supersymmetric transformation generated by

$$\delta\eta = \begin{pmatrix} \delta\eta_1 \\ \delta\eta_2 \end{pmatrix}$$

and δt, namely

$$\begin{aligned}\delta x^i &= \dot{x}^i\, \delta t + \mathrm{i}\xi^i_\alpha\, \delta\eta_\alpha, \\ \delta\xi^i_\alpha &= (\mathrm{d}\xi^i_\alpha/\mathrm{d}t)\delta t + \dot{x}^i\, \delta\eta_\alpha + \mathrm{i}\Gamma^i{}_{jk}\xi^j_\alpha\, \xi^k_\beta\, \delta\eta_\beta,\end{aligned} \tag{10.29}$$

where the summation convention applies also to the repeated Greek indices

$$\xi_\alpha\, \delta\eta_\alpha = \xi_1\, \delta\eta_1 + \xi_2\, \delta\eta_2 \text{ etc.} \tag{10.30}$$

and δt and $\delta\eta$ are of compact support in $\mathbb{T}$. Therefore the action $S(x, \xi)$ is supersymmetric.

The hamiltonian $H = \tfrac{1}{2}g_{ij}\dot{x}^i\dot{x}^j - \tfrac{1}{8}R_{ijkl}\xi^i_\alpha\xi^j_\alpha\xi^k_\beta\xi^l_\beta$ derived from the action functional (10.27) is precisely equal to half the laplacian operator Δ on forms [6, p. 319]. Therefore, using (10.6) and (10.14),

$$\chi(\mathbb{X}) = \operatorname{Str} \exp(-\mathrm{i}Ht). \tag{10.31}$$

We shall show that, for m even,[5]

$$\operatorname{Str} \exp(-\mathrm{i}Ht) = \frac{1}{(8\pi)^{m/2}(m/2)!} \int_\mathbb{X} \mathrm{d}^m x\; g^{-1/2} \epsilon^{i_1 \ldots i_m} \epsilon^{j_1 \ldots j_m}$$

$$\times R_{i_1 i_2 j_1 j_2} \cdots R_{i_{m-1} i_m j_{m-1} j_m} \tag{10.32}$$

and

$$\operatorname{Str} \exp(-\mathrm{i}Ht) = 0 \quad \text{for } m \text{ odd.} \tag{10.33}$$

That is, we shall show that the Gauss–Bonnet–Chern–Avez formula for the Euler–Poincaré characteristic can be obtained by computing a supersymmetric path integral. Since $\operatorname{Str} \exp(-\mathrm{i}Ht)$ is independent of the

---

[5] We set $g := \det(g_{\mu\nu})$.

10.1 The Euler–Poincaré characteristic

magnitude of t, we compute it for an infinitesimal time interval ϵ. The path integral reduces to an ordinary integral, and the reader may question the word "path integral" in the title of this section. The reason is that the path-integral formalism simplifies the calculation considerably since it uses the action rather than the hamiltonian.

To spell out $\operatorname{Str}\exp(-iHt)$ one needs a basis in the super Hilbert space $\mathcal{H}$ on which the hamiltonian H, i.e. the laplacian Δ, operates. A convenient basis is the coherent-states basis defined as follows. For more on its property see [5, p. 381]. Let $|x', t\rangle$ be a basis for the bosonic sector

$$x^i(t)|x', t\rangle = x'^i |x', t\rangle. \tag{10.34}$$

Replace the a-type dynamical variables

$$\xi = \begin{pmatrix} \xi_1 \\ \xi_2 \end{pmatrix}$$

by

$$z^i := \frac{1}{\sqrt{2}}(\xi_1^i - i\xi_2^i), \qquad z^{i\dagger} := \frac{1}{\sqrt{2}}(\xi_1^i + i\xi_2^i). \tag{10.35}$$

The superjacobian of this transformation is

$$\frac{\partial(z^\dagger, z)}{\partial(\xi_1, \xi_2)} = \left(\operatorname{sdet}\begin{pmatrix} 1/\sqrt{2} & i/\sqrt{2} \\ 1/\sqrt{2} & -i/\sqrt{2} \end{pmatrix}\right)^m = i^m. \tag{10.36}$$

Set

$$z_i = g_{ij} z^j, \qquad z_i^\dagger = g_{ij} z^{j\dagger}. \tag{10.37}$$

The new variables satisfy

$$\left[z_i, z_j^\dagger\right]_+ = g_{ij}, \tag{10.38}$$

and the other graded commutators between the zs vanish.

Define supervectors in $\mathcal{H}$:

$$|x', z', t\rangle := \exp\left(-\frac{1}{2} z_i'^\dagger z'^i + z_i^\dagger(t) z'^i\right) |x', t\rangle \tag{10.39}$$

and

$$\langle x', z'^\dagger, t| := |x', z', t\rangle^\dagger = \langle x', t| \exp\left(\frac{1}{2} z_i'^\dagger z'^i + z_i'^\dagger z^i(t)\right). \tag{10.40}$$

This basis is called a coherent-states basis because the $|x', z', t\rangle$ are right eigenvectors of the $z^i(t)$ while the $\langle x', z'^\dagger, t|$ are left eigenvectors of the

182 *Grassmann analysis: applications*

$z^{i\dagger}(t)$. In terms of this basis

$$\operatorname{Str}\exp(-\mathrm{i}Ht) = \frac{1}{(2\pi\mathrm{i})^m}\int\mathrm{d}^m x' \prod_{j=1}^{m}(\delta z_j'^{\dagger}\,\delta z_j')g^{-1}(x')$$
$$\times \langle x',z'^{\dagger},t'|e^{-\mathrm{i}Ht}|x',z',t'\rangle. \qquad (10.41)$$

We need not compute the hamiltonian; it is sufficient to note that it is a time-translation operator. Therefore

$$\langle x',z'^{\dagger},t'|\exp(-\mathrm{i}Ht)|x',z',t'\rangle = \langle x',z'^{\dagger},t'+t|x',z',t'\rangle. \qquad (10.42)$$

Two circumstances simplify the path-integral representation of this probability amplitude.

- It is a trace; therefore the paths are loops beginning and ending at the same point in the supermanifold.
- The supertrace is scale-invariant; therefore, the time interval t can be taken arbitrarily small.

It follows that the only term in the action functional (10.27) contributing to the supertrace is the Riemann tensor integral,

$$\frac{1}{8}\int_{\mathbb{T}}\mathrm{d}t\,R_{ijkl}\xi_\alpha^i\xi_\alpha^j\xi_\beta^k\xi_\beta^l = \frac{1}{2}\int_{\mathbb{T}}\mathrm{d}t\,R_{ijkl}z^{i\dagger}z^j z^{k\dagger}z^l. \qquad (10.43)$$

For an infinitesimal time interval ϵ,

$$\operatorname{Str}\exp(-\mathrm{i}H\epsilon) = (2\pi\mathrm{i}\epsilon)^{-m/2}(2\pi)^{-m}$$
$$\times \int_{\mathbb{R}^{m|2m}}\exp\left(\frac{1}{4}\mathrm{i}R_{ijkl}(x')\xi_1^i\xi_1^j\xi_2^k\xi_2^l\epsilon\right)g^{-1/2}(x')\mathrm{d}^m x'\prod_{i=1}^{m}\delta\xi_1^i\,\delta\xi_2^i. \qquad (10.44)$$

The detour by the z-variables was useful for constructing a supervector basis. The return to the ξ-variables simplifies the Berezin integrals. On expanding the exponential in (10.44), one sees that the integral vanishes for m odd, and for m even is equal to[6]

$$\operatorname{Str}\exp(-\mathrm{i}H\epsilon) = \frac{(2\pi)^{-3m/2}}{4^{m/2}(m/2)!}\int_{\mathbb{R}^{m|2m}}\left(R_{ijkl}(x)\xi_1^i\xi_1^j\xi_2^k\xi_2^l\right)^{m/2}$$
$$\times g^{-1/2}(x)\mathrm{d}^m x\,\delta\xi_1^1\,\delta\xi_2^1\ldots\delta\xi_1^m\,\delta\xi_2^m. \qquad (10.45)$$

---

[6] The detailed calculation is carried out in [5, pp. 388–389]. There the normalization of the Berezin integral is not (9.51) but $C^2 = 2\pi\mathrm{i}$; however, it can be shown that (10.45) does not depend on the normalization.

Finally,

$$\chi(\mathbb{X}) = \text{Str} \exp(-iH\epsilon)$$
$$= \frac{1}{(8\pi)^{m/2}(m/2)!} \int_X g^{-1/2} \epsilon^{i_1...i_m} \epsilon^{j_1...j_m} R_{i_1 i_2 j_1 j_2} \cdots R_{i_{m-1} i_m j_{m-1} j_m} \mathrm{d}^m x. \tag{10.46}$$

□

In two dimensions, $m = 2$, one obtains the well-known formula for the Euler number

$$\begin{aligned}\chi(\mathbb{X}) &= \frac{1}{8\pi} \int_X g^{-1/2} \epsilon^{ij} \epsilon^{kl} R_{ijkl} \mathrm{d}^2 x \\ &= \frac{1}{8\pi} \int_X g^{1/2} (g^{ik} g^{jl} - g^{il} g^{jk}) R_{ijkl} \, \mathrm{d}^2 x \\ &= \frac{1}{2\pi} \int g^{1/2} R \, \mathrm{d}^2 x. \end{aligned} \tag{10.47}$$

Again one can check that the r.h.s. is invariant under a scale transformation of the metric. When $g \mapsto cg$ with a constant c, the Christoffel symbols are invariant, the Riemann scalar $R = R_{ij} g^{ij}$ goes into Rc^{-1} and $g^{1/2}$ goes into $c^{m/2} g^{1/2}$. Therefore in two dimensions $g^{1/2} R$ goes into $g^{1/2} R$.

Starting from a superclassical system more general than (10.27), A. Mostafazadeh [7] has used supersymmetric path integrals for deriving the index of the Dirac operator formula and the Atiyah–Singer index theorem.

10.2 Supersymmetric quantum field theory

It is often, but erroneously, stated that the "classical limits" of Fermi fields take their values in a Grassmann algebra because "their anticommutators vanish when $\hbar = 0$." Leaving aside the dubious concept of a "Fermi field's classical limit," we note that in fact the canonical anticommutators of Fermi fields do *not* vanish when $\hbar = 0$ because they do not depend on $\hbar$: given the canonical quantization

$$[\Phi(x), \Pi(y)]_- = i\hbar \delta(x-y) \quad \text{for a bosonic system,} \tag{10.48}$$
$$[\psi(x), \pi(y)]_+ = i\hbar \delta(x-y) \quad \text{for a fermionic system,} \tag{10.49}$$

and the Dirac lagrangian

$$\mathcal{L} = \bar{\psi}(-p_\mu \gamma^\mu - mc)\psi = i\hbar \bar{\psi} \gamma^\mu \partial_\mu \psi - mc\bar{\psi}\psi, \tag{10.50}$$

the conjugate momentum $\pi(x) = \delta \mathcal{L}/\delta \dot{\psi}$ is proportional to $\hbar$. The net result is that the graded commutator is independent of $\hbar$. □

Clearing up the above fallacy does not mean that Grassmann algebra plays no role in fermionic systems. Grassmann analysis *is* necessary for a consistent and unified functional approach to quantum field theory; the functional integrals are integrals over functions of Grassmann variables.

Supersymmetry in quantum field theory is a symmetry that unites particles of integer and half-integer spin in common symmetry multiplets, called *supermultiplets*.

Supersymmetry in physics is too complex to be thoroughly addressed in this book. We refer the reader to works by Martin (S. Martin, "A supersymmetry primer," hep-ph/9709356), Weinberg (S. Weinberg, *The Quantum Theory of Fields Volume III: Supersymmetry*, Cambridge, Cambridge University Press, 2000), and Wess and Bagger (J. Wess and J. Bagger, *Supersymmetry and Supergravity*, Princeton, NJ, Princeton University Press, 2nd edn., 1992).

Here we mention only supersymmetric Fock spaces, i.e. spaces of states that carry a representation of a supersymmetric algebra – that is, an algebra of bosonic and fermionic creation and annihilation operators.

Representations of fermionic and bosonic creation and annihilation operators are easily constructed on the ascending complex of forms (Sections 9.2 and 9.4)

- on $\mathbb{M}^D$, for the fermionic case; and
- on $\mathbb{R}^{0|\nu}$, for the bosonic case.

They provide representations of operators on supersymmetric Fock spaces. The operator e defines creation operators; and the operator i defines annihilation operators. On $\mathbb{M}^D$, let f be a scalar function $f : \mathbb{M}^D \to \mathbb{R}$,

$$e(f) : \mathcal{A}^p \to \mathcal{A}^{p+1} \qquad \text{by } \omega \mapsto df \wedge \omega. \tag{10.51}$$

Let X be a vector field on $\mathbb{M}^D$,

$$i(X) : \mathcal{A}^p \to \mathcal{A}^{p-1} \qquad \text{by contraction with the vector field } X. \tag{10.52}$$

Let ϕ be a scalar function on $\mathbb{R}^{0|\nu}$ such that $\phi : \mathbb{R}^{0|\nu} \to \mathbb{R}$,

$$e(\phi) : \mathcal{A}^p \to \mathcal{A}^{p+1} \qquad \text{by } \omega \mapsto d\phi \wedge \omega. \tag{10.53}$$

Let Ξ be a vector field on $\mathbb{R}^{0|\nu}$,

$$i(\Xi) : \mathcal{A}^p \to \mathcal{A}^{p-1} \qquad \text{by contraction with the vector field } \Xi. \tag{10.54}$$

Representations of fermionic and bosonic creation and annihilation operators can also be constructed on a descending complex of densities. They can be read off from Section 9.2 (a compendium of Grassmann analysis). They are given explicitly in [8].

A physical example of fermionic operators: Dirac fields

The second set of Maxwell's equations (see e.g. [6, p. 336]),

$$\delta F + J = 0,$$

together with some initial data, gives the electromagnetic field F created by a current J. Dirac gave an expression for the current J:

$$J_\mu = e c \bar{\psi} \gamma_\mu \psi, \quad \text{with } \bar{\psi} \text{ such that } \bar{\psi}\psi \text{ is a scalar,} \quad (10.55)$$

e is the electric charge, the $\{\gamma_\mu\}$s are the Dirac matrices, and ψ is a Dirac field that obeys Dirac's equation. The structural elements of quantum electrodynamics are the electromagnetic field and the Dirac fields. Their quanta are photons, electrons, and positrons, which are viewed as particles. The Dirac field ψ is an operator on a Fock space. It is a linear combination of an electron-annihilation operator a and a positron-creation operator $b^\dagger$, constructed so as to satisfy the causality principle, namely, the requirement that supercommutators of field operators vanish when the points at which the operators are evaluated are separated by a space-like interval (see [10, Vol. I] and [11]).

The Dirac field describes particles other than electrons and antiparticles other than positrons, generically called fermions and antifermions. There are several representations of Dirac fields that depend on the following:

- the signature of the metric tensor;
- whether the fields are real [9] or complex; and
- the choice of Dirac pinors (one set of four complex components) or Weyl spinors (two sets of two complex components).

For example [11],

$$\Psi(x) = \int \frac{d^3 p}{(2\pi)^3} \frac{1}{\sqrt{2E_\mathbf{p}}} \sum_s \left(a_\mathbf{p}^s u^s(p) e^{-ip\cdot x} + b_\mathbf{p}^{s\dagger} v^s(p) e^{ip\cdot x} \right)$$

and

$$\bar{\Psi}(x) = \int \frac{d^3 p}{(2\pi)^3} \frac{1}{\sqrt{2E_\mathbf{p}}} \sum_s \left(b_\mathbf{p}^s \bar{v}^s(p) e^{-ip\cdot x} + a_\mathbf{p}^{s\dagger} \bar{u}^s(p) e^{ip\cdot x} \right)$$

represent a fermion and an antifermion, respectively.

In quantum electrodynamics, the operator $a_\mathbf{p}^{s\dagger}$ creates electrons with energy $E_\mathbf{p}$, momentum $\mathbf{p}$, spin $1/2$, charge $+1$ (in units of e, hence $e < 0$), and polarization determined by the (s)pinor u^s. The operator $b_\mathbf{p}^{s\dagger}$ creates positrons with energy $E_\mathbf{p}$, momentum $\mathbf{p}$, spin $1/2$, charge -1, and polarization opposite to that of u^s.

The creation and annihilation operators are normalized so that

$$\{\Psi_a(\mathbf{x}), \Psi_b^\dagger(\mathbf{y})\} = \delta^3(\mathbf{x}-\mathbf{y})\delta_{ab}$$

with all other anticommutators equal to zero. The (s)pinors u^s and v^s obey Dirac's equation. The term $\mathrm{d}^3 p/\sqrt{2E_\mathbf{p}}$ is Lorentz-invariant; it is the positive-energy part of $\mathrm{d}^4 p\, \delta(p^2 - m^2)$.

There are many [8] constructions of supermanifolds and many representations of bosonic and fermionic algebras, that is many possibilities for a framework suitable for supersymmetric quantum field theory.

10.3 The Dirac operator and Dirac matrices

The Dirac operator is the operator on a Pin bundle,

$$\not{\partial} := \gamma^a \nabla_a,$$

where ∇_a is the covariant derivative and $\{\gamma^a\}$ are the Dirac matrices. The Dirac operator is the square root of the laplacian

$$\Delta = g^{ab} \nabla_a \nabla_b.$$

The operator Q on differential forms, (10.8),

$$Q = \mathrm{d} + \delta,$$

is also a square root of the laplacian. We shall develop a connection between the Dirac operator $\not{\partial}$ and the Q-operator acting on superfunctions.

Consider a D-dimensional real vector space V with a scalar product. Introducing a basis $e_1, \ldots, e_D$, we represent a vector by its components $v = e_a v^a$. The scalar product reads

$$g(v, w) = g_{ab} v^a w^b. \tag{10.56}$$

Let $C(V)$ be the corresponding Clifford algebra generated by $\gamma_1, \ldots, \gamma_D$ satisfying the relations

$$\gamma_a \gamma_b + \gamma_b \gamma_a = 2 g_{ab}. \tag{10.57}$$

The dual generators are given by $\gamma^a = g^{ab} \gamma_b$ and

$$\gamma^a \gamma^b + \gamma^b \gamma^a = 2 g^{ab}, \tag{10.58}$$

where $g^{ab} g_{bc} = \delta^a{}_c$ as usual.

We define now a representation of the Clifford algebra $C(V)$ by operators acting on a Grassmann algebra. Introduce Grassmann variables

10.3 The Dirac operator and Dirac matrices

$\xi^1, \ldots, \xi^D$ and put[7]

$$\gamma^a = \xi^a + g^{ab}\frac{\partial}{\partial \xi^b}. \tag{10.59}$$

Then the relations (10.58) hold. In more intrinsic terms we consider the exterior algebra ΛV^* built on the dual V^* of V with a basis $(\xi^1, \ldots, \xi^D)$ dual to the basis $(e_1, \ldots, e_D)$ of V. The scalar product g defines an isomorphism $v \mapsto I_g v$ of V with V^* characterized by

$$\langle I_g v, w \rangle = g(v, w). \tag{10.60}$$

Then we define the operator $\gamma(v)$ acting on ΛV^* as follows:

$$\gamma(v) \cdot \omega = I_g v \wedge \omega + i(v)\omega, \tag{10.61}$$

where the contraction operator $i(v)$ satisfies

$$i(v)(\omega_1 \wedge \ldots \wedge \omega_p) = \sum_{j=1}^{p}(-1)^{j-1}\langle \omega_j, v\rangle \omega_1 \wedge \ldots \wedge \hat{\omega}_j \wedge \ldots \wedge \omega_p \tag{10.62}$$

for $\omega_1, \ldots, \omega_p$ in V^*. (The hat ˆ means that the corresponding factor is omitted.) An easy calculation gives

$$\gamma(v)\gamma(w) + \gamma(w)\gamma(v) = 2g(v, w). \tag{10.63}$$

We recover $\gamma_a = \gamma(e_a)$, hence $\gamma^a = g^{ab}\gamma_b$.

The representation constructed in this manner is not the spinor representation since it is of dimension 2^D. Assume that D is even, $D = 2n$, for simplicity. Hence ΛV^* is of dimension $2^D = (2^n)^2$, and *the spinor representation should be a "square root" of ΛV^**.

Consider the operator J on ΛV^* given by

$$J(\omega_1 \wedge \ldots \wedge \omega_p) = \omega_p \wedge \ldots \wedge \omega_1 = (-1)^{p(p-1)/2}\omega_1 \wedge \ldots \wedge \omega_p \tag{10.64}$$

for $\omega_1, \ldots, \omega_p$ in V^*. Introduce the operator

$$\gamma^o(v) = J\gamma(v)J. \tag{10.65}$$

In components $\gamma^o(v) = v^a\gamma_a^o$, where $\gamma_a^o = J\gamma_a J$. Since $J^2 = 1$, they satisfy the Clifford relations

$$\gamma^o(v)\gamma^o(w) + \gamma^o(w)\gamma^o(v) = 2g(v, w). \tag{10.66}$$

The interesting point is the commutation property[8]

$$\gamma(v) \text{ and } \gamma^o(w) \text{ commute for all } v, w.$$

---

[7] Here again $\partial/\partial \xi^b$ denotes the left derivation operator, which is often denoted by $\overleftarrow{\partial}/\partial \xi^b$.

[8] This construction is reminiscent of Connes' description of the standard model in [12].

According to the standard wisdom of quantum theory, the degrees of freedom associated with the γ_a decouple from those for the γ_a^o. Assume that the scalars are complex numbers, hence the Clifford algebra is isomorphic to the algebra of matrices of type $2^n \times 2^n$. Then ΛV^* can be decomposed as a tensor square

$$\Lambda V^* = S \otimes S \qquad (10.67)$$

with the $\gamma(v)$ acting on the first factor only, and the $\gamma^o(v)$ acting on the second factor in the same way:

$$\gamma(v)(\psi \otimes \psi') = \Gamma(v)\psi \otimes \psi', \qquad (10.68)$$

$$\gamma^o(v)(\psi \otimes \psi') = v \otimes \Gamma(v)\psi'. \qquad (10.69)$$

The operator J is then the exchange

$$J(\psi \otimes \psi') = \psi' \otimes \psi. \qquad (10.70)$$

The decomposition $S \otimes S = \Lambda V^*$ corresponds to the formula

$$c_{i_1 \ldots i_p} = \bar\psi \gamma_{[i_1} \ldots \gamma_{i_p]} \psi \qquad (0 \le p \le D) \qquad (10.71)$$

for the currents[9] $c_{i_1 \ldots i_p}$ (by $[\ldots]$ we denote antisymmetrization).

In differential geometric terms, let $(\mathbb{M}^D, g)$ be a (pseudo)riemannian manifold. The Grassmann algebra ΛV^* is replaced by the graded algebra $\mathcal{A}(\mathbb{M}^D)$ of differential forms. The Clifford operators are given by

$$\gamma(f)\omega = df \wedge \omega + i(\nabla f)\omega \qquad (10.72)$$

(∇f is the gradient of f with respect to the metric g, a vector field). In components $\gamma(f) = \partial_\mu f \cdot \gamma^\mu$ with

$$\gamma^\mu = e(x^\mu) + g^{\mu\nu} i\left(\frac{\partial}{\partial x^\nu}\right). \qquad (10.73)$$

The operator J satisfies

$$J(\omega) = (-1)^{p(p-1)/2}\omega \qquad (10.74)$$

for a p-form ω. To give a spinor structure on the riemannian manifold $(\mathbb{M}^D, g)$ (in the case of D even) is to give a splitting[10]

$$\Lambda T_\mathbb{C}^* \mathbb{M}^D \simeq S \otimes S \qquad (10.75)$$

---

[9] For $n = 4$, this gives a scalar, a vector, a bivector, a pseudo-vector and a pseudo-scalar.

[10] $T_\mathbb{C}^* \mathbb{M}^D$ is the complexification of the cotangent bundle. We perform this complexification in order to avoid irrelevant discussions on the signature of the metric.

satisfying the analog of relations (10.68) and (10.70). The Dirac operator $\slashed{\partial}$ is then characterized by the fact that $\slashed{\partial} \times 1$ acting on bispinor fields (sections of $S \otimes S$ on $\mathbb{M}^D$) corresponds to $d + \delta$ acting on (complex) differential forms; that is, on (complex) superfunctions on $\Pi T \mathbb{M}^D$.

References

[1] S.S. Chern (1979). "From triangles to manifolds," *Am. Math. Monthly* **86**, 339–349.

[2] Y. Choquet-Bruhat and C. DeWitt-Morette (2000). *Analysis, Manifolds, and Physics, Part II*, revised and enlarged edn., pp. 4 and 57–64.

[3] E. Witten (1982). "Supersymmetry and Morse theory," *J. Diff. Geom.* **17**, 661–692.

[4] E. Getzler (1985). "Atiyah–Singer index theorem," in Les Houches Proceedings *Critical Phenomena, Random Systems, Gauge Theories*, eds. K. Osterwalder and R. Stora (Amsterdam, North Holland).

[5] B. DeWitt (1992). *Supermanifolds*, 2nd edn. (Cambridge, Cambridge University Press).

[6] Y. Choquet-Bruhat and C. DeWitt-Morette, with M. Dillard-Bleick (1996). *Analysis, Manifolds, and Physics, Part I: Basics*, revised edn. (Amsterdam, North Holland).

[7] A. Mostafazadeh (1994). "Supersymmetry and the Atiyah–Singer index theorem. I. Peierls brackets, Green's functions, and a proof of the index theorem via Gaussian superdeterminants," *J. Math. Phys.* **35**, 1095–1124.
A. Mostafazadeh (1994). "Supersymmetry and the Atiyah–Singer index theorem. II. The scalar curvature factor in the Schrödinger equation," *J. Math. Phys.* **35**, 1125–1138.

[8] P. Cartier, C. DeWitt-Morette, M. Ihl, and Ch. Sämann, with an appendix by M.E. Bell (2002). "Applications to supersymmetry" (arXiv: math-ph/0202026 v1 19 Feb 2002); also in Michael Marinov Memorial Volume *Multiple Facets of Quantization and Supersymmetry*, eds. M. Olshanetsky and A. Vainshtein (River Edge, NJ, World Scientific, 2002), pp. 412–457.

[9] B. DeWitt (2003). *The Global Approach to Quantum Field Theory* (Oxford, Oxford University Press; with corrections, 2004).

[10] S. Weinberg (1995, 1996, 2000). *The Quantum Theory of Fields*, Vols. I–III (Cambridge, Cambridge University Press).

[11] M.E. Peskin and D.V. Schroeder (1995). *An Introduction to Quantum Field Theory* (Reading, Perseus Books).

[12] A. Connes (1994). *Noncommutative Geometry* (San Diego, CA, Academic Press), pp. 418–427.

Other references that have inspired this chapter

N. Berline, E. Getzler, and M. Vergne (1992). *Heat Kernels and Dirac Operators* (Berlin, Springer).

B. Booss and D. D. Bleecker (1985). *Topology and Analysis, the Atiyah–Singer Index Formula and Gauge-Theoretic Physics* (Berlin, Springer), Part IV.

N. Bourbaki (1969). *Eléments de mathématiques, intégration* (Paris, Hermann), Ch. 9, p. 39. English translation by S. K. Berberian, *Integration II* (Berlin, Springer, 2004).

H. Cartan (1950). "Notions d'algèbre différentielle; application aux groupes de Lie et aux variétés où opère un groupe de Lie," and "La transgression dans un groupe de Lie et dans un espace fibré principal," in *Henri Cartan Œuvres* (Berlin, Springer), Vol. III, pp. 1255–1267 and 1268–1282.

W. Greub, S. Halperin, and R. Vanstone (1973). *Connections, Curvature, and Cohomology*, Vol. II (New York, Academic Press).

D. A. Leites (1980). "Introduction to the theory of supermanifolds," *Russian Mathematical Surveys* **35**, 1–64.

A. Salam and J. Strathdee (1974). "Super-gauge transformations," *Nucl. Phys.* **B76**, 477–482, reprinted in *Supersymmetry*, ed. S. Ferrara (Amsterdam/Singapore, North Holland/World Scientific, 1987).

Seminar on Supermanifolds, known as SoS, over 2000 pages written by D. Leites and his colleagues, students, and collaborators from 1977 to 2000 (expected to be available electronically from arXiv).

11
Volume elements, divergences, gradients

$$\int_{\mathrm{M}^D} \mathcal{L}_X \omega^D = 0 \qquad \mathcal{L}_X \omega = \mathrm{Div}_\omega(X) \cdot \omega$$

11.1 Introduction. Divergences

So far we have constructed the following volume elements.

- Chapter 2. An integral definition of gaussian volume elements on Banach spaces, (2.29) and (2.30).
- Chapter 4. A class of ratios of infinite-dimensional determinants that are equal to finite determinants.
- Chapter 7. A mapping from spaces of pointed paths on $\mathbb{R}^D$ to spaces of pointed paths on riemannian manifolds $\mathbb{N}^D$ that makes it possible to use the results of Chapter 2 in nonlinear spaces.
- Chapter 9. A differential definition of volume elements, in terms of scalar densities, that is useful for integration over finite-dimensional spaces of Grassmann variables (Section 9.4).

In this chapter we exploit the triptych volume elements–divergences–gradients on nonlinear, infinite-dimensional spaces.

Lessons from finite-dimensional spaces

Differential calculus on Banach spaces and differential geometry on Banach manifolds are natural generalizations of their finite-dimensional counterparts. Therefore, we review differential definitions of volume elements on D-dimensional manifolds, which can be generalized to infinite-dimensional spaces.

Top-forms and divergences

Let ω be a D-form on $\mathbb{M}^D$, i.e. a top-form.[1] Let X be a vector field on $\mathbb{M}^D$. Koszul [1] has introduced the following definition of divergence, henceforth abbreviated to "Div," which generalizes "div":

$$\mathcal{L}_X \omega =: \operatorname{Div}_\omega(X) \cdot \omega. \tag{11.1}$$

According to this formula, the divergence of a vector X is the rate of change of a volume element ω under a transformation generated by the vector field X. The rate of change of a volume element is easier to comprehend than the volume element. This situation is reminiscent of other situations. For example, it is easier to comprehend, and measure, energy differences rather than energies. Another example: ratios of infinite-dimensional determinants may have meaning even when each determinant alone is meaningless.

We shall show in (11.20) that this formula applies to the volume element ω_g on a riemannian manifold $(\mathbb{M}, g)$,

$$\omega_g(x) := |\det g_{\alpha\beta}(x)|^{1/2} \, \mathrm{d}x^1 \wedge \ldots \wedge \mathrm{d}x^D, \tag{11.2}$$

and in (11.30) that it applies to the volume element ω_Ω on a symplectic manifold $(\mathbb{M}^{2N}, \Omega)$, where $2N = D$ and the symplectic form Ω is a nondegenerate closed 2-form

$$\omega_\Omega(x) = \frac{1}{N!} \Omega \wedge \ldots \wedge \Omega \qquad (N \text{ factors}). \tag{11.3}$$

In canonical coordinates (p, q), the symplectic form is

$$\Omega = \sum_\alpha \mathrm{d}p_\alpha \wedge \mathrm{d}q^\alpha \tag{11.4}$$

and the volume element is

$$\omega_\Omega(p, q) = \mathrm{d}p_1 \wedge \mathrm{d}q^1 \wedge \ldots \wedge \mathrm{d}p_N \wedge \mathrm{d}q^N. \tag{11.5}$$

It is surprising that ω_g and ω_Ω obey equations with the same structure, namely

$$\mathcal{L}_X \omega = D(X) \cdot \omega, \tag{11.6}$$

---

[1] The set of top-forms on $\mathbb{M}^D$ is an interesting subset of the set of closed forms on $\mathbb{M}^D$. Top-forms satisfy a property not shared with arbitrary closed forms, namely

$$\mathcal{L}_{fX} \omega = \mathcal{L}_X(f\omega), \qquad \text{where } f \text{ is a scalar function on } \mathbb{M}^D.$$

11.1 Introduction. Divergences

because riemannian geometry and symplectic geometry are notoriously different [2]. For instance,

Riemannian geometry	**Symplectic geometry**
line element	surface area
$\int ds$	$\int \Omega$
geodesics	minimal surface areas
$\mathcal{L}_X g = 0 \Rightarrow X$ is Killing.	$\mathcal{L}_X \Omega = 0 \Rightarrow X$ is hamiltonian.
Killings are few.	Hamiltonians are many.

Riemannian manifolds $(\mathbb{M}^D, g)$ [3]

We want to show that the equation (11.6) with

$$D(X) := \frac{1}{2} \mathrm{Tr}(g^{-1}\mathcal{L}_X g) \tag{11.7}$$

characterizes the volume element ω_g up to a multiplicative constant. Indeed, let

$$\omega(x) = \mu(x) \mathrm{d}^D x. \tag{11.8}$$

By the Leibniz rule

$$\mathcal{L}_X(\mu \mathrm{d}^D x) = \mathcal{L}_X(\mu) \mathrm{d}^D x + \mu \mathcal{L}_X(\mathrm{d}^D x). \tag{11.9}$$

Since $\mathrm{d}^D x$ is a top-form on $\mathbb{M}^D$,

$$\mathcal{L}_X(\mathrm{d}^D x) = \mathrm{d}(i_X \mathrm{d}^D x) = \partial_\alpha X^\alpha \, \mathrm{d}^D x = X^\alpha{}_{,\alpha} \, \mathrm{d}^D x. \tag{11.10}$$

Hence

$$\mathcal{L}_X(\mu \mathrm{d}^D x) = \left(X^\alpha \mu_{,\alpha} + \mu X^\alpha{}_{,\alpha}\right) \mu^{-1} \cdot \mu \mathrm{d}^D x. \tag{11.11}$$

On the other hand,

$$(\mathcal{L}_X g)_{\alpha\beta} = X^\gamma g_{\alpha\beta,\gamma} + g_{\gamma\beta} X^\gamma{}_{,\alpha} + g_{\alpha\gamma} X^\gamma{}_{,\beta}, \tag{11.12}$$

that is $(\mathcal{L}_X g)_{\alpha\beta} = X_{\alpha;\beta} + X_{\beta;\alpha}$. Hence

$$D(X) = \frac{1}{2} \mathrm{Tr}(g^{-1}\mathcal{L}_X g) = \frac{1}{2} g^{\beta\alpha} X^\gamma g_{\alpha\beta,\gamma} + X^\alpha{}_{,\alpha}. \tag{11.13}$$

The equation

$$\mathcal{L}_X \omega = D(X) \cdot \omega \tag{11.14}$$

is satisfied if and only if

$$(X^\gamma \mu_{,\gamma} + \mu X^\alpha{}_{,\alpha})\mu^{-1} = \frac{1}{2}g^{\alpha\beta}X^\gamma g_{\alpha\beta,\gamma} + X^\alpha{}_{,\alpha}, \qquad (11.15)$$

i.e. if and only if

$$\partial_\gamma \ln \mu = \frac{\mu_{,\gamma}}{\mu} = \frac{1}{2}g^{\alpha\beta}g_{\alpha\beta,\gamma} = \frac{1}{2}\partial_\gamma \ln|\det g|, \qquad (11.16)$$

that is

$$\mu(x) = C|\det g(x)|^{1/2}, \qquad (11.17)$$

where C is a constant. The quantity $D(X) = \frac{1}{2}\operatorname{Tr}(g^{-1}\mathcal{L}_X g)$ is the covariant divergence

$$\operatorname{Div}_g(X) := X^\alpha{}_{;\alpha} := X^\alpha{}_{,\alpha} + \Gamma^\beta{}_{\beta\alpha}X^\alpha \qquad (11.18)$$

$$= X^\alpha{}_{,\alpha} + \frac{1}{2}g^{\beta\gamma}g_{\gamma\beta,\alpha}X^\alpha$$

$$= \frac{1}{2}\operatorname{Tr}(g^{-1}\mathcal{L}_X g). \qquad (11.19)$$

In conclusion,

$$\mathcal{L}_X \omega_g = X^\alpha{}_{;\alpha}\omega_g = \operatorname{Div}_g X \cdot \omega_g. \qquad (11.20)$$

Remark. If X is a Killing vector field with respect to isometries, then $\mathcal{L}_X g = 0$, $\mathcal{L}_X \omega_g = 0$, $X_{\alpha;\beta} + X_{\beta;\alpha} = 0$, $X^\alpha{}_{;\alpha} = 0$, and equation (11.20) is trivially satisfied.

Remark. On $\mathbb{R}^D$ the gaussian volume element $d\Gamma_Q$ has the same structure as ω_g:

$$d\Gamma_Q(x) := |\det Q|^{1/2}\exp(-\pi Q(x))dx^1 \wedge \ldots \wedge dx^D. \qquad (11.21)$$

In the infinite-dimensional version (2.19) of (11.21), we have regrouped the terms in order to introduce $\mathcal{D}_{s,Q}(x)$, a dimensionless translation-invariant volume element on a Banach space,

$$d\Gamma_{s,Q}(x) \stackrel{f}{=} \mathcal{D}_{s,Q}(x)\exp\left(-\frac{\pi}{s}Q(x)\right). \qquad (11.22)$$

Remark. For historical reasons, different notation is used for volume elements. For instance, in the above remark we use different notations when we say "$d\Gamma$ has the same structure as ω." Why not use "$d\omega$"? Integrals were introduced with the notation $\int f(x)dx$. Much later, $f(x)dx$ was identified as a differential one-form, say ω, $\int f(x)dx = \int \omega$. The symbol $d\omega$ is used for the differential of ω, i.e. for a two-form.

11.1 Introduction. Divergences

Symplectic manifolds $(\mathbb{M}^D, \Omega)$, $D = 2N$

We shall show that the symplectic volume element ω_Ω satisfies the structural equation

$$\mathcal{L}_X \omega = D(X) \cdot \omega \tag{11.23}$$

with $D(X) = \text{Div}_\Omega(X)$ defined by (11.29) if and only if[2]

$$\omega_\Omega = \frac{1}{N!} \Omega^{\wedge N} \tag{11.24}$$
$$= |\det \Omega_{\alpha\beta}|^{1/2} \, d^D x =: \text{Pf}(\Omega_{\alpha\beta}) d^D x$$

up to a multiplicative constant. Pf is a pfaffian. We define Ω^{-1} and calculate $\text{Tr}(\Omega^{-1} \mathcal{L}_X \Omega)$.

- The symplectic form Ω defines an isomorphism from the tangent bundle TM to the cotangent bundle T^*M by

$$\Omega : X \mapsto i_X \Omega. \tag{11.25}$$

We can then define

$$X_\alpha := X^\beta \Omega_{\beta\alpha}.$$

The inverse $\Omega^{-1} : T^*M \to TM$ of Ω is given by

$$X^\alpha = X_\beta \Omega^{\beta\alpha},$$

where

$$\Omega^{\alpha\beta} \Omega_{\beta\gamma} = \delta^\alpha_\gamma. \tag{11.26}$$

Note that in strict components, i.e. with $\Omega = \Omega_{AB}\, dx^A \wedge dx^B$ for $A < B$, X_A is not equal to $X^B \Omega_{BA}$.
- We compute

$$(\mathcal{L}_X \Omega)_{\alpha\beta} = X^\gamma \Omega_{\alpha\beta,\gamma} + \Omega_{\gamma\beta} X^\gamma{}_{,\alpha} + \Omega_{\alpha\gamma} X^\gamma{}_{,\beta}$$
$$= X_{\beta,\alpha} - X_{\alpha,\beta} \tag{11.27}$$

using $d\Omega = 0$, also written as $\Omega_{\beta\gamma,\alpha} + \Omega_{\gamma\alpha,\beta} + \Omega_{\alpha\beta,\gamma} = 0$. Hence

$$(\Omega^{-1} \mathcal{L}_X \Omega)^\gamma_\beta = \Omega^{\gamma\alpha}(X_{\beta,\alpha} - X_{\alpha,\beta}) \tag{11.28}$$

and

$$\frac{1}{2} \text{Tr}(\Omega^{-1} \mathcal{L}_X \Omega) = \Omega^{\gamma\alpha} X_{\gamma,\alpha}$$
$$=: \text{Div}_\Omega(X). \tag{11.29}$$

---

[2] The symplectic form Ω is given in coordinates as $\Omega = \frac{1}{2} \Omega_{\alpha\beta}\, dx^\alpha \wedge dx^\beta$ with $\Omega_{\alpha\beta} = -\Omega_{\beta\alpha}$.

- We compute $\mathcal{L}_X \omega_\Omega$. According to Darboux' theorem, there is a coordinate system (x^α) in which the volume form $\omega_\Omega = (1/N!)\Omega^{\wedge N}$ is

$$\omega_\Omega = dx^1 \wedge \ldots \wedge dx^{2N}$$

and $\Omega = \Omega_{\alpha\beta}\, dx^\alpha \otimes dx^\beta$ with *constant* coefficients $\Omega_{\alpha\beta}$. The inverse matrix $\Omega^{\beta\alpha}$ of $\Omega_{\alpha\beta}$ is also made up of constants, hence $\Omega^{\beta\alpha}{}_{,\gamma} = 0$. In these coordinates

$$\begin{aligned}
\mathcal{L}_X \omega_\Omega &= X^\alpha{}_{,\alpha} \omega_\Omega \\
&= (X_\beta \Omega^{\beta\alpha})_{,\alpha} \omega_\Omega \\
&= (X_{\beta,\alpha} \Omega^{\beta\alpha} + X_\beta \Omega^{\beta\alpha}{}_{,\alpha}) \omega_\Omega \\
&= X_{\beta,\alpha} \Omega^{\beta\alpha} \omega_\Omega \\
&= \mathrm{Div}_\Omega(X) \cdot \omega_\Omega,
\end{aligned} \qquad (11.30)$$

and equation (11.23) is satisfied.
- The uniqueness is proved as in the riemannian case (see equations (11.15)–(11.17)).

Remark. If X is a hamiltonian vector field, then $\mathcal{L}_X \Omega = 0$, $\mathcal{L}_X \omega_\Omega = 0$, and $\mathrm{Div}_\Omega(X) = 0$. The basic equation (11.23) is trivially satisfied.

Supervector spaces $\mathbb{R}^{n|\nu}$ [4]

Let x be a point in the supervector space (Section 9.2) $\mathbb{R}^{n|\nu}$ with coordinates

$$x^A = (x^a, \xi^\alpha) \qquad \begin{cases} a \in \{1, \ldots, n\}, \\ \alpha \in \{1, \ldots, \nu\}, \end{cases} \qquad (11.31)$$

where x^a is a bosonic variable, and ξ^α is a fermionic variable. Let X be a vector field on $\mathbb{R}^{n|\nu}$,

$$X = X^A\, \partial/\partial x^A,$$

and let ω be a top-form. The divergence $\mathrm{Div}(X)$ defined, up to an invertible "volume density" f, by the Koszul formula

$$\mathcal{L}_X \omega = \mathrm{Div}(X) \cdot \omega \qquad (11.32)$$

is [4]

$$\mathrm{Div}(X) = \frac{1}{f}(-1)^{\tilde{A}(1+\tilde{X})} \frac{\partial}{\partial X^A}(X^A f), \qquad (11.33)$$

where $\tilde{A}$ and $\tilde{X}$ are the parities of A and X, respectively.

The general case $\mathcal{L}_X \omega = D(X) \cdot \omega$

Two properties of $D(X)$ dictated by properties of $\mathcal{L}_X \omega$ follow:

(i)
$$\mathcal{L}_{[X,Y]} = \mathcal{L}_X \mathcal{L}_Y - \mathcal{L}_Y \mathcal{L}_X \Leftrightarrow D([X,Y]) = \mathcal{L}_X D(Y) - \mathcal{L}_Y D(X); \tag{11.34}$$

(ii) since ω is a top-form,
$$\mathcal{L}_X \omega = d i_X \omega,$$
and
$$\mathcal{L}_X(f\omega) = d i_X(f\omega) = d i_{fX}\omega = \mathcal{L}_{fX}\omega.$$

Therefore, when acting on a top-form (Project 19.5.1),
$$\mathcal{L}_{fX} = f\mathcal{L}_X + \mathcal{L}_X(f) \Leftrightarrow D(fX) = fD(X) + \mathcal{L}_X(f). \tag{11.35}$$

11.2 Comparing volume elements

Volume elements and determinants of quadratic forms are offsprings of integration theory. Their values are somewhat elusive because they depend on the choice of coordinates. For instance, consider a quadratic form Q on some finite-dimensional space with coordinates $x^1, \ldots, x^D$, namely

$$Q(x) = h_{ab} x^a x^b. \tag{11.36}$$

In another system of coordinates defined by

$$x^a = u_\ell^a \bar{x}^\ell, \tag{11.37}$$

the quadratic form

$$Q(x) = \bar{h}_{\ell m} \bar{x}^\ell \bar{x}^m \tag{11.38}$$

introduces a new kernel,

$$\bar{h}_{\ell m} = u_\ell^a u_m^b h_{ab}. \tag{11.39}$$

There is no such thing as "the determinant of a quadratic form" because it scales with a change of coordinates:

$$\det \bar{h}_{\ell m} = \det h_{ab} \cdot \left(\det u_\ell^a\right)^2. \tag{11.40}$$

Ratios of the determinants of these forms, on the other hand, have an intrinsic meaning. Consider two quadratic forms Q_0 and Q_1,

$$Q_0 = h_{ab}^{(0)} x^a x^b, \qquad Q_1 = h_{ab}^{(1)} x^a x^b. \tag{11.41}$$

Denote by $\det(Q_1/Q_0)$ the ratio of the determinants of their kernels,

$$\det(Q_1/Q_0) := \det\bigl(h_{ab}^{(1)}\bigr)/\det\bigl(h_{ab}^{(0)}\bigr). \tag{11.42}$$

This ratio is invariant under a change of coordinates. Therefore, one expects that ratios of infinite-dimensional determinants can be defined using projective systems [5] or similar techniques.

Ratios of infinite-dimensional determinants

Consider two continuous quadratic forms Q_0 and Q_1 on a Banach space $\mathbb{X}$. Assume Q_0 to be invertible in the following sense. Let D_0 be a continuous, linear map from $\mathbb{X}$ into its dual $\mathbb{X}'$ such that

$$Q_0(x) = \langle D_0 x, x \rangle, \qquad \langle D_0 x, y \rangle = \langle D_0 y, x \rangle. \tag{11.43}$$

The form Q_0 is said to be invertible if the map D_0 is invertible, i.e. if there exists a unique[3] inverse G of D_0

$$G \circ D_0 = \mathbf{1}. \tag{11.44}$$

Let D_1 be defined similarly, but without the requirement of invertibility. There exists a unique continuous operator U on $\mathbb{X}$ such that

$$D_1 = D_0 \circ U, \tag{11.45}$$

that is $U = G \circ D_1$. If $U - \mathbf{1}$ is nuclear (see equations (11.47)–(11.63)), then the determinant of U is defined [6]. Let us denote the determinant of U as $\mathrm{Det}(Q_1/Q_0)$. It can be calculated as follows. Let V be a finite-dimensional subspace of $\mathbb{X}$, and let $Q_{0,V}$ and $Q_{1,V}$ be the restrictions of Q_0 and Q_1 to V. Assume that V runs through an increasing sequence of subspaces, whose union is dense in $\mathbb{X}$, and that $Q_{0,V}$ is invertible for every V. Then

$$\mathrm{Det}(Q_1/Q_0) = \lim_V \det(Q_{1,V}/Q_{0,V}). \tag{11.46}$$

The fundamental trace–determinant relation

The fundamental relation between the trace and the determinant of a matrix A in $\mathbb{R}^D$ is

$$\mathrm{d}\ln\det A = \mathrm{tr}(A^{-1}\,\mathrm{d}A), \tag{11.47}$$

which is also written

$$\det\exp A = \exp\mathrm{tr}\,A. \tag{11.48}$$

---

[3] The inverse G of D_0 is uniquely determined either by restricting $\mathbb{X}$ or by choosing W in (2.28) and (2.29).

11.2 Comparing volume elements

Indeed, the trace and the determinant of A are invariant under similarity transformations; the matrix A can be made triangular by a similarity transformation, and the above formula is easy to prove for triangular matrices [7, Part I, p. 174].

The fundamental relation (11.47) is valid for operators on nuclear spaces [6, 8].

Let $\mathbb{X}$ be a Banach space, and let Q be an invertible[4] positive-definite quadratic form on $\mathbb{X}$. The quadratic form $Q(x)$ defines a norm on $\mathbb{X}$, namely

$$\|x\|^2 = Q(x), \tag{11.49}$$

and a dual norm $\|x'\|$ on $\mathbb{X}'$, as usual.

According to Grothendieck, an operator T on $\mathbb{X}$ is *nuclear* if it admits a representation of the form

$$Tx = \sum_{n \geq 0} \langle x'_n, x \rangle x_n \tag{11.50}$$

with elements x_n in $\mathbb{X}$ and x'_n in $\mathbb{X}'$ such that $\sum_{n \geq 0} \|x_n\| \cdot \|x'_n\|$ is finite. The greatest lower bound of all such sums $\sum_n \|x_n\| \cdot \|x'_n\|$ is called the *nuclear norm* of T, denoted by $\|T\|_1$. The nuclear operators on $\mathbb{X}$ form a Banach space, denoted by $\mathcal{L}^1(\mathbb{X})$, with norm $\|\cdot\|_1$. On $\mathcal{L}^1(\mathbb{X})$, there exists a continuous linear form, the *trace*, such that

$$\mathrm{Tr}(T) = \sum_{n \geq 0} \langle x'_n, x_n \rangle \tag{11.51}$$

for an operator T given by (11.50).

We now introduce a power series in λ, namely

$$\sum_{p \geq 0} \sigma_p(T) \lambda^p := \exp\left(\lambda \,\mathrm{Tr}(T) - \frac{\lambda^2}{2} \mathrm{Tr}(T^2) + \frac{\lambda^3}{3} \mathrm{Tr}(T^3) - \cdots \right). \tag{11.52}$$

From Hadamard's inequality of determinants, we obtain the *basic estimate*

$$|\sigma_p(T)| \leq p^{p/2} \|T\|_1^p / p!. \tag{11.53}$$

---

[4] A positive-definite continuous quadratic form is not necessarily invertible. For instance, let $\mathbb{X}$ be the space ℓ^2 of sequences $(x_1, x_2, \ldots)$ of real numbers with $\sum_{n=1}^{\infty} (x_n)^2 < \infty$ and define the norm by $\|x\|^2 = \sum_{n=1}^{\infty} (x_n)^2$. We can identify $\mathbb{X}$ with its dual $\mathbb{X}'$, where the scalar product is given by $\sum_{n=1}^{\infty} x'_n x_n$. The quadratic form $Q(x) = \sum_{n=1}^{\infty} x_n^2/n$ corresponds to the map $D : \mathbb{X} \to \mathbb{X}'$ that takes $(x_1, x_2, \ldots)$ into $(x_1/1, x_2/2, \ldots)$. The inverse of D does not exist as a map from ℓ^2 into ℓ^2 since the sequence $1, 2, 3, \ldots$ is unbounded.

It follows that the power series $\sum_{p\geq 0} \sigma_p(T)\lambda^p$ has an infinite radius of convergence. We can therefore define the determinant as

$$\mathrm{Det}(1+T) := \sum_{p\geq 0} \sigma_p(T) \tag{11.54}$$

for any nuclear operator T. Given the definition (11.52) of σ_p, we obtain the more general definition

$$\mathrm{Det}(1+\lambda T) = \sum_{p\geq 0} \sigma_p(T)\lambda^p. \tag{11.55}$$

The fundamental property of determinants is, as expected, the *multiplicative rule*

$$\mathrm{Det}(U_1 \circ U_2) = \mathrm{Det}(U_1)\mathrm{Det}(U_2), \tag{11.56}$$

where U_i is of the form $1 + T_i$, with T_i nuclear (for $i = 1, 2$). From equation (11.56) and the relation $\sigma_1(T) = \mathrm{Tr}(T)$, we find a *variation formula* (for nuclear operators $U - 1$ and δU)

$$\frac{\mathrm{Det}(U + \delta U)}{\mathrm{Det}(U)} = 1 + \mathrm{Tr}(U^{-1} \cdot \delta U) + \mathcal{O}(\|\delta U\|_1^2). \tag{11.57}$$

In other words, if $U(\nu)$ is an operator of the form $1 + T(\nu)$, where $T(\nu)$ is nuclear and depends smoothly on the parameter ν in the Banach space $\mathcal{L}^1(\mathbb{X})$, we obtain the *derivation formula*

$$\frac{\mathrm{d}}{\mathrm{d}\nu} \ln \mathrm{Det}(U(\nu)) = \mathrm{Tr}\left(U(\nu)^{-1} \frac{\mathrm{d}}{\mathrm{d}\nu} U(\nu)\right). \tag{11.58}$$

Remark. For any other norm $\|\cdot\|^1$ defining the topology of $\mathbb{X}$, we have an estimate

$$C^{-1}\|x\| \leq \|x\|^1 \leq C\|x\|, \tag{11.59}$$

for a finite numerical constant $C > 0$. It follows easily from (11.59) that the previous definitions are independent of the choice of the particular norm $\|x\| = Q(x)^{1/2}$ in $\mathbb{X}$.

Explicit formulas

Introduce a basis $(e_n)_{n\geq 1}$ of $\mathbb{X}$ that is orthonormal for the quadratic form Q. Therefore $Q(\sum_n t_n e_n) = \sum_n t_n^2$. An operator T in $\mathbb{X}$ has a matrix representation (t_{mn}) such that

$$Te_n = \sum_m e_m \cdot t_{mn}. \tag{11.60}$$

Assume that T is nuclear. Then the series $\sum_n t_{nn}$ of diagonal terms in the matrix converges absolutely, and the trace $\mathrm{Tr}(T)$ is equal to $\sum_n t_{nn}$,

11.2 Comparing volume elements

as expected. Furthermore, $\sigma_p(T)$ is the sum of the series consisting of the principal minors of order p,

$$\sigma_p(T) = \sum_{i_1 < \cdots < i_p} \det(t_{i_\alpha, i_\beta})_{1 \leq \alpha \leq p, 1 \leq \beta \leq p}. \tag{11.61}$$

The determinant of the operator $U = 1 + T$, whose matrix has elements $u_{mn} = \delta_{mn} + t_{mn}$, is a *limit of finite-size determinants*:

$$\text{Det}(U) = \lim_{N=\infty} \det(u_{mn})_{1 \leq m \leq N, 1 \leq n \leq N}. \tag{11.62}$$

As a special case, suppose that the basic vectors e_n are *eigenvectors* of T,

$$T e_n = \lambda_n e_n. \tag{11.63}$$

Then

$$\text{Tr}(T) = \sum_n \lambda_n \quad \text{and} \quad \text{Det}(1+T) = \prod_n (1 + \lambda_n),$$

where both the series and the infinite product converge absolutely.

The nuclear norm $\|T\|_1$ can also be computed as follows: there exists an orthonormal basis (e_n) such that the vectors $T e_n$ are mutually orthogonal (for the quadratic form Q). Then $\|T\|_1 = \sum_n \|T e_n\|$.

Remark. Let T be a continuous linear operator in $\mathbb{X}$. Assume that the series of diagonal terms $\sum_n t_{nn}$ converges absolutely for *every* orthonormal basis. Then T is nuclear. When T is symmetric and positive, it suffices to assume that this statement holds for *one* given orthonormal basis. Then it holds for all. Counterexamples exist for the case in which T is not symmetric and positive [6].

Comparing divergences

A divergence is a trace. For example[5]

$$\text{Div}_A(X) = \frac{1}{2} \text{tr}(A^{-1} \mathcal{L}_X A) \tag{11.64}$$

regardless of whether A is the metric tensor g or the symplectic form Ω (in both cases, an invertible bilinear form on the tangent vectors). It follows

---

[5] Notice the analogy between the formulas (11.2) and (11.24) for the volume elements,

$$\omega_g = |\det g_{\mu\nu}|^{1/2} \, dx^1 \wedge \ldots \wedge dx^D,$$
$$\omega_\Omega = |\det \Omega_{\alpha\beta}|^{1/2} \, dx^1 \wedge \ldots \wedge dx^D,$$

for the metric $g = g_{\mu\nu} \, dx^\mu \, dx^\nu$ and the symplectic form $\Omega = \frac{1}{2} \Omega_{\alpha\beta} \, dx^\alpha \wedge dx^\beta = \Omega_{\alpha\beta} \, dx^\alpha \otimes dx^\beta$.

from the fundamental relation between trace and determinant,
$$d \ln \det A = \operatorname{tr}(A^{-1} dA), \tag{11.65}$$
that
$$\operatorname{Div}_A(X) - \operatorname{Div}_B(X) = \mathcal{L}_X \ln(\det(A/B)^{1/2}). \tag{11.66}$$
Given Koszul's definition of divergence (11.1) in terms of volume elements, namely
$$\mathcal{L}_X \omega = \operatorname{Div}_\omega(X) \cdot \omega, \tag{11.67}$$
equation (11.66) gives ratios of volume elements in terms of ratios of determinants.

Example. Let Φ be a manifold equipped with two riemannian metrics g_A and g_B. Let
$$P : \Phi \to \Phi \tag{11.68}$$
be a map transforming g_A into g_B. Let ω_A and ω_B be the corresponding volume elements on Φ such that, by P,
$$\omega_A \mapsto \omega_B = \rho \omega_A. \tag{11.69}$$
Then
$$\operatorname{Div}_{\omega_B}(X) - \operatorname{Div}_{\omega_A}(X) = \mathcal{L}_X \ln \rho \tag{11.70}$$
and
$$\omega_B / \omega_A = \det(g_B/g_A)^{1/2}. \tag{11.71}$$
$\rho(x)$ is the determinant of the jacobian matrix of the map P. The proof is given in Appendix F, equations (F.23)–(F.25). It can also be done by an explicit calculation of the change of coordinates defined by P.

11.3 Integration by parts

Integration theory is unthinkable without integration by parts, not only because it is a useful technique but also because it has its roots in the fundamental requirements of definite integrals, namely (see Section 9.3)
$$DI = 0, \qquad ID = 0, \tag{11.72}$$
where D is a derivative operator and I an integral operator. We refer the reader to Chapter 9 for the meaning and some of the uses of the fundamental requirements (11.72).

Already in the early years of path integrals Feynman was promoting integration by parts in functional integration, jokingly and seriously. Here

11.3 Integration by parts

we use integration by parts for relating divergences and gradients; this relation completes the triptych "volume elements–divergences–gradients."

Divergences and gradients

In $\mathbb{R}^3$ the concept of a "gradient" is intuitive and easy to define: it measures the steepness of a climb. Mathematically, the gradient (or nabla, $\nabla \equiv \nabla_{g^{-1}}$) is a *contravariant* vector:

$$\nabla^i := g^{ij} \frac{\partial}{\partial x^j}. \tag{11.73}$$

The divergence $\operatorname{Div} V$ of a vector field V is its scalar product with the gradient vector:

$$(\nabla | V)_g = g_{ij} \nabla^i V^j = g_{ij} g^{ik} \frac{\partial}{\partial x^k} V^j = V^j{}_{,j}. \tag{11.74}$$

Divergence and gradient are related by integration by parts. Indeed,

$$(V|\nabla f)_g(x) = g_{ij} V^i g^{jk} \partial f/\partial x^k = V^i f_{,i}(x)$$
$$= \mathcal{L}_V f(x), \tag{11.75}$$
$$(\operatorname{div} V | f)(x) = V^i{}_{,i} f(x), \tag{11.76}$$

and

$$\int d^3x \big((V|\nabla f)_g(x) + (\operatorname{div} V|f)(x) \big) = \int d^3x \, \frac{\partial}{\partial x^k}(fV^k). \tag{11.77}$$

Assume that fV vanishes on the boundary of the domain of integration, and integrate the right-hand side by parts. The right-hand side vanishes because the volume element d^3x is invariant under translation, hence

$$\int_{\mathbb{R}^3} d^3x (V|\nabla f)(x) = - \int_{\mathbb{R}^3} d^3x (\operatorname{div} V|f)(x); \tag{11.78}$$

the *functional scalar products* satisfy modulo a sign the adjoint relation

$$(V|\nabla f) = -(\operatorname{div} V|f). \tag{11.79}$$

The generalizations of (11.78) and (11.74) to spaces other than $\mathbb{R}^3$ face two difficulties:

- in contrast to d^3x, generic volume elements are not invariant under translation; and
- traces in infinite-dimensional spaces are notoriously sources of problems. Two examples follow.

The infinite-dimensional matrix

$$M = \operatorname{diag}(1, 1/2, 1/3, \ldots)$$

has the good properties of a Hilbert–Schmidt operator (the sum of the squares of its elements is finite), but its trace is infinite.

Closed loops in quantum field theory introduce traces; they have to be set apart by one technique or another. For instance Wick products (see Appendix D) serve this purpose among others.

The definition of divergence provided by the Koszul formula (11.1),

$$\mathcal{L}_V \omega =: \text{Div}_\omega(V) \cdot \omega, \tag{11.80}$$

bypasses both difficulties mentioned above: the definition (11.80) is not restricted to translation-invariant volume elements and it is meaningful in infinite-dimensional spaces.

The definition (11.73) of the gradient vector as a contravariant vector requires the existence of a metric tensor g. Let $\mathcal{A}^p$ be the space of p-forms on $\mathbb{M}^D$ and $\mathcal{X}$ the space of contravariant vector fields on $\mathbb{M}^D$:

$$\mathcal{A}^0 \xrightarrow{d} \mathcal{A}^1 \underset{g}{\overset{g^{-1}}{\rightleftarrows}} \mathcal{X}, \tag{11.81}$$

$$\nabla_{g^{-1}} = g^{-1} \circ d. \tag{11.82}$$

On the other hand the scalar product (11.75),

$$(V | \nabla_{g^{-1}} f)_g = \mathcal{L}_V f,$$

is simply the Lie derivative with respect to V of the scalar function f, hence it is independent of the metric and easy to generalize to scalar functionals.

From the definitions (11.80) and (11.82), one sees that

- the volume element defines the divergence and
- the metric tensor defines the gradient because it provides a canonical isomorphism between covariant and contravariant vectors.

If it happens that the volume element ω_g is defined by the metric tensor g, then one uses the explicit formulas (11.2) and (11.18).

With the Koszul definition (11.80) and the property (11.75), one can derive the grad–div relationship as follows:

$$\int_{\mathbb{M}^D} (V | \nabla_{g^{-1}} f)_g(x) \cdot \omega = \int_{\mathbb{M}^D} \mathcal{L}_V f(x) \cdot \omega \quad \text{by the property (11.75)} \tag{11.83}$$

$$= -\int_{\mathbb{M}^D} f(x) \cdot \mathcal{L}_V \omega \quad \text{by integration by parts}$$

$$= -\int_{\mathbb{M}^D} f(x) \text{Div}_\omega(V) \cdot \omega \tag{11.84}$$

by virtue of the Koszul formula, and hence, finally,
$$(V|\nabla f)_\omega = -(\text{Div}_\omega V|f)_\omega. \tag{11.85}$$

Divergence and gradient in function spaces

The basic ingredients in constructing the grad–div relationship are Lie derivatives, the Koszul formula, scalar products of functions and scalar products of contravariant vectors. They can be generalized in function spaces $\mathbb{X}$, as follows.

- Lie derivatives. As usual on function spaces, one introduces a one-parameter family of paths $\{x_\lambda\}_\lambda$, $\lambda \in [0,1]$:
$$x_\lambda : \mathbb{T} \longrightarrow \mathbb{M}^D, \qquad x_\lambda(t) \equiv x(\lambda, t), \tag{11.86}$$
$$\dot{x}(\lambda, t) := \frac{d}{dt} x(\lambda, t), \tag{11.87}$$
$$x'(\lambda, t) := \frac{d}{d\lambda} x(\lambda, t). \tag{11.88}$$

For $\lambda = 0$, x_0 is abbreviated to x and
$$x'(\lambda, t)\big|_{\lambda=0} =: V_x(t); \tag{11.89}$$

V_x is a vector at[6] $x \in \mathbb{X}$, tangent to the one-parameter family $\{x_\lambda\}$.

Let F be a scalar functional on the function space $\mathbb{X}$, then
$$\mathcal{L}_V F(x) = \frac{d}{d\lambda} F(x_\lambda)\bigg|_{\lambda=0}, \qquad x \in \mathbb{X}. \tag{11.90}$$

- The Koszul formula (11.1) defines the divergence of a vector field V as the rate of change of a volume element ω under the group of transformations generated by the vector field V:
$$\mathcal{L}_V \omega =: \text{Div}_\omega(V) \cdot \omega.$$

We adopt the Koszul formula as the definition of divergence in function space.

- Scalar products of real-valued functionals:
$$(F_1|F_2)_\omega = \int_\mathbb{X} \omega F_1 F_2. \tag{11.91}$$

- Scalar products of contravariant vectors. In the finite-dimensional case, such a scalar product requires the existence of a metric tensor defining a canonical isomorphism between the dual spaces $\mathbb{R}^D$ and $\mathbb{R}_D$. In the infinite-dimensional case, a gaussian volume element on $\mathbb{X}$,
$$d\Gamma_Q(x) \stackrel{f}{=} \mathcal{D}_Q(x) \cdot \exp\left(-\frac{\pi}{s} Q(x)\right) =: \omega_Q \tag{11.92}$$

---

[6] More precisely, for each epoch t in $\mathbb{T}$, $V_x(t)$ is in the space $T_{x(t)} \mathbb{M}^D$.

defined by (2.30) and (2.30)$_s$, does provide a canonical isomorphism between $\mathbb{X}$ and its dual $\mathbb{X}'$, namely the pair (D, G) defined by Q and W, respectively,

$$Q(x) = \langle Dx, x \rangle \quad \text{and} \quad W(x') = \langle x', Gx' \rangle, \tag{11.93}$$

$$\mathbb{X}' \underset{D}{\overset{G}{\rightleftarrows}} \mathbb{X}, \quad GD = 1, \quad DG = 1. \tag{11.94}$$

The role of the pair (G, D) as defining a canonical isomorphism between $\mathbb{X}$ and $\mathbb{X}'$ is interesting but is not necessary for generalizing the grad–div relation (11.85): indeed, the scalar product of a contravariant vector V with the gradient of a scalar function F does not depend on the metric tensor (11.75); it is simply the Lie derivative of F in the V-direction.

- The functional grad–div relation. At $x \in \mathbb{X}$,

$$(V|\nabla F)(x) = \mathcal{L}_V F(x), \quad x \in \mathbb{X}; \tag{11.95}$$

upon integration on $\mathbb{X}$ with respect to the gaussian (11.92)

$$\int_{\mathbb{X}} \omega_Q \mathcal{L}_V F(x) = - \int_{\mathbb{X}} \mathcal{L}_V \omega_Q \cdot F(x) \tag{11.96}$$

$$= - \int_{\mathbb{X}} \text{Div}_{\omega_Q}(V) \cdot \omega_Q F(x).$$

The functional grad–div relation is the global scalar product

$$(V|\nabla F)_{\omega_Q} = -(\text{Div}_{\omega_Q}(V)|F)_{\omega_Q}. \tag{11.97}$$

Translation-invariant symbols

The symbol "dx" for $x \in \mathbb{R}$ is translation-invariant:

$$\mathrm{d}(x + a) = \mathrm{d}x \quad \text{for } a, \text{ a fixed point in } \mathbb{R}. \tag{11.98}$$

Equivalently, one can characterize the translation invariance of dx by the integral

$$\int_{\mathbb{R}} \mathrm{d}x \, \frac{\mathrm{d}}{\mathrm{d}x} f(x) = 0, \tag{11.99}$$

provided that f is a function vanishing on the boundary of the domain of integration. In order to generalize (11.99) on $\mathbb{R}^D$, introduce a vector field V in $\mathbb{R}^D$; equation (11.99) becomes

$$\int \mathrm{d}^D x \, \partial_\alpha V^\alpha(x) = 0, \quad \partial_\alpha = \partial/\partial x^\alpha,$$

and

$$\int \mathrm{d}^D x \, \partial_\alpha (fV^\alpha)(x) = \int \mathrm{d}^D x (\partial_\alpha f \cdot V^\alpha + f \partial_\alpha V^\alpha) = 0.$$

11.3 Integration by parts

Hence

$$\int d^D x (\mathcal{L}_V f + f \operatorname{div} V) = 0.$$

This calculation reproduces (11.78) in a more familiar notation.

Let $\mathcal{F}$ be a class of functionals F on the space $\mathbb{X}$ of functions x. We shall formally generalize the characterization of translation-invariant symbols $\mathcal{D}$ on linear spaces $\mathbb{X}$, namely

$$\mathcal{D}(x + x_0) = \mathcal{D}x \qquad \text{for } x_0 \text{ a fixed function,} \qquad (11.100)$$

by imposing the following requirement on $\mathcal{D}x$:

$$\int_\mathbb{X} \mathcal{D}x \, \frac{\delta F}{\delta x(t)} = 0. \qquad (11.101)$$

The characterization (11.101) is meaningful only for F in a class $\mathcal{F}$ such that

$$\int_\mathbb{X} \frac{\delta}{\delta x(t)} (\mathcal{D}x \, F) = 0. \qquad (11.102)$$

Although work remains to be done for an operational definition of $\mathcal{F}$, we note that it is coherent with the fundamental requirement (11.72), $ID = 0$. We shall assume that F satisfies (11.102) and exploit the triptych

gradient–divergence–volume element

linked by the grad–div relation (11.97) and the Koszul equation (11.1).

The Koszul formula applied to the translation-invariant symbol $\mathcal{D}x$ is a straightforward generalization of (11.10). Namely

$$\mathcal{L}_V d^D x = d(i_V d^D x) = V^i{}_{,i} d^D x = \operatorname{div}(V) d^D x \qquad (11.103)$$

generalizes to

$$\mathcal{L}_V \mathcal{D}x = \int_\mathbb{T} dt \, \frac{\delta V(x,t)}{\delta x(t)} \mathcal{D}x =: \operatorname{div}(V) \mathcal{D}x, \qquad (11.104)$$

where the Lie derivative $\mathcal{L}_V$ is defined by (11.90), i.e. by a one-parameter family of paths $\{x_\lambda\}$. Recall that deriving the coordinate expression of $\operatorname{div}(V)$ by computing $\mathcal{L}_V d^D x$ is not a trivial exercise on D-forms. We take for granted its naive generalization

$$\operatorname{div}(V) = \int_\mathbb{T} dt \, \frac{\delta V(x,t)}{\delta x(t)} \qquad (11.105)$$

and bypass the elusive concept of a "top-form on $\mathbb{X}$."

Example. The divergence $\text{Div}_Q(V)$ of the gaussian volume element,

$$\omega_Q(x) := \mathrm{d}\Gamma_Q(x) := \exp\left(-\frac{\pi}{s}Q(x)\right)\mathcal{D}_Q x, \qquad (11.106)$$

is given by the Koszul formula

$$\mathcal{L}_V \omega_Q = \mathcal{L}_V\!\left(\exp\!\left(-\frac{\pi}{s}Q(x)\right)\right)\mathcal{D}_Q x + \exp\!\left(-\frac{\pi}{s}Q(x)\right)\mathcal{L}_V \mathcal{D}_Q x$$
$$=: \text{Div}_Q(V)\omega_Q.$$

The gaussian divergence is the sum of the "naive" divergence (11.105) and $-(\pi/s)\mathcal{L}_V Q(x)$, that is, with the notation (11.93),

$$\text{Div}_Q(V) = -\frac{2\pi}{s}\langle Dx, V(x)\rangle + \text{div}(V). \qquad (11.107)$$

We thereby recover a well-known result of Malliavin calculus [9].

Remark. Two translation-invariant symbols are frequently used: the dimensionless volume elements $\mathcal{D}_Q x$ defined by (2.30) and the formal product

$$\mathrm{d}[x] = \prod_t \mathrm{d}x(t). \qquad (11.108)$$

In D dimensions,

$$\mathcal{D}_Q(x) = |\det Q|^{1/2}\, \mathrm{d}x^1 \ldots \mathrm{d}x^D; \qquad (11.109)$$

in function spaces, we write symbolically

$$\mathcal{D}_Q(x) = |\text{Det}\, Q|^{1/2}\, \mathrm{d}[x]. \qquad (11.110)$$

Application: the Schwinger variational principle

In the equation (11.101) which defines the translation-invariant symbol $\mathcal{D}x$, replace $F(x)$ by

$$F(x)\mu(x)\exp\left(\frac{\mathrm{i}}{\hbar}S_{\text{cl}}(x)\right). \qquad (11.111)$$

Then

$$\int_{\mathbb{X}} \mathcal{D}x \left(\frac{\delta F}{\delta x(t)} + F\,\frac{\delta \ln \mu(x)}{\delta x(t)} + \frac{\mathrm{i}}{\hbar}F\,\frac{\delta S_{\text{cl}}}{\delta x(t)}\right)\mu(x)\exp\left(\frac{\mathrm{i}}{\hbar}S_{\text{cl}}(x)\right) = 0. \qquad (11.112)$$

On being translated into the operator formalism (6.80), this equation gives, for $\mathbb{X} = \mathcal{P}_{a,b}$,

$$\left\langle b\left|\widehat{\frac{\delta F}{\delta x(t)}} + F\,\widehat{\frac{\delta \ln \mu}{\delta x(t)}} + F\frac{\mathrm{i}}{\hbar}\,\widehat{\frac{\delta S_{\text{cl}}}{\delta x(t)}}\right|a\right\rangle = 0, \qquad (11.113)$$

11.3 Integration by parts

where $\widehat{O}$ is the time-ordered product of operators corresponding to the functional O. For $F = 1$, equation (11.113) becomes

$$\left\langle b \left| \widehat{\frac{\delta \ln \mu}{\delta x(t)}} + \frac{i}{\hbar} \widehat{\frac{\delta S_{\mathrm{cl}}}{\delta x(t)}} \right| a \right\rangle = 0, \qquad (11.114)$$

which gives, for fixed initial and final states, the same quantum dynamics as the Schwinger variational principle,

$$0 = \delta\langle b | a \rangle = \left\langle b \left| \frac{i}{\hbar} \widehat{\delta S_{\mathrm{q}}} \right| a \right\rangle \qquad (11.115)$$

where the quantum action functional

$$S_{\mathrm{q}} = S_{\mathrm{cl}} + \frac{\hbar}{i} \ln \mu$$

corresponds to the Dirac quantum action function [10] up to order $\hbar^2$.

Remark. In Chapter 6 we derived the Schwinger variational principle by varying the potential, i.e. by varying the action functional.

Group-invariant symbols

LaChapelle (private communication) has investigated the use of integration by parts when there is a group action on the domain of integration other than translation. We have generalized [6, 11] the gaussian volume element defined in Section 2.3 on Banach spaces as follows. Let Θ and Z be two given continuous bounded functionals defined on a Banach space $\mathbb{X}$ and its dual $\mathbb{X}'$, respectively, by

$$\Theta : \mathbb{X} \times \mathbb{X}' \to \mathbb{C}, \qquad Z : \mathbb{X}' \to \mathbb{C}. \qquad (11.116)$$

Define a volume element $\mathcal{D}_{\Theta,Z}$ by

$$\int_{\mathbb{X}} \mathcal{D}_{\Theta,Z}(x) \Theta(x, x') = Z(x'). \qquad (11.117)$$

There is a class $\mathcal{F}$ of functionals on $\mathbb{X}$ that are integrable with respect to $\mathcal{D}_{\Theta,Z}$ defined [11] as follows:

$$F_\mu \in \mathcal{F} \Leftrightarrow F_\mu(x) = \int_{\mathbb{X}'} \Theta(x, x') d\mu(x'), \qquad (11.118)$$

where μ is a bounded measure on $\mathbb{X}'$. Although μ is not necessarily defined by $F = F_\mu$, it can be proved that $\int_{\mathbb{X}} F(x) \mathcal{D}_{\Theta,Z}(x)$ is defined. Moreover, it is not necessary to identify μ in order to compute $\int_{\mathbb{X}} F(x) \mathcal{D}_{\Theta,Z}(x)$. The class $\mathcal{F}$ generalizes the class chosen by Albeverio and Høegh-Krohn [12].

Let σ_g be a group action on $\mathbb{X}$ and σ'_g be a group action on $\mathbb{X}'$ such that
$$\langle \sigma'_g x', x \rangle = \langle x', \sigma_{g^{-1}} x \rangle. \tag{11.119}$$

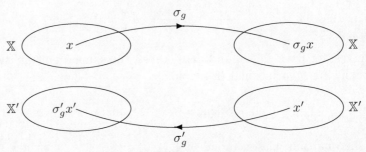

The volume element $\mathcal{D}_{\Theta,Z}$ defined by (11.117) is invariant under the group action if Z and Θ are invariant, namely if
$$Z(\sigma'_g x') = Z(x')$$
$$\Theta(\sigma_g x, \sigma'_g x') = \Theta(x, x'). \tag{11.120}$$

Then
$$\int_\mathbb{X} F_\mu(\sigma_g x) \mathcal{D}_{\Theta,Z}(x) = \int_\mathbb{X} F_\mu(x) \mathcal{D}_{\Theta,Z}(x). \tag{11.121}$$

Let V be an infinitesimal generator of the group of transformations $\{\sigma_g\}$, then
$$\int_\mathbb{X} \mathcal{L}_V F_\mu(x) \cdot \mathcal{D}_{\Theta,Z}(x) = -\int_\mathbb{X} F_\mu(x) \cdot \mathcal{L}_V \mathcal{D}_{\Theta,Z}(x) = 0. \tag{11.122}$$

This equation is to the group-invariant symbol $\mathcal{D}_{\Theta,Z}$ what the following equation is to the translation-invariant symbol $\mathcal{D}_Q$:
$$\int_\mathbb{X} \frac{\delta F}{\delta x(t)} \mathcal{D}_Q(x) = -\int_\mathbb{X} F \frac{\delta}{\delta x(t)} \mathcal{D}_Q(x) = 0. \tag{11.123}$$

References

[1] J. L. Koszul (1985). "Crochet de Schouten–Nijenhuis et cohomologie," in *The Mathematical Heritage of Elie Cartan*, Astérisque, Numéro Hors Série, 257–271.
[2] D. McDuff (1998). "Symplectic structures," *Notices of AMS* **45**, 952–960.
[3] P. Cartier, M. Berg, C. DeWitt-Morette, and A. Wurm (2001). "Characterizing volume forms," in *Fluctuating Paths and Fields*, ed. A. Pelster (Singapore, World Scientific), pp. 139–156.
[4] T. Voronov (1992). "Geometric integration theory on supermanifolds," *Sov. Sci. Rev. C: Math. Phys.* **9**, 1–138.

[5] See for instance C. DeWitt-Morette, A. Makeshwari, and B. Nelson (1979). "Path integration in nonrelativistic quantum mechanics," *Phys. Rep.* **50**, 255–372.
[6] P. Cartier and C. DeWitt-Morette (1995). "A new perspective on functional integration," *J. Math. Phys.* **36**, 2237–2312. See pp. 2295–2301 "Infinite-dimensional determinants."
[7] Y. Choquet-Bruhat and C. DeWitt-Morette (1996 and 2000). *Analysis, Manifolds, and Physics*, Part I: Basics (with M. Dillard-Bleick), revised edn.; Part II, revised and enlarged edn. (Amsterdam, North Holland).
[8] P. Cartier (1989). "A course on determinants," in *Conformal Invariance and String Theory*, eds. P. Dita and V. Georgescu (New York, Academic Press).
[9] D. Nualart (1995). *The Malliavin Calculus and Related Topics (Probability and its Applications)* (Berlin, Springer).
[10] P. A. M. Dirac (1947). *The Principles of Quantum Mechanics* (Oxford, Clarendon Press).
[11] P. Cartier and C. DeWitt-Morette (1993). "Intégration fonctionnelle; éléments d'axiomatique," *C. R. Acad. Sci. Paris* **316 Série II**, 733–738.
[12] S. A. Albeverio and R. J. Høegh-Krohn (1976). *Mathematical Theory of Feynman Path Integrals* (Berlin, Springer).

Other references that have inspired this chapter

J. LaChapelle (1997). "Path integral solution of the Dirichlet problem," *Ann. Phys.* **254**, 397–418.

J. LaChapelle (2004). "Path integral solution of linear second order partial differential equations I. The general construction," *Ann. Phys.* **314**, 362–395.

Part IV
Non-gaussian applications

12
Poisson processes in physics

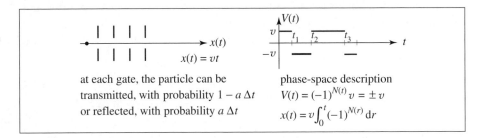

at each gate, the particle can be transmitted, with probability $1 - a\,\Delta t$ or reflected, with probability $a\,\Delta t$

phase-space description
$V(t) = (-1)^{N(t)} v = \pm v$
$x(t) = v \int_0^t (-1)^{N(r)}\, dr$

Chapters 12 and 13 belong together. Chapter 13 gives the foundation for Chapter 12. The problems in Chapter 12, namely the telegraph equation and a two-state system interacting with its environment, were the motivation for Chapter 13; they were worked out before the mathematical theory in Chapter 13 had been elaborated. Minor duplications in these two chapters allow the reader to begin with either chapter.

12.1 The telegraph equation

In the previous chapters, functional integrals served as solutions to parabolic equations. They were defined as averages of generalized brownian paths. In Chapters 12 and 13, we consider functional integrals that average Poisson paths. These "path integrals" are solutions to hyperbolic equations and Dirac equations. Path integrals as a tool for studying the telegraph equation were introduced in the Colloquium Lectures delivered in 1956 by Mark Kac at the Magnolia Petroleum Company [1]. In this section we present an outline of Kac's analysis of the telegraph equation.

When should one use Monte Carlo calculation?

The telegraph equation is the simplest wave equation with dissipation: here a and v are positive constants, and

$$\frac{1}{v^2}\frac{\partial^2 F}{\partial t^2} + \frac{2a}{v^2}\frac{\partial F}{\partial t} - \frac{\partial^2 F}{\partial x^2} = 0, \tag{12.1}$$

$$\left.\begin{array}{r}F(x,0) = \phi(x) \\ \dfrac{\partial}{\partial t}F(x,t)\bigg|_{t=0} = 0\end{array}\right\} \tag{12.2}$$

where $\phi(x)$ is an "arbitrary" function.

A stochastic solution of the telegraph equation first appeared in a paper by Sidney Goldstein [2] following a model proposed by G.I. Taylor. The solution is an average over the set of random walks defined by the following: consider a lattice of points on $\mathbb{R}$ with lattice spacing Δx, and a particle starting at time $t = 0$ at $x_0 = x(0)$ moving with constant speed v in the positive or negative direction. Each step Δx is of duration Δt,

$$\Delta x = v\,\Delta t. \tag{12.3}$$

At each lattice point,

$$\begin{array}{l}\text{the probability of reversing direction is } \;a\,\Delta t; \\ \text{the probability of maintaining direction is } \;1 - a\,\Delta t.\end{array} \tag{12.4}$$

Let vS_n be the displacement of the particle after n steps, i.e. at time

$$t = n\,\Delta t. \tag{12.5}$$

Let the random variable ϵ_i be defined such that

$$\text{for } i = 1, \ldots, n, \quad \epsilon_i = \begin{cases} 0 & \text{for no change of direction,} \\ 1 & \text{for a change of direction.} \end{cases} \tag{12.6}$$

The random variables ϵ_i are stochastically independent and have the property[1]

$$\begin{aligned}\Pr[\epsilon_i = 0] &= 1 - a\,\Delta t, \\ \Pr[\epsilon_i = 1] &= a\,\Delta t.\end{aligned} \tag{12.7}$$

The number of changes of direction after i steps is

$$N_i = \epsilon_1 + \cdots + \epsilon_{i-1} \tag{12.8}$$

---

[1] In this chapter and the following we denote by $\Pr[\ldots]$ the probability of the event inside the brackets and by $\langle\ldots\rangle$ the average (or mean) value.

for $i \geq 1$. We assume the convention $N_1 = 0$, and write the displacement after n steps as[2]

$$vS_n = v\,\Delta t \sum_{i=1}^{n} (-1)^{N_i}. \qquad (12.9)$$

The quantity $S_n \in [-t, t]$ is a "randomized time," and is not necessarily positive. The solution $F_a(x_0, t)$ of the telegraph equation, (12.1) and (12.2), is given by an average [1]

$$F_a(x_0, t) = \lim_{n \to \infty} \frac{1}{2}\langle \phi(x_0 + vS_n) \rangle + \frac{1}{2}\langle \phi(x_0 - vS_n) \rangle. \qquad (12.10)$$

Without dissipation, $a = 0$, and we obtain $S_n = t$. The solution

$$F_0(x_0, t) = \frac{1}{2}\phi(x_0 + vt) + \frac{1}{2}\phi(x_0 - vt) \qquad (12.11)$$

is a superposition of right- and left-moving pulses. With dissipation, F_a is the following average:

$$F_a(x_0, t) = \lim_{n \to \infty} \langle F_0(x_0, S_n) \rangle. \qquad (12.12)$$

The symmetry $F_0(x_0, t) = F_0(x_0, -t)$ is broken when $a \neq 0$.

Remark. The nonrelativistic limit of the telegraph equation. When the conditions $v = \infty$ and $a = \infty$ occur together such that $2a/v^2$ remains constant, the limiting case of the telegraph equation is the diffusion equation

$$\frac{1}{D}\frac{\partial F}{\partial t} = \frac{\partial^2 F}{\partial x^2}, \qquad (12.13)$$

where we have let $2a/v^2 = 1/D$, for example.

A Monte Carlo calculation of (12.10)

Position a large number of particles at x_0. Impart to half of them a velocity in one direction, and to half a velocity in the opposite direction. At every time-step, "flip a coin" using random numbers weighted by the probability law (12.4). At each time t, record the position of each particle, compute ϕ for each particle, and average over the number of particles. As Kac said, "this is a completely ridiculous scheme": Δt has to be very small, making direction reversal a very unlikely event, and one needs an enormously

---

[2] We tacitly assume that the initial speed was in the positive direction. In the opposite case, replace v by $-v$.

large number of particles. Kac concludes that "we have to be clever" to find a more efficient calculation.

A better stochastic solution of the telegraph equation

Equation (12.10) is the discretized version of a path-integral solution of the telegraph equation. Kac did away with discretization by introducing the random variable $S(t)$, the continuous counterpart of S_n, defined such that

$$S(t) = \int_0^t (-1)^{N(\tau)} d\tau, \qquad S(t) \in]-t, t]. \tag{12.14}$$

The random variable $N(\tau)$ is defined with the Poisson distribution of parameter $a\tau$, i.e. with probability

$$\Pr[N(\tau) = k] = e^{-a\tau}(a\tau)^k/k!, \qquad k = 0, 1, \ldots \tag{12.15}$$

Given a finite number of epochs, t_i, where

$$0 = t_0 < t_1 < t_2 < \cdots < t_n,$$

the increments $N(t_j) - N(t_{j-1})$ are stochastically independent for $j = 1, \ldots, n$.

In the previous random walk, the discrete probability $P_k(t)$ of k reversals during the time interval $[0, t]$ satisfies

$$P_k(t) = (a \Delta t) P_{k-1}(t - \Delta t) + (1 - a \Delta t) P_k(t - \Delta t),$$

with the convention $P_{-1}(t) = 0$. To lowest order in Δt this becomes

$$[P_k(t) - P_k(t - \Delta t)]/\Delta t = a[P_{k-1}(t) - P_k(t)].$$

On replacing the left-hand side by a derivative with respect to t we get a set of coupled first-order ordinary differential equations for the $P_k(t)$. The solution with the right boundary condition at $t = 0$ ($P_k(0) = 0$ for $k \geq 1$, $P_0(0) = 1$) is (12.15), $P_k(t) = e^{-at}(at)^k/k!$ in the limit $\Delta t = 0$.

In terms of the random variable $S(t)$, Kac's solution of the telegraph equation is [1]

$$F_a(x_0, t) = \frac{1}{2}\langle \phi(x_0 + vS(t)) \rangle + \frac{1}{2}\langle \phi(x_0 - vS(t)) \rangle. \tag{12.16}$$

The Monte Carlo calculation of this equation is considerably faster than one that involves small probabilities. For evaluating a sample process $S(t)$ one needs a machine, or a source of radioactive material, which produces a Poisson process. One computes the value of ϕ for each sample process and averages over the number of samples. This number is not necessarily enormous; $n = 100$ is sufficient.

Random times

The introduction of the random variable $S(t)$ is accompanied by a new concept: $S(t)$ randomizes time in the solution (12.11) of the equation without dissipation ((12.1) when $a = 0$). Kac notes that "this amusing observation persists for all equations of this form in *any number* of dimensions." Given a solution $\phi(x,t)$ of the wave equation without dissipation,

$$\frac{1}{v^2}\frac{\partial^2 \phi}{\partial t^2} - \Delta_x \phi = 0, \tag{12.17}$$

where $\phi(x,0)$ is "arbitrary" and

$$\left.\frac{\partial \phi}{\partial t}\right|_{t=0} = 0, \tag{12.18}$$

define

$$F(x,t) = \langle \phi(x, S(t)) \rangle. \tag{12.19}$$

By exploiting the symmetry $\phi(x,t) = \phi(x,-t)$, the right-hand side of (12.19) may be written

$$\frac{1}{2}\langle \phi(x, S(t)) \rangle + \frac{1}{2}\langle \phi(x, -S(t)) \rangle,$$

and can be shown to be $F(x,t)$, the solution of the wave equation with dissipation,

$$\frac{1}{v^2}\frac{\partial^2 F}{\partial t^2} + \frac{2a}{v^2}\frac{\partial F}{\partial t} - \Delta_x F = 0, \tag{12.20}$$

where

$$F(x,0) = \phi(x,0), \qquad \left.\frac{\partial F}{\partial t}\right|_{t=0} = 0. \tag{12.21}$$

Equation (12.14) can be thought of as a path-dependent time reparametrization. Indeed, $|S(t)|$ is the time required for a particle moving with velocity v without reversal to reach the point $x(t) = v|S(t)|$.

Kac's solution (12.16) of the telegraph equation suggests that the stochastic process $S(t)$ is of seminal importance for the construction and calculation of path-integral solutions of the wave equation. First, we need to compute the probability density $g(t,r)$ of $S(t)$,

$$g(t,r)dr = \Pr[r < S(t) < r + dr], \qquad -t < r < t,$$

and the probability density $h(t,r)$ of $|S(t)|$,

$$h(t,r)\mathrm{d}r = \Pr[r < |S(t)| < r + \mathrm{d}r], \qquad 0 < r < t.$$

These probabilities are related by

$$h(t,r) = g(t,r) + g(t,-r).$$

A nontrivial calculation, based on Laplace transforms [3], yields

$$h(t,r) = \mathrm{e}^{-at}\delta(t-r) + a\mathrm{e}^{-at}\theta(t-r)$$
$$\times \left[I_0\!\left(a(t^2-r^2)^{1/2}\right) + \frac{t}{(t^2-r^2)^{1/2}} I_1\!\left(a(t^2-r^2)^{1/2}\right) \right] \quad (12.22)$$

for $r \geq 0$, where I_0 and I_1 are the standard Bessel functions [4], and θ is the Heaviside step function.

The solution (12.19) of the telegraph equation (12.20) can then be given explicitly in terms of the solution ϕ of the wave equation without dissipation (12.17) by a quadrature,

$$F(x,t) = \int_{-t}^{t} \mathrm{d}r\, \phi(x,r) g(t,r)$$
$$= \int_{0}^{t} \mathrm{d}r\, \phi(x,r) h(t,r). \quad (12.23)$$

See-Kit Foong [5] has exploited and generalized this procedure, and has applied his results to a variety of problems. Among these is the design and restoration of wave fronts propagating in random media, including the case in which the probability $a\,\Delta t$ is time-dependent: $a(t)\Delta t$.

Random times will be used extensively in Chapter 14 on "exit times."

12.2 Klein–Gordon and Dirac equations

The telegraph equation versus the Klein–Gordon equation

Recall the telegraph equation

$$\frac{1}{v^2}\frac{\partial^2 F}{\partial t^2} + \frac{2a}{v^2}\frac{\partial F}{\partial t} - \frac{\partial^2 F}{\partial x^2} = 0. \quad (12.24)$$

If $F = F(x,t)$ is a solution of this equation then the function $G(x,t) = \mathrm{e}^{at} F(x,t)$ satisfies the relation

$$\frac{1}{v^2}\frac{\partial^2 G}{\partial t^2} - \frac{\partial^2 G}{\partial x^2} - \frac{a^2}{v^2} G = 0. \quad (12.25)$$

In order to interpret $a\,\Delta t$ as the probability of reversing the speed, one requires the constant a to be real and positive. In equation (12.25), let

12.2 Klein–Gordon and Dirac equations

$v = c$, where c is the speed of light. Let a be *purely imaginary*,

$$a = -imc^2/\hbar; \tag{12.26}$$

the equation (12.25) is now the *Klein–Gordon equation*

$$\frac{1}{c^2}\frac{\partial^2 G}{\partial t^2} - \frac{\partial^2 G}{\partial x^2} + \frac{m^2 c^2}{\hbar^2} G = 0. \tag{12.27}$$

The transition from the telegraph equation to the Klein–Gordon equation is achieved by introducing the functional transformation $G(x,t) = e^{at} F(x,t)$ and the analytic continuation in a from $a > 0$ to $a = -imc^2/\hbar$.

As remarked already, if we let a and v approach infinity in such a way that $2a/v^2$ remains constant and equal to $1/D$, the telegraph equation becomes the diffusion equation

$$\frac{1}{D}\frac{\partial F}{\partial t} = \frac{\partial^2 F}{\partial x^2}. \tag{12.28}$$

If $v = c$ and $a = -imc^2/\hbar$, then $D = i\hbar/(2m)$. Analytic continuation in a transforms the diffusion equation (12.28) into the Schrödinger equation

$$i\hbar\frac{\partial F}{\partial t} = -\frac{\hbar^2}{2m}\frac{\partial^2 F}{\partial x^2}. \tag{12.29}$$

These calculations reaffirm that, for any solution $G(x,t)$ of the Klein–Gordon equation, $e^{imc^2 t/\hbar} G(x,t)$ is a solution of the Schrödinger equation in the non-relativistic limit $c = \infty$. We summarize these results in the following chart:

$$
\begin{array}{ccc}
\text{Telegraph} & \xrightarrow{v=\infty} & \text{Diffusion} \\
\updownarrow \text{analytic continuation} & & \updownarrow \text{analytic continuation} \\
\text{Klein–Gordon} & \xrightarrow{c=\infty} & \text{Schrödinger}
\end{array}
$$

The two-dimensional Dirac equation

To achieve the analytic continuation from the telegraph equation to the Klein–Gordon equation, we follow B. Gaveau, T. Jacobson, M. Kac, and L. S. Schulman [6] in replacing one second-order differential equation by two first-order differential equations.

This process had already been accomplished for the telegraph equation by Mark Kac [1]; here is his method in probabilistic terms. Let $(N(t))_{t \geq 0}$ be a Poisson process with intensity a. The process $(N(t))$ is markovian, but the process

$$X(t) = v \int_0^t (-1)^{N(\tau)} d\tau, \tag{12.30}$$

introduced by Kac, is not. Let $P(x,t)$ be the probability density of the random variable $X(t)$:
$$\Pr[x < X(t) < x + \mathrm{d}x] = P(x,t)\mathrm{d}x. \tag{12.31}$$
The probability density $P(x,t)$ satisfies the second-order differential equation
$$\frac{1}{v^2}\frac{\partial^2 P}{\partial t^2} + \frac{2a}{v^2}\frac{\partial P}{\partial t} - \frac{\partial^2 P}{\partial x^2} = 0. \tag{12.32}$$
To reduce (12.32) to two first-order equations we must relate it to a markovian process. The simplest method is to introduce a phase-space description $(X(t), V(t))$, where the velocity is $V(t) = v(-1)^{N(t)}$. The process moves along the lines $L_\pm = \{V = \pm v\}$. Let $P_\pm(x,t)$ be the probability density on the line $L_\pm$; by an obvious projection we get
$$P(x,t) = P_+(x,t) + P_-(x,t). \tag{12.33}$$
During an infinitesimal time interval $[t, t+\Delta t]$, the processes move along the lines $L_\pm$ by the amount $\pm v\,\Delta t$. The probability that the speed reverses, i.e. that the process jumps from $L_\pm$ to $L_\mp$, is $a\,\Delta t$. The probability of continuing along the same line is $1 - a\,\Delta t$. From this description, one derives the so-called "master equation"
$$\begin{pmatrix} P_+(x, t+\Delta t) \\ P_-(x, t+\Delta t) \end{pmatrix} = \begin{pmatrix} 1 - a\,\Delta t & a\,\Delta t \\ a\,\Delta t & 1 - a\,\Delta t \end{pmatrix} \begin{pmatrix} P_+(x - v\,\Delta t, t) \\ P_-(x + v\,\Delta t, t) \end{pmatrix}. \tag{12.34}$$
In the limit $\Delta t = 0$, we obtain the differential equation
$$\frac{\partial}{\partial t}\begin{pmatrix} P_+ \\ P_- \end{pmatrix} = a\begin{pmatrix} -1 & 1 \\ 1 & -1 \end{pmatrix}\begin{pmatrix} P_+ \\ P_- \end{pmatrix} - v\frac{\partial}{\partial x}\begin{pmatrix} 1 & 0 \\ 0 & -1 \end{pmatrix}\begin{pmatrix} P_+ \\ P_- \end{pmatrix}. \tag{12.35}$$
It is straightforward to show that both components, P_+ and P_-, and hence their sum $P = P_+ + P_-$, satisfy the telegraph equation.

In a spacetime of dimension $2m$ or $2m+1$, spinors have 2^m components: in two or three dimensions spinors have two components, and in four dimensions spinors have four components. In a spacetime with coordinates x, t, the spinors have two components; the Dirac equation reads
$$i\hbar\frac{\partial \psi}{\partial t} = mc^2\beta\psi - i\hbar c\alpha\frac{\partial \psi}{\partial x}, \tag{12.36}$$
where the 2×2 hermitian matrices α and β satisfy
$$\alpha^2 = \beta^2 = 1_2, \qquad \alpha\beta = -\beta\alpha. \tag{12.37}$$
Recall that both components of a solution to the Dirac equation satisfy the Klein–Gordon equation. To compare the equations (12.35) and (12.36),

12.2 Klein–Gordon and Dirac equations

let
$$u(x,t) = e^{imc^2 t/\hbar} \psi(x,t) \qquad (12.38)$$

with components
$$u = \begin{pmatrix} u_+ \\ u_- \end{pmatrix}.$$

Assume the Weyl matrices α and β:
$$\alpha = \begin{pmatrix} 1 & 0 \\ 0 & -1 \end{pmatrix}, \qquad \beta = \begin{pmatrix} 0 & 1 \\ 1 & 0 \end{pmatrix}.^3 \qquad (12.39)$$

The system satisfied by u_+ and u_- reduces to (12.35) with
$$a = -imc^2/\hbar, \qquad v = c, \qquad (12.40)$$

as before. This substitution provides the analytic continuation from the telegraph to the Klein–Gordon equation. Analytic continuation can also be implemented in imaginary time, by means of an imaginary Planck constant.

Feynman's checkerboard

Rather than performing the analytic continuation at the level of differential equations, it is useful to perform it at the level of functional integration. The random walk, or Poisson process, at the heart of Kac's solution of the telegraph equation should be replaced by a path integral in Feynman's sense: one with complex amplitudes.

Feynman proposed in [7] a solution of the two-dimensional Dirac equation based on his checkerboard model: particles move at the speed of light on a spacetime checkerboard (figure 12.1). Particles move to the right or to the left with an amplitude (imaginary probability) of a change of direction at each step. (See Section 13.3 and Appendix H for a rigorous mathematical description.)

Four-dimensional spacetime

The checkerboard model has inspired a number of physicists (see references in [6]). Let us consider now the familiar four-dimensional spacetime. The telegraph equation takes the form
$$\frac{\partial^2 P_\pm}{\partial t^2} + 2a \frac{\partial P_\pm}{\partial t} - v^2 \Delta_x P_\pm = 0, \qquad (12.41)$$

---

[3] Using Pauli matrices this is $\alpha = \sigma_3$, $\beta = \sigma_1$.

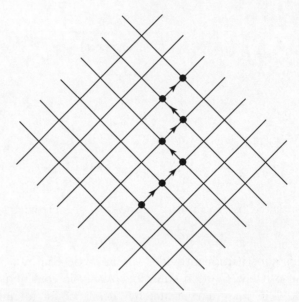

Fig. 12.1 Feynman's checkerboard

where Δ_x is the three-dimensional laplacian. Upon introduction of the vector $\vec{\sigma} = (\sigma_1, \sigma_2, \sigma_3)$ whose components are the Pauli matrices, Δ_x becomes the square of the operator $\vec{\sigma} \cdot \vec{\nabla}_x$. The second-order equation (12.41) is now replaced by two first-order equations:

$$\frac{\partial}{\partial t}\begin{pmatrix} \Pi_+ \\ \Pi_- \end{pmatrix} = a \begin{pmatrix} -1 & 1 \\ 1 & -1 \end{pmatrix} \begin{pmatrix} \Pi_+ \\ \Pi_- \end{pmatrix} - v \begin{pmatrix} \vec{\sigma} \cdot \vec{\nabla} & 0 \\ 0 & -\vec{\sigma} \cdot \vec{\nabla} \end{pmatrix} \begin{pmatrix} \Pi_+ \\ \Pi_- \end{pmatrix} \tag{12.42}$$

in analogy with (12.35). In equation (12.42) Π_+ and Π_- are 2-spinors. The change of parameters $a = -imc^2/\hbar$, $v = c$ as in (12.38) and (12.40) transforms the last equation into the four-component Dirac equation

$$i\hbar \frac{\partial \psi}{\partial t} = mc^2 \beta \psi - i\hbar c \alpha_j \nabla_j \psi, \tag{12.43}$$

where the 4×4 matrices α_j and β are defined in Weyl's normalization:

$$\beta = \begin{pmatrix} 0 & 1_2 \\ 1_2 & 0 \end{pmatrix}, \quad \alpha_j = \begin{pmatrix} \sigma_j & 0 \\ 0 & -\sigma_j \end{pmatrix}. \tag{12.44}$$

Unfortunately there is no straightforward generalization of the checkerboard model. In [8] Jacobson constructed a spinor-chain path-integral solution of the Dirac equation, but there is no known stochastic process related by analytic continuation to Jacobson's spinor-chain path integral.

12.3 Two-state systems interacting with their environment[4]

A two-state system interacting with its environment is a good model for a variety of physical systems, including an ammonia molecule, a benzene molecule, the H_2^+ ion, and a nucleus of spin 1/2 in a magnetic field. For further investigation, see the book of C. Cohen-Tannoudji, B. Diu, and F. Laloë [9], and the Ph.D. dissertation of D. Collins and references therein [10].

The influence functional

In 1963 Feynman and Vernon [11] introduced the concept of the influence functional for describing the effects of the environment on a system, when both the system and its environment are quantum-mechanical systems with a classical limit. They considered a total action

$$S_{\text{tot}}(x, q) = S_{\text{syst}}(x) + S_{\text{int}}(x, q) + S_{\text{env}}(q),$$

where x are the system paths and q are the environment paths. Both sets of paths are gaussian distributed and the interaction is approximately linear in q. They integrated out the q variables, and obtained the influence of the environment on the system; the influence functional is given in terms of the paths x of the system. They worked out several explicit examples, and derived a fluctuation–dissipation theorem.

Functional integration is our tool of choice for integrating out, or averaging, contributions that affect the system of interest but are not of primary interest. (See in particular Chapter 16.) In this section the system of interest is a two-state system that is Poisson distributed and has no classical analog.

Isolated two-state systems

The description of a quantum system consists of a vector space over the complex numbers called *state space*, and self-adjoint operators on the state space called *observables*. The state space of most familiar quantum systems is infinite-dimensional. There are also quantum systems whose state spaces are finite-dimensional. The paradigm is the two-state system.

Let $\mathcal{H}_{\text{syst}}$ be the Hilbert space of a two-state system. Let

$$|\psi(t)\rangle \in \mathcal{H}_{\text{syst}} \qquad (12.45)$$

---

[4] This section is a summary of David Collins' Ph.D. dissertation [10].

be a normalized state vector, and

$$|\psi(t)\rangle = \sum_{i=\pm} \psi_i(t)|i\rangle \tag{12.46}$$

be the decomposition of $|\psi(t)\rangle$ in the basis $\{|+\rangle, |-\rangle\}$,

$$\psi_i(t) := \langle i|\psi(t)\rangle. \tag{12.47}$$

The matrix elements of an operator A on $\mathcal{H}_{\text{syst}}$ in the basis $\{|i\rangle\}$ are $A_{ij} = \langle i|A|j\rangle$;

$$A = \sum_{i,j=\pm} |i\rangle A_{ij} \langle j|. \tag{12.48}$$

The operator A is an observable if and only if

$$A = a^0 \mathbf{1} + a^j \sigma_j, \tag{12.49}$$

where $a^\mu \in \mathbb{R}$, and $\{\sigma_j\}$ are the Pauli matrices

$$\sigma_j \sigma_k = \delta_{jk} \mathbf{1} + i\epsilon_{jk\ell} \sigma_\ell. \tag{12.50}$$

The representation of an operator by Pauli matrices gives readily its *site (diagonal) component* and its *hopping component*. For example, let the operator A be a hamiltonian H. Equation (12.49) reads

$$H = \begin{pmatrix} H^0 + H^3 & H^1 - iH^2 \\ H^1 + iH^2 & H^0 - H^3 \end{pmatrix}. \tag{12.51}$$

The H^0 terms may be discarded by shifting the overall energy scale so that $H^0 = 0$. The site (diagonal) component of the hamiltonian is

$$H_{\text{s}} := H^3 \sigma_3. \tag{12.52}$$

Its hopping component is

$$H_{\text{h}} := H^1 \sigma_1 + H^2 \sigma_2. \tag{12.53}$$

When $H^0 = 0$,

$$H = H_{\text{s}} + H_{\text{h}} = H^j \sigma_j. \tag{12.54}$$

In situations in which the hamiltonian is time-dependent, or in which it interacts with its environment, it is simpler to work in the *interaction picture* defined as follows:

$$|\psi_{\text{int}}(T)\rangle := \exp(i(T - t_0) H_{\text{s}}/\hbar) |\psi(T)\rangle. \tag{12.55}$$

The Schrödinger equation in the interaction picture is

$$i\hbar \frac{\partial}{\partial T} |\psi_{\text{int}}(T)\rangle = H_{\text{h int}}(T) |\psi_{\text{int}}(T)\rangle, \tag{12.56}$$

where $H_{\text{h int}}$ is the hopping hamiltonian in the interaction picture defined by

$$H_{\text{h int}}(T) := e^{i(T-t_0)H_s/\hbar} H_h e^{-i(T-t_0)H_s/\hbar}. \tag{12.57}$$

The evolution operator $U_{\text{int}}(T, t_0)$, defined by

$$\psi_{\text{int}}(T) = U_{\text{int}}(T, t_0)\psi_{\text{int}}(t_0), \tag{12.58}$$

satisfies the equations

$$\frac{\partial}{\partial T} U_{\text{int}}(T, t_0) = -\frac{i}{\hbar} H_{\text{h int}}(T) U_{\text{int}}(T, t_0), \tag{12.59}$$

$$U_{\text{int}}(t_0, t_0) = \mathbf{1}.$$

Two-state systems interacting with their environment[5]

Let $\mathcal{H}_{\text{tot}}$ be the Hilbert space for the total system,

$$\mathcal{H}_{\text{tot}} = \mathcal{H}_{\text{syst}} \otimes \mathcal{H}_{\text{env}} \tag{12.60}$$

with orthonormal basis

$$|j, n\rangle := |j\rangle \otimes |n\rangle, \qquad j \in \{+, -\}, \; n \in \mathbb{N}, \tag{12.61}$$

and state-vector decomposition

$$|\psi\rangle_{\text{tot}} = \sum_{j=\pm} \sum_{n=0}^{\infty} \psi_{nj} |j, n\rangle. \tag{12.62}$$

The system components are

$$|\psi_j\rangle := \sum_{n=0}^{\infty} \psi_{nj} |n\rangle \in \mathcal{H}_{\text{env}} \tag{12.63}$$

and

$$|\psi\rangle_{\text{tot}} = \sum_{j=\pm} |j\rangle \otimes |\psi_j\rangle. \tag{12.64}$$

There is a version of (12.49) for the system plus environment: an operator A is an observable if and only if

$$A = A^0 \mathbf{1}_{\text{syst}} + A^j \sigma_j, \tag{12.65}$$

where A^μ is a self-adjoint operator on $\mathcal{H}_{\text{env}}$.

---

[5] For interaction with many-environment systems and related problems see [10].

For the two-state interacting system, the hamiltonian is

$$H_{\text{tot}} = H_{\text{syst}} \otimes \mathbf{1}_{\text{env}} + \mathbf{1}_{\text{syst}} \otimes H_{\text{env}} + H_{\text{int}}, \tag{12.66}$$

which is abbreviated to

$$H_{\text{tot}} = H_s + H_h + H_e. \tag{12.67}$$

In (12.67) the identity operators are omitted and[6]

$$H_s + H_h = H_{\text{syst}}, \qquad H_e = H_{\text{env}} + H_{\text{int}}.$$

The evolution operator in the interaction picture satisfies (12.59). Now $H_{h\,\text{int}}$ is obtained from H_h by substituting $H_s + H_e$ for H_s in (12.57).

Poisson functional integrals[7]

The goal of this subsection is to construct a path-integral solution of the differential equation (12.59).

Let a Poisson path N be characterized by its number $N_\mathbb{T}$ of speed reversals during a given finite time interval $\mathbb{T} = [t_a, t_b]$ of duration $T = t_b - t_a$. Let $\mathcal{P}_n$ be the space of paths N that jump n times. A space $\mathcal{P}_n$ is parametrized by the set of jump times $\{t_1, \ldots, t_n\}$. A path $N \in \mathcal{P}_n$ can be represented by the sum

$$N \leftrightarrow \delta_{t_1} + \delta_{t_2} + \cdots + \delta_{t_n}. \tag{12.68}$$

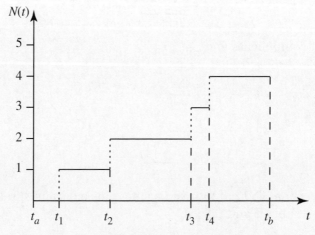

Fig. 12.2 A path $N \in \mathcal{P}_4$

---

[6] Here H_s and H_h are, respectively, the site and hopping components of $H_{\text{syst}} = H^0\mathbf{1} + H^j \sigma_j$.
[7] See Chapter 13 for a rigorous mathematical description.

12.3 Two-state systems interacting with their environment

The probability of N jumping n times during the interval $\mathbb{T}$ is

$$\Pr[N_{\mathbb{T}} = n] = e^{-\lambda} \frac{\lambda^n}{n!}, \qquad (12.69)$$

where $\lambda = aT$:

$$\lambda = \int_{t_a}^{t_b} d\lambda_a(t), \qquad (12.70)$$

where a is the decay constant and $d\lambda_a(t) = a\, dt$.

The space $\mathcal{P}$ of Poisson paths N is the disjoint union of the spaces $\mathcal{P}_n$ parametrized by n jump times $\{t_1, \ldots, t_n\}$,

$$\mathcal{P} = \bigcup_n \mathcal{P}_n. \qquad (12.71)$$

Therefore a functional

$$F : \mathcal{P} \to \text{ linear operators on a vector space (e.g. } \mathbb{C}) \qquad (12.72)$$

is completely determined by its components

$$F_n := F\big|_{\mathcal{P}_n}. \qquad (12.73)$$

The Poisson expectation value of a functional F on $\mathcal{P}$ is

$$\langle F \rangle = \sum_{n=0}^{\infty} \int_{\mathcal{P}_n} d\mu_a^n(t_1, \ldots, t_n) F_n(t_1, \ldots, t_n), \qquad (12.74)$$

where

$$d\mu_a^n(t_1, \ldots, t_n) = e^{-aT} d\lambda_a(t_n) \ldots d\lambda_a(t_1). \qquad (12.75)$$

The expectation value $\langle F \rangle$ can also be written

$$\langle F \rangle = \int_{\mathcal{P}} \mathcal{D}_{a,T} N \cdot F(N), \qquad (12.76)$$

where $\mathcal{D}_{a,T} N$ can be characterized by its Laplace transform[8]

$$\int_{\mathcal{P}} \mathcal{D}_{a,T} N \cdot \exp(\langle f, N \rangle) = \exp \int_{\mathbb{T}} \left(e^{f(t)} - 1\right) d\lambda_a(t). \qquad (12.77)$$

In order to construct a path-integral solution of (12.59) we consider the case in which $\mathbb{T} = [0, T]$. The components of F are

$$\begin{aligned} F_n(t_1, \ldots, t_n) &= Q(t_n) \ldots Q(t_1) \qquad \text{for } n \geq 1, \\ F_0 &= 1, \end{aligned} \qquad (12.78)$$

---

[8] Since we identified N with $\delta_{t_1} + \cdots + \delta_{t_n}$, $\langle f, N \rangle$ is equal to $f(t_1) + \cdots + f(t_n)$ for any function f on $\mathbb{T}$; it is a random variable for each given f. In Section 13.1, we denote $\langle f, N \rangle$ by $\int_{\mathbb{T}} f \cdot dN$ (see formula (13.27)).

with
$$0 < t_1 < \ldots < t_n < T.$$

Theorem (see proof in Chapter 13). *The Poisson path integral*
$$U_a(T) := e^{aT} \int_{\mathcal{P}} \mathcal{D}_{a,T} N \cdot F(N) \qquad (12.79)$$
is the solution of the equations
$$\frac{\partial U_a(T)}{\partial T} = aQ(T)U_a(T),$$
$$U_a(0) = 1. \qquad (12.80)$$

In order to apply this theorem to equation (12.59), we need to identify the decay constant a and the integrand $F(N)$ for the interacting two-state system with total hamiltonian (12.67),
$$H_{\text{tot}} = H_{\text{s}} + H_{\text{h}} + H_{\text{e}}. \qquad (12.81)$$

The hopping hamiltonian H_{h} (12.53) is often simplified by setting $H^2 = 0$ in (12.51). The commonly used notation for the nonzero components of the hamiltonian (12.51) is
$$H^1 =: \Delta, \qquad H^3 =: -\varepsilon.$$

The total hamiltonian is the sum (12.81) with
$$H_{\text{s}} = \begin{pmatrix} -\varepsilon & 0 \\ 0 & \varepsilon \end{pmatrix}, \qquad H_{\text{h}} = \begin{pmatrix} 0 & \Delta \\ \Delta & 0 \end{pmatrix},$$
$$H_{\text{e}} = H_{\text{env}}\mathbf{1} + H_{\text{int}}^1 \sigma_1 + H_{\text{int}}^3 \sigma_3 \qquad (\text{when } H^2 = 0).$$

The decay constant is complex (see Chapter 13),
$$a = -\mathrm{i}\Delta/\hbar. \qquad (12.82)$$

The components of the integrand $F(N)$ are
$$F_0 = \mathbf{1}_{\text{syst}},$$
$$F_n(t_1, \ldots, t_n) = Q_{\text{h int}}(t_n) \ldots Q_{\text{h int}}(t_1), \qquad (12.83)$$
where
$$Q_{\text{h int}}(t) = \frac{1}{\Delta} H_{\text{h int}}(t). \qquad (12.84)$$

The evolution operator $U_{\text{int}}(T, t_0)$ that satisfies (12.59) can therefore be represented by the functional integral
$$U_{\text{int}}(T, t_0) = \exp(-\mathrm{i}\Delta(T-t_0)/\hbar) \cdot \int_{\mathcal{P}} \mathcal{D}_{-\mathrm{i}\Delta/\hbar, T-t_0} N \cdot F(N). \qquad (12.85)$$

The influence operator

The functional integral (12.85) provides information about the evolution of the environment as well as the system; however, in general we care only about the environment's effects on the system. Collins has developed a method for computing an influence operator for the interacting two-state system, which is similar to the Feynman–Vernon influence-functional method.

Collins has applied the influence-operator method to the spin-boson system. This system consists of a two-state system in an environment composed of a collection of quasi-independent harmonic oscillators coupled diagonally to the two-state system. The total hamiltonian is

$$H_{\text{tot}} = H_{\text{s}} + H_{\text{h}} + \sum_\alpha \hbar \omega_\alpha \left(a_\alpha^\dagger a_\alpha + \frac{1}{2} \right) + \sum_\alpha \mu_\alpha (a_\alpha^\dagger + a_\alpha) \sigma_3, \quad (12.86)$$

where α labels the oscillators in the bath, ω_α is the frequency of the α-oscillator, $a_\alpha^\dagger$ is the corresponding creation operator, a_α is the corresponding annihilation operator, and $\mu_\alpha \in \mathbb{C}$ is the coupling strength. The computation of the influence operator for the system (12.86) is interesting because it provides an *exact* influence functional. The answer is too unwieldy to be given here; it can be found in [10].

Other examples involving different environments and different initial states of the two-state system are worked out in [10] and compared with other calculations.

Poisson functional integrals are relatively new and not as developed as gaussian functional integrals. The analysis in Chapter 13 is a solid point of departure.

References

[1] M. Kac (1974). *Some Stochastic Problems in Physics and Mathematics*, lectures given in 1956 at the Magnolia Petroleum Co.; partially reproduced in *Rocky Mountain J. Math.* **4**, 497–509.

[2] S. Goldstein (1951). "On diffusion by discontinuous movements, and the telegraph equation," *Quart. J. Mech. Appl. Math.* **4**, 129–156.

[3] C. DeWitt-Morette and S.-K. Foong (1990). "Kac's solution of the telegrapher's equation revisited: part I," *Annals Israel Physical Society*, **9**, 351–366, Nathan Rosen Jubilee Volume, eds. F. Cooperstock, L. P. Horowitz, and J. Rosen (Jerusalem, Israel Physical Society, 1989).
C. DeWitt-Morette and S.-K. Foong (1989). "Path integral solutions of wave equation with dissipation," *Phys. Rev. Lett.* **62**, 2201–2204.

[4] M. Abramowitz and I. A. Stegun (eds.) (1970). *Handbook of Mathematical Functions*, 9th edn. (New York, Dover), p. 375, formula 9.6.10.

[5] S.-K. Foong (1990). "Kac's solution of the telegrapher's equation revisited: Part II," *Annals Israel Physical Society*, **9**, 367–377.

S.-K. Foong (1993). "Path integral solution for telegrapher's equation," in *Lectures on Path Integration, Trieste 1991*, eds. H. A. Cerdeira, S. Lundqvist, D. Mugnai, A. Ranfagni, V. Sayakanit, and L. S. Schulman (Singapore, World Scientific).

[6] B. Gaveau, T. Jacobson, M. Kac, and L. S. Schulman (1984). "Relativistic extension of the analogy between quantum mechanics and Brownian motion," *Phys. Rev. Lett.* **53**, 419–422.

T. Jacobson and L. S. Schulman (1984). "Quantum stochastics: the passage from a relativistic to a non-relativistic path integral," *J. Phys. A. Math. Gen.* **17**, 375–383.

[7] R. P. Feynman and A. R. Hibbs (1965). *Quantum Mechanics and Path Integrals* (New York, McGraw Hill), Problem 2-6, pp. 34–36.

[8] T. Jacobson (1984). "Spinor chain path integral for the Dirac equation," *J. Phys. A. Math. Gen.* **17**, 2433–2451.

[9] C. Cohen-Tannoudji, B. Diu, and F. Laloë (1977). *Quantum Mechanics*, Vol. 1 (New York, Wiley-Interscience).

[10] D. Collins (1997). Two-state quantum systems interacting with their environments: a functional integral approach, unpublished Ph.D. Dissertation (University of Texas at Austin).

D. Collins (1997). "Functional integration over complex Poisson processes," appendix to "A rigorous mathematical foundation of functional integration" by P. Cartier and C. DeWitt-Morette in *Functional Integration: Basics and Applications*, eds. C. DeWitt-Morette, P. Cartier, and A. Folacci (New York, Plenum).

[11] R. P. Feynman and F. L. Vernon (1963). "Theory of a general quantum system interacting with a linear dissipative system," *Ann. Phys.* **24**, 118–173.

13
A mathematical theory of Poisson processes

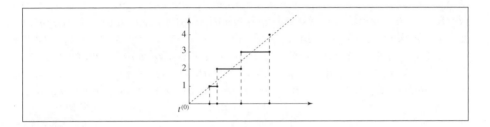

In the previous chapter, we described various physical problems in which Poisson processes appear quite naturally. We aim now at a general and rigorous mathematical theory of such processes and their use in solving a certain class of differential equations.

In Section 13.1, we review the classical theory of the Poisson random processes. We give a dual description:

- in terms of *waiting times* between two consecutive records; and
- in terms of the *counting process* based on the number of records in any given time interval.

We conclude this section by introducing a generating functional,

$$Z(v) = \left\langle \exp \int v \cdot dN \right\rangle,$$

which is a kind of Laplace transform. This functional plays a crucial role in the rest of this chapter.

In Section 13.2, we generalize the previous results in three different ways:

- the *decay constant* a (which is also the temporal density of records) may be a complex number; it is no longer restricted to a positive real number as in the probabilistic interpretation;

234   *A mathematical theory of Poisson processes*

- the one-dimensional time axis is replaced by a multi-dimensional space, and records become *hits* (corresponding to random point processes (also known as random hits)); and
- the case of an *unbounded domain* is covered in addition to the case of a bounded one.

The mathematical tools necessary for these generalizations are the complex bounded Radon measures and the promeasures (also called cylindrical measures).

We conclude this chapter (in Sections 13.3 and 13.4) with a discussion of a quite general class of differential equations. The formula for the propagator that we obtain is a variant of Dyson's formula in the so-called interaction picture. The advantage of our formulation is that it *applies equally well to parabolic diffusion equations that are not time-reversible.* Also, it is one instance in which Feynman's idea of representing a propagator as a "sum over histories" is vindicated in a completely rigorous mathematical fashion, while retaining the conceptual simplicity of the idea. As a corollary, we give *a transparent proof of Kac's solution of the telegraph equation and a rigorous formulation of the process of analytic continuation from the telegraph equation to the Klein–Gordon equation.*

13.1 Poisson stochastic processes

Basic properties of Poisson random variables

A Poisson random variable is a random variable N taking values in the set $\mathbb{N}$ of nonnegative integers, $0, 1, 2, \ldots$, such that the probability p_n that $N = n$ is equal to $e^{-\lambda} \lambda^n / n!$ (and in particular $p_0 = e^{-\lambda}$). Here λ is a nonnegative parameter, so that $p_n \geq 0$. The important requirement is that p_n is proportional to $\lambda^n / n!$, the normalizing constant $e^{-\lambda}$ arising then from the relation $\sum_{n=0}^{\infty} p_n = 1$. The mean value $\langle e^{vN} \rangle$ is given by $\sum_{n=0}^{\infty} p_n e^{nv}$, hence

$$\langle e^{vN} \rangle = e^{\lambda(e^v - 1)}. \tag{13.1}$$

The goal of this chapter is the generalization of this formula, which characterizes Poisson probability laws. The limiting case, $\lambda = 0$, corresponds to $N = 0$ (with probability $p_0 = 1$).

More generally, we obtain

$$\langle (1+u)^N \rangle = e^{\lambda u}, \tag{13.2}$$

from which (13.1) follows by the change of variable $u = e^v - 1$. Upon expanding $(1+u)^N$ by means of the binomial series and identifying the coefficients of like powers of u, we find

$$\langle N(N-1)\ldots(N-k+1)\rangle = \lambda^k \qquad \text{for } k \geq 1, \tag{13.3}$$

and, in particular,

$$\langle N \rangle = \lambda, \qquad \langle N(N-1) \rangle = \lambda^2. \tag{13.4}$$

It follows that the *mean value* $\langle N \rangle$ is equal to λ, and that the standard deviation $\sigma = \sqrt{\langle N^2 \rangle - \langle N \rangle^2}$ is equal to $\sqrt{\lambda}$.

Consider a family $(N_1, \ldots, N_r)$ of independent random variables, where N_i is a Poisson variable with mean value λ_i. From

$$\exp\left(\sum_{i=1}^{r} v_i N_i\right) = \prod_{i=1}^{r} \exp(v_i N_i) \tag{13.5}$$

we deduce by independence of the N_i

$$\left\langle \exp\left(\sum_{i=1}^{r} v_i N_i\right) \right\rangle = \prod_{i=1}^{r} \langle \exp(v_i N_i) \rangle, \tag{13.6}$$

that is

$$\left\langle \exp\left(\sum_{i=1}^{r} v_i N_i\right) \right\rangle = \exp\left(\sum_{i=1}^{r} \lambda_i (e^{v_i} - 1)\right). \tag{13.7}$$

Conversely, the independence of the N_i follows from this formula, and, after reduction to (13.1), we conclude that each variable N_i is of Poisson type with mean value λ_i.

On specializing to the case in which $v_1 = \ldots = v_r = v$ in (13.7), we find that $N = N_1 + \cdots + N_r$ is a Poisson variable with mean value $\lambda = \lambda_1 + \cdots + \lambda_r$. In other words,

$$\langle N \rangle = \langle N_1 \rangle + \cdots + \langle N_r \rangle$$

as required by the additivity of mean values.

A definition of Poisson processes

A Poisson process is a good model for many physical processes, including radioactive decay. Let the observations begin at time $t^{(0)}$ (also denoted t_0). We make *records* of a fortuitous event, such as the emission of an α-particle from the radioactive sample. The times of these records are random, forming a sequence

$$t^{(0)} < T_1 < T_2 < \ldots < T_k < T_{k+1} < \ldots$$

that increases without bounds and has the following properties:

(a) *the waiting times*

$$W_1 = T_1 - t^{(0)}, \qquad W_2 = T_2 - T_1, \ldots, \qquad W_k = T_k - T_{k-1}, \ldots$$

are stochastically independent; and
(b) *the probability law of W_k is exponential,*

$$\Pr[W_k > t] = e^{-at} \qquad \text{for } t \geq 0. \tag{13.8}$$

The parameter $a > 0$ is called the *decay constant*; since at is a pure number, a has the physical dimension of inverse time (or frequency).

The *probability density* of W_k, defined by

$$\Pr[t < W_k < t + dt] = p_a(t)dt, \tag{13.9}$$

is given by

$$p_a(t) = ae^{-at}\theta(t), \tag{13.10}$$

where $\theta(t)$ is the *Heaviside function*, normalized such that $\theta(t) = 1$ for $t \geq 0$, and $\theta(t) = 0$ for $t < 0$.

The sample space Ω corresponding to this process[1] is the set of sequences

$$\omega = \left(t^{(0)} < t_1 < \ldots < t_k < t_{k+1} < \ldots\right)$$

or, in terms of the coordinates $w_k = t_k - t_{k-1}$, the set of sequences

$$\mathbf{w} = (w_1, w_2, \ldots)$$

of elements of the half-line $L =]0, +\infty[$. Denoting the probability measure $p_a(t)dt = ae^{-at}dt$ on L by $d\Pi_a(t)$, the probability law Π_a on Ω is defined by

$$d\Pi_a(\mathbf{w}) = d\Pi_a(w_1)d\Pi_a(w_2)\ldots \tag{13.11}$$

One calculates the probability that $T_k > c$ explicitly by means of the integral

$$I(k,c) := \int_{w_1+\ldots+w_k > c} dw_1 \ldots dw_k \, a^k e^{-a(w_1+\ldots+w_k)}\theta(w_1)\ldots\theta(w_k)$$

$$= e^{-ac}\left(1 + ac + \frac{1}{2!}(ac)^2 + \cdots + \frac{1}{(k-1)!}(ac)^{k-1}\right). \tag{13.12}$$

It follows that $\lim_{k=\infty} I(k,c) = 1$ for each $c > 0$. Hence with probability unity, we find that $\lim_{k=\infty} T_k = +\infty$. We can therefore replace the space

---

[1] See Section 1.6 for a review of the foundations of probability theory.

13.1 Poisson stochastic processes

Ω by the subspace $\mathcal{P}$ of sequences ω such that $\lim_{k=\infty} t_k = +\infty$; hence $\mathcal{P}$ is our basic probability space, with measure Π_a.

The *joint probability density* for the finite subsequence $(T_1, \ldots, T_k)$ is given by

$$p_k^a(t_1, \ldots, t_k) = p_a(t_1 - t_0)p_a(t_2 - t_1)\ldots p_a(t_k - t_{k-1})$$

$$= a^k e^{-a(t_k - t_0)} \prod_{i=1}^{k} \theta(t_i - t_{i-1}). \qquad (13.13)$$

Notice the *normalization*

$$\int_{\Omega_k} dt_1 \ldots dt_k \, p_k^a(t_1, \ldots, t_k) = 1, \qquad (13.14)$$

where Ω_k is the subset of $\mathbb{R}^k$ defined by the inequalities $t_0 < t_1 < \ldots < t_k$. Notice also the *consistency relation*

$$\int_{t_k}^{\infty} dt_{k+1} \, p_{k+1}^a(t_1, \ldots, t_k, t_{k+1}) = p_k^a(t_1, \ldots, t_k) \qquad (13.15)$$

for $k \geq 0$.

Let $X = u(T_1, \ldots, T_k)$ be a random variable depending on the observation of $T_1, \ldots, T_k$ alone; its mean value is given by

$$\langle X \rangle = a^k \int_{t_0}^{\infty} dt_1 \int_{t_1}^{\infty} dt_2 \ldots \int_{t_{k-1}}^{\infty} dt_k \, e^{-a(t_k - t_0)} u(t_1, \ldots, t_k). \qquad (13.16)$$

The counting process

Let $N(t)$, for $t > t^{(0)}$, be the number of records occurring at a time less than or equal to time t; this is the random variable

$$N(t) := \sum_{k=1}^{\infty} \theta(t - T_k). \qquad (13.17)$$

Here, $N(t)$ is finite on the path space $\mathcal{P}$, since $\lim_{k=\infty} T_k = +\infty$ on $\mathcal{P}$. The *random function* $(N(t))_{t > t^{(0)}}$ is a step function with unit jumps occurring at the *random times* $T_1, T_2, \ldots, T_k, \ldots$ (see figure 13.1).

Since $N(t)$ is a step function, its derivative is a sum of Dirac δ-functions,

$$\frac{d}{dt} N(t) = \sum_{k=1}^{\infty} \delta(t - T_k),$$

which leads to the (Stieltjes) integral

$$\int_{t_0}^{\infty} v(t) \cdot dN(t) = \sum_{k=1}^{\infty} v(T_k). \qquad (13.18)$$

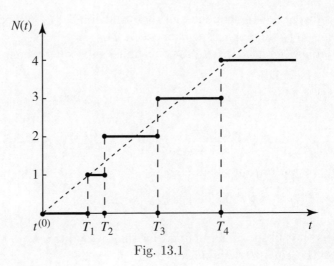

Fig. 13.1

This summation is finite if the function v vanishes outside the bounded time interval $]t^{(0)}, t^{(1)}]$.

Let $\mathbb{T} =]t', t'']$ be a finite time interval (with $t^{(0)} \leq t' < t''$). The increment $N_{\mathbb{T}} = N(t'') - N(t')$ is the number of records occurring during $\mathbb{T}$. The following properties give an alternative description of the Poisson process.

(A) *For each finite time interval $\mathbb{T} =]t', t'']$, the random variable $N_{\mathbb{T}}$ follows a Poisson law with mean value $\lambda_a(\mathbb{T}) = a(t'' - t')$.*
(B) *For mutually disjoint time intervals $\mathbb{T}^{(1)}, \ldots, \mathbb{T}^{(r)}$, the random variables $N_{\mathbb{T}^{(1)}}, \ldots, N_{\mathbb{T}^{(r)}}$ are stochastically independent.*

We begin the proof by calculating the distribution of $N(t)$ for $t > t_0$. For $k = 0, 1, 2, \ldots$, the event $\{N(t) = k\}$ is equivalent to $T_k \leq t < T_{k+1}$. If A is the subset of $\mathbb{R}^{k+1}$ defined by the inequalities

$$t_0 < t_1 < \ldots < t_k \leq t < t_{k+1},$$

then the probability $\Pr[N(t) = k]$ is given by the integral over A of the function $p_{k+1}^a(t_1, \ldots, t_k, t_{k+1}) = a^{k+1} e^{-a(t_{k+1} - t_0)}$. This integral factors into the volume of the set

$$\{t_0 < t_1 < \ldots < t_k < t\}$$

in $\mathbb{R}^k$, which is equal to $(t - t_0)^k / k!$, and the integral

$$\int_t^\infty dt_{k+1} \, a^{k+1} e^{-a(t_{k+1} - t_0)} = a^k e^{-a(t - t_0)}. \tag{13.19}$$

Using the definition

$$\varphi_k^a(t) := (at)^k e^{-at} / k!, \tag{13.20}$$

13.1 Poisson stochastic processes

we conclude that $\Pr[N(t) = k] = \varphi_k^a(t - t_0)$. In other words, $N(t)$ *follows a Poisson law with mean value* $a(t - t_0)$.

Let now $\mathbb{T}^{(1)} =]t^{(0)}, t^{(1)}]$ and $\mathbb{T}^{(2)} =]t^{(1)}, t^{(2)}]$ be two consecutive time intervals. To calculate the probability of the event $\{N_{\mathbb{T}^{(1)}} = k_1, N_{\mathbb{T}^{(2)}} = k_2\}$, we have to integrate the function

$$p_{k_1+k_2+1}^a(t_1, \ldots, t_{k_1+k_2+1}) = a^{k_1+k_2+1} e^{-a(t_{k_1+k_2+1} - t_0)} \tag{13.21}$$

over the subset B of $\mathbb{R}^{k_1+k_2+1}$ defined by the inequalities

$$t^{(0)} < t_1 < \ldots < t_{k_1} \leq t^{(1)}, \tag{13.22}$$
$$t^{(1)} < t_{k_1+1} < \ldots < t_{k_1+k_2} \leq t^{(2)}, \tag{13.23}$$
$$t^{(2)} < t_{k_1+k_2+1}. \tag{13.24}$$

Because this integration domain B is a cartesian product, the integral splits into three factors:

(a) the volume of the set defined by (13.22), which is $(t^{(1)} - t^{(0)})^{k_1}/k_1!$;
(b) the volume $(t^{(2)} - t^{(1)})^{k_2}/k_2!$ corresponding to the domain defined by (13.23); and
(c) the integral

$$\int_{t^{(2)}}^{\infty} dt_{k_1+k_2+1}\, a^{k_1+k_2+1} e^{-a(t_{k_1+k_2+1} - t_0)}$$
$$= a^{k_1+k_2} e^{-a(t^{(2)} - t^{(0)})}$$
$$= a^{k_1} e^{-a(t^{(1)} - t^{(0)})} \cdot a^{k_2} e^{-a(t^{(2)} - t^{(1)})}.$$

By combining these factors, we obtain finally

$$\Pr[N_{\mathbb{T}^{(1)}} = k_1, N_{\mathbb{T}^{(2)}} = k_2] = \varphi_{k_1}^a(t^{(1)} - t^{(0)})\varphi_{k_2}^a(t^{(2)} - t^{(0)}). \tag{13.25}$$

In other words, $N_{\mathbb{T}^{(1)}}$ and $N_{\mathbb{T}^{(2)}}$ are independent, and each follows a Poisson law with mean value

$$\langle N_{\mathbb{T}^{(i)}} \rangle = \lambda_a(\mathbb{T}^{(i)}) \qquad \text{for } i \in \{1, 2\}. \tag{13.26}$$

The generalization from two time intervals to a finite collection of consecutive time intervals, namely

$$\mathbb{T}^{(1)} =]t^{(0)}, t^{(1)}], \ldots, \mathbb{T}^{(r)} =]t^{(r-1)}, t^{(r)}],$$

is obvious; hence $(N_{\mathbb{T}^{(1)}}, \ldots, N_{\mathbb{T}^{(r)}})$ is a collection of stochastically independent Poisson variables with mean value $\langle N_{\mathbb{T}^{(i)}} \rangle = \lambda_a(\mathbb{T}^{(i)})$. Since any collection of mutually disjoint time intervals can be extended to a collection like $\mathbb{T}^{(1)}, \ldots, \mathbb{T}^{(r)}$, assertions (A) and (B) follow immediately.

Remark. From our previous calculations, we find that

$$\langle N(t) \rangle = a(t - t^{(0)}).$$

The random step function $(N(t))_{t>t^{(0)}}$ averages to the continuous linear function $a(t - t^{(0)})$, which corresponds to the dotted line in figure 13.1. In other words, the discrete set of records, due to the fluctuations, gives rise to a *density of events*, which is equal to a. It would require few changes to accommodate a time-varying decay constant $a(t)$, where now $\lambda(\mathbb{T}) = \int_{t'}^{t''} dt \, a(t)$ for $\mathbb{T} =]t', t'']$ (see Section 13.2).

A generating functional

Let us fix a finite time interval $\mathbb{T} =]t^{(0)}, t^{(1)}]$ and a complex-valued function[2] v on $\mathbb{T}$. We are interested in the distribution of the random variable

$$J = \int_{\mathbb{T}} v(t) \cdot dN(t), \qquad (13.27)$$

abbreviated $J = \int v \cdot dN$. It suffices to calculate the Laplace transform $\langle e^{pJ} \rangle$, and, since $pJ = \int (pv) \cdot dN$, it suffices to consider the case $p = 1$ and therefore $\langle e^{J} \rangle$.

The event $\{N_{\mathbb{T}} = k\}$ is equal to $\{t^{(0)} < T_1 < \ldots < T_k \leq t^{(1)} < T_{k+1}\}$. When it occurs $\int v \cdot dN$ is equal to $v(T_1) + \cdots + v(T_k)$, and $e^{J} = u(T_1) \ldots u(T_k)$, where $u = e^v$. Under the same circumstances, the joint probability density of $(T_1, \ldots, T_k)$ is equal to

$$\int_{t^{(1)}}^{\infty} dt_{k+1} \, p_{k+1}^a(t_1, \ldots, t_k, t_{k+1}),$$

which is constant and equal to $a^k e^{-a(t^{(1)} - t^{(0)})}$. Therefore, if I_k is a random variable that is equal to 1 if $N_{\mathbb{T}} = k$ and to 0 otherwise, we find that

$$\langle I_k \cdot e^J \rangle = a^k e^{-a(t^{(1)} - t^{(0)})} \int_{A_k} dt_1 \ldots dt_k \, u(t_1) \ldots u(t_k), \qquad (13.28)$$

where A_k is the domain defined by the inequalities

$$t^{(0)} < t_1 < \ldots < t_k < t^{(1)}.$$

The function to be integrated is symmetric in $(t_1, \ldots, t_k)$; its integral over the domain A_k is therefore $1/k!$ times the integral over the cube

$$\mathbb{T}^k = \{t^{(0)} < t_1 < t^{(1)}, \ldots, t^{(0)} < t_k < t^{(1)}\}.$$

By combining these factors, we obtain

$$\langle I_k \cdot e^J \rangle = a^k e^{-\lambda_a(\mathbb{T})} \left(\int_{\mathbb{T}} dt \cdot u(t) \right)^k \Big/ k!. \qquad (13.29)$$

---

[2] It suffices to assume that v is measurable and bounded.

On summing over $k = 0, 1, 2, \ldots$ we obtain the formula

$$\left\langle \exp\left(\int_\mathbb{T} v \cdot dN\right) \right\rangle = \exp\left(\int_\mathbb{T} d\lambda_a \cdot (e^v - 1)\right), \qquad (13.30)$$

where $d\lambda_a(t) := a\, dt$.

Remarks. (1) As noted already, at is a pure number if t is a time. Hence the measure λ_a on $\mathbb{T}$ defined by $d\lambda_a(t) = a \cdot dt$ is without physical dimension, and the integral $\lambda_a(\mathbb{T}) = \int_\mathbb{T} d\lambda_a(t)$ is a pure number. Indeed, $\lambda_a(\mathbb{T})$ is the mean value of the integer-valued random variable $N_\mathbb{T}$ (the number of records occurring during the time interval $\mathbb{T}$).

(2) Suppose that v is a step function. If the interval $\mathbb{T}$ is subdivided into suitable subintervals $\mathbb{T}^{(1)}, \ldots, \mathbb{T}^{(r)}$, the function v assumes a constant value v_i on $\mathbb{T}^{(i)}$. Thus we find that

$$J = \sum_{i=1}^r v_i N_{\mathbb{T}^{(i)}}, \qquad \int_\mathbb{T} d\lambda_a \cdot (e^v - 1) = \sum_{i=1}^r \lambda_a(\mathbb{T}^{(i)})(e^{v_i} - 1).$$

Specializing (13.30) to this case yields

$$\left\langle \exp\left(\sum_{i=1}^r v_i N_{\mathbb{T}^{(i)}}\right) \right\rangle = \exp\left(\sum_{i=1}^r \lambda_a(\mathbb{T}^{(i)})(e^{v_i} - 1)\right).$$

From the multivariate generating series (13.7), we recover the basic properties (A) and (B) above. Therefore, *formula (13.30) gives a characterization of the probability law of a Poisson process.*

13.2 Spaces of Poisson paths

Statement of the problem

Our goal is to solve the *Klein–Gordon* (or *Dirac*) equation by means of functional integration. We recall the problem from Section 12.2[3]

$$\frac{1}{c^2}\frac{\partial^2 G}{\partial t^2} - \frac{\partial^2 G}{\partial x^2} - \frac{a^2}{c^2}G = 0, \qquad (13.31)$$

$$G(x, 0) = \phi(x), \qquad \frac{\partial}{\partial t}G(x, t)\Big|_{t=0} = a\phi(x). \qquad (13.32)$$

The data are the numerical constants a and c and the function ϕ. We incorporate ϕ and c into the function

$$G_0(x, t) = \frac{1}{2}(\phi(x + ct) + \phi(x - ct)), \qquad (13.33)$$

---

[3] The Cauchy data (13.32) are adjusted so that the solution $G_a(x, t)$ is given in the form $e^{at} F(x, t)$, where $F(x, t)$ is a solution of the Cauchy problem (12.1) and (12.2).

which is the solution of the *wave equation*

$$\frac{1}{c^2}\frac{\partial^2 G_0}{\partial t^2} - \frac{\partial^2 G_0}{\partial x^2} = 0, \tag{13.34}$$

$$G_0(x,0) = \phi(x), \qquad \frac{\partial}{\partial t}G_0(x,t)\Big|_{t=0} = 0. \tag{13.35}$$

When a is real and positive, Kac's stochastic solution is given by

$$G(x,t) = e^{at}\langle G_0(x, S(t))\rangle_a \qquad \text{for } t \geq 0, \tag{13.36}$$

where the dependence of the averaging process on a has been made explicit. The precise meaning of Kac's solution is as follows.

- The space $\mathcal{P}$ of Poisson paths consists of the sequences (of record times)

$$\omega = (0 < t_1 < t_2 < \ldots < t_k < t_{k+1} < \ldots),$$

where $\lim_{k=\infty} t_k = +\infty$.
- For a given $t \geq 0$, $S(t)$ is a functional on $\mathcal{P}$, and is defined by the relations

$$N(t,\omega) = \sum_{k=1}^{\infty} \theta(t - t_k),$$

$$S(t,\omega) = \int_0^t (-1)^{N(\tau,\omega)} d\tau. \tag{13.37}$$

- The averaging process for a functional X on $\mathcal{P}$ is defined by

$$\langle X \rangle_a = \int_{\mathcal{P}} d\Pi_a(\omega) X(\omega), \tag{13.38}$$

where the measure Π_a on $\mathcal{P}$ is defined by (13.11).

In the Klein–Gordon equation, a is purely imaginary,

$$a = -imc^2/\hbar, \tag{13.39}$$

and we would like *to perform the analytic continuation on a at the level of the averaging*, i.e. to define $\langle X \rangle_a$ for a complex parameter a and suitable functionals X on $\mathcal{P}$. On the basis of our analysis of the way in which the properties of the Poisson process are used in Kac's solution, we require the following relation:

$$\left\langle \exp\left(\int_0^\infty v(t) \cdot dN(t)\right) \right\rangle_a = \exp\left(a \int_0^\infty dt \cdot \left(e^{v(t)} - 1\right)\right) \tag{13.40}$$

for any (piecewise) continuous function $v(t)$ defined for $t > 0$ and vanishing for all t sufficiently large.

13.2 Spaces of Poisson paths

A sketch of the method

An initial approach might be to extend the validity of (13.11) and to define a *complex measure* $\mathbf{\Pi}_a$ on $\mathcal{P}$ by

$$\mathrm{d}\mathbf{\Pi}_a(\mathbf{w}) = \prod_{k=1}^{\infty} \mathrm{d}\Pi_a(w_k). \qquad (13.41)$$

The complex measure Π_a on the half-line $]0,+\infty[$ is defined by $\mathrm{d}\Pi_a(t) = ae^{-at}\,\mathrm{d}t$ as before; its total variation is finite only if $\operatorname{Re} a > 0$. Then

$$\operatorname{Var}(\Pi_a) = \int_0^{\infty} |ae^{-at}|\,\mathrm{d}t = \frac{|a|}{\operatorname{Re} a}. \qquad (13.42)$$

To define an infinite product of complex measures Θ_n, we have to assume the finiteness[4] of

$$\prod_{n=1}^{\infty} \operatorname{Var}(\Theta_n).$$

Here $\Theta_n = \Pi_a$ for every $n \geq 1$, and the finiteness of the infinite product requires $\operatorname{Var}(\Pi_a) = 1$. A straightforward calculation shows that this is true only if $a > 0$.

In order to solve the Klein–Gordon equation for times t in $[0,T]$, we need to know $S(t)$ for $t \in [0,T]$, and this quantity depends on the $N(t)$ values for $0 \leq t \leq T$ alone. Let $\mathcal{P}_{[0,T]}$ be the space of the Poisson paths $(N(t))_{0 \leq t \leq T}$ over a finite interval. We shall construct a complex measure $\mathbf{\Pi}_{a,T}$ on $\mathcal{P}_{[0,T]}$ of finite total variation, which, for $a > 0$, will be the projection of $\mathbf{\Pi}_a$ under the natural projection $\mathcal{P} \to \mathcal{P}_{[0,T]}$. The integral of a functional X on $\mathcal{P}_{[0,T]}$ with respect to this measure $\mathbf{\Pi}_a$ shall be denoted as $\langle X \rangle_{a,T}$ or $\int \mathcal{D}_{a,T} N \cdot X(N)$.

In the last step, we let T approach $+\infty$. We prove a consistency relation between $\mathbf{\Pi}_{a,T}$ and $\mathbf{\Pi}_{a,T'}$ for $0 < T < T'$. The collection of the measures $(\mathbf{\Pi}_{a,T})_{T>0}$ on the various spaces $\mathcal{P}_{[0,T]}$ defines a *promeasure* $\mathbf{\Pi}_a$ on the space $\mathcal{P}$ in the sense of [1] and as presented in Section 1.6. Denoting the corresponding average by $\langle X \rangle_a$, we prove (13.40) for any complex number a.

We postpone the application to differential equations until Section 13.3.

Basic assumptions

Let $\mathbb{M}$ be any D-dimensional manifold, which serves as a configuration space. For simplicity, we assume $\mathbb{M}$ to be oriented, Hausdorff, and covered

---

[4] See (1.19)–(1.21) for an account of product measures.

by a countable family of charts. A subset $\mathbb{B}$ of $\mathbb{M}$ shall be called *bounded* if its closure is a compact subset of $\mathbb{M}$. Furthermore, we denote a (complex) volume element on $\mathbb{M}$ by λ; in the domain of a coordinate system $(t^1, \ldots, t^D)$, the analytic form of λ is

$$d\lambda(\mathbf{t}) = \lambda(t^1, \ldots, t^D) dt^1 \ldots dt^D. \tag{13.43}$$

If $\mathbb{B} \subset \mathbb{M}$ is bounded, its (complex) volume is $\lambda(\mathbb{B}) = \int_{\mathbb{B}} d\lambda(\mathbf{t})$, which is a well-defined number.

We denote the space consisting of the closed discrete subsets of $\mathbb{M}$ by $\mathcal{P}$, i.e. subsets N such that $N \cap \mathbb{B}$ is finite for each bounded set $\mathbb{B}$ in $\mathbb{M}$. The elements of N are called *hits*. We are looking for a model of *collections of random hits*. If $\mathbb{B}$ is a bounded set, we denote the number of points in $N \cap \mathbb{B}$ by $N(\mathbb{B})$. For any function v on $\mathbb{B}$, we denote the sum of the values $v(\mathbf{t})$ for $\mathbf{t}$ running over the $N(\mathbb{B})$ points in $N \cap \mathbb{B}$ by $\int_{\mathbb{B}} v \cdot dN$. Our purpose is to define an averaging process (or volume element on $\mathcal{P}$) satisfying the characteristic property

$$\left\langle \exp\left(\int_{\mathbb{B}} v \cdot dN\right) \right\rangle = \exp\left(\int_{\mathbb{B}} d\lambda \cdot (e^v - 1)\right) \tag{13.44}$$

for any bounded (measurable) set $\mathbb{B}$ in $\mathbb{M}$ and any bounded (measurable) function v on $\mathbb{B}$.

The bounded case

Fix a compact set $\mathbb{B}$ in $\mathbb{M}$. We denote by $\mathcal{P}_k(\mathbb{B})$ (for $k = 0, 1, 2, \ldots$) the class of subsets of $\mathbb{B}$ consisting of k elements. Define $\mathcal{P}(\mathbb{B})$ as the class of finite subsets of $\mathbb{B}$; $\mathcal{P}(\mathbb{B})$ is the disjoint union of the sets $\mathcal{P}_k(\mathbb{B})$ for $k = 0, 1, 2, \ldots$

We require a parametrization of $\mathcal{P}_k(\mathbb{B})$: indeed, let $\mathbb{B}^k$ be the usual cartesian product of k copies of $\mathbb{B}$, endowed with the product topology and the volume element $d\lambda(\mathbf{t}_1) \ldots d\lambda(\mathbf{t}_k)$. Remove from $\mathbb{B}^k$ the set of sequences $(\mathbf{t}_1, \ldots, \mathbf{t}_k)$ where at least one equality of type $\mathbf{t}_i = \mathbf{t}_j$ occurs. We arrive at a locally compact space Ω_k that carries the full measure of $\mathbb{B}^k$. On the space Ω_k, the group S_k of the $k!$ permutations of $\{1, 2, \ldots, k\}$ acts by permutation of the components $\mathbf{t}_1, \ldots, \mathbf{t}_k$. The space of orbits Ω_k / S_k can be identified with $\mathcal{P}_k(\mathbb{B})$ by mapping the orbit of the ordered sequence $(\mathbf{t}_1, \ldots, \mathbf{t}_k)$ onto the (unordered) finite subset $\{\mathbf{t}_1, \ldots, \mathbf{t}_k\}$. We define the measure $d\mu^k$ on $\mathcal{P}_k(\mathbb{B})$ as the image of the measure

$$d\lambda^k(\mathbf{t}_1, \ldots, \mathbf{t}_k) = \frac{1}{k!} e^{-\lambda(\mathbb{B})} d\lambda(\mathbf{t}_1) \ldots d\lambda(\mathbf{t}_k) \tag{13.45}$$

on Ω_k by the projection $\Omega_k \to \mathcal{P}_k(\mathbb{B})$.

13.2 Spaces of Poisson paths

A functional X on $\mathcal{P}$ is given by a collection of functions $X_k : \Omega_k \to \mathbb{C}$, namely

$$X_k(\mathbf{t}_1, \ldots, \mathbf{t}_k) = X(N), \qquad \text{where } N = \{\mathbf{t}_1, \ldots, \mathbf{t}_k\}. \tag{13.46}$$

These functions X_k are symmetric in their arguments. The mean value is defined by

$$\langle X \rangle := e^{-\lambda(\mathbb{B})} \sum_{k=0}^{\infty} \frac{1}{k!} \int_{\mathbb{B}} \cdots \int_{\mathbb{B}} d\lambda(\mathbf{t}_1) \ldots d\lambda(\mathbf{t}_k) X_k(\mathbf{t}_1, \ldots, \mathbf{t}_k). \tag{13.47}$$

Let v be a function on $\mathbb{B}$, which is measurable and bounded. Consider the functional $J(N) = \exp \int_{\mathbb{B}} v \cdot dN$ on $\mathcal{P}$. Since $\int_{\mathbb{B}} v \cdot dN = v(\mathbf{t}_1) + \cdots + v(\mathbf{t}_k)$ for $N = \{\mathbf{t}_1, \ldots, \mathbf{t}_k\}$, the functional J is multiplicative; for each k, we find that

$$J_k(\mathbf{t}_1, \ldots, \mathbf{t}_k) = u(\mathbf{t}_1) \ldots u(\mathbf{t}_k), \tag{13.48}$$

where $u = \exp v$. From equation (13.47), we obtain

$$\langle J \rangle = e^{-\lambda(\mathbb{B})} \sum_{k=0}^{\infty} \frac{1}{k!} \int_{\mathbb{B}} d\lambda(\mathbf{t}_1) u(\mathbf{t}_1) \ldots \int_{\mathbb{B}} d\lambda(\mathbf{t}_k) u(\mathbf{t}_k)$$

$$= e^{-\lambda(\mathbb{B})} \exp\left(\int_{\mathbb{B}} d\lambda(\mathbf{t}) \cdot u(\mathbf{t})\right)$$

and finally

$$\langle J \rangle = \exp\left(\int_{\mathbb{B}} d\lambda(\mathbf{t}) \cdot (u(\mathbf{t}) - 1)\right). \tag{13.49}$$

Since $u = e^v$, equation (13.44) is proved.

Since $\mathcal{P}(\mathbb{B})$ is the disjoint union of the sets Ω_k/S_k, and each Ω_k is locally compact, $\mathcal{P}(\mathbb{B})$ is endowed with a natural locally compact topology, and each μ^k is a complex bounded Radon measure (see Section 1.5). We denote by $\mathcal{D}_{\lambda,\mathbb{B}} N$ (abbreviated as $\mathcal{D}N$) the measure on $\mathcal{P}(\mathbb{B})$ that induces μ^k on each $\mathcal{P}_k(\mathbb{B})$. The total variation of μ^k on $\mathcal{P}_k(\mathbb{B})$ is $|e^{-\lambda(\mathbb{B})}| |\lambda|(\mathbb{B})^k/k!$; therefore the total variation of $\mathcal{D}N$, that is

$$\int_{\mathcal{P}(\mathbb{B})} |\mathcal{D}_{\lambda,\mathbb{B}} N| = |e^{-\lambda(\mathbb{B})}| \sum_{k=0}^{\infty} |\lambda|(\mathbb{B})^k/k!$$

$$= \exp(|\lambda|(\mathbb{B}) - \operatorname{Re} \lambda(\mathbb{B})),$$

is a finite number.

246   *A mathematical theory of Poisson processes*

It follows that $\mathcal{D}_{\lambda,\mathbb{B}}N$ *is a bounded complex Radon measure on the locally compact Polish space* $\mathcal{P}(\mathbb{B})$, *and the averaging process is*

$$\langle X \rangle = \int_{\mathcal{P}(\mathbb{B})} \mathcal{D}_{\lambda,\mathbb{B}} N \cdot X(N). \tag{13.50}$$

To avoid confusion, we can write $\langle X \rangle_{\lambda,\mathbb{B}}$ instead of $\langle X \rangle$.

An interpretation

We shall interpret equation (13.44) in terms of a Fourier transformation on a Banach space. Indeed, let $E(\mathbb{B})$ denote the Banach space consisting of the continuous real-valued functions v on $\mathbb{B}$, with the norm

$$\|v\| = \sup_{\mathbf{t} \in \mathbb{B}} |v(\mathbf{t})|. \tag{13.51}$$

We endow the space $E(\mathbb{B})'$, dual to $E(\mathbb{B})$, with the weak topology.[5] If one associates with any $N = \{\mathbf{t}_1, \ldots, \mathbf{t}_k\}$ in $\mathcal{P}(\mathbb{B})$ the linear form $v \to \int_\mathbb{B} v \cdot dN = v(\mathbf{t}_1) + \cdots + v(\mathbf{t}_k)$, one identifies $\mathcal{P}(\mathbb{B})$ with a subspace of $E(\mathbb{B})'$. The bounded complex measure $\mathcal{D}N = \mathcal{D}_{\lambda,\mathbb{B}}N$ on $\mathcal{P}(\mathbb{B})$ can be extended to a measure[6] with the same symbol on $E(\mathbb{B})'$, which is identically 0 outside $\mathcal{P}(\mathbb{B})$. Hence

$$\int_{E(\mathbb{B})'} \mathcal{D}N \cdot X(N) = \int_{\mathcal{P}(\mathbb{B})} \mathcal{D}N \cdot X(N) \tag{13.52}$$

for any functional X on $E(\mathbb{B})'$.

The *Fourier transform* of the bounded measure $\mathcal{D}_{\lambda,\mathbb{B}}N$ on $E(\mathbb{B})'$ is the function on $E(\mathbb{B})$ defined by

$$Z_\mathbb{B}(v) = \int_{E(\mathbb{B})'} \mathcal{D}_{\lambda,\mathbb{B}} N \cdot e^{2\pi i \int v \cdot dN}. \tag{13.53}$$

By applying equation (13.44), we find that

$$Z_\mathbb{B}(v) = \exp\left(\int_\mathbb{B} d\lambda \cdot (e^{2\pi i v} - 1) \right). \tag{13.54}$$

---

[5] For any ξ_0 in $E(\mathbb{B})'$ and $v_1, \ldots, v_r$ in $E(\mathbb{B})$, let $V(\xi_0; v_1, \ldots, v_r)$ consist of the elements ξ such that $|\xi(v_i) - \xi_0(v_i)| < 1$ for $1 \leq i \leq r$. Then these sets $V(\xi_0; v_1, \ldots, v_r)$ for varying systems $(v_1, \ldots, v_r)$ form a basis of neighborhoods of ξ_0 in $E(\mathbb{B})'$ for the weak topology.

[6] The space $E(\mathbb{B})'$ with the weak topology is *not* a Polish space. Nevertheless, it can be decomposed as a disjoint union $C_0 \cup C_1 \cup C_2 \cup \ldots$, where each piece C_n consists of the elements ξ in $E(\mathbb{B})'$ with $n \leq \|\xi\| < n+1$ and is therefore a Polish space. Integrating over $E(\mathbb{B})'$ is as easy as it would be if this space were Polish.

Since the Banach space $E(\mathbb{B})$ is separable, we have a uniqueness theorem for Fourier transformation. Therefore $\mathcal{D}_{\lambda,\mathbb{B}} N$ is the complex bounded measure on $E(\mathbb{B})'$ whose Fourier transform is given by equation (13.54).

The unbounded case

To generalize the previous constructions from the bounded subset $\mathbb{B}$ to the unbounded space $\mathbb{M}$, we shall imitate a known strategy of quantum field theory, local algebras, using the mathematical tool of "promeasures" (see [1] and Section 1.6).

We view the elements N of $\mathcal{P}$ as point distributions, a kind of random field, for example a shoal of herring distributed randomly in the sea. A functional $X(N)$ on $\mathcal{P}$ is supported by a compact subset $\mathbb{B}$ of $\mathbb{M}$ if $X(N)$ depends on $N \cap \mathbb{B}$ alone; it is obtained from a functional Ξ on $\mathcal{P}(\mathbb{B})$ by the rule

$$X(N) = \Xi(N \cap \mathbb{B}). \tag{13.55}$$

These functionals on $\mathcal{P}$, under some weak regularity assumption, for example that Ξ is bounded and measurable on the locally compact Polish space $\mathcal{P}(\mathbb{B})$, form an algebra $\mathcal{O}(\mathbb{B})$ for the pointwise operations. It is quite obvious that $\mathcal{O}(\mathbb{B})$ is contained in $\mathcal{O}(\mathbb{B}')$ for $\mathbb{B} \subset \mathbb{B}' \subset \mathbb{M}$. Any functional in some algebra $\mathcal{O}(\mathbb{B})$ shall be called *local*. Our goal is to define the mean values of local functionals.

Let X be a local functional, let $\mathbb{B}$ be a compact subset of $\mathbb{M}$ such that X belongs to $\mathcal{O}(\mathbb{B})$, and let Ξ be the corresponding functional on $\mathcal{P}(\mathbb{B})$ (see equation (13.55)). *We claim that*

$$I(\mathbb{B}) := \int_{\mathcal{P}(\mathbb{B})} \mathcal{D}_{\lambda,\mathbb{B}} N \cdot \Xi(N) \tag{13.56}$$

is independent of $\mathbb{B}$. Once this has been proved, we shall denote the previous integral by $\int_\mathcal{P} \mathcal{D}_\lambda N \cdot X(N)$, or $\langle X \rangle_\lambda$, or simply $\langle X \rangle$.

Proof. (a) We have to prove the equality $I(\mathbb{B}) = I(\mathbb{B}')$ for any pair of compact subsets $\mathbb{B}$ and $\mathbb{B}'$ in $\mathbb{M}$. Since the union of two compact subsets is compact, it suffices to prove the equality under the assumption that $\mathbb{B} \subset \mathbb{B}'$.

(b) Assuming that $\mathbb{B} \subset \mathbb{B}'$, consider the map $\rho: N \to N \cap \mathbb{B}$ from $\mathcal{P}(\mathbb{B}')$ to $\mathcal{P}(\mathbb{B})$. The spaces $\mathcal{P}(\mathbb{B})$ and $\mathcal{P}(\mathbb{B}')$ are locally compact, *but ρ is not necessarily continuous*. We leave it as an exercise for the reader to construct a decomposition of each $\mathcal{P}_k(\mathbb{B}')$ into a finite number of pieces, each of which is the difference between two compact subsets (for example $K \backslash L$, with

$L \subset K$ compact), such that the restriction of ρ to each piece is continuous. It follows that the map ρ is measurable.

(c) Since ρ is measurable, the image under ρ of the measure $\mathcal{D}_{\lambda,\mathbb{B}'}N$ on $\mathcal{P}(\mathbb{B}')$ is defined as a (complex, bounded) measure on $\mathcal{P}(\mathbb{B})$, and is denoted by $\overline{\mathcal{D}}_{\lambda,\mathbb{B}}N$. The characteristic property of $\overline{\mathcal{D}}_{\lambda,\mathbb{B}}N$ is

$$\int_{\mathcal{P}(\mathbb{B})} \overline{\mathcal{D}}_{\lambda,\mathbb{B}}N \cdot X(N) = \int_{\mathcal{P}(\mathbb{B}')} \mathcal{D}_{\lambda,\mathbb{B}'}N' \cdot X(N' \cap \mathbb{B}) \tag{13.57}$$

for every (measurable and bounded) functional X on $\mathcal{P}(\mathbb{B})$. The consistency relation $I(\mathbb{B}) = I(\mathbb{B}')$ can be reformulated as

$$\int_{\mathcal{P}(\mathbb{B})} \mathcal{D}_{\lambda,\mathbb{B}}N \cdot X(N) = \int_{\mathcal{P}(\mathbb{B}')} \mathcal{D}_{\lambda,\mathbb{B}'}N' \cdot X(N' \cap \mathbb{B}). \tag{13.58}$$

Using equation (13.57), this reduces to the equality $\overline{\mathcal{D}}_{\lambda,\mathbb{B}}N = \mathcal{D}_{\lambda,\mathbb{B}}N$.

(d) To prove that these measures on $\mathcal{P}(\mathbb{B})$ are equal, we use the uniqueness of the Fourier transformation; we have to prove that $\overline{\mathcal{D}}_{\lambda,\mathbb{B}}N$ and $\mathcal{D}_{\lambda,\mathbb{B}}N$ give the same integral for a functional X of the form

$$X(N) = \exp\left(i \int_{\mathbb{B}} v \cdot dN\right) \tag{13.59}$$

for any continuous function v on $\mathbb{B}$. Hence, we are reduced to the proof of equation (13.58) for the functionals X of type (13.59). The left-hand side of (13.58) is then

$$\int_{\mathcal{P}(\mathbb{B})} \mathcal{D}_{\lambda,\mathbb{B}}N \cdot \exp\left(i \int_{\mathbb{B}} v \cdot dN\right) = \exp\left(\int_{\mathbb{B}} d\lambda(\mathbf{t}) \cdot \left(e^{iv(\mathbf{t})} - 1\right)\right). \tag{13.60}$$

Notice the formula

$$\int_{\mathbb{B}} v \cdot d(N' \cap \mathbb{B}) = \int_{\mathbb{B}'} v' \cdot dN', \tag{13.61}$$

where v' is the extension[7] of v to $\mathbb{B}'$ which is identically 0 outside $\mathbb{B}$. The right-hand side of (13.58) is therefore equal to

$$\int_{\mathcal{P}(\mathbb{B}')} \mathcal{D}_{\lambda,\mathbb{B}'}N' \cdot \exp\left(i \int_{\mathbb{B}'} v' \cdot dN'\right) = \exp\left(\int_{\mathbb{B}'} d\lambda(\mathbf{t}) \cdot \left(e^{iv'(\mathbf{t})} - 1\right)\right). \tag{13.62}$$

To conclude, notice that, since $v' = 0$ outside $\mathbb{B}$, the function $e^{iv'} - 1$ is also 0 outside $\mathbb{B}$. Integrating either $e^{iv} - 1$ over $\mathbb{B}$ or $e^{iv'} - 1$ over $\mathbb{B}'$ gives the same result. □

---

[7] This function is not always continuous, but equation (13.44) remains true for bounded measurable functions, and v' is measurable!

We know that for any (bounded measurable) function v on $\mathbb{M}$, vanishing outside a bounded subset $\mathbb{B}$, the functional $\int_{\mathbb{M}} v \cdot dN$, and therefore $\exp(\int_{\mathbb{M}} v \cdot dN)$, belongs to $\mathcal{O}(\mathbb{B})$. From equation (13.60), we obtain immediately[8]

$$\int_{\mathcal{P}} \mathcal{D}_\lambda N \cdot \exp\left(\int_{\mathbb{M}} v \cdot dN\right) = \exp\left(\int_{\mathbb{M}} d\lambda \cdot (e^v - 1)\right). \qquad (13.63)$$

As a corollary, let v be a step function; there exist disjoint bounded subsets $\mathbb{B}_1, \ldots, \mathbb{B}_r$ and constants $v_1, \ldots, v_r$ such that $v(\mathbf{t}) = v_i$ for $\mathbf{t}$ in $\mathbb{B}_i$ ($1 \leq i \leq r$) and $v(\mathbf{t}) = 0$ for $\mathbf{t}$ not in $\mathbb{B}_1 \cup \ldots \cup \mathbb{B}_r$. Equation (13.63) reduces to

$$\int \mathcal{D}_\lambda N \cdot \exp\left(\sum_{i=1}^{r} v_i N(\mathbb{B}_i)\right) = \exp\left(\sum_{i=1}^{r} \lambda(\mathbb{B}_i)(e^{v_i} - 1)\right). \qquad (13.64)$$

Application to probability theory

Assume that λ is nonnegative, and therefore that its density $\lambda(t^1, \ldots, t^D)$ is nonnegative in any coordinate system $(t^1, \ldots, t^D)$ of $\mathbb{M}$. Then, for each compact subset $\mathbb{B}$, $\mathcal{D}_{\lambda,\mathbb{B}} N$ is a positive measure of total mass 1 on $\mathcal{P}(\mathbb{B})$, i.e. a probability measure. In this case, *there is a probability measure $\mathcal{D}_\lambda N$ on $\mathcal{P}$, such that the mean value of a local functional $X(N)$ is the integral of $X(N)$ with respect to this measure* (which justifies our notation!). This enables us to find the mean value of some nonlocal functionals.

To prove our claim, we introduce a sequence of compact subsets

$$\mathbb{B}_1 \subset \mathbb{B}_2 \subset \mathbb{B}_3 \subset \ldots \subset \mathbb{B}_n \subset \ldots \subset \mathbb{M}$$

exhausting $\mathbb{M}$, i.e.

$$\mathbb{M} = \bigcup_n \mathbb{B}_n.$$

Denote by $\mathcal{D}_n N$ the probability measure $\mathcal{D}_{\lambda,\mathbb{B}_n} N$ on $\mathcal{P}(\mathbb{B}_n)$. We proved earlier that the restriction map $N_n \mapsto N_n \cap \mathbb{B}_{n-1}$ of $\mathcal{P}(\mathbb{B}_n)$ into $\mathcal{P}(\mathbb{B}_{n-1})$ maps the measure $\mathcal{D}_n N_n$ into $\mathcal{D}_{n-1} N_{n-1}$. Furthermore, elements N of $\mathcal{P}$ correspond, via the equation $N_n = N \cap \mathbb{B}_n$, to sequences N_1 in $\mathcal{P}(\mathbb{B}_1)$, N_2 in $\mathcal{P}(\mathbb{B}_2), \ldots$ such that $N_n \cap \mathbb{B}_{n-1} = N_{n-1}$ for $n \geq 2$. From a well-known theorem of Kolmogoroff [2], we know that there exists a unique probability measure $\mathcal{D}_\lambda N$ on $\mathcal{P}$ mapping onto $\mathcal{D}_n N$ for each $n \geq 1$. Since

---

[8] Notice that the function $e^v - 1$ is bounded and measurable on $\mathbb{M}$ and vanishes outside the bounded set $\mathbb{B}$; hence the integral $\int_{\mathbb{M}} d\lambda \cdot (e^v - 1)$ is well defined, and equal to $\int_{\mathbb{B}} d\lambda \cdot (e^v - 1)$. Furthermore, equation (13.63) is valid *a priori* for v purely imaginary. To get the general case, use analytic continuation by writing v as $\lambda v_1 + iv_2|_{\lambda=1}$ with a real parameter λ.

every compact subset of $\mathbb{M}$ is contained in some $\mathbb{B}_n$, every local functional belongs to $\mathcal{O}(\mathbb{B}_n)$ for some n; hence the promeasure $\mathcal{D}_\lambda N$ is indeed a probability measure on $\mathcal{P}$.

To conclude, *when λ is nonnegative, $\mathcal{D}_\lambda N$ is the probability law of a random point process on $\mathbb{M}$.*

A special case

Let $\mathbb{M}$ be now the real line $\mathbb{R}$, with the standard topology; a subset of $\mathbb{R}$ is bounded iff it is contained in a finite interval $\mathbb{T} = [t_a, t_b]$. The measure λ shall be λ_a defined by $d\lambda_a(t) = a\, dt$, where a is a *complex constant*. The measure $\lambda_a(\mathbb{T})$ of the interval $\mathbb{T} = [t_a, t_b]$ is aT, where $T = t_b - t_a$ is the length of $\mathbb{T}$. The only notable change from the general theory occurs in the description of $\mathcal{P}_k(\mathbb{T})$; by taking advantage of the natural ordering on the line, an element N of $\mathcal{P}_k(\mathbb{T})$ can be uniquely enumerated as $N = \{t_1, \ldots, t_k\}$ with

$$t_a \leq t_1 < \ldots < t_k \leq t_b. \tag{13.65}$$

The volume element on $\mathcal{P}_k(\mathbb{T})$ is

$$\begin{aligned} d\mu_a^k(t_1, \ldots, t_k) &= e^{-\lambda(\mathbb{T})} d\lambda_a(t_1) \ldots d\lambda_a(t_k) \\ &= a^k e^{-aT}\, dt_1 \ldots dt_k. \end{aligned} \tag{13.66}$$

Recall that the hypercube $\mathbb{T} \times \ldots \times \mathbb{T}$ (k factors) of size T is decomposed into $k!$ simplices derived from $\mathcal{P}_k(\mathbb{T})$ by the permutations of the coordinates $t_1, \ldots, t_k$, and recall that $dt_1 \ldots dt_k$ is invariant under such permutations. Therefore the total mass $\mu_a^k(\mathcal{P}_k(\mathbb{T}))$ is equal to $e^{-\lambda(\mathbb{T})} \lambda(\mathbb{T})^k / k!$ as it should be. Hence μ_a^k is properly normalized.

For a functional X on $\mathcal{P}(\mathbb{T})$, with restriction X_k to $\mathcal{P}_k(\mathbb{T})$, the mean value is given by

$$\langle X \rangle = e^{-aT} \sum_{k=0}^{\infty} a^k \int_{t_a \leq t_1 < \ldots < t_k \leq t_b} dt_1 \ldots dt_k\, X_k(t_1, \ldots, t_k). \tag{13.67}$$

The volume element on $\mathcal{P}(\mathbb{T})$ is denoted by $\mathcal{D}_{a,\mathbb{T}} N$. The volume element on $\mathcal{P}$ shall be denoted by $\mathcal{D}_a N$. When $a > 0$, this is a probability measure on $\mathcal{P}(\mathbb{R})$; for complex a, it is, in general, a promeasure.

The first part of our program has been completed: *the probability measure $\mathcal{D}_a N$ on $\mathcal{P}(\mathbb{R})$ describing the Poisson random process has been continued analytically to a promeasure $\mathcal{D}_a N$ for every complex a, enabling us to take mean values of functionals supported by a finite time interval.*

It remains to solve differential equations.

13.3 Stochastic solutions of differential equations

A general theorem

We want to solve a differential equation of the form

$$\frac{d}{dt}\psi(t) = A(t)\psi(t). \qquad (13.68)$$

The operator $A(t)$ is *time-dependent*, and decomposes as

$$A(t) = A_0(t) + aA_1(t), \qquad (13.69)$$

where $A_0(t)$ and $A_1(t)$ are operators, and a is a complex constant.

As usual, we are seeking the *propagator* $U(t_b, t_a)$, which is defined for $t_a \leq t_b$ and satisfies

$$\frac{\partial}{\partial t_b} U(t_b, t_a) = A(t_b) U(t_b, t_a),$$
$$U(t_a, t_a) = \mathbf{1}, \qquad (13.70)$$

from which it follows (for $t_a \leq t_b \leq t_c$) that

$$U(t_c, t_a) = U(t_c, t_b) U(t_b, t_a). \qquad (13.71)$$

The solution of equation (13.68) with initial value $\psi(t_0) = \psi_0$ is then given by $\psi(t) = U(t, t_0)\psi_0$.

The essential result is as follows. Let $U_0(t_b, t_a)$ be the propagator for the unperturbed equation

$$\frac{d}{dt}\psi(t) = A_0(t)\psi(t). \qquad (13.72)$$

Assume that $t_b \geq t_a$ and define the operator-valued functions

$$V_k(t_b, t_a | t_1, \ldots, t_k) = U_0(t_b, t_k) A_1(t_k) U_0(t_k, t_{k-1}) A_1(t_{k-1}) \cdots$$
$$\cdots A_1(t_2) U_0(t_2, t_1) A_1(t_1) U_0(t_1, t_a). \qquad (13.73)$$

Then the "perturbed propagator" is given by

$$U(t_b, t_a) = \sum_{k=0}^{\infty} a^k \int_{\Delta_k} dt_1 \ldots dt_k \, V_k(t_b, t_a | t_1, \ldots, t_k), \qquad (13.74)$$

where the domain of integration Δ_k is defined by the inequalities

$$t_a \leq t_1 < t_2 < \ldots < t_k \leq t_b. \qquad (13.75)$$

Two particular cases are noteworthy.

(a) If $A_0(t) = 0$ and the inequalities (13.75) are satisfied, then $U_0(t_b, t_a) = \mathbf{1}$, and V_k reduces to

$$V_k(t_b, t_a) = A_1(t_k) \ldots A_1(t_1). \qquad (13.76)$$

In this case, equation (13.74) reduces to *Dyson's time-ordered exponential formula*.

(b) Suppose that $A_0(t)$ and $A_1(t)$ are *time-independent*, i.e. $A_0(t) = A_0$, $A_1(t) = A_1$. Let $U_0(t) = \exp(tA_0)$. We find

$$U_0(t_b, t_a) = U_0(t_b - t_a) \tag{13.77}$$

for the free propagator. Assume now that the unperturbed equation (13.72) is time reversible; therefore $U_0(t)$ is defined for negative t as well. With the definitions

$$\begin{aligned} A_{\text{int}}(t) &:= U_0(-t) A_1 U_0(t), \\ U_{\text{int}}(t_b, t_a) &:= U_0(-t_b) U(t_b, t_a) U_0(t_a), \end{aligned} \tag{13.78}$$

equation (13.74) takes the form

$$U_{\text{int}}(t_b, t_a) = \sum_{k=0}^{\infty} a^k \int_{\Delta_k} dt_1 \ldots dt_k\, A_{\text{int}}(t_k) \ldots A_{\text{int}}(t_1). \tag{13.79}$$

A comparison with the special case (a) leads to the conclusion that $U_{\text{int}}(t_b, t_a)$ is the propagator for the equation

$$\frac{d}{dt} \psi_{\text{int}}(t) = a A_{\text{int}}(t) \psi_{\text{int}}(t). \tag{13.80}$$

The solutions of equation (13.68),

$$\frac{d}{dt} \psi(t) = (A_0 + a A_1) \psi(t), \tag{13.81}$$

and the solutions of equation (13.80) are connected by the transformation

$$\psi_{\text{int}}(t) = U_0(-t) \psi(t). \tag{13.82}$$

In summary, equations (13.78)–(13.82) describe the well-known method of *interaction representation* for solving a "perturbed" stationary differential equation (see Section 12.3).

Proof of the main theorem

(A) Assume that $t_a = t_b$. The domain of integration Δ_k consists of one point $t_1 = \ldots = t_k = t_a$ for $k \geq 1$, and therefore the integral over Δ_k is 0 for $k \geq 1$. Taking into account the limiting value for $k = 0$, namely

$$V_0(t_b, t_a |\ ) = U_0(t_b, t_a), \tag{13.83}$$

we find that

$$V_0(t_a, t_a |\ ) = U_0(t_a, t_a) = \mathbf{1}. \tag{13.84}$$

13.3 Stochastic solutions of differential equations

Formula (13.74) reduces to
$$U(t_a, t_a) = V_0(t_a, t_a|) = 1 \qquad (13.85)$$
since the integral is 0 for $k \geq 1$.

(B) The inductive form of definition (13.73) is
$$V_k(t_b, t_a|t_1, \ldots, t_k) = U_0(t_b, t_k) A_1(t_k) V_{k-1}(t_k, t_a|t_1, \ldots, t_{k-1}) \qquad (13.86)$$
for $k \geq 1$; the initial value for $k = 0$ is derived from equation (13.83). Fix t_a, and let $t_b > t_a$ be variable. Denote by $\Phi_k(t_b)$ the coefficient of a^k in equation (13.74), and define $\Phi(t_b)$ as the sum
$$\sum_{k=0}^{\infty} a^k \Phi_k(t_b).$$

The result of (A) can be expressed as
$$\Phi(t_a) = 1. \qquad (13.87)$$

By integrating equation (13.86) over $t_1, \ldots, t_k$, we obtain the inductive relation (for $k \geq 1$)
$$\Phi_k(t_b) = \int_{t_a}^{t_b} dt_k \, U_0(t_b, t_k) A_1(t_k) \Phi_{k-1}(t_k). \qquad (13.88)$$

The integration variable t_k can be written as τ. Taking into account the initial value, we find that
$$\Phi_0(t_b) = U_0(t_b, t_a). \qquad (13.89)$$

On multiplying both sides of equation (13.88) by a^k and summing over k, we obtain by virtue of equations (13.87) and (13.89)
$$\Phi(t_b) = U_0(t_b, t_a) \Phi(t_a) + \int_{t_a}^{t_b} d\tau \, U_0(t_b, \tau) \Psi(\tau) \qquad (13.90)$$

with $\Psi(\tau) = a A_1(\tau) \Phi(\tau)$. Any function $\Phi(t)$ satisfying (13.90) is a solution of the inhomogeneous differential equation
$$\left(\frac{d}{dt_b} - A_0(t_b) \right) \Phi(t_b) = \Psi(t_b). \qquad (13.91)$$

Using the definition of $\Psi(t_b)$, we find that $\Phi(t_b)$ is the solution of the differential equation
$$\frac{d}{dt_b} \Phi(t_b) = (A_0(t_b) + a A_1(t_b)) \Phi(t_b) \qquad (13.92)$$

with the initial value $\Phi(t_a) = 1$, and therefore $\Phi(t_b) = U(t_b, t_a)$ as requested.

The relation with functional integration

Fix t_a and t_b such that $t_b \geq t_a$ and let $T = t_b - t_a$. A functional $M(t_b, t_a)$ on $\mathcal{P}$ that is supported by the interval $\mathbb{T} = [t_b, t_a]$ is defined by

$$M(t_b, t_a|N) = e^{aT} V_k(t_b, t_a|t_1, \ldots, t_k) \tag{13.93}$$

for $N = \{t_1, \ldots, t_k\}$ in $\mathcal{P}_k(\mathbb{T})$ and $t_a \leq t_1 < \ldots < t_k \leq t_b$. Equation (13.74) now reads

$$U(t_b, t_a) = \int_{\mathcal{P}} \mathcal{D}_a N \cdot M(t_b, t_a|N). \tag{13.94}$$

When $A_0(t) = 0$ and $A_1(t) = Q(t)$, the functional $M(t_b, t_a)$ is given by

$$M(t_b, t_a|N) = e^{aT} Q(t_k) \ldots Q(t_1). \tag{13.95}$$

Formula (13.94) then becomes the theorem of D. Collins referred to in Section 12.3.

The functional $M(t_b, t_a)$ is an example of a *multiplicative functional*,[9] in other words, it satisfies the equation

$$M(t_c, t_a) = M(t_c, t_b) M(t_b, t_a) \tag{13.96}$$

for $t_a < t_b < t_c$. More explicitly, the relation

$$V_{k+\ell}(t_c, t_a|t_1, \ldots, t_{k+\ell}) = V_\ell(t_c, t_b|t_{k+1}, \ldots, t_{k+\ell}) V_k(t_b, t_a|t_1, \ldots, t_k) \tag{13.97}$$

for[10]

$$t_a \leq t_1 < \ldots < t_k \leq t_b \leq t_{k+1} < \ldots < t_{k+\ell} \leq t_c$$

follows from the definition (13.73) and the property

$$U_0(t_{k+1}, t_k) = U_0(t_{k+1}, t_b) U_0(t_b, t_k) \tag{13.98}$$

for the "free propagator" $U_0(t_b, t_a)$.

The main formula (13.74) can be written as

$$\langle M(t_b, t_a) \rangle_a = U(t_b, t_a). \tag{13.99}$$

This result exactly validates Feynman's insight that the propagator $U(t_b, t_a)$ should be expressed as a "sum over histories." The propagation property

$$U(t_c, t_a) = U(t_c, t_b) U(t_b, t_a) \tag{13.100}$$

---

[9] This notion has been used extensively in probability theory after the discovery of the Feynman–Kac formula. See for instance P. A. Meyer in [3].

[10] Here and in similar situations, we do not need to distinguish between weak and strong inequalities in $t_k \leq t_b \leq t_{k+1}$ since we eventually integrate over the variables t_k and t_{k+1}, and a single point contributes nothing to a simple integral.

13.3 Stochastic solutions of differential equations

for $t_c > t_b > t_a$ can be restated as

$$\langle M(t_c, t_b) \cdot M(t_b, t_a) \rangle_a = \langle M(t_c, t_b) \rangle_a \langle M(t_b, t_a) \rangle_a \qquad (13.101)$$

using equations (13.96) and (13.99). We now give a proof using functional integration.

Formula (13.101) is a special case of a general formula[11]

$$\langle M \cdot M' \rangle_a = \langle M \rangle_a \langle M' \rangle_a \qquad (13.102)$$

for functionals M and M' supported by adjacent intervals: M by $[t_b, t_c]$ and M' by $[t_a, t_b]$. It suffices to prove (13.102) for the case

$$M = \exp\left(\int_{t_b}^{t_c} v \cdot dN\right), \qquad M' = \exp\left(\int_{t_a}^{t_b} v \cdot dN\right),$$

using the methods of Section 13.2. In this case, $M \cdot M'$ is equal to $\exp\left(\int_{t_a}^{t_c} v \cdot dN\right)$, and this equation reduces to

$$\exp\left(a \int_{t_a}^{t_c} dt \cdot \left(e^{v(t)} - 1\right)\right) = \exp\left(a \int_{t_a}^{t_b} dt \cdot \left(e^{v(t)} - 1\right)\right)$$
$$\times \exp\left(a \int_{t_b}^{t_c} dt \cdot \left(e^{v(t)} - 1\right)\right), \qquad (13.103)$$

which is a consequence of the additivity property of the integral

$$\int_{t_a}^{t_c} = \int_{t_a}^{t_b} + \int_{t_b}^{t_c}.$$

The short-time propagator

Beginning with the definition of the propagator in terms of a sum over histories (see equation (13.99)), we derived the propagator property (13.100). To complete the picture, we must establish the value of the *short-time propagator*

$$U(t + \Delta t, t) = \mathbf{1} + A(t) \cdot \Delta t + \mathcal{O}((\Delta t)^2), \qquad (13.104)$$

where the error is of the order of $(\Delta t)^2$.

We first recall *how to derive the differential equation from equations (13.100) and (13.104)*. Given any initial value, ψ_0, at time t_0, for the state vector, we define

$$\psi(t) = U(t, t_0)\psi_0 \qquad \text{for } t \geq t_0. \qquad (13.105)$$

---

[11] One derives in the same manner the same property for M and M' when these functionals are supported by disjoint intervals.

Since $U(t_0, t_0) = \mathbf{1}$, $\psi(t_0) = \psi_0$. Furthermore, from the propagator property (13.100), we find that

$$\psi(t') = U(t', t)\psi(t) \qquad \text{for } t' \geq t \geq t_0. \tag{13.106}$$

Upon inserting the short-time propagator (13.104) into (13.106), we obtain

$$\psi(t + \Delta t) = \psi(t) + A(t)\psi(t)\Delta t + \mathcal{O}((\Delta t)^2), \tag{13.107}$$

and therefore

$$\frac{\mathrm{d}\psi(t)}{\mathrm{d}t} = A(t)\psi(t) \tag{13.108}$$

on taking the limit $\Delta t = 0$.

To compute the short-time propagator, let us return to equation (13.74) for $t_a = t$, $t_b = t + \Delta t$. Denote by I_k the coefficient of a^k on the right-hand side of equation (13.74).

- For $k = 0$, this contribution is

$$I_0 = U_0(t + \Delta t, t) = 1 + A_0(t)\Delta t + \mathcal{O}((\Delta t)^2). \tag{13.109}$$

- For $k = 1$, this contribution is $I_1 = \int_t^{t+\Delta t} \mathrm{d}\tau\, \Phi(\tau)$, where

$$\Phi(\tau) = U_0(t + \Delta t, \tau) A_1(\tau) U_0(\tau, t). \tag{13.110}$$

By applying well-known infinitesimal principles, we find that the integral I_1 is equal to $\Phi(t)\Delta t$, up to an error of order $(\Delta t)^2$; but

$$\Phi(t) = U_0(t + \Delta t, t) A_1(t) U_0(t, t) = A_1(t) + \mathcal{O}(\Delta t).$$

We conclude that

$$I_1 = A_1(t)\Delta t + \mathcal{O}((\Delta t)^2). \tag{13.111}$$

- For $k \geq 2$, we are integrating over a part of a hypercube of volume $(\Delta t)^k$, hence $I_k = \mathcal{O}((\Delta t)^k)$. Each term I_k can be neglected individually. Uniformity is easy to achieve. By continuity, $A_1(\tau)$ and $U_0(\tau', \tau)$ are bounded around $\tau = \tau' = t$, hence there exist constants B and C such that $|I_k| \leq C(B\,\Delta t)^k$ for $k \geq 2$. Therefore, if Δt is small enough, then

$$\left| \sum_{k \geq 2} a^k I_k \right| \leq \frac{a^2 C(B\,\Delta t)^2}{1 - aB\,\Delta t}, \tag{13.112}$$

and consequently

$$\sum_{k \geq 2} a^k I_k = \mathcal{O}((\Delta t)^2).$$

13.3 Stochastic solutions of differential equations

In summary, we find that

$$U(t+\Delta t, t) = I_0 + aI_1 + \mathcal{O}((\Delta t)^2) = 1 + A_0(t)\Delta t + aA_1(t)\Delta t + \mathcal{O}((\Delta t)^2),$$

hence the short-time propagator (13.104).

Remark. The summation over k amounts to a summation over the (random) number of hits $N_{[t,t+\Delta t]}$ in the short time interval $[t, t+\Delta t]$. Hence I_0 and I_1 correspond to $N_{[t,t+\Delta t]} = 0$ and $N_{[t,t+\Delta t]} = 1$, respectively. The *main point is that the probability of occurrence of more than one hit is of the order of* $(\Delta t)^2$; hence this case does not contribute to the short-time propagator.

A remark

We want to show that *our scheme of functional integration is the only one for which the propagator of any differential equation*

$$\frac{d}{dt}\psi(t) = (A_0(t) + aA_1(t))\psi(t) \tag{13.113}$$

is given by

$$U(t_b, t_a) = \langle M(t_b, t_a)\rangle_a \tag{13.114}$$

for $t_b \geq t_a$ (see equations (13.73) and (13.93) for the definition of $M(t_b, t_a)$).

As a special case, consider the ordinary differential equation

$$\frac{d}{dt}\psi(t) = a\,u(t)\psi(t), \tag{13.115}$$

where $\psi(t)$ and $u(t)$ are scalar-valued functions. The propagator is given by

$$U(t_b, t_a) = \exp\left(a\int_{t_a}^{t_b} dt\, u(t)\right), \tag{13.116}$$

and the functional $M(t_b, t_a)$ is defined as

$$M(t_b, t_a|N) = e^{a(t_b - t_a)} u(t_1)\ldots u(t_k) \tag{13.117}$$

for $N = \{t_1, \ldots, t_k\}$ included in $[t_a, t_b]$. For $u = \exp v$, the functional $M(t_b, t_a)$ is $e^{a(t_b-t_a) + \int v \cdot dN}$, and the formula $\langle M(t_b, t_a)\rangle_a = U(t_b, t_a)$ is equivalent to

$$\left\langle \exp\left(\int_{t_a}^{t_b} v \cdot dN\right)\right\rangle_a = \exp\left(a\int_{t_a}^{t_b} dt \cdot \left(e^{v(t)} - 1\right)\right), \tag{13.118}$$

which is the characteristic property (13.44).

Solution of the Dirac equation

The equation that we wish to solve is

$$\frac{d\psi}{dt} = A\psi$$

with

$$A = \begin{pmatrix} -v\partial_x & a \\ a & v\partial_x \end{pmatrix}. \qquad (13.119)$$

This matrix may be decomposed as $A = A_0 + aA_1$, where

$$A_0 = \begin{pmatrix} -v\partial_x & 0 \\ 0 & v\partial_x \end{pmatrix}, \qquad A_1 = \begin{pmatrix} 0 & 1 \\ 1 & 0 \end{pmatrix}. \qquad (13.120)$$

The vector ψ, which is a function of t, is of the form

$$\begin{pmatrix} \psi_+(x) \\ \psi_-(x) \end{pmatrix}$$

for fixed t, where $\psi_\pm(x)$ are functions of the space variable x, and $\partial_x = \partial/\partial x$ as usual.

We are working with the time-independent version of equations (13.68) and (13.69); the one-parameter group $U_0(t) = \exp(tA_0)$ is given explicitly by

$$U_0(t)\begin{pmatrix} \psi_+(x) \\ \psi_-(x) \end{pmatrix} = \begin{pmatrix} \psi_+(x - vt) \\ \psi_-(x + vt) \end{pmatrix}, \qquad (13.121)$$

whereas

$$A_1\begin{pmatrix} \psi_+(x) \\ \psi_-(x) \end{pmatrix} = \begin{pmatrix} \psi_-(x) \\ \psi_+(x) \end{pmatrix}. \qquad (13.122)$$

According to equation (13.73), we need to calculate the operator $V_k(T, 0|t_1, \ldots, t_k)$, abbreviated as V_k, which is defined as the product

$$\left[\prod_{i=0}^{k-1} U_0(t_{k+1-i} - t_{k-i})A_1\right] \cdot U_0(t_1 - t_0). \qquad (13.123)$$

For simplicity, we write $t_0 = 0$, $t_{k+1} = T$. It is straightforward to show that

$$V_k\begin{pmatrix} \psi_+(x) \\ \psi_-(x) \end{pmatrix} = \begin{cases} \begin{pmatrix} \psi_+(x - vS_k) \\ \psi_-(x + vS_k) \end{pmatrix} & \text{if } k \text{ is even,} \\ \begin{pmatrix} \psi_-(x + vS_k) \\ \psi_+(x - vS_k) \end{pmatrix} & \text{if } k \text{ is odd,} \end{cases} \qquad (13.124)$$

13.3 Stochastic solutions of differential equations

with the inductive definition

$$S_0 = t_1 - t_0, \qquad (13.125)$$
$$S_k = S_{k-1} + (-1)^k(t_{k+1} - t_k). \qquad (13.126)$$

In other words,

$$S_k = \sum_{i=0}^{k}(-1)^i(t_{i+1} - t_i). \qquad (13.127)$$

We can subsume both cases in equation (13.124) by

$$V_k \begin{pmatrix} \psi_+(x) \\ \psi_-(x) \end{pmatrix} = A_1^k \begin{pmatrix} \psi_+(x - vS_k) \\ \psi_-(x + vS_k) \end{pmatrix}. \qquad (13.128)$$

We give now the stochastic interpretation of this formula. Let us define a step function $(N(t))$ by

$$N(t) = \sum_{i=1}^{k} \theta(t - t_i) \qquad (13.129)$$

with unit jumps at $t_1, \ldots, t_k$ such that $0 < t_1 < \ldots < t_k < T$. Then $k = N(T)$ and $S_k = S(T)$ with

$$S(T) = \int_0^T (-1)^{N(\tau)} \, d\tau. \qquad (13.130)$$

We treat the operators $e^{aT} V_k(T, 0|t_1, \ldots, t_k)$ as the components of a functional $M(T, 0|N)$ on $\mathcal{P}_{[0,T]}$, and we rewrite equation (13.128) as

$$M(T, 0|N) \begin{pmatrix} \psi_+(x) \\ \psi_-(x) \end{pmatrix} = e^{aT} A_1^{N(T)} \begin{pmatrix} \psi_+(x - vS(T)) \\ \psi_-(x + vS(T)) \end{pmatrix}. \qquad (13.131)$$

According to equation (13.99), the propagator $U(T) = e^{TA}$ can be calculated as a functional integral, namely

$$U(T) = \int_{\mathcal{P}_{[0,T]}} \mathcal{D}_{a,T} N \cdot M(T, 0|N). \qquad (13.132)$$

To summarize, *the functional integral*

$$\begin{pmatrix} \psi_+(x, T) \\ \psi_-(x, T) \end{pmatrix} = e^{aT} \int_{\mathcal{P}_{[0,T]}} \mathcal{D}_{a,T} N \cdot A_1^{N(T)} \begin{pmatrix} \psi_+(x - vS(T)) \\ \psi_-(x + vS(T)) \end{pmatrix} \qquad (13.133)$$

is the solution of the differential equation

$$\partial_T \begin{pmatrix} \psi_+(x, T) \\ \psi_-(x, T) \end{pmatrix} = \begin{pmatrix} -v\partial_x & a \\ a & v\partial_x \end{pmatrix} \begin{pmatrix} \psi_+(x, T) \\ \psi_-(x, T) \end{pmatrix} \qquad (13.134)$$

with the initial value $\psi_\pm(x, 0) = \psi_\pm(x)$.

Remark. The matrix
$$A_1 = \begin{pmatrix} 0 & 1 \\ 1 & 0 \end{pmatrix}$$
satisfies $A_1^2 = \mathbf{1}_2$, hence $A_1^k = \mathbf{1}_2$ for k even, and $A_1^k = A_1$ for k odd. An explicit formula is
$$A_1^k = \frac{1}{2}\begin{pmatrix} 1+(-1)^k & 1-(-1)^k \\ 1-(-1)^k & 1+(-1)^k \end{pmatrix}. \tag{13.135}$$

Some special cases

As mentioned previously, when $v = c$ and $a = -imc^2/\hbar$, equation (13.134) is the Dirac equation in a two-dimensional spacetime. Our theory of complex path-integrals (in the Poisson case) gives the path-integral representation (13.133) of the solution of the Dirac equation.

If we omit the factor e^{aT} in equation (13.133), we immediately conclude that the solution to the differential equation
$$\partial_T \begin{pmatrix} u_+(x,T) \\ u_-(x,T) \end{pmatrix} = \begin{pmatrix} -v\,\partial_x - a & a \\ a & v\,\partial_x - a \end{pmatrix} \begin{pmatrix} u_+(x,T) \\ u_-(x,T) \end{pmatrix} \tag{13.136}$$
with initial value $u_\pm(x,0) = w_\pm(x)$ is given by
$$\begin{pmatrix} u_+(x,T) \\ u_-(x,T) \end{pmatrix} = \int_{\mathcal{P}_{[0,T]}} \mathcal{D}_{a,T} N \cdot A_1^{N(T)} \begin{pmatrix} w_+(x - vS(T)) \\ w_-(x + vS(T)) \end{pmatrix}. \tag{13.137}$$

Let us return to the case of second-order differential equations. Let
$$M = \begin{pmatrix} p & q \\ r & s \end{pmatrix}$$
be any matrix of pairwise commuting operators, with trace $\tau = p + s$ and determinant $\Delta = ps - qr$. Define $M' = \tau \cdot \mathbf{1}_2 - M$, i.e.
$$M' = \begin{pmatrix} s & -q \\ -r & p \end{pmatrix};$$
then
$$MM' = \begin{pmatrix} ps - qr & 0 \\ 0 & ps - qr \end{pmatrix} = \begin{pmatrix} \Delta & 0 \\ 0 & \Delta \end{pmatrix}$$
by the commutativities $pq = qp$ etc. In other words, M satisfies the so-called Hamilton–Cayley equation
$$M^2 - \tau M + \Delta \cdot \mathbf{1}_2 = 0. \tag{13.138}$$

13.3 Stochastic solutions of differential equations

A corollary: if M is time-independent, any solution of the first-order equation

$$\partial_t \psi = M\psi \qquad (13.139)$$

satisfies the second-order equation

$$(\partial_t^2 - \tau \partial_t + \Delta)\psi = 0. \qquad (13.140)$$

An application: for

$$M = \begin{pmatrix} -c\partial_x & -imc^2/\hbar \\ -imc^2/\hbar & c\partial_x \end{pmatrix},$$

one finds that $\tau = 0$, and

$$\Delta = -c^2 \partial_x^2 + \frac{m^2 c^4}{\hbar^2}.$$

Any solution of the Dirac equation is also a solution of the Klein–Gordon equation.

If M is the matrix that appears in equation (13.136), then $\tau = -2a$ and $\Delta = -v^2 \partial_x^2$. Equation (13.140) therefore specializes to

$$(\partial_t^2 + 2a\,\partial_t - v^2 \partial_x^2)\psi = 0, \qquad (13.141)$$

which is the telegraph equation; in equation (13.137), both $u_+(x,T)$ and $u_-(x,T)$ are solutions of the telegraph equation. Notice that the operator A_1 interchanges the two components of a vector

$$\begin{pmatrix} w_+ \\ w_- \end{pmatrix},$$

and therefore leaves invariant the sum $w_+ + w_-$. From equation (13.137), we find that the formula

$$u(x,T) := u_+(x,T) + u_-(x,T)$$
$$= \int_{\mathcal{P}_{[0,T]}} \mathcal{D}_{a,T} N \cdot [w_+(x - vS(T)) + w_-(x + vS(T))]$$
$$= \langle w_+(x - vS(T)) \rangle_a + \langle w_-(x + vS(T)) \rangle_a \qquad (13.142)$$

defines a solution of the telegraph equation. The initial values of this solution are given by

$$u(x,0) = w_+(x) + w_-(x), \qquad (13.143)$$
$$\partial_T u(x,T)|_{T=0} = -v\,\partial_x(w_+(x) - w_-(x)). \qquad (13.144)$$

When $w_+ = w_- = \tfrac{1}{2}\phi$, we recover Kac's solution of the telegraph equation.

13.4 Differential equations: explicit solutions

The wave equation

Recall the well-known method of solving hyperbolic equations introduced by Hadamard.

We want to solve the wave equation

$$\frac{1}{v^2}\frac{\partial^2 G}{\partial t^2} - \frac{\partial^2 G}{\partial x^2} = 0 \tag{13.145}$$

with Cauchy data

$$G(x,0) = \phi_0(x), \tag{13.146}$$

$$\frac{\partial}{\partial t}G(x,t)|_{t=0} = \phi_1(x). \tag{13.147}$$

Consider the differential operator

$$L = v^{-2}\partial_t^2 - \partial_x^2, \tag{13.148}$$

where $\partial_t := \partial/\partial t$ and $\partial_x := \partial/\partial x$. We are seeking a (generalized) function $E(x,t)$ with the following properties:

- it satisfies the *differential equation* $L \cdot E(x,t) = \delta(x)\delta(t)$; and
- it satisfies the *support condition* $E(x,t) = 0$ for $t < 0$.

This function is called the elementary solution (or retarded propagator).

Let $G(x,t)$ be a solution of the wave equation $L \cdot G = 0$ in the spacetime $\mathbb{R}^2$ with coordinates x,t. Let us truncate G so that it is only a function over positive times, i.e.

$$G^+(x,t) = \theta(t)G(x,t), \tag{13.149}$$

where $\theta(t)$ is the Heaviside step function. It is straightforward to show that

$$L \cdot G^+(x,t) = \theta(t) L \cdot G(x,t) + v^{-2}[\phi_0(x)\delta'(t) + \phi_1(x)\delta(t)]. \tag{13.150}$$

This formula simplifies since we have assumed that $L \cdot G = 0$. Since the operator L is translation-invariant, its inverse is given by the convolution with the elementary solution $E(x,t)$. We can therefore invert equation (13.150), and we find that

$$G^+(x,t) = v^{-2}\int\int d\xi\, d\tau\, E(x-\xi, t-\tau)[\phi_0(\xi)\delta'(\tau) + \phi_1(\xi)\delta(\tau)]. \tag{13.151}$$

13.4 Differential equations: explicit solutions

By performing the time integration (with respect to τ), we obtain

$$G^+(x,t) = v^{-2} \int d\xi\, E(x-\xi, t)\phi_1(\xi) + v^{-2} \int d\xi\, \partial_t E(x-\xi, t)\phi_0(\xi). \tag{13.152}$$

Our task is now to determine the elementary solution. The simplest method is to use so-called lightcone coordinates

$$x_+ = vt - x, \qquad x_- = vt + x. \tag{13.153}$$

On writing $\partial_\pm$ for $\partial/\partial x_\pm$, one obtains the transformation formulas

$$\partial_t = v(\partial_+ + \partial_-), \qquad \partial_x = \partial_- - \partial_+, \tag{13.154}$$

and therefore

$$L = 4\partial_+ \partial_-. \tag{13.155}$$

The volume element is $dx\, dt = [1/(2v)]dx_+\, dx_-$, and, since $\delta(x)\delta(t)dx\, dt$ is invariant under the coordinate changes, we obtain

$$\delta(x)\delta(t) = 2v\delta(x_+)\delta(x_-). \tag{13.156}$$

Since

$$\partial_+ \partial_- [\theta(x_+)\theta(x_-)] = \delta(x_+)\delta(x_-) \tag{13.157}$$

we find that

$$E(x,t) = \frac{v}{2}\theta(vt-x)\theta(vt+x), \tag{13.158}$$

and therefore

$$\partial_t E(x,t) = \frac{v^2}{2}\theta(t)[\delta(x+vt) + \delta(x-vt)] \tag{13.159}$$

for the retarded propagator. Upon inserting this solution into equation (13.152), we obtain the well-known solution[12] to the wave equation (13.145) with Cauchy data (13.146) and (13.147):

$$G(x,t) = \frac{1}{2}[\phi_0(x+vt) + \phi_0(x-vt)] + \frac{1}{2v}\int_{x-vt}^{x+vt} d\xi\, \phi_1(\xi) \tag{13.160}$$

for $t \geq 0$. The last term can also be written as

$$\frac{1}{2}\int_{-t}^{t} dr\, \phi_1(x+vr).$$

---

[12] Taking $\phi_1 = 0$, we recover the solution (13.33) to the equations (13.34) and (13.35).

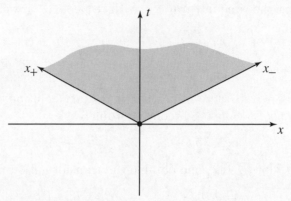

Fig. 13.2 The shaded region is the support of the elementary solution

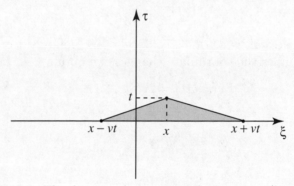

Fig. 13.3 The domain of integration for equation (13.151)

The Klein–Gordon equation

We extend the previous method to the Klein–Gordon equation, written as $L_a \cdot G = 0$, with the differential operator

$$L_a := L - a^2/v^2 = v^{-2}\partial_t^2 - \partial_x^2 - a^2 v^{-2}. \tag{13.161}$$

By comparison with equation (13.158), we construct the following *Ansatz* for the retarded propagator:

$$E_a(x,t) = \frac{v}{2}\theta(x_+)\theta(x_-)H_a(x_+, x_-). \tag{13.162}$$

The function H_a must satisfy the differential equation

$$\partial_+ \partial_- H_a(x_+, x_-) = \frac{a^2}{4v^2} H_a(x_+, x_-), \tag{13.163}$$

whose solution is given by

$$H_a(x_+, x_-) = \sum_{m=0}^{\infty} \left(\frac{a^2 x_+ x_-}{4v^2}\right)^m \bigg/ (m!)^2. \tag{13.164}$$

13.4 Differential equations: explicit solutions

On reverting to the variables x and t, we obtain

$$E_a(x,t) = \frac{v}{2}\theta(vt-x)\theta(vt+x)I_0\left(a\sqrt{t^2-x^2/v^2}\right), \qquad (13.165)$$

where I_0 is the modified Bessel function [4]. From equation (13.165), one finds that

$$\partial_t E_a(x,t) = \frac{v^2}{2}\theta(t)[\delta(x+vt)+\delta(x-vt)] \qquad (13.166)$$

$$+ \frac{avt}{2\sqrt{t^2-x^2/v^2}}\theta(vt-x)\theta(vt+x)I_1\left(a\sqrt{t^2-x^2/v^2}\right),$$

since the modified Bessel function I_1 is the derivative of I_0 [5].

By insertion, one finds an explicit solution for the Klein–Gordon equation

$$\left(\frac{1}{v^2}\frac{\partial^2}{\partial t^2} - \frac{\partial^2}{\partial x^2} - \frac{a^2}{v^2}\right)G_a(x,t) = 0, \qquad (13.167)$$

$$G_a(x,0) = \phi_0(x), \qquad \frac{\partial}{\partial t}G_a(x,t)|_{t=0} = \phi_1(x), \qquad (13.168)$$

namely (see equation (13.152))

$$G_a(x,t) = \frac{1}{2}[\phi_0(x+vt)+\phi_0(x-vt)]$$

$$+ \frac{1}{2}\int_{x-vt}^{x+vt} d\xi \cdot \phi_0(\xi) \frac{at}{\sqrt{v^2t^2-(x-\xi)^2}} I_1\left(a\sqrt{t^2-\frac{(x-\xi)^2}{v^2}}\right)$$

$$+ \frac{1}{2v}\int_{x-vt}^{x+vt} d\xi \cdot \phi_1(\xi) I_0\left(a\sqrt{t^2-\frac{(x-\xi)^2}{v^2}}\right). \qquad (13.169)$$

The telegraph equation

Recall the problem

$$\frac{1}{v^2}\frac{\partial^2 F_a}{\partial t^2} + \frac{2a}{v^2}\frac{\partial^2 F_a}{\partial t} - \frac{\partial^2 F_a}{\partial x^2} = 0, \qquad (13.170)$$

$$F_a(x,0) = \phi_0(x), \qquad \frac{\partial}{\partial t}F_a(x,t)|_{t=0} = 0, \qquad (13.171)$$

from Section 12.1. It is obvious that, for any solution $F_a(x,t)$ to this problem, the function $G_a(x,t) = e^{at}F_a(x,t)$ is a solution of the Klein–Gordon equation (13.167) with Cauchy data

$$\phi_0(x) = \phi(x), \qquad \phi_1(x) = a\phi(x). \qquad (13.172)$$

Given the solution to the Klein–Gordon equation in formula (13.169), and given the change of integration variable $\xi = x + vr$, we find the explicit solution to the telegraph equation:

$$F_a(x,t) = \frac{e^{-at}}{2}[\phi(x+vt) + \phi(x-vt)] + \int_{-t}^{t} dr\, \phi(x-vr) p_a(t,r), \tag{13.173}$$

with

$$p_a(t,r) := \frac{a}{2} e^{-at} I_0\left(a\sqrt{t^2-r^2}\right) + \frac{a}{2} \frac{e^{-at} t}{\sqrt{t^2-r^2}} I_1\left(a\sqrt{t^2-r^2}\right). \tag{13.174}$$

Comparison with the stochastic solution

Assume now that $a > 0$. Recall Kac's stochastic solution (12.16):

$$F_a(x,t) = \frac{1}{2}\langle \phi(x + v\,S(t))\rangle_a + \frac{1}{2}\langle \phi(x - v\,S(t))\rangle_a. \tag{13.175}$$

Here the random variable $S(t)$ is defined on the sample space of the Poisson process $N(t)$ with intensity a, i.e. $\langle N(t)\rangle_a = at$. We introduce the probability density $g_a(t,r)$ of $S(t)$ and deduce the analytic form of equation (13.175):

$$F_a(x,t) = \int_{-\infty}^{+\infty} dr\, \phi(x+vr) \frac{1}{2}(g_a(t,r) + g_a(t,-r)). \tag{13.176}$$

By comparing equation (13.176) with equation (13.174), we conclude that, for $t \geq 0$,

$$g_a(t,r) + g_a(t,-r) = e^{-at}[\delta(t+r) + \delta(t-r)] + ae^{-at}\theta(t-|r|)$$
$$\times \left[I_0\left(a\sqrt{t^2-r^2}\right) + \frac{t}{\sqrt{t^2-r^2}} I_1\left(a\sqrt{t^2-r^2}\right)\right]. \tag{13.177}$$

We thereby recover the result of [6], which was described by equation (12.22) in Section 12.1. For the case of the Dirac equation, we refer the reader to Appendix H.

References

[1] N. Bourbaki (1969). *Eléments de mathématiques, intégration*, Chapter 9 (Paris, Hermann). English translation by S. K. Berberian, *Integration II* (Berlin, Springer, 2004).

[2] N. Bourbaki (1969). *Eléments de mathématiques, intégration*, Chapter 9 (Paris, Hermann), p. 119. English translation by S. K. Berberian, *Integration II* (Berlin, Springer, 2004).

[3] P. A. Meyer (1966). *Probabilités et potentiel* (Paris, Hermann).

[4] M. Abramowitz and I. A. Stegun (1970). *Handbook of Mathematical Functions*, 9th edn. (New York, Dover Publications), p. 375, formula 9.6.12.
H. Bateman and A. Erdelyi (1953). *Higher Transcendental Functions*, Vol. II, (New York, McGraw-Hill), p. 5, equation (12).

[5] H. Bateman and A. Erdelyi (1953). *Higher Transcendental Functions*, Vol. II, (New York, McGraw-Hill), p. 79, formula (20).
M. Abramowitz and I. A. Stegun (1970). *Handbook of Mathematical Functions*, 9th edn. (New York, Dover Publications), p. 376, formula 9.6.27.

[6] C. DeWitt-Morette and S.-K. Foong (1990). "Kac's solution of the telegrapher's equation revisited: part I," *Annals Israel Physical Society*, **9**, 351–366. Nathan Rosen Jubilee Volume, eds. F. Cooperstock. L. P. Horowitz, and J. Rosen (Israel Physical Society, Jerusalem, 1989).
C. DeWitt-Morette and S.-K. Foong (1989). "Path integral solutions of wave equation with dissipation," *Phys. Rev. Lett.* **62**, 2201–2204.
S.-K. Foong (1990). "Kac's solution of the telegrapher's equation revisited: part II," *Annals Israel Physical Society* **9**, 367–377.
S.-K. Foong (1993). "Path integral solution for telegrapher's equation," in *Lectures on Path Integration, Trieste 1991*, eds. H. A. Cerdeira, S. Lundqvist, D. Mugnai, A. Ranfagni, V. Sayakanit, and L. S. Schulman (Singapore, World Scientific).

14
The first exit time; energy problems

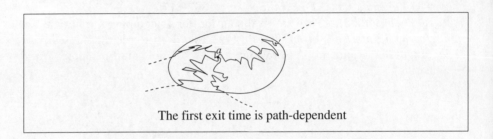

The first exit time is path-dependent

14.1 Introduction: fixed-energy Green's function

In previous chapters, we constructed functional integral representations of the time evolution of solutions of Schrödinger's equation and Dirac's equation with scattering amplitudes such as

$$\langle b|\exp(-iH(t_b - t_a)/\hbar)|a\rangle =: \langle b, t_b|a, t_b\rangle \qquad (14.1)$$
$$=: \langle b|a\rangle_T.$$

In this section we construct solutions of the time-independent Schrödinger equation, which is an elliptic equation of the form

$$H\psi_n = E_n\psi_n \quad \text{(no summation)}, \qquad (14.2)$$

where $\{\psi_n\}$ and $\{E_n\}$ are the set of energy eigenfunctions and the spectrum of a hamiltonian operator H, respectively. We recall their construction by means of a resolvent operator $\lim_{\epsilon=0}(E - H + i\epsilon)$.

The amplitude $\langle b|a\rangle_T$, restricted to positive T, can be written [1, p. 88]

$$\langle b|a\rangle_T = \theta(T)\sum_n \psi_n(b)\psi_n^*(a)\exp(-iE_nT/\hbar), \qquad (14.3)$$

14.1 Introduction: fixed-energy Green's function

where $T = t_b - t_a$. The Heaviside step function $\theta(T)$ is responsible for the term $i\epsilon$ in the resolvent operator. Let $G(b, a; E)$ be the Green function of the resolvent operator

$$(E - H + i\epsilon)G(b, a; E) = \delta(b - a). \tag{14.4}$$

Its spectral decomposition is

$$G(b, a; E) = \left\langle b \left| \frac{1}{E - H + i\epsilon} \right| a \right\rangle \tag{14.5}$$

$$= \sum_n \langle b|\psi_n\rangle \frac{1}{E - E_n + i\epsilon} \langle \psi_n|a\rangle.$$

$G(b, a; E)$ is a fixed-energy transition amplitude. The time elapsed when the system goes from the state $|a\rangle$ to the state $|b\rangle$ is irrelevant, but the system is constrained to have a fixed-energy E at all times. We sometimes use a notation similar to (14.1) and set

$$G(b, a; E) =: \langle b|a\rangle_E. \tag{14.6}$$

Classical mechanics

For formulating variational principles of energy-constrained systems, Abraham and Marsden [2] introduced a one-parameter family of functions on a fixed time interval $[t_a, t_b]$:

$$\tau : [0, 1] \times [t_a, t_b] \to \mathbb{R},$$
$$\tau(\mu, t) =: \tau_\mu(t), \qquad \mu \in [0, 1]; \tag{14.7}$$

the functions τ_μ are chosen to be monotonically increasing,

$$d\tau_\mu(t)/dt =: \lambda_\mu(t) > 0. \tag{14.8}$$

Let $\{x_{(\mu)}\}_\mu$, abbreviated to $\{x_\mu\}_\mu$, be a one-parameter family of paths. Each path x_μ is defined on its own time interval $[\tau_\mu(t_a), \tau_\mu(t_b)]$,

$$x_\mu : [\tau_\mu(t_a), \tau_\mu(t_b)] \to \mathbb{M}^D \tag{14.9}$$

with fixed end points

$$\begin{aligned} x_\mu(\tau_\mu(t_a)) &= a, \\ x_\mu(\tau_\mu(t_b)) &= b. \end{aligned} \tag{14.10}$$

Systems with constant energy being invariant under time translation, we set

$$\tau_\mu(t_a) = 0.$$

Each path x_μ reaches b at its own time $\tau_\mu(t_b)$.

Proposition. The action functional for fixed-energy systems is defined over the function space $\{\tau_\mu, x_\mu\}$ and is given by

$$S(\tau_\mu, x_\mu; E) = \int_0^{\tau_\mu(t_b)} d\tau (L(x_\mu(\tau), \dot{x}_\mu(\tau)) + E), \qquad (14.11)$$

or, emphasizing the fact that μ is a parameter[1], we can write the action functional

$$S(\tau_\mu, x_\mu; E) = \int_0^{\tau(\mu, t_b)} d\tau (L(x(\mu, \tau), \dot{x}(\mu, \tau)) + E).$$

Proof. Let (τ_0, x_0) be the critical point(s) of the action functional, namely the solutions of

$$\left. \frac{dS}{d\mu} \right|_{\mu_0} = 0. \qquad (14.12)$$

The condition $x_\mu(\tau_\mu(t_b)) = b$ implies

$$\frac{dx_\mu}{d\mu}(\tau_\mu(t_b)) = \frac{dx_\mu}{d\tau_\mu(t_b)} \frac{d\tau_\mu(t_b)}{d\mu} = 0. \qquad (14.13)$$

The condition (14.12) for the critical point(s) (τ_0, x_0) implies

$$\frac{\partial L}{\partial x_0(\tau)} - \frac{d}{d\tau} \frac{\partial L}{\partial \dot{x}_0(\tau)} = 0, \qquad (14.14)$$

$$\tau_{\mu_0}(\tau) = \tau, \qquad (14.15)$$

and

$$\left. \left(L + E - \frac{\partial L}{\partial \dot{x}_0} \dot{x}_0 \right) \right|_{t_a}^{t_b} = 0. \qquad (14.16)$$

The first two equations identify x_0 as a classical path, and $t_b - t_a$ as the time elapsed when traveling along x_0 from a to b. The third equation states that E is the constant energy of the classical path. □

For path x with values in $\mathbb{M}^D$, $D > 1$, the condition $x_\mu(\tau_\mu(t_b)) = b$ becomes

$$x_\mu(\tau_\mu(t_b)) \in \partial \mathbb{M}^D. \qquad (14.17)$$

$\tau_\mu(t_b)$ is called *the first exit time* of the path x_μ out of the domain $\mathbb{M}^D$.

---

[1] Hence, for any given value of the parameter μ, the variable τ runs over the interval $[0, \tau_\mu(t_b)]$ and $x[\mu, \tau]$ is a variable point in $\mathbb{M}^D$, such that $x(\mu, 0) = a$ and $x(\mu, \tau_\mu(t_b)) = b$.

14.1 Introduction: fixed-energy Green's function

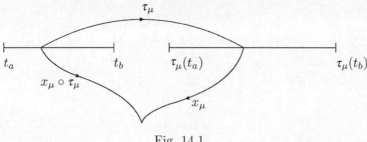

Fig. 14.1

One-parameter families of paths are sufficient for computing the critical points of the action functional but they are not sufficient for constructing domains of integration of path integrals.

Spaces of (random) paths are constructed by probabilists who introduce a probability space Ω where elements $\omega \in \Omega$ are path parameters. Here a path x_μ and its time τ_μ would be parametrized by the same element $\omega \in \Omega$:

$$(x_\mu, \tau_\mu) \quad \text{becomes} \quad (x(\omega), \tau(\omega)). \tag{14.18}$$

A probability space is a useful concept, but not the most appropriate in quantum physics. Therefore we shall reexpress the pair (x_μ, τ_μ) by a function $x_\mu \circ \tau_\mu$ defined on a fixed interval $[t_a, t_b]$ and the action functional (14.11) as an integral over $[t_a = 0, t_b]$ by the change of variable $\tau = \tau_\mu(t)$:

$$S(\tau_\mu, x_\mu, E) = \int_0^{t_b} dt \left(L\left((x_\mu \circ \tau_\mu)(t), \frac{d}{dt}(x_\mu \circ \tau_\mu)(t) \right) + E \right) \frac{d\tau_\mu(t)}{dt}. \tag{14.19}$$

The path x_μ and its own time τ_μ, together, make a new function $x_\mu \circ \tau_\mu$ (figure 14.1), which need not be restricted to a one-parameter family of paths, so we consider action functionals of the form

$$S_\lambda(q, E) = \int_0^{t_b} dt \left(L\left(q(t), \frac{d}{dt} q(t) \right) + E \right) \lambda(t), \tag{14.20}$$

where $q := x \circ \tau$ and

$$\lambda(t) := d\tau(t)/dt. \tag{14.21}$$

As expected when dealing with energy as a variable, the phase-space formalism is more convenient than the configuration-space formalism. The phase-space action functional for energy-constrained systems is given by

$$S_\lambda(q,p;E) = \int_{q_a}^{q_b} dq\, p(q,E) + \int_{t_a}^{t_b} dt(E - H(q(t),p(t)))\lambda(t), \quad (14.22)$$

where $q = x \circ \tau$ and

$$p = \partial L \Big/ \partial\left(\frac{d}{dt} x \circ \tau\right).$$

We use the same symbol S_λ in (14.20) and (14.22) for different functionals that are easily identifiable by their arguments.

Quantum mechanics

By introducing paths more general than critical points of an action functional, we have introduced a new function $\lambda(t) = d\tau(t)/dt$. When $\lambda(t)$ is a constant function,

$$\lambda(t) = \text{constant}, \quad \text{also labeled } \lambda, \quad (14.23)$$

then

$$\int d\lambda\, \exp\left(\frac{2\pi i}{h}\lambda \int_{t_a}^{t_b} dt(E - H(q(t),p(t)))\right)$$
$$= \delta\left(h^{-1} \int_{t_a}^{t_b} dt(E - H(q(t),p(t)))\right). \quad (14.24)$$

This ordinary integral over λ gives an *average* energy constraint

$$\frac{1}{T}\int_{t_a}^{t_b} dt\, H(q(t),p(t)) = E. \quad (14.25)$$

On the other hand, a functional integral over the *function* λ is expected to give a delta-functional (see Project 19.1.2)

$$\Delta((E - H(q,p))/h), \quad (14.26)$$

i.e.

$$H(q(t),p(t)) = E \quad \text{for every } t \in [t_a, t_b]. \quad (14.27)$$

Thus the path (q,p) is constrained to an energy surface.

Remark. The fixed-energy amplitude was first computed by C. Garrod [3], who assumed the function $\lambda(t)$ in (14.22) to be a constant. The "fixed energy" is then only an "average fixed energy."

In conclusion, the energy-constraint idea and the first-exit-time concept are dual expressions of the same requirements. In Abraham and Marsden's

14.2 The path integral for a fixed-energy amplitude

words, "they balance" each other. They are both useful for solving Dirichlet problems.

14.2 The path integral for a fixed-energy amplitude

The energy amplitude is the Fourier transform of the time amplitude

$$\langle b|a\rangle_E = -2\pi i h^{-1} \int_{-\infty}^{\infty} dT \, \exp\left(\frac{2\pi i}{h} ET\right) \langle b|a\rangle_T. \tag{14.28}$$

Proof. Use the definition (14.6) of $\langle b|a\rangle_E$ and the definition (14.3) of $\langle b|a\rangle_T$,

$$\langle b|a\rangle_T = \theta(T)\left\langle b\left|\exp\left(-\frac{2\pi i}{h}HT\right)\right|a\right\rangle. \tag{14.29}$$

□

Remark. $\langle b|a\rangle_E$ is not truly a probability amplitude. It does not satisfy a combination law. It is nevertheless a useful concept; see e.g. applications in Gutzwiller [4].

We begin with $\langle b|a\rangle_E^{\text{WKB}}$, the stationary-phase approximation of (14.28). From a WKB approximation one can read the action *function* of the system and a finite-dimensional determinant equal to the ratio of the infinite-dimensional determinants of the quadratic forms which characterize the system and its path integral (Section 4.2). Therefore the WKB approximation of any proposed path integral for $\langle b|a\rangle_E$ has to be equal to the stationary-phase approximation of (14.28), which is easy to compute.

The WKB approximation

The WKB approximation of the point-to-point amplitude is

$$\langle b|a\rangle_T = \langle b, t_b|a, t_a\rangle$$
$$= h^{-D/2} \sum_j \left|\det\left(\frac{\partial^2 \mathcal{S}_j(b,a;T)}{\partial a^\alpha \, \partial b^\beta}\right)\right|^{1/2} \cdot \exp\left(\frac{2\pi i}{h}\mathcal{S}_j(b,a;T) - \frac{\pi i}{2}\lambda_j\right),$$
$$\tag{14.30}$$

where the sum is over all classical paths j from a to b, with a and b not being conjugate to each other; $\mathcal{S}_j$ is the action function for the classical path j going from a to b in time T; and the Morse index λ_j of the hessian of the action functional is equal to the number of points conjugate to a along the j-path, each conjugate point being counted with its multiplicity.

Let τ_j be the value of T which minimizes the exponent of the integrand in (14.28),

$$\partial \mathcal{S}_j(b,a;T)/\partial T|_{T=\tau_j} + E = 0. \tag{14.31}$$

Then

$$\langle b|a\rangle_E^{\text{WKB}} = \sum_j |D_{W_j}(b,a;E)|^{1/2} \exp\left(\frac{2\pi\mathrm{i}}{h}W_j(b,a;E) - \frac{\pi\mathrm{i}}{2}(\lambda_j + p_j)\right) \tag{14.32}$$

with

$$W_j(b,a;E) = \mathcal{S}_j(b,a;\tau_j) + E\tau_j = \int_{j\text{-path}} \mathrm{d}q\, p(q,E) \tag{14.33}$$

and

$$D_{W_j}(b,a;E) = \det\left(\frac{\partial^2 \mathcal{S}_j(b,a;\tau_j)}{\partial a^\alpha\, \partial b^\beta}\right) \bigg/ \left(\frac{\partial^2 \mathcal{S}_j(b,a;T)}{\partial T^2}\right)\bigg|_{T=\tau_j}. \tag{14.34}$$

A nontrivial calculation (see e.g. [5, pp. 339–340]) shows that

$$D_{W_j}(b,a;E) = \det\begin{pmatrix} \partial^2 W_j/\partial a^\alpha\, \partial b^\beta & \partial^2 W_j/\partial a^\alpha\, \partial E \\ \partial^2 W_j/\partial E\, \partial b^\beta & \partial^2 W_j/\partial E^2 \end{pmatrix}. \tag{14.35}$$

The phase gain p_j is equal to the number of turning points (libration points) which occur each time $\partial^2 W_j/\partial E^2$ changes sign (recall that τ_j is a function of E). The phase gain λ_j comes from the fixed-time point-to-point amplitude (14.30).

A nondynamical degree of freedom
(contributed by John LaChapelle)

The concept "to each path its own time" can be codified by a pair (x_μ, τ_μ) or a pair $(x(\omega), \tau(\omega))$. When the pair (x_μ, τ_μ) is replaced by one function $x_\mu \circ \tau_\mu$ together with a new function $\mathrm{d}\tau_\mu/\mathrm{d}t$, and the action functional (14.19) is replaced by the action functional (14.20), a new (nondynamical) degree of freedom λ is introduced and the concept of path-dependent time-parametrization is eliminated.

The price to be paid is an integration over the function space Λ of the possible λ, and one does not know a priori what volume element $\mathcal{D}\lambda$ should be used. Proceed heuristically for guidance: assume initially that $\mathcal{D}\lambda$ is characterized by a quadratic form $Q(\lambda)$. Since λ is nondynamical, its kinetic term should vanish. This can be effected by allowing $s \in \mathbb{C}^+$ in equation (1.101) and taking the limit $|s| \to \infty$. The resulting volume element is called a "Dirac" volume element. It encodes the functional analog of the statement "the Fourier transform of 1 is a delta function."

14.2 The path integral for a fixed-energy amplitude

The other limit, $|s| \to 0$, also leads to a "Dirac" volume element that encodes the functional analog of "the Fourier transform of a delta function is 1" [6, 7]. In this chapter we use only the equation

$$\int \mathcal{D}\lambda \, \exp\left(\frac{2\pi i}{h} \int_T dt \, \lambda(t) \, (E - H(q(t), p(t)))\right)$$
$$= \int \mathcal{D}\lambda \, \exp(2\pi i \langle \lambda, (E - H(q,p))/h \rangle)$$
$$= \Delta((E - H(q,p))/h), \qquad (14.36)$$

where $\mathcal{D}\lambda$ is assumed to be a Dirac volume element and Δ is a delta-functional (see Project 19.1.2).

The path-integral representation of the fixed-energy amplitude is a gaussian integral over (q, p) defined in Section 3.4 and an integral over λ satisfying (14.36). Let $\int \mathcal{D}q \, \mathcal{D}p$ be the gaussian in phase space defined by the quadratic forms Q and W (equations (3.96) and (3.97)). Then

$$\langle b|a\rangle_E = \int_{\mathbb{X}_b} \mathcal{D}q \, \mathcal{D}p \int \mathcal{D}\lambda \, \exp\left(\frac{2\pi i}{h} S_\lambda(q, p; E)\right), \qquad (14.37)$$

where $q(t) = (x \circ \tau)(t)$ and

$$p(t) = \partial L \Big/ \partial\left(\frac{d}{dt} x \circ \tau\right).$$

The domain of integration $\mathbb{X}_b$ is the space of paths with fixed end point $q(t_b) = b$, defined on the interval $[t_a, t_b]$, but with no restriction on p. The WKB approximation $\langle b|a\rangle_E^{\text{WKB}}$ of (14.37) is easy to compute; it is a straightforward application of the technique presented in Section 4.2. In brief outline, it is done as follows.

- In the WKB approximation, the delta-functional is given by (see e.g. (14.36))

$$\Delta^{\text{WKB}}((E - H(q,p))/h) = \Delta((E - H(q_{\text{cl}}, p_{\text{cl}}))/h), \qquad (14.38)$$

where $(q_{\text{cl}}, p_{\text{cl}}) \in \mathbb{X}_b$ is a classical path on the interval $[t_a, t_b]$, such that $q_{\text{cl}}(t_b) = b$. Recall that $q_{\text{cl}} = x_0 \circ \mathbf{1}$ (see e.g. (14.15)). The WKB approximation Δ^{WKB} is independent of the variables of integration (q, p).

- The phase-space integral over $\mathbb{X}_b$ proceeds along standard lines.

  Parametrize the space of paths q by a space of paths vanishing at t_b.

  Expand the action functional around its value, on classical paths up to its second variation. The hessian defines a quadratic form

$$Q = Q_0 + Q_1,$$

where Q_0 comes from the free action functional and is used for defining the volume element.

Perform the gaussian integration. The result obtained is proportional to $(\text{Det}((Q_0 + Q_1)/Q_0))^{-1/2}$, which is equal to the finite determinant of the hessian of the corresponding action function.

This sequence of calculations is too long to include here. It can be found in [6, 7]. The result is identical to (14.32). Applying this procedure to the computation of $\langle b|a\rangle_E$ for the harmonic oscillator is of interest; it is worked out in detail in [8].

Incidently, it is possible to treat τ as the nondynamical degree of freedom instead of λ. However, identifying the relevant volume element is trickier. It turns out to be a "gamma" volume element. The gamma volume element is to gamma probability distributions what a gaussian volume element is to gaussian probability distributions. Gamma volume elements play a key role in path-integral solutions of linear second-order PDEs in bounded regions [7]. We leave to Chapter 19 (projects) a study of gamma volume elements.

14.3 Periodic and quasiperiodic orbits
The Bohr–Sommerfeld rule

The relation between periodic orbits of a classical system and the energy levels of the corresponding quantum system was discovered by Einstein, Bohr, and Sommerfeld (see e.g. [9, pp. 6–7]). J. B. Keller [10], M. C. Gutzwiller [11], A. Voros [12], and W. J. Miller [13] generalized the Bohr–Sommerfeld quantum formula by introducing the Morse index, or the Maslov index, and using the characteristic exponents of celestial mechanics.

The characteristic exponents introduce complex-valued classical trajectories and complex energies into the Bohr–Sommerfeld formula. Complex-valued classical trajectories were first introduced by Keller [14] in his "geometrical theory of diffraction." With McLaughlin [15] he showed how classical paths of all types – including the classical diffracted path – enter the WKB approximation. Balian and Bloch [16] have systematically investigated the complex Hamilton–Jacobi equation when the potential is analytic and developed quantum mechanics in terms of complex classical paths. Balian, Parisi, and Voros [17] have shown in an example how asymptotic expansions can fail if the classical complex trajectories are not included in the WKB approximation.

What is the contribution of the path-integral formalism to the study of the energy levels of a quantum system? The trace of the fixed-energy amplitude, $\langle b|a\rangle_E$, can be used for computing the bound-state energy

14.3 Periodic and quasiperiodic orbits

spectrum; it incorporates naturally the characteristic exponents and the Morse index, and it provides a simple proof [4] of the Gutzwiller–Voros result. In brief, it *yields* the generalized Bohr–Sommerfeld formula.

The density of energy states

The density of energy states of a bound system can be obtained from the fixed-energy propagator $\langle b|a\rangle_E$ (see e.g. (14.37)). The WKB approximation follows easily from the WKB approximation $\langle b|a\rangle_E^{\text{WKB}}$.

By definition, the density of energy states is

$$\rho(E) := \sum_n \delta(E - E_n), \tag{14.39}$$

where E_n are the eigenvalues of the time-independent Schrödiger equation (14.2),

$$H\psi_n = E_n\psi_n. \tag{14.40}$$

The density $\rho(E)$ can be obtained from the Fourier transform of the trace of the fixed-time propagator (14.3). Indeed,

$$\langle b|a\rangle_T = \theta(T)\sum_n \psi_n(b)\psi_n^*(a)\exp(-iE_nT/\hbar),$$

$$\text{tr}\langle a|a\rangle_T = \theta(T)\sum_n \exp(-iE_nT/\hbar),$$

provided that the eigenfunctions $\{\psi_n\}$ are normalized to unity. Let $g(E)$ be the Fourier transform of $\text{tr}\langle a|a\rangle_T$:

$$g(E) = \sum_n \mathcal{F}(\theta)((E_n - E)/\hbar)$$

$$= \sum_n (P(E - E_n)^{-1} - i\pi\delta(E - E_n))$$

$$= \sum_n (E - E_n + i\epsilon)^{-1},$$

where P stands for principal value. Therefore

$$\rho(E) := \sum_n \delta(E - E_n) = \frac{1}{i\pi}\left(\sum_n P(E - E_n)^{-1} - g(E)\right). \tag{14.41}$$

□

The function $g(E)$ is the Fourier transform of $\text{tr}\langle a|a\rangle_T$; therefore (see (14.28)) it is equal to

$$g(E) = \text{tr}\langle a|a\rangle_E. \tag{14.42}$$

The WKB approximation of $g(E)$

The WKB approximation $g(E)^{\text{WKB}}$ can be obtained by the stationary-phase approximation of the trace of the WKB approximation of the fixed-energy propagator $\langle b|a \rangle_E^{\text{WKB}}$ given explicitly by (14.32). We compute the trace for a, b in $\mathbb{R}^D$ for simplicity. When a and b are points of a riemannian manifold $\mathbb{M}^D$, we refer the reader to [4] for the generalization of the following equation:

$$g(E)^{\text{WKB}} := \text{stationary-phase approximation}_{b=a} \int_{\mathbb{R}^D} d^D a \sum_j \left| D_{W_j}(b, a; E) \right|^{1/2}$$

$$\times \exp\left(\frac{2\pi i}{h} W_j(b, a; E) - \frac{\pi i}{2} (\lambda_j + p_j) \right) \quad (14.43)$$

in which we have used the notation of (14.32). Let a^* be a value of a that minimizes the exponent in the integral (14.43),

$$0 = \left(\partial W_j(b, a; E)/\partial a + \partial W_j(b, a; E)/\partial b \right)\big|_{a=b=a^*}$$
$$= -p_{\text{in}}(a^*, E) + p_{\text{fin}}(a^*, E), \quad (14.44)$$

where p_{in} and p_{fin} are the initial and final momenta of the stationary path that starts at a^* and ends up at a^*. A closed stationary path that satisfies (14.44) is a periodic orbit. We recall the stationary-phase formula in its simplest form (5.36). The asymptotic approximation for large λ of the integral

$$F(\lambda) = \int_{\mathbb{M}^D} d\mu(x) h(x) \exp(i\lambda f(x)) \quad (14.45)$$

is

$$F(\lambda) \approx \mathcal{O}(\lambda^{-N}) \quad \text{for any } N \text{ if } f \text{ has no critical point in the support of } h, \quad (14.46)$$

$$F(\lambda) \approx \mathcal{O}(\lambda^{-D/2}) \quad \text{if } f \text{ has a finite number of nondegenerate critical points on the support of } h, \quad (14.47)$$

where h is a real-valued smooth function of compact support on the riemannian manifold $\mathbb{M}^D$ with metric g and volume form $d\mu(x)$. Here we assume that there is only one critical point $y \in \mathbb{M}^D$ of f (there is a unique solution y to $\nabla f(y) = 0$) and that it is nondegenerate (the hessian Hf does not vanish at y):

$$|Hf(x)| := |\det \partial^2 f/\partial x^i\, \partial x^j| |\det g_{ij}|^{-1} \quad (14.48)$$

14.3 Periodic and quasiperiodic orbits

does not vanish at $x = y$. Then

$$F(\lambda) = \left(\frac{2\pi}{\lambda}\right)^{D/2} \exp(i\lambda f(y))\exp(i\pi\,\text{sign}(Hf(y))/4)\frac{h(y)}{|Hf(y)|^{1/2}} + o\left(\lambda^{-D/2}\right), \tag{14.49}$$

where $\text{sign}(Hf) = D - 2p$ and p is the Morse index of the hessian. In order to apply (14.49) to (14.43), we need the value of the exponent $W(a, b; E)$ at the unique critical point $a = b = a^*$. Set

$$W(E) = W(a^*, a^*; E). \tag{14.50}$$

We shall show that the hessian HW is related to the Poincaré map $R(\tau)$ which gives the deviation from a periodic orbit after a period τ has elapsed. Let $(\bar{q}, \bar{p})$ be the phase-space coordinate of a periodic orbit. The phase-space Jacobi operator is

$$\mathcal{J}(\bar{q}, \bar{p}) = \begin{pmatrix} -\partial^2 H/\partial \bar{q}^\alpha \partial \bar{q}^\beta & -\nabla_t - \partial^2 H/\partial \bar{q}^\alpha \partial \bar{p}_\beta \\ \nabla_t - \partial^2 H/\partial \bar{p}_\alpha \partial \bar{q}^\beta & -\partial^2 H/\partial \bar{p}_\alpha \partial \bar{p}_\beta \end{pmatrix}; \tag{14.51}$$

the Jacobi fields are the solutions

$$k(t) = \begin{pmatrix} h^\beta(t) \\ j_\beta(t) \end{pmatrix} \tag{14.52}$$

of the Jacobi equation

$$\mathcal{J}(\bar{q}, \bar{p})k(t) = 0. \tag{14.53}$$

In general there are $2D$ linearly independent solutions of the form

$$k_n(t) = \exp(\alpha_n t)S_n(t), \quad n = \pm 1, \ldots, \pm D, \tag{14.54}$$

where the functions $\{S_n(t)\}$ are periodic in t with period τ and the $2D$ constants are the characteristic exponents (stability angles).

Remark. Poincaré writes the Jacobi equation as follows:

$$(\nabla_t \mathbf{1} - \mathcal{H}(\bar{q}, \bar{p}))k(t) = \left(\begin{pmatrix} \nabla_t & 0 \\ 0 & \nabla_t \end{pmatrix} + \begin{pmatrix} -\dfrac{\partial^2 H}{\partial \bar{p}_\alpha \partial \bar{q}_\beta} & -\dfrac{\partial^2 H}{\partial \bar{p}_\alpha \partial \bar{p}_\beta} \\ \dfrac{\partial^2 H}{\partial \bar{q}^\alpha \partial \bar{q}^\beta} & \dfrac{\partial^2 H}{\partial \bar{q}^\alpha \partial \bar{p}^\beta} \end{pmatrix}\right)\begin{pmatrix} h^\beta(t) \\ j_\beta(t) \end{pmatrix}$$
$$= 0. \tag{14.55}$$

The Poincaré map $R(\tau)$ is the $2D \times 2D$ matrix defined by

$$k(t + \tau) = R(\tau)k(t). \tag{14.56}$$

The Jacobi fields $k = (h, j)$ can be obtained by variation through stationary paths (see Chapter 4). To first order in (h, j) we have

$$p(t_a) + j(t_a) = -\partial W(b + h(t_b), a + h(t_a); E)/\partial a,$$
$$p(t_b) + j(t_b) = \partial W(b + h(t_b), a + h(t_a); E)/\partial b.$$

Hence

$$j_\alpha(t_a) = -\frac{\partial^2 W}{\partial b^\beta \, \partial a^\alpha} h^\beta(t_b) - \frac{\partial^2 W}{\partial a^\beta \, \partial a^\alpha} h^\beta(t_a),$$
$$j_\alpha(t_b) = \frac{\partial^2 W}{\partial b^\beta \, \partial b^\alpha} h^\beta(t_b) + \frac{\partial^2 W}{\partial a^\beta \, \partial b^\alpha} h^\beta(t_a). \quad (14.57)$$

When $a = b = a^*$, the pair $(h^\alpha(t_a + \tau), j_\alpha(t_a + \tau))$ is obtained from $(h^\alpha(t_a), j_\alpha(t_a))$ by a linear transformation function of period τ. *If the hessian HW of the energy function (14.50) is not degenerate*, one can show [4] (an easy calculation in one dimension, more involved when $D > 1$) that

$$\begin{pmatrix} h^\alpha(t_a + \tau) \\ j_\alpha(t_a + \tau) \end{pmatrix} = R(\tau) \begin{pmatrix} h^\alpha(t_a) \\ j_\alpha(t_a) \end{pmatrix} \quad (14.58)$$

and

$$-\det(\partial^2 W/\partial b^\alpha \, \partial a^\beta)|_{a=b=a^*} = \det R(\tau) = 1. \quad (14.59)$$

However, *the hessian HW is degenerate* because the system has at least one constant of the motion, namely the energy (14.27). The degeneracy of hessians caused by conservation laws is worked out in Section 5.2 on constants of the motion, as are the implications of the degeneracy. The strategy for handling these degeneracies consists of block diagonalizing the hessian into a vanishing matrix and a matrix with nonvanishing determinant. The degrees of freedom corresponding to the vanishing matrix can be integrated explicitly. Let $\widehat{\det}$ be the determinant of the nonvanishing submatrix. It can be shown [4] that

$$-\widehat{\det}(\partial^2 W/\partial b^\alpha \, \partial a^\beta)|_{a=b=a^*} = \widehat{\det} R(\tau) = 1. \quad (14.60)$$

We refer the reader to [4] for a full calculation (14.43) of $g(E)^{\text{WKB}}$. The result, when the energy is the only constant of the motion, is

$$g(E)^{\text{WKB}} = \frac{2\pi i}{h} \sum_j \tau_j(E) |\widehat{\det}(R(\tau) - \mathbf{1})|^{-1/2}$$

$$\times \exp\left(\frac{2\pi i}{h} W_j(E) - \frac{\pi i}{2}(\lambda_j + p_j + k_j)\right), \quad (14.61)$$

where j labels a periodic orbit; the term k_j is the number of negative eigenvalues of $(R(\tau) - \mathbf{1})$. The other terms are defined in (14.32). The

14.4 Intrinsic and tuned times of a process

poles of $g(E)^{\mathrm{WKB}}$ are the energy eigenvalues of the system. They yield the generalized Bohr–Sommerfeld formula [4].

Path-integral representations

By introducing a new nondynamical degree of freedom λ we have constructed a path-integral representation (14.37) of $\langle b|a\rangle_E$. The trace $g(E)$ of $\langle b|a\rangle_E$ is given in the WKB approximation by (14.61). It is tantalizing to think of (14.61) as the WKB approximation of the path-integral representation of $g(E)$, which would bypass the path-integral representation of $\langle b|a\rangle_E$. To the best of our knowledge, however, it has not yet been constructed.

14.4 Intrinsic and tuned times of a process

In the previous section, we replaced the path-dependent parametrization τ_μ of a path x_μ by a path-independent parametrization of the path $x_\mu \circ \tau_\mu$ and a new function λ. In this section we return to a path-dependent time parametrization and a reparametrization, which are powerful tools in stochastic calculus and suggest techniques valid in other spaces [18, and references therein, 19]. We introduce the concepts of tuned-time and intrinsic-time substitution in stochastic processes with one-dimensional examples. We refer the reader to H. P. McKean Jr.'s book [20] for a study of the notions introduced here.

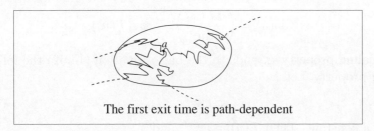

Fig. 14.2 When a path $x(t) \in \mathbb{R}$, constrained to be of energy E, starts at a at time t_a, it leaves the interval $[a,b] \in \mathbb{R}$ at a path-dependent time. Paths $x(t) \in \mathbb{R}^D$, constrained to be of energy E, leave the boundary $\partial \mathrm{M}^D$ of a domain $\mathrm{M}^D \subset \mathbb{R}^D$ at their own individual times.

Tuned-time substitutions

Let z be a brownian process and x a stochastic process such that

$$\mathrm{d}x(t) = X(x(t))\mathrm{d}z(t), \qquad x(0) = 0. \qquad (14.62)$$

The expectation value of $z(t)^2$ is
$$\mathbb{E}[z(t)^2] = t \tag{14.63}$$
and the expectation value of $x(t)^2$ is
$$\mathbb{E}[x(t)^2] = \mathbb{E}\left[\int_0^t X^2(x(s))\,\mathrm{d}s\right]. \tag{14.64}$$

Let $u \in \mathbb{R}$ be defined by a stochastic stopping time $\mathcal{T}(u)$,
$$u =: \int_0^{\mathcal{T}(u)} X^2(x(t))\,\mathrm{d}t. \tag{14.65}$$

Then the process defined by b, namely
$$b(u) := x(\mathcal{T}(u)), \tag{14.66}$$

is brownian. The time substitution $t \mapsto \mathcal{T}(u)$ such that $b(u)$ is brownian is said to be *tuned to the process* x.

From the definition (14.66) one derives a scaling property for a path-dependent time substitution:
$$\mathrm{d}b(u) = \mathrm{d}x(\mathcal{T}(u)) = X(x(\mathcal{T}(u)))\mathrm{d}z(\mathcal{T}(u))$$
$$= X(b(u))\mathrm{d}z(\mathcal{T}(u)). \tag{14.67}$$

It follows from the definition (14.65) of $\mathcal{T}(u)$ that
$$\mathrm{d}u = \mathrm{d}\mathcal{T}(u)X^2(x(\mathcal{T}(u))). \tag{14.68}$$

Therefore
$$\mathrm{d}b(u) = \left(\frac{\mathrm{d}\mathcal{T}(u)}{\mathrm{d}u}\right)^{-1/2} \mathrm{d}z(\mathcal{T}(u)). \tag{14.69}$$

This scaling property extends to the basic equation (14.62) the following scaling property. Let
$$x(t) = c\,z(t) \tag{14.70}$$
with c a constant. Then $\mathcal{T}(u) = u/c^2$, and
$$b(u) = cz(u/c^2). \tag{14.71}$$

The intrinsic time of a process

Let z be a brownian process and x a stochastic process such that
$$\mathrm{d}x(t) = X(x(t))\mathrm{d}z(t), \qquad x(0) = 0. \tag{14.72}$$
By definition, $U(t)$ is the intrinsic time of x if the process a defined by
$$\mathrm{d}x(t) =: \mathrm{d}a(U(t)) \tag{14.73}$$

14.4 Intrinsic and tuned times of a process

is brownian. The scaling property (14.69) can be used for relating $U(t)$ to $X(x(t))$. Indeed, according to (14.69) the brownian process z can be written as another scaled brownian process:

$$dz(t) = \left(\frac{dU(t)}{dt}\right)^{-1/2} da(U(t)). \tag{14.74}$$

Insertion of (14.74) into (14.72) yields the required condition (14.73) if

$$dU(t) = X^2(x(t))dt. \tag{14.75}$$

In summary, time substitutions have given us two new brownian processes, a and b, related to the process $dx(t) = X(x(t))dz(t)$:

- the tuned-time substitution gives

$$b(u) := x(\mathcal{T}(u)); \tag{14.76}$$

- the intrinsic time defines

$$x(t) =: a(U(t)). \tag{14.77}$$

These results are not limited to one-dimensional equations. They also apply to processes x defined by

$$dx(t) = \sum_\alpha X_{(\alpha)}(x(t))dz^\alpha(t). \tag{14.78}$$

One can define arbitrary time substitutions other than the ones above. Time substitutions can be used to relate different equations having the generic form

$$dx(t) = X_{(\alpha)}(x(t))dz^\alpha(t) + Y(x(t))dt. \tag{14.79}$$

Ipso facto, they relate different parabolic equations.

Kustaanheimo–Stiefel transformations

Kustaanheimo–Stiefel (KS) transformations [21], which are well known in celestial mechanics, map a dynamical system with coordinates $(x \in \mathbb{R}^N, t)$ into a dynamical system with coordinates $(y \in \mathbb{R}^n, u)$ for the following pairs

$$\begin{array}{c|cccc} N & 1 & 2 & 4 & 8 \\ n & 1 & 2 & 3 & 4 \end{array} \tag{14.80}$$

Under the inverse ($n = 3$, $N = 4$) KS transformation the equation of motion for a particle in $\mathbb{R}^3$ in a Newtonian potential is mapped into the equation of motion for a harmonic oscillator in $\mathbb{R}^4$, which remains regular at the center of attraction. Duru and Kleinert [22] introduced the KS transformation for computing path-integral representations of the energy levels and wave functions of a particle in a Coulomb potential. Blanchard and Sirugue [23] have applied KS transformations to stochastic processes and identified the change of process underlying the Duru–Kleinert calculation.

References

[1] R. P. Feynman and A. R. Hibbs (1965). *Quantum Mechanics and Path Integrals* (New York, McGraw-Hill).

[2] R. Abraham and J. E. Marsden (1985). *Foundations of Mechanics* (New York, Addison-Wesley).

[3] C. Garrod (1966). "Hamiltonian path integral methods," *Rev. Mod. Phys.* **38**, 483–494.

[4] M. C. Gutzwiller (1995). *Chaos in Classical and Quantum Mechanics (Interdisciplinary Applied Mathematics, Vol. 1)* (Berlin, Springer).

[5] C. DeWitt-Morette, A. Maheshwari, and B. Nelson (1979). "Path integration in nonrelativistic quantum mechanics," *Phys. Rep.* **50**, 266–372.

[6] J. LaChapelle (1995). Functional integration on symplectic manifolds, unpublished Ph.D. dissertation, University of Texas at Austin.
J. LaChapelle (1997). "Path integral solution of the Dirichlet problem," *Ann. Phys.* **254**, 397–418 and references therein.

[7] J. LaChapelle (2004). "Path integral solution of linear second order partial differential equations, I. The general construction, II. Elliptic, parabolic, and hyperbolic cases," *Ann. Phys.* **314**, 362–395 and 396–424.

[8] A. Wurm (1997). The Cartier/DeWitt path integral formalism and its extension to fixed energy Green's function, unpublished Diplomarbeit, Julius-Maximilians-Universität, Würzburg.

[9] M. Born (1927). *The Mechanics of the Atom* (London, G. Bell and Sons).

[10] J. B. Keller (1958). "Corrected Bohr–Sommerfeld quantum conditions for nonseparable systems," *Ann. Phys.* **4**, 180–188.

[11] M. C. Gutzwiller (1967). "Phase-integral approximation in momentum space and the bound state of an atom," *J. Math. Phys.* **8**, 1979–2000.

[12] A. Voros (1974). "The WKB–Maslov method for non-separable systems," in *Géometrie symplectique et physique mathématique* (Paris, CNRS), pp. 277–287.

[13] W. J. Miller (1975). "Semi-classical quantization of non-separable systems: a new look at periodic orbit theory," *J. Chem. Phys.* **63**, 996–999.

[14] J. B. Keller (1958). "A geometrical theory of diffraction," in *Calculus of Variations and its Applications* (New York, McGraw-Hill), pp. 27–52.

[15] J. B. Keller and D. W. McLaughlin (1975). "The Feynman integral," *Am. Math. Monthly* **82**, 457–465.

[16] R. Balian and C. Bloch (1971). "Analytical evaluation of the Green function for large quantum numbers," *Ann. Phys.* **63**, 592–600.
R. Balian and C. Bloch (1974). "Solution of the Schrödinger equation in terms of classical paths," *Ann. Phys.* **85**, 514–545.

[17] R. Balian, G. Parisi, and A. Voros (1978). "Discrepancies from asymptotic series and their relation to complex classical trajectories," *Phys. Rev. Lett.* **41**, 1141–1144; erratum *Phys. Rev. Lett.* **41**, 1627.
[18] A. Young and C. DeWitt-Morette (1986). "Time substitutions in stochastic processes as a tool in path integration," *Ann. Phys.* **169**, 140–166.
[19] H. Kleinert (2004). *Path Integrals in Quantum Mechanics, Statistics, and Polymer Physics*, 3rd edn. (Singapore, World Scientific).
[20] H. P. McKean Jr. (1969). *Stochastic Integrals* (New York, Academic Press).
[21] P. Kustaanheimo and E. Stiefel (1965). "Perturbation theory of Kepler motion based on spinor regularization," *J. Reine Angew. Math.* **218**, 204–219.
[22] I. H. Duru and H. Kleinert (1982). "Quantum mechanics of H-atoms from path integrals," *Fortschr. Phys.* **30**, 401–435.
I. H. Duru and H. Kleinert (1979). "Solution of the path integral for the H-atom," *Phys. Lett. B* **84**, 185–188.
S. N. Storchak (1989). "Rheonomic homogeneous point transformations and reparametrization in path integrals," *Phys. Lett. A* **135**, 77–85.
[23] Ph. Blanchard and M. Sirugue (1981). "Treatment of some singular potentials by change of variables in Wiener integrals," *J. Math. Phys.* **22**, 1372–1376.

Part V
Problems in quantum field theory

15
Renormalization 1: an introduction

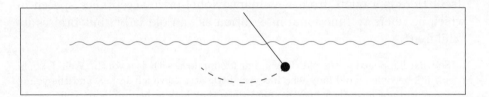

15.1 Introduction

The fundamental difference between quantum mechanics (systems with a finite number of degrees of freedom) and quantum field theory (systems with an infinite number of degrees of freedom) can be said to be "radiative corrections." In quantum mechanics "a particle is a particle" characterized, for instance, by mass, charge, and spin. In quantum field theory the concept of a "particle" is intrinsically associated to the concept of a "field." The particle is affected by its own field. Its mass and charge are modified by the surrounding fields, namely its own and other fields interacting with it.

In the previous chapters we have developed path integration and its applications to quantum mechanics. How much of the previous analysis can be generalized to functional integration and its applications to quantum field theory? Gaussians, time-ordering, and variational methods, as presented in the previous chapters, are prototypes that are easy to adapt to quantum field theory. We shall carry out this adaptation in Section 15.2.

However, the lessons learned from quantum mechanics do not address all the issues that arise in quantum field theory. We introduce them in Section 15.3. We use the example, due to G. Green, of pendulum motion modified by a surrounding fluid [1]. In the equation of motion the mass of the pendulum is modified by the fluid. Nowadays we say the mass is "renormalized." The remarkable fact is that the equation of motion remains valid, provided that one replaces the "bare" mass by the renormalized mass.

In Chapter 16, we work out renormalization by a scaling method developed by D. C. Brydges, J. Dimock, and T. R. Hurd. This method proceeds via scale-dependent effective action functionals – in contrast to other methods, which proceed via counterterms in the lagrangian. The scaling method exploits *directly* the power of functional integration. Other renormalization techniques work with Feynman diagrams, an *offspring* of functional integration. Diagrams are so powerful and yield so many results in renormalization theory and elsewhere that there was a time, in the sixties and seventies, when they overshadowed functional integration itself. Here is a quote from Feynman's Nobel-prize acceptance speech in which he refers to functional integration as "an old lady" but praises its "children":

> So what happened to the old theory that I fell in love with as a youth? Well, I would say it's become an old lady who has very little attractive left in her, and the young today will not have their hearts pound when they look at her anymore. But, we can say the best we can for any old woman, that she has been a very good mother and has given birth to some very good children. And I thank the Swedish Academy of Sciences for complimenting one of them. Thank you.[2][1]

The children are alive and well. The combinatorics of diagram expansions, worked out by A. Connes and D. Kreimer, is an amazing chapter of calculus by graphic methods. We present in Chapter 17 the diagram combinatorics worked out by M. Berg and P. Cartier. They have developed a Lie algebra of renormalization related to the Hopf algebra governing the work of Connes and Kreimer.

Regularization

Renormalization is usually difficult to work out because the Green functions in field theory are singular (see for instance the difference between equations (4.58) and (4.59)). This problem is addressed by regularization and presented in Section 15.4.

Although regularization and renormalization are different concepts, they are often linked together. The following handwaving argument could

---

[1] To C. DeWitt, who wrote to him that "She is beautiful in her own right," Feynman replied "You only dressed her up."

15.2 From paths to fields

lead to the expectation that radiative corrections (renormalization) regularize the theory: we have noted (2.40) that the Green functions (covariances) G are the values of the variance W evaluated at pointlike sources. In general, given a source J,

$$W(J) = \langle J, GJ \rangle; \qquad (15.1)$$

it is singular only if J is pointlike. On the other hand, radiative corrections "surrounding" a point particle can be thought of as an extension of the particle. Unfortunately calculations do not bear out this expectation in general. In addition to inserting radiative corrections into "tree" diagrams, one needs to introduce a regularization technique to control the divergences of the Green functions.

15.2 From paths to fields

Gaussian integrals
(See Chapter 2)

The results of Section 2.3 on gaussians in Banach spaces apply equally well to Banach spaces of paths $x \in \mathbb{X}$ and to Banach spaces of fields $\phi \in \Phi$. A gaussian volume element

$$\mathcal{D}_{s,Q}\phi \cdot \exp\left(-\frac{\pi}{s}Q(\phi)\right) \stackrel{f}{=} \mathrm{d}\Gamma_{s,Q}(\phi) \qquad (15.2)$$

is defined by its Fourier transform

$$\int_\Phi \mathcal{D}_{s,Q}\phi \cdot \exp\left(-\frac{\pi}{s}Q(\phi)\right)\exp(-2\pi\mathrm{i}\langle J, \phi\rangle) = \exp(-\pi s W(J)), \qquad (15.3)$$

where s is real or complex.

- For s real, the quadratic form Q/s is positive.
- For s complex, the real part of Q/s is positive.
- J is in the dual Φ' of Φ, and is often called a source.
- Q and W are inverses of each other in the following sense. Set

$$Q(\phi) = \langle D\phi, \phi \rangle \quad \text{and} \quad W(J) = \langle J, GJ \rangle,$$

where $D : \Phi \to \Phi'$ and $G : \Phi' \to \Phi$ are symmetric linear maps; then

$$DG = 1_{\Phi'} \quad \text{and} \quad GD = 1_{\Phi}.$$

We refer the reader to Chapter 2 and to the last paragraphs of Section 4.3, as well as to the titles listed in Section 1.4, for properties and uses of gaussian functional integrals.

Operator causal ordering
(see time-ordering in Chapter 6)

In path integrals, the variable of integration $x \in \mathbb{X}$ is a *parametrized* path. The parameter creates an ordering in the values of the variable of integration. We call it "time-ordering" even when the parameter is not time. We have shown in Section 6.5 that a path integral representing the matrix element of an operator product on a Hilbert space "attaches time-ordering labels" to the operator product.

For example, the power of path integrals for time-ordering shows up already in its most elementary formula:

$$\langle b, t_b | T(\mathbf{x}(t)\mathbf{x}(t')) | a, t_a \rangle_Q = \int_{\mathbb{X}_{a,b}} \mathcal{D}_{i,Q}(x) \exp(\pi i Q(x)) x(t) x(t'), \quad (15.4)$$

where x is a function and $\mathbf{x}(t)$ is an operator operating on the state $|a, t_a\rangle$. We recall the normalization obtained in formula (3.39), namely

$$\int_{\mathbb{X}_{a,b}} \mathcal{D}_{i,Q}(x) \cdot \exp(\pi i Q(x)) = (\det(-iG_b(t_a, t_a)))^{-1/2},$$

where $G_b(t, t')$ is the covariance associated with the quadratic form $Q(x)$ on $\mathbb{X}_b$.

The path integral (15.4) gives the two-point function for a free particle of mass m in one dimension (see Section 3.1 and equation (2.41)) when

$$Q(x) = \frac{m}{2\pi\hbar} \int_{\mathbb{T}} dt\, \dot{x}(t)^2$$

$$\langle 0, t_b | T(\mathbf{x}(t)\mathbf{x}(t')) | 0, t_a \rangle = \begin{cases} \dfrac{i\hbar}{m}(t' - t_a) & \text{if } t > t', \\ \dfrac{i\hbar}{m}(t - t_a) & \text{if } t < t'. \end{cases} \quad (15.5)$$

The path integral on the right-hand side of (15.4) has time-ordered the matrix element on the left-hand side of (15.4).

In field theory time-ordering is replaced by causal ordering dictated by lightcones: if a point x_j is in the future lightcone of x_i, written

$$j > i,$$

then the causal (time, or chronological) ordering T of the field operators $\phi(x_j)\phi(x_i)$ is the symmetric function

$$T(\phi(x_j)\phi(x_i)) = T(\phi(x_i)\phi(x_j)) \quad (15.6)$$

which satisfies the equation

$$T(\phi(x_j)\phi(x_i)) = \begin{cases} \phi(x_j)\phi(x_i) & \text{for } j > i, \\ \phi(x_i)\phi(x_j) & \text{for } i > j. \end{cases} \quad (15.7)$$

15.2 From paths to fields

In general

$$\langle \text{out}|T(\mathcal{F}(\phi))|\text{in}\rangle_S = \int_{\Phi_{\text{in,out}}} \mathcal{F}(\phi)\omega_S(\phi), \qquad (15.8)$$

where $\mathcal{F}$ is a functional of the field ϕ (or a collection of fields ϕ_i), S is the action functional of a system, and $\omega_S(\phi)$ is a volume element that will be required to satisfy certain basic properties identified in Chapters 6 and 11.

In equation (15.8) the causal ordering on the left-hand side is dictated by the functional integral. In the work of Bryce DeWitt, causal ordering of operators follows from the Peierls bracket. The functional integral on the right-hand side is then constructed by requiring the left-hand side to obey Schwinger's variational principle.

Remark. A path can be parametrized by a parameter other than time, for instance by a scaling parameter. All previous discussions and properties of path integrals apply to any parametrized path – including "Schrödinger equations" satisfied by path integrals.

Variational methods
(see Chapters 6 and 11)

We summarize two variational methods that apply readily to quantum field theory: the one developed in Chapter 6 for gaussian integrals and the one developed in Chapter 11 for integrals more general than gaussian.

In Chapter 6 we varied the action functional (6.52),

$$S(\gamma) = S_0(\gamma) + eS_1(\gamma), \qquad (15.9)$$

i.e. the sum of a kinetic term

$$S_0(\gamma) := \frac{m}{2} \int_\gamma \frac{(\mathrm{d}x(t))^2}{\mathrm{d}t} \qquad (15.10)$$

and a potential term

$$S_1(\gamma) = -\int_\gamma \mathrm{d}t\, V(x(t), t), \qquad (15.11)$$

where γ is a path $x : [t_a, t_b] \to \mathbb{R}^D$. Calling U_S the Schrödinger propagator, i.e. the solution of

$$\begin{aligned} i\hbar\, \partial_t U_S(t, t_a) &= H U_S(t, t_a), \quad \text{with } H = -\frac{\hbar^2}{2m}\Delta + eV, \\ U_S(t_a, t_a) &= \mathbf{1}, \end{aligned} \qquad (15.12)$$

294   *Renormalization 1: an introduction*

and calling $U_F(t_b, t_a)$ the path integral[2]

$$U_F(t_b, t_a) = \int_{\mathcal{P}_{a,b}} \mathcal{D}\gamma \cdot \exp\left(\frac{i}{\hbar}S(\gamma)\right), \qquad (15.13)$$

we showed, (6.64) and (6.68), that

$$\frac{\delta U_S}{\delta V(x(t), t)} = \frac{\delta U_F}{\delta V(x(t), t)}. \qquad (15.14)$$

We had previously proved that $U_S = U_F$ when $V = 0$. Therefore it follows from (15.14) that $U_S = U_F$ when $V \neq 0$.

In Chapter 6 we worked with the gaussian volume element on a space $\mathbb{X}$ of paths

$$d\Gamma(x) \stackrel{f}{:=} \mathcal{D}x \cdot \exp\left(\frac{i}{\hbar}S_0(x)\right). \qquad (15.15)$$

In Chapter 11 we introduced a class of volume elements ω_S more general than gaussians, namely

$$\omega_S(x) := \mathcal{D}x \cdot \exp\left(\frac{i}{\hbar}S(x)\right) \simeq \mathcal{D}x \cdot \mu(x)\exp\left(\frac{i}{\hbar}S_{cl}(x)\right), \qquad (15.16)$$

where S_{cl} is the classical action functional of a system and

$$S(x) = S_{cl}(x) - i\hbar \ln \mu(x) + \mathcal{O}(\hbar^2) \qquad (15.17)$$

is the quantum action functional suggested by Dirac [3]. The symbol "$\simeq$" in equation (15.16) means that the terms of order $\hbar^2$ in $S(x)$ are neglected.

The symbol $\mathcal{D}x$ is a translation-invariant symbol,

$$\mathcal{D}(x + x_0) = \mathcal{D}x \quad \text{for } x_0 \text{ a fixed function.} \qquad (15.18)$$

It can be normalized by extracting a quadratic term Q in the classical action functional S:

$$\int_{\mathbb{X}} \mathcal{D}_Q(x) \exp(i\pi Q(x)) = 1. \qquad (15.19)$$

This equation defines a normalized gaussian volume element

$$\omega_Q(x) := \mathcal{D}_Q(x)\exp(i\pi Q(x)), \qquad (15.20)$$

which can serve as "background" volume element for $\omega_S(x)$. It was shown in Sections 11.3 and 4.2 and Appendix E that ratios of volume elements,

---

[2] Recall that the volume element $\mathcal{D}\gamma$ is normalized by

$$\int_{\mathcal{P}_b} \mathcal{D}\gamma \cdot \exp\left(\frac{i}{\hbar}S_0(\gamma)\right) = 1,$$

where $\mathcal{P}_b$ is the set of paths with fixed final position x_b at time t_b.

15.2 From paths to fields

rather than volume elements themselves, are the important concept in functional integration. Therefore background volume elements have a role to play in explicit calculations.

The translation invariance of the symbol $\mathcal{D}x$, together with an integration by parts, provides a variational equation for the volume element $\omega_S(x)$. Let F be a functional on $\mathbb{X}$ such that

$$\int_{\mathbb{X}} \frac{\delta}{\delta x(t)}(\mathcal{D}x \cdot F(x)) = 0; \qquad (15.21)$$

then the translation invariance of $\mathcal{D}x$ implies

$$\int_{\mathbb{X}} \mathcal{D}x \cdot \frac{\delta F(x)}{\delta x(t)} = 0. \qquad (15.22)$$

If one uses $F(x)\exp(iS(x)/\hbar)$ rather than $F(x)$ in (15.21), then (15.22) reads

$$\int_{\mathbb{X}} \omega_S(x)\left(\frac{\delta F(x)}{\delta x(t)} + \frac{i}{\hbar}F(x)\frac{\delta S(x)}{\delta x(t)}\right) = 0, \qquad (15.23)$$

where

$$\omega_S(x) = \mathcal{D}x \cdot \exp(iS(x)\hbar). \qquad (15.24)$$

Equation (15.23) gives the relationship between the volume element ω_S and the action functional S. Equation (15.23) can also be written as a variational equation:

$$\int_{\mathbb{X}} \omega_S(x)\frac{\delta F(x)}{\delta x(t)} = -\frac{i}{\hbar}\int_{\mathbb{X}} \omega_S(x)F(x)\frac{\delta S(x)}{\delta x(t)}. \qquad (15.25)$$

Equation (15.23) exemplifies a useful property of $\omega_S(x)$, namely a generalization of the Malliavin formula [4]

$$\int_{\mathbb{X}} \omega_S(x)\frac{\delta F(x)}{\delta x(t)} = -\int \frac{\delta \omega_S(x)}{\delta x(t)}F(x). \qquad (15.26)$$

When $F = 1$, equation (15.23) together with (15.8) gives the (Schwinger) quantum version of the classical equation of motion:

$$\langle\text{out}|T(\delta S(\mathbf{x})/\delta \mathbf{x}(t))|\text{in}\rangle = 0. \qquad (15.27)$$

Equation (15.23) does *not* define the volume element $\omega_S(x)$ because the quantum action functional S is not determined by the classical action functional S_{cl}. They differ by the unknown term $i\hbar \ln \mu(x)$ and terms of the order of $\hbar^2$. Nevertheless, equation (15.23) is a cornerstone of functional integration because it is a direct consequence of integration by parts (15.22), which is one of the fundamental requirements of integration theory, (9.38)–(9.41).

In conclusion, equations (15.15)–(15.27) apply to fields $\phi(x)$ as well as to paths $x(t)$.

Remark. The term $\mu(\phi)$ used by Bryce DeWitt in Chapter 18 is not the equivalent of $\mu(x)$ in (15.15) because the translation-invariant symbol in (18.2) is not equivalent to the translation-invariant symbol $\mathcal{D}x$. They differ by a determinant that makes $\mathcal{D}x$ dimensionless.

A functional differential equation

The Fourier transform $Z(J)$ of the volume element

$$\omega_S(\phi) = \mathcal{D}\phi \cdot \exp(\mathrm{i}S(\phi)/\hbar)$$

is

$$Z(J) = \int_\Phi \omega_S(\phi) \exp(-2\pi\mathrm{i}\langle J, \phi\rangle). \tag{15.28}$$

We shall obtain a functional differential equation for $Z(J)$ by applying (15.23) to $F(\phi) = \exp(-2\pi\mathrm{i}\langle J, \phi\rangle)$.

Use of the fundamental properties of integration (9.38) leads to the following equations:

$$\frac{\delta Z(J)}{\delta \phi(x)} = 0 \tag{15.29}$$

and

$$\int_\Phi \omega_S(\phi)\left(\frac{\delta F(\phi)}{\delta\phi(x)} + \frac{\mathrm{i}}{\hbar}F(\phi)\frac{\delta S(\phi)}{\delta\phi(x)}\right) = 0, \tag{15.30}$$

i.e.

$$\int \omega_S(\phi)\left(-2\pi\mathrm{i}J(x) + \frac{\mathrm{i}}{\hbar}\frac{\delta S(\phi)}{\delta\phi(x)}\right)\exp(-2\pi\mathrm{i}\langle J, \phi\rangle) = 0. \tag{15.31}$$

If we assume the quantum action functional to be of the form

$$S(\phi) = -\frac{1}{2}\langle D\phi, \phi\rangle + P(\phi), \quad \text{i.e.} \quad \frac{\delta S(\phi)}{\delta\phi(x)} = -D\phi + P'(\phi), \tag{15.32}$$

where D is a differential operator and P a polynomial in ϕ, we can exploit the properties of Fourier transforms to rewrite (15.31) as a functional differential equation for $Z(J)$. Given the following property of Fourier transforms,

$$\int_\Phi \omega_S(\phi)\phi(x)\exp(-2\pi\mathrm{i}\langle J,\phi\rangle) = \int_\Phi \omega_S(\phi)\frac{-1}{2\pi\mathrm{i}}\frac{\delta}{\delta J(x)}\exp(-2\pi\mathrm{i}\langle J,\phi\rangle), \tag{15.33}$$

it follows from (15.31) and (15.32) that

$$-2\pi i J(x)Z(J) - \frac{i}{\hbar}D\left(-\frac{1}{2\pi i}\frac{\delta}{\delta J(x)}Z(J)\right) + \frac{i}{\hbar}P'\left(-\frac{1}{2\pi i}\frac{\delta}{\delta J(x)}\right)Z(J) = 0, \quad (15.34)$$

which can be simplified by using h rather than $\hbar$. One may normalize $Z(J)$ by dividing it by $Z(0)$:

$$i\hbar \ln(Z(J)/Z(0)) =: W(J). \quad (15.35)$$

It has been shown (e.g. [5]) that the diagram expansion of $Z(J)$ contains disconnected diagrams, but the diagram expansion of $W(J)$ contains only connected diagrams.

In conclusion, we have shown that the lessons learned from path integrals on gaussian volume elements (15.3), operator time-ordering (15.8), and integration by parts (15.22) apply readily to functional integrals.

15.3 Green's example

Alain Connes introduced us[3] to a delightful paper [1] by Green that serves as an excellent introduction to modern renormalization in quantum field theory. Green derived the harmonic motion of an ellipsoid in fluid media. This system can be generalized and described by the lagrangian of an arbitrary solid C in an incompressible fluid F, under the influence of an arbitrary potential $V(x)$, $x \in F$. In vacuum (i.e. when the density ρ_F of the fluid is negligible compared with the density ρ_C of the solid) the lagrangian of the solid moving with velocity $\vec{v}$ under $V(x)$ is

$$L_C := \int_C d^3x \left(\frac{1}{2}\rho_C|\vec{v}|^2 - \rho_C V(x)\right). \quad (15.36)$$

In the static case, $\vec{v} = 0$, Archimedes' principle can be stated as the renormalization of the potential energy of a solid immersed in a fluid. The potential energy of the system fluid plus solid is

$$E_{\text{pot}} := \int_{\mathbb{R}^3 \setminus C} d^3x\, \rho_F V(x) + \int_C d^3x\, \rho_C V(x) \quad (15.37)$$

$$= \int_{\mathbb{R}^3} d^3x\, \rho_F V(x) + \int_C d^3x (\rho_C - \rho_F) V(x). \quad (15.38)$$

The potential energy of the immersed solid C is

$$E_{\text{pot}} = \int_C d^3x (\rho_C - \rho_F) V(x). \quad (15.39)$$

---

[3] The UT librarian, Molly White, found the reference.

In modern parlance ρ_C is the bare coupling constant, and $\rho_C - \rho_F$ is the coupling constant renormalized by the immersion. We ignore the infinite term $\int_{\mathbb{R}^3} d^3x\, \rho_F V(x)$ because energy difference, not energy, is the physically meaningful concept.

In the dynamic case, $\vec{v} \neq 0$, Green's calculation can be shown to be the renormalization of the mass of a solid body by immersion in a fluid. The kinetic energy of the system fluid plus solid is

$$E_{\text{kin}} := \frac{1}{2}\int_F d^3x\, \rho_F |\vec{\nabla}\Phi|^2 + \frac{1}{2}\int_C d^3x\, \rho_C |\vec{v}|^2, \tag{15.40}$$

where Φ is the velocity potential of the fluid,

$$\vec{v}_F = \vec{\nabla}\Phi. \tag{15.41}$$

If the fluid is incompressible, then

$$\Delta\Phi = 0.$$

If the fluid is nonviscous, then the relative velocity $\vec{v}_F - \vec{v}$ is tangent to the boundary ∂C of the solid. Let $\vec{n}$ be the unit normal to ∂C, and $\partial_n \Phi := \vec{n} \cdot \vec{\nabla}\Phi$. Then $\vec{v}_F - \vec{v}$ is tangent to ∂C iff

$$\partial_n \Phi = \vec{v} \cdot \vec{n} \qquad \text{on } \partial C = \partial F. \tag{15.42}$$

Moreover, Φ and $\vec{\nabla}\Phi$ must vanish at infinity sufficiently rapidly for $\int_F d^3x |\vec{\nabla}\Phi|^2$ to be finite. Altogether Φ satisfies a Neumann problem

$$\Delta\Phi = 0 \text{ on } F, \qquad \Phi(\infty) = 0, \qquad \partial_n \Phi = \vec{v} \cdot \vec{n} \text{ on } \partial F, \tag{15.43}$$

where $\vec{n}$ is the inward normal to ∂F. If ρ_F is constant, then Green's formula

$$\int_F d^3x |\vec{\nabla}\Phi|^2 = -\int_{\partial F} d^2\sigma\, \Phi \cdot \partial_n \Phi \tag{15.44}$$

together with the boundary condition (15.42) simplifies the calculation of $E_{\text{kin}}(F)$:

$$E_{\text{kin}}(F) = \frac{1}{2}\rho_F \int_{\partial F} d^2\sigma(-\Phi)\vec{v} \cdot \vec{n}. \tag{15.45}$$

Example. Let C be a ball of radius R centered at the origin. The dipole potential

$$\Phi(\vec{r}) = -\frac{1}{2}(\vec{v} \cdot \vec{r})(R/r)^3 \tag{15.46}$$

satisfies the Neumann conditions (15.43). On the sphere $\partial C = \partial F$,

$$\Phi(\vec{r}) = -\frac{1}{2}R\vec{v} \cdot \vec{n} \tag{15.47}$$

and
$$E_{\text{kin}}(F) = \frac{1}{2}\rho_F \frac{R}{2} \int_{\partial C} d^2\sigma (\vec{v}\cdot\vec{n})^2. \tag{15.48}$$

By homogeneity and rotational invariance
$$\int_{\partial C} d^2\sigma(\vec{v}\cdot\vec{n})^2 = \frac{1}{3}|\vec{v}|^2 \int_{\partial C} d^2\sigma = \frac{1}{3}|\vec{v}|^2 4\pi R^2$$

and
$$E_{\text{kin}}(F) = \frac{1}{2}\rho_F \cdot \text{vol}(C) \cdot \frac{1}{2}|\vec{v}|^2 \tag{15.49}$$

with $\text{vol}(C) = \frac{4}{3}\pi R^3$. Finally, the total kinetic energy (15.40) is
$$E_{\text{kin}} = \left(\rho_C + \frac{1}{2}\rho_F\right) \cdot \text{vol}(C) \cdot \frac{1}{2}|\vec{v}|^2. \tag{15.50}$$

The mass density ρ_C is renormalized to $\rho_C + \frac{1}{2}\rho_F$ by immersion.

In conclusion, one can say that the bare lagrangian (15.36)
$$\int_C d^3x \left(\frac{1}{2}\rho_c|\vec{v}|^2 - \rho_c V(x)\right) \tag{15.51}$$

is renormalized by immersion to
$$\int_C d^3x \left(\frac{1}{2}\left(\rho_C + \frac{1}{2}\rho_F\right)|\vec{v}|^2 - (\rho_C - \rho_F)V(x)\right), \tag{15.52}$$

where the renormalized inertial mass density (15.50) is different from the gravitational mass density (15.39). Green's calculation for a vibrating ellipsoid in a fluid media leads to a similar conclusion. As in (15.50) the density of the vibrating body is increased by a term proportional to the density of the fluid, which is obviously more complicated than $\rho_F/2$.

Remark. Different renormalizations for the inertial and gravitational masses. An example. Let the lagrangian (15.36) be the lagrangian of a simple pendulum of length l
$$L = \frac{1}{2}m_i v^2 + m_g g x(t),$$

where m_i is the inertial mass and m_g the gravitational mass, which are equal but correspond to different concepts. In polar coordinates in the plane of the pendulum motion
$$L = \frac{1}{2}m_i l^2 \dot\theta^2 + m_g g l(1 - \cos\theta).$$

The period, for small oscillations
$$T = 2\pi \left(\frac{m_i l}{m_g g}\right)^{1/2},$$

is independent of the mass, but when the pendulum is immersed in a fluid, equation (15.52) shows that m_i and m_g are not renormalized identically; their ratio is not equal to 1. In Project 19.7.1 we raise the issue of the principle of equivalence of inertial and gravitational masses in classical and quantum field theory.

From Archimedes' principle, to Green's example, to the lagrangian formulation, one sees the evolution of the concepts used in classical mechanics: Archimedes gives the *force* felt by an immersed solid, Green gives the harmonic *motion* of an ellipsoid in fluid media, and the lagrangian gives the kinetic and the potential *energy* of the system [6]. In quantum physics, action functionals defined in infinite-dimensional spaces of functions play a more important role than do lagrangians or hamiltonians defined on finite-dimensional spaces. Renormalization is often done by renormalizing the lagrangian. Renormalization in quantum field theory by scaling (Chapter 16), on the other hand, uses the action functional.

15.4 Dimensional regularization

The $\lambda\phi^4$ model

The $\lambda\phi^4$ system is a self-interacting relativistic scalar field in a spacetime of dimension $1+3$ described by a lagrangian density

$$\mathcal{L}(\phi) = \frac{\hbar}{2}\eta^{\mu\nu}\partial_\mu\phi \cdot \partial_\nu\phi - \frac{m^2c^2}{2\hbar}\phi^2 - \frac{\hbar\lambda}{4!}\phi^4. \qquad (15.53)$$

Our notation is as follows: vectors in the Minkowski space $\mathbb{M}^4$ have covariant coordinates a_0, a_1, a_2, a_3 and contravariant coordinates a^0, a^1, a^2, a^3 related by

$$a^0 = a_0, \qquad a^1 = -a_1, \qquad a^2 = -a_2, \qquad a^3 = -a_3. \qquad (15.54)$$

The dot product $a \cdot b$ is given by the following expressions:

$$a \cdot b = a^\mu b_\mu = \eta^{\mu\nu}a_\mu b_\nu = \eta_{\mu\nu}a^\mu b^\nu, \qquad (15.55)$$

where $(\eta^{\mu\nu})$ and $(\eta_{\mu\nu})$ are diagonal matrices with entries $+1, -1, -1, -1$. The 4-vector x giving the location of a point at place $\vec{x}$ and time t is given by $(x^0, \vec{x})$, where

$$x^0 = ct, \qquad (x^1, x^2, x^3) = \vec{x}, \qquad (15.56)$$

and the 4-momentum p is given by

$$p_0 = E/c, \qquad (p_1, p_2, p_3) = \vec{p} \qquad (15.57)$$

for the energy E and momentum $\vec{p}$. The volume element $d^4x = dx^0 \cdot d^3\vec{x}$. Finally, m is the mass of the particle described by the field ϕ, and λ

15.4 Dimensional regularization

is the coupling constant. Here, in contrast to in Section 16.2, we keep explicitly the Planck constant $\hbar$, but we hide the speed of light c by introducing the parameter $\mu = mc/\hbar$. Hence μ is an inverse length, as well as the values $\phi(x)$ of the field ϕ, and λ is dimensionless. The action is given by

$$S(\phi) = \int_{\mathbb{M}^4} d^4x \, \mathcal{L}(\phi(x)). \tag{15.58}$$

The n-point function

Consider any quantum field theory described by an action functional $S(\phi)$. We denote by $\hat{\phi}(x)$ the operator quantizing the value $\phi(x)$ of the field at the point x in spacetime. The *causal ordering* is defined by the following properties:

- $T(\hat{\phi}(x_1)\ldots\hat{\phi}(x_n))$ is an operator symmetric in the spacetime points $x_1, \ldots, x_n$; and
- if u is any timelike vector in $\mathbb{M}^4$, and $i_1, \ldots, i_n$ is the permutation of $1, 2, \ldots, n$ such that

$$u \cdot x_{i_1} > u \cdot x_{i_2} > \cdots > u \cdot x_{i_n},$$

then

$$T(\hat{\phi}(x_1)\ldots\hat{\phi}(x_n)) = \hat{\phi}(x_{i_1})\ldots\hat{\phi}(x_{i_n}) \tag{15.59}$$

(time increasing from right to left).

The scattering matrix (S-matrix) is usually computed using the n-point functions

$$G_n(x_1, \ldots, x_n) = \langle \text{out}|T(\hat{\phi}(x_1)\ldots\hat{\phi}(x_n))|\text{in}\rangle. \tag{15.60}$$

There is a functional integral representation of these functions:

$$G_n(x_1, \ldots, x_n) = \frac{\int \mathcal{D}\phi \cdot e^{iS(\phi)/\hbar} \phi(x_1)\ldots\phi(x_n)}{\int \mathcal{D}\phi \cdot e^{iS(\phi)/\hbar}}. \tag{15.61}$$

The volume element $\mathcal{D}\phi$ on the space of classical fields ϕ is translation-invariant, $\mathcal{D}(\phi + \phi_0) = \mathcal{D}\phi$ for a fixed function ϕ_0. Its normalization doesn't matter because of the division in equation (15.61). The corresponding generating series is given by

$$\sum_{n \geq 0} \frac{(-2\pi i)^n}{n!} \int d^4x_1 \ldots d^4x_n \, G_n(x_1, \ldots, x_n) J(x_1) \ldots J(x_n) \tag{15.62}$$

evaluated as $Z(J)/Z(0)$, where $Z(J)$ is defined as in (15.28)

$$Z(J) = \int \mathcal{D}\phi \cdot e^{iS(\phi)/\hbar} e^{-2\pi i \langle J, \phi \rangle}. \tag{15.63}$$

Hence, we have

$$Z(0) = \int \mathcal{D}\phi \cdot e^{iS(\phi)/\hbar} \tag{15.64}$$

and $Z(J)$ is obtained by replacing $S(\phi)$ by $S(\phi) - 2\pi\hbar\langle J, \phi\rangle$ in the definition of $Z(0)$. The term $\langle J, \phi\rangle$ is considered as a perturbation coming from a "source" J.

In the $\lambda\varphi^4$ model, we have

$$S(\phi) = S_0(\phi) + \lambda S_{\text{int}}(\phi), \tag{15.65}$$

where

$$S_0(\phi) = \frac{\hbar}{2} \int d^4x [\eta^{\mu\nu} \partial_\mu \phi(x) \partial_\nu \phi(x) - \mu^2 \phi(x)^2], \tag{15.66}$$

$$S_{\text{int}}(\phi) = -\frac{\hbar}{4!} \int d^4x\, \phi(x)^4. \tag{15.67}$$

We can therefore expand $e^{iS(\phi)/\hbar}$ as a power series in λ, namely

$$e^{iS(\phi)/\hbar} = e^{iS_0(\phi)/\hbar} \sum_{m \geq 0} \frac{(-\lambda/4!)^m}{m!} \int d^4y_1 \ldots d^4y_m\, \phi(y_1)^4 \ldots \phi(y_m)^4. \tag{15.68}$$

After integrating term by term in equations (15.61), (15.63), and (15.64), we obtain an expansion of $G_n(x_1, \ldots, x_n)$, $Z(J)$, and $Z(0)$ as a power series in λ, which is known as a perturbation expansion.

The free field

The quadratic part $S_0(\phi)$ in the action $S(\phi)$ for $\lambda\varphi^4$ is obtained by putting $\lambda = 0$ in $S(\phi)$ and corresponds to a free field, since the classical equation of motion $\delta S_0(\phi)/\delta\phi(x) = 0$ reads as

$$\Box \phi + \mu^2 \phi = 0 \tag{15.69}$$

with the d'Alembertian operator $\Box = \eta^{\mu\nu}\partial_\mu\partial_\nu$. The advantage of the perturbation expansion is that each term in the λ-series is a functional integral over a free field. In this subsection, we adapt the results of Section 2.3 to our case. From (15.66), we derive $S_0(\phi)/\hbar = \pi Q(\phi)$ with

$$Q(\phi) = \frac{1}{2\pi} \int d^4x [\eta^{\mu\nu} \partial_\mu \phi(x) \partial_\nu \phi(x) - \mu^2 \phi(x)^2], \tag{15.70}$$

15.4 Dimensional regularization

that is

$$Q(\phi) = -\frac{1}{2\pi}\langle \Box\phi + \mu^2\phi, \phi\rangle. \tag{15.71}$$

The two-point function for the free field is given by[4]

$$G_0(x,y) = \int \mathcal{D}\phi \cdot e^{\pi i Q(\phi)} \phi(x)\phi(y). \tag{15.72}$$

Using equations (2.35)–(2.37) (with $s = i$), we derive

$$G_0(x,y) = g(x-y), \tag{15.73}$$

where

$$(\Box + \mu^2)g(x) = -i\delta(x). \tag{15.74}$$

Using a Fourier transform, we end up with[5]

$$G_0(x,y) = \frac{i}{(2\pi)^4\hbar^2} \int d^4p \, \frac{e^{ip\cdot(x-y)/\hbar}}{p \cdot p - m^2c^2}. \tag{15.75}$$

The generating function for the free field

$$Z_0(J) = \int \mathcal{D}\phi \cdot e^{-iS_0(\phi)/\hbar} e^{-2\pi i \langle J, \phi\rangle} \tag{15.76}$$

is given by

$$Z_0(J) = e^{-2\pi^2 W_0(J)} \tag{15.77}$$

with

$$W_0(J) = \iint d^4x \cdot d^4y \, G_0(x,y) J(x) J(y). \tag{15.78}$$

By expanding both sides of (15.77) in powers of J, we can evaluate the n-point functions for the free field as

$$G_{0,n}(x_1,\ldots,x_n) = 0 \quad \text{for } n \text{ odd}, \tag{15.79}$$

$$G_{0,2m}(x_1,\ldots,x_{2m}) = \sum G_0(x_{i_1}, x_{j_1}) G_0(x_{i_2}, x_{j_2}) \ldots G_0(x_{i_m}, x_{j_m}) \tag{15.80}$$

(summation over the permutations $i_1 \ldots i_m j_1 \ldots j_m$ of $1, 2 \ldots, 2m$ such that $i_1 < i_2 < \ldots < i_m, i_1 < j_1, \ldots, i_m < j_m$ (see equation (2.47)).

---

[4] We normalize $\mathcal{D}\phi$ by $\int \mathcal{D}\phi \cdot e^{\pi i Q(\phi)} = 1$, that is $\mathcal{D}\phi = \mathcal{D}_{i,Q}\phi$ with the notation of Section 2.3.

[5] Notice our convention $p \cdot p = (p_0)^2 - (p_1)^2 - (p_2)^2 - (p_3)^2$. In quantum field theory, most authors define $p \cdot p$ as $(-p_0)^2 + (p_1)^2 + (p_2)^2 + (p_3)^2$ and find $p \cdot p + m^2c^2$ in the denominator of (15.75).

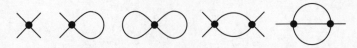

Fig. 15.1 Some examples of diagrams

Diagrams

To define a diagram D of the kind considered here, we first introduce a set V of *vertices* and we attach to each vertex v a set H_v of four *half-edges*. The disjoint union H of the sets H_v consists of the half-edges. We then select a subset H_{ext} of H, consisting of the *external half-edges* (also called *legs*). The set $H_{\text{int}} = H \setminus H_{\text{ext}}$ of internal half-edges is then partitioned into a collection E of two-element subsets.[6] Hence E consists of *edges*, each edge being composed of two half-edges. To orient an edge, choose one of its two half-edges. Notice that loops and multiple edges are permitted. As usual, a diagram is connected iff any two vertices can be joined by a sequence of edges. Otherwise it is decomposed uniquely into a collection of connected diagrams, called *its connected components*.

The action functional will be written as

$$S_\lambda(\phi) = S_0(\phi) + \lambda S_{\text{int}}(\phi) \tag{15.81}$$

to make explicit the dependence on λ. Accordingly we set

$$Z_\lambda(J) = \int \mathcal{D}\phi \cdot e^{iS_\lambda(\phi)/\hbar} e^{-2\pi i \langle J, \phi \rangle} \tag{15.82}$$

and we define $W_\lambda(J)$ by[7]

$$Z_\lambda(J)/Z_\lambda(0) = e^{-2\pi^2 W_\lambda(J)}. \tag{15.83}$$

We have a double expansion

$$W_\lambda(J) = W_0(J) + \sum_{m \geq 1} \sum_{n \geq 2} \lambda^m W_{m,n}(J), \tag{15.84}$$

where

$$W_{m,n}(J) = \int d^4 x_1 \ldots d^4 x_n\, W_{m,n}(x_1, \ldots, x_n) F(x_1) \ldots F(x_n) \tag{15.85}$$

and $F = G_0 J$, that is

$$F(x) = \int d^4 y\, G_0(x,y) J(y). \tag{15.86}$$

---

[6] Since external legs are half-edges, we are describing what is usually called an *amputated diagram*.
[7] Notice that $Z_0(0) = 1$ by our normalization of $\mathcal{D}\phi$; hence $W_\lambda(J)$ reduces to $W_0(J)$ for $\lambda = 0$.

15.4 Dimensional regularization

From (15.62), (15.83), (15.84), and (15.85) we see that *the task of computing the n-point functions is reduced to the determination of the functions* $W_{m,n}(x_1,\ldots,x_n)$. The main combinatorial result is that *these functions can be expressed as a sum* $\Sigma_D W_D(x_1,\ldots,x_n)$, *where D runs over the connected diagrams* with m vertices and n legs labeled by the points $x_1,\ldots,x_n$ in spacetime.

Regularization

The theory has one drawback. The two-point function $G_0(x,y)$ is singular for $x = y$, and the interaction

$$S_{\text{int}}(\phi) = -\frac{\hbar}{4!}\int d^4x\,\phi(x)^4$$

refers to coinciding points. When evaluating gaussian free-field integrals according to (15.80) we shall meet singular factors of the type $G_0(x,x)$.

Fig. 15.2

To cure the singularities, one may proceed as follows.

- Express the kernels $W_D(x_1,\ldots,x_n)$ as Fourier transforms of functions $W_D(p_1,\ldots,p_n)$ depending on n 4-momenta $p_1,\ldots,p_n$ labeling the legs of the diagram D. For instance, for the diagram depicted in figure 15.2, the conservation of energy-momentum requires

$$p_1 + p_2 = p_3 + p_4 \quad (=p \text{ say}). \tag{15.87}$$

- Carry out a Wick rotation, that is construct an analytic continuation replacing the component p_0 of p by ip_0 and introducing the corresponding euclidean momentum $p_E = (p_0, p_1, p_2, p_3)$. Hence $p\cdot p = -p_E\cdot p_E$. For the diagram above we are reduced to the evaluation of the integral

$$I(p_E) := \int \frac{d^4k_E}{(|k_E|^2 + m^2c^2)(|k_E + p_E|^2 + m^2c^2)}, \tag{15.88}$$

where $|k_E|$ is the length $\sqrt{k_E\cdot k_E}$ of the euclidean vector k_E. This integral is logarithmically divergent since the integrand behaves like $1/|k_E|^4$ for $|k_E|$ large.

- Introduce the Schwinger parameters α_1 and α_2 associated with the *edges* of D by

$$(|\pi_i|^2 + m^2c^2)^{-1} = \int_0^\infty d\alpha_i\,e^{-\alpha_i(|\pi_i|^2 + m^2c^2)} \tag{15.89}$$

(here $\pi_1 = k_E$ and $\pi_2 = k_E + p_E$).

- Plug (15.89) into (15.88), and interchange the integrations with respect to k_E and α_1, α_2. We end up with

$$I(p_E) = \pi^2 \int_0^\infty \int_0^\infty \frac{d\alpha_1\, d\alpha_2}{(\alpha_1 + \alpha_2)^2} e^{-\Phi(\alpha_1, \alpha_2, p_E)}, \qquad (15.90)$$

where

$$\Phi(\alpha_1, \alpha_2, p_E) = (\alpha_1 + \alpha_2) m^2 c^2 + \frac{\alpha_1 \alpha_2}{\alpha_1 + \alpha_2} |p_E|^2. \qquad (15.91)$$

The integral (15.90) diverges in the neighborhood of the origin $\alpha_1 = \alpha_2 = 0$ in the α-plane, because of the exponent in $(\alpha_1 + \alpha_2)^2$. This exponent originates from an intermediate integral,

$$\int d^4 k_E\, e^{-(\alpha_1+\alpha_2)|k_E|^2} = \left(\frac{\pi}{\alpha_1 + \alpha_2}\right)^2. \qquad (15.92)$$

On repeating the same calculation for vectors k_E in a euclidean space of dimension D, the previous integral would become $(\pi/(\alpha_1+\alpha_2))^{D/2}$. On writing D as $4 - 2z$, we end up with

$$I_z(p_E) = \pi^{2-z} \int_0^\infty \int_0^\infty \frac{d\alpha_1\, d\alpha_2}{(\alpha_1 + \alpha_2)^2} (\alpha_1 + \alpha_2)^z e^{-\Phi(\alpha_1,\alpha_2,p_E)}. \qquad (15.93)$$

Consider now z as a complex parameter and *pretend that we are working in a space of $D = 4 - 2z$ dimensions*. Owing to the factor $(\alpha_1+\alpha_2)^z$, the integral $I_z(p_E)$ is now convergent for Re $z > 0$ and has an analytic continuation as a meromorphic function of z.

For a general diagram D, we end up with a slightly more complicated expression:

$$I_D(p_1, \ldots, p_n) = \int_{[0,\infty[^e} d^e \alpha\, \frac{e^{-\Phi(\alpha,p)}}{U_D(\alpha)^2}. \qquad (15.94)$$

Here $\alpha = (\alpha_i)$ is a set of Schwinger parameters, one for each of the e edges of D, $U_D(\alpha)$ is a certain combinatorial polynomial (Kirchhoff polynomial), and $\Phi(\alpha, p)$ is the sum $m^2 c^2 \sum_i \alpha_i + V_D(\alpha, p)/U_D(\alpha)$, where $V_D(\alpha, p)$ is a polynomial in the α_i and p_j, which is quadratic in the 4-momenta p_j.

Conclusion

The dimensional regularization amounts to inserting a convergence factor $U_D(\alpha)^z$ into the integral (15.94). Thereby the integral converges for Re $z > 0$ and extends as a meromorphic function of z.

There is a drawback, since $U_D(\alpha)$ has a physical dimension. To cure the disease, use the fact that $\alpha m^2 c^2$ is dimensionless and introduce a *mass scale μ* (or dually a length scale as in Chapter 16). This mass scale is crucial in the Wilsonian approach to renormalization. With this mass scale, we can introduce the dimensionless convergence factor $U_D(\alpha\mu^2 c^2)^z$ in the divergent integral (15.94).

We don't have to invent a euclidean space of dimension $4 - 2z$. However, A. Connes recently suggested replacing the Minkowski space $\mathbb{M}^4$ by $\mathbb{M}^4 \times P_z$, where P_z is a suitable *noncommutative space* (spectral triple) of dimension $-2z$.

Neutrix calculus [7] developed in connection with asymptotic series can be used to obtain finite renormalizations for quantum field theory.

References

[1] G. Green (1836). "Researches on the vibration of pendulums in fluid media," *Roy. Soc. Edinburgh Trans.*, 9 pp. Reprinted in *Mathematical Papers* (Paris, Hermann, 1903), pp. 313–324.
For more information, see D. M. Cannell (1993). *George Green Mathematician and Physicist 1793–1841* (London, The Athlone Press).

[2] R. P. Feynman (1966). "The development of the space-time view of quantum electrodynamics," *Phys. Today* (August), 31–44.

[3] P. A. M Dirac (1947). *The Principles of Quantum Mechanics* (Oxford, Clarendon Press).

[4] P. M. Malliavin (1997). *Stochastic Analysis* (Berlin, Springer).

[5] L. H. Ryder (1996). *Quantum Field Theory*, 2nd edn. (Cambridge, Cambridge University Press).
B. DeWitt (2003). *The Global Approach to Quantum Field Theory* (Oxford, Oxford University Press; with corrections 2004).

[6] In the 1877 preface of his book *Matter and Motion* (New York, Dover Publications, 1991) James Clerk Maxwell recalls the evolution of physical science from forces acting between one body and another to the energy of a material system.

[7] Y. Jack Ng and H. van Dam (2004). "Neutrix calculus and finite quantum field theory," arXiv:hep-th/0410285, v3 30 November 2004.

Other references that inspired this chapter

R. Balian and J. Zinn-Justin, eds. (1976). *Methods in Field Theory* (Amsterdam, North-Holland).

M. Berg (2001). Geometry, renormalization, and supersymmetry, unpublished Ph.D. Dissertation, University of Texas at Austin.

A. Das (1993). *Field Theory, A Path Integral Approach* (Singapore, World Scientific).

M. E. Peskin and D. V. Schroeder (1995). *An Introduction to Quantum Field Theory* (Reading, Perseus Books).

F. Ravndal (1976). *Scaling and Renormalization Groups* (Copenhagen, Nordita).

A. Wurm (2002). Renormalization group applications in area-preserving nontwist maps and relativistic quantum field theory, unpublished Ph.D. Dissertation, University of Texas at Austin.

16
Renormalization 2: scaling

16.1 The renormalization group

The aim of the renormalization group is to describe how the dynamics of a system evolves as one changes the scale of the phenomena being observed.

D. J. Gross [1]

The basic physical idea underlying the renormalization group is that many length or energy scales are locally coupled.

K. J. Wilson [2]

While working out estimates on the renormalization-group transformations, D. C. Brydges, J. Dimock, and T. R. Hurd [3] developed, cleaned up, and simplified scaling techniques. Their coarse-graining techniques have a wide range of applications and are presented in Section 2.5. A gaussian covariance can be decomposed into scale-dependent contributions (2.79) and (2.82). Indeed, a gaussian $\mu_{s,G}$, abbreviated to μ_G, of covariance G and variance W, is defined by

$$\int_\Phi \mathrm{d}\mu_G(\phi)\exp(-2\pi\mathrm{i}\langle J,\phi\rangle) := \exp(-\pi s W(J)), \tag{16.1}$$

$$W(J) =: \langle J, GJ\rangle. \tag{16.2}$$

16.1 The renormalization group

It follows that, if $W = W_1 + W_2$, the gaussian μ_G can be decomposed into two gaussians, μ_{G_1} and μ_{G_2}:

$$\mu_G = \mu_{G_1} * \mu_{G_2} = \mu_{G_2} * \mu_{G_1}, \quad (16.3)$$

or, more explicitly,

$$\int_\Phi d\mu_G(\phi)\exp(-2\pi i\langle J, \phi\rangle) = \int_\Phi d\mu_{G_2}(\phi_2) \int_\Phi d\mu_{G_1}(\phi_1)$$
$$\times \exp(-2\pi i\langle J, \phi_1 + \phi_2\rangle), \quad (16.4)$$

where

$$G = G_1 + G_2, \quad (16.5)$$
$$\phi = \phi_1 + \phi_2. \quad (16.6)$$

The convolution property (16.3) and the additive properties (16.5) and (16.6) make it possible to decompose a gaussian covariance into scale-dependent contributions as follows.

- Introduce an independent scale variable $l \in\,]0, \infty[$. A gaussian covariance is a homogeneous two-point function on $\mathbb{M}^D$ (a D-dimensional euclidean or minkowskian space) defined by

$$\frac{s}{2\pi} G(|x-y|) := \int_\Phi d\mu_G(\phi)\phi(x)\phi(y). \quad (16.7)$$

It follows that the covariance can be represented by an integral

$$G(\xi) = \int_0^\infty d^\times l\, S_l u(\xi), \quad \text{where } d^\times l = dl/l,\ \xi = |x-y|, \quad (16.8)$$

for some dimensionless function u. The scaling operator S_l is defined by an extension of (2.73),

$$S_l u(\xi) = l^{2[\phi]} u(\xi/l), \quad (16.9)$$

where $[\phi]$ is the length dimension of ϕ. For example, see (2.78)–(2.86):

$$G(\xi) = C_D/\xi^{D-2}. \quad (16.10)$$

- Break the domain of integration of the scaling variable l into subdomains $[2^j l_0, 2^{j+1} l_0[$,

$$G(\xi) = \sum_{j=-\infty}^{+\infty} \int_{2^j l_0}^{2^{j+1} l_0} d^\times l\, S_l u(\xi) \quad (16.11)$$

and set

$$G(\xi) =: \sum_{j=-\infty}^{+\infty} G_j(l_0, \xi) \quad \text{abbreviated to } \sum_j G_j(\xi). \quad (16.12)$$

The contributions G_j to the covariance G are self-similar. Indeed, set $l = (2^j l_0)k$, then

$$\begin{aligned} G_j(\xi) &= (2^j l_0)^{2[\phi]} \int_1^2 d^\times k \, k^{[2\phi]} u(\xi/2^j l_0 k) \\ &= (2^j)^{2[\phi]} G_0(\xi/2^j) \\ &= S_{2^j} G_0(\xi). \end{aligned} \qquad (16.13)$$

Another equivalent formulation can be found in Section 2.5, equation (2.89).

The corresponding field decomposition (2.81) is

$$\phi = \sum_{j=-\infty}^{+\infty} \phi_j,$$

which is also written

$$\phi(x) = \sum_j \phi_j(l_0, x). \qquad (16.14)$$

The subdomains $[2^j l_0, 2^{j+1} l_0[$ are exponentially increasing for $j \geq 0$, and the subdomains $[2^{j-1} l_0, 2^j l_0[$ are exponentially decreasing for $j \leq 0$. A scale-dependent covariance defines a scale-dependent gaussian, a scale-dependent functional laplacian (2.63), and scale-dependent Bargmann–Segal and Wick transforms (2.64) and (2.65).

Remark. Brydges uses a scale variable $k \in [1, \infty[$. The domain decomposition

$$\sum_{j=-\infty}^{+\infty} [2^j l_0, 2^{j+1} l_0[$$

is then reorganized as

$$\sum_{j=-\infty}^{+\infty} = \sum_{j<0} + \sum_{j \geq 0},$$

and l_0 is set equal to 1, i.e.

$$]0, \infty[= [0, 1[\cup [1, \infty[\qquad (16.15)$$

with $k \in [1, \infty[$ and $k^{-1} \in [0, 1[$. Note that

$$\int_0^1 d^\times l \, S_l u(\xi) = \int_1^\infty d^\times k \, S_{k^{-1}} u(\xi). \qquad (16.16)$$

Remark. The Mellin transform. Equation (16.8) defines the Mellin transform $\tilde{f}(\alpha)$ of a function $f(t)$ decreasing sufficiently rapidly at infinity,

$$\tilde{f}(\alpha) := \int_0^\infty d^\times t \, f(t) t^\alpha < \infty$$

$$= a^\alpha \int_0^\infty d^\times l \, f\left(\frac{a}{l}\right) l^{-a},$$

where $l = a/t$. For example,

$$\frac{1}{a^{2-\epsilon}} = \frac{1}{2} \int_0^\infty d^\times l \, \exp\left(-\frac{a^2}{4l^2}\right) l^{-2+\epsilon}. \tag{16.17}$$

Scale-dependent gaussians can be used to define effective actions as follows: break the action functional S into a free (quadratic) term $\frac{1}{2}Q$ and an interacting term S_{int}:

$$S = \frac{1}{2}Q + S_{\text{int}}. \tag{16.18}$$

In this context a free action is defined as an action that can be decomposed into scale-dependent contributions that are self-similar. The quadratic form in (16.18) defines a differential operator D,

$$Q(\phi) = \langle D\phi, \phi \rangle.$$

Provided that the domain Φ of ϕ is properly restricted (e.g. by the problem of interest) the operator D has a unique Green function G,

$$DG = 1, \tag{16.19}$$

which serves as the covariance of the gaussian μ_G defined by (16.1) and (16.2). If D is a linear operator with constant coefficients then $G(x,y)$ is a function of $|x - y| =: \xi$ with scale decomposition (16.12),

$$G(\xi) = \sum_j G_j(l_0, \xi). \tag{16.20}$$

The contributions G_j satisfy the self-similarity condition (16.13). The corresponding action functional Q is a "free" action.

To construct effective actions one splits the domain of the scale variable into two domains,

$$[\Lambda, \infty[\, = [\Lambda, L[\, \cup \, [L, \infty[, \tag{16.21}$$

where Λ is a short-distance (high-energy) cut-off.[1] The variable L is the renormalization scale.

---

[1] The cut-off Λ is used for handling divergences at the spacetime origin in euclidean quantum field theory, or on the lightcone at the origin in the minkowskian case.

The quantity to be computed, which is formally written

$$\int \mathcal{D}\phi \, \exp\left(\frac{i}{\hbar} S(\phi)\right),$$

is the limit $\Lambda \to 0$ of

$$I_\Lambda := \int_\Phi \mathrm{d}\mu_{[\Lambda,\infty[}(\phi) \exp\left(\frac{i}{\hbar} S_{\text{int}}(\phi)\right) \qquad (16.22)$$

$$\equiv \left\langle \mu_{[\Lambda,\infty[}, \exp\left(\frac{i}{\hbar} S_{\text{int}}\right)\right\rangle,$$

$$\left\langle \mu_{[\Lambda,\infty[}, \exp\left(\frac{i}{\hbar} S_{\text{int}}\right)\right\rangle = \left\langle \mu_{[L,\infty[}, \mu_{[\Lambda,L[} * \exp\left(\frac{i}{\hbar} S_{\text{int}}\right)\right\rangle. \qquad (16.23)$$

The integrand on the right-hand side is an effective action at scales greater than L and is obtained by integrating short-distance degrees of freedom over the range $[\Lambda, L[$. Equation (16.23) transforms S_{int} into an effective action.

Equation (16.23) is conceptually beautiful: one integrates an effective action at scales greater than L over the range $[L, \infty[$, but it is difficult to compute. It has been rewritten by Brydges *et al.* as a coarse-graining transformation (2.94) in terms of the coarse-graining operator P_l:

$$P_l := S_{l/l_0} \mu_{[l_0, l[} *, \qquad (16.24)$$

where the scaling operator S_{l/l_0} is defined by (2.73) and (2.74). The coarse-graining operator P_l is an element of a semigroup (2.95). The semigroup property is necessary to derive the scale evolution equation of the effective action, and to simplify the writing we now require that $l_0 = 1$ by choosing a proper unit of length. The coarse-graining operator is then

$$P_l := S_l \mu_{[1,l[} * \qquad (16.25)$$

and the domain of the scale variable is split at $L = 1$:

$$[\Lambda, \infty[= [\Lambda, 1[\cup [1, \infty[,$$

$$\left\langle \mu_{[\Lambda,\infty[}, \exp\left(\frac{i}{\hbar} S_{\text{int}}\right)\right\rangle = \left\langle \mu_{[\Lambda,1[}, \mu_{[1,\infty[} * \exp\left(\frac{i}{\hbar} S_{\text{int}}\right)\right\rangle. \qquad (16.26)$$

The new integral on the right-hand side can then be written as the scale-independent gaussian of a coarse-grained integrand (2.112):

$$\left\langle \mu_{[1,\infty[}, \exp\left(\frac{i}{\hbar} S_{\text{int}}\right)\right\rangle = \left\langle \mu_{[l,\infty[}, \mu_{[1,l[} * \exp\left(\frac{i}{\hbar} S_{\text{int}}\right)\right\rangle$$

$$= \left\langle \mu_{[1,\infty[}, S_l \mu_{[1,l[} * \exp\left(\frac{i}{\hbar} S_{\text{int}}\right)\right\rangle$$

$$= \left\langle \mu_{[1,\infty[}, P_l \exp\left(\frac{i}{\hbar} S_{\text{int}}\right)\right\rangle.$$

16.1 The renormalization group

It follows that

$$\left\langle \mu_{[\Lambda,\infty[}, \exp\left(\frac{i}{\hbar}S_{\text{int}}\right)\right\rangle = \left\langle \mu_{[\Lambda,1[}, \mu_{[1,\infty[} * P_l \exp\left(\frac{i}{\hbar}S_{\text{int}}\right)\right\rangle. \quad (16.27)$$

The scale evolution of a coarse-grained quantity $P_l A(\phi)$ is given by (2.98),

$$\left(\frac{\partial^\times}{\partial l} - H\right) P_l A(\phi) = 0, \quad (16.28)$$

where

$$H = \dot{S} + \frac{s}{4\pi}\dot{\Delta} \quad (16.29)$$

and

$$\dot{S} := \left.\frac{d^\times S}{dl}\right|_{l=l_0}.$$

The functional laplacian $\dot{\Delta}$ is defined by (2.101).

We are interested in the scale evolution of $P_l \exp((i/\hbar)S_{\text{int}}(\phi))$. This term appears in the evolution equation (16.28) when $A(\phi)$ is an exponential. The derivation (2.102)–(2.110) is repeated, and differs only in the calculation of[2]

$$\int_\Phi d\mu_{[l_0,l[}(\psi)(\exp B(\phi))''\psi\psi = (\exp B(\phi))''\frac{s}{2\pi}G_{[l_0,l[}$$

$$= \int dvol x \int dvol y \frac{s}{2\pi}G_{[l_0,l[}(|x-y|)$$

$$\times \frac{\delta^2}{\delta\phi(x)\delta\phi(y)}\exp B(\phi), \quad (16.30)$$

$$\frac{\delta^2}{\delta\phi(x)\delta\phi(y)}\exp B(\phi) = \left(\frac{\delta^2 B(\phi)}{\delta\phi(x)\delta\phi(y)} + \frac{\delta B(\phi)}{\delta\phi(x)}\frac{\delta B(\phi)}{\delta\phi(y)}\right)\exp B(\phi). \quad (16.31)$$

Finally, $P_l B(\phi)$ satisfies the equation

$$\left(\frac{\partial^\times}{\partial l} - H\right) P_l B(\phi) = \frac{s}{4\pi} B(\phi) \overset{\leftrightarrow}{\Delta} B(\phi) \quad (16.32)$$

with

$$B(\phi) \overset{\leftrightarrow}{\Delta} B(\phi) = \int dvol x \int dvol y\, G_{[l_0,l[}(|x-y|)\frac{\delta B(\phi)}{\delta\phi(x)}\frac{\delta B(\phi)}{\delta\phi(y)}. \quad (16.33)$$

---

[2] We use $dvol x$ instead of $d^D x$ because the formulas are valid if there is a nontrivial metric and for $s = 1$ or i.

Given (16.11), it follows that

$$G_{[1,l[}(\xi) = \int_1^l d^\times s\, S_s u(\xi), \tag{16.34}$$

$$\left.\frac{\partial^\times}{\partial l} G_{[1,l[}(\xi)\right|_{l=1} = u(\xi), \tag{16.35}$$

and

$$B(\phi) \overset{\leftrightarrow}{\Delta}\, B(\phi) = \int d\text{vol}x \int d\text{vol}y\, u(|x-y|)\, \frac{\delta B(\phi)}{\delta \phi(x)}\frac{\delta B(\phi)}{\delta \phi(y)}. \tag{16.36}$$

In conclusion, we have derived two *exact* scale-evolution equations (16.28), for $P_l S_{\text{int}}(\phi)$ and (16.32), for $P_l \exp((i/\hbar)S_{\text{int}}(\phi))$. Now let us define

$$S_{[l]} := P_l S_{\text{int}}(\phi). \tag{16.37}$$

The coarse-grained interaction $S_{[l]}$ (not to be confused with S_l) is often called the naive scaling of the interaction, or its scaling by "engineering dimension," hence the use of the square bracket. $S_{[l]}$ satisfies (16.28):

$$\left(\frac{\partial^\times}{\partial l} - H\right) S_{[l]} = 0. \tag{16.38}$$

Let $S(l, \phi)$ be the effective action defined by

$$\exp S(l, \phi) := P_l \exp\left(\frac{i}{\hbar} S_{\text{int}}(\phi)\right). \tag{16.39}$$

It follows that

$$\left(\frac{\partial^\times}{\partial l} - H\right) S(l,\phi) = \frac{s}{4\pi} S(l,\phi) \overset{\leftrightarrow}{\Delta}\, S(l,\phi). \tag{16.40}$$

Approximate solutions to the exact scale-evolution equation of the effective action are worked out for the $\lambda \phi^4$ system in the next section. The approximate solutions include divergent terms. The strategy is *not* to drop divergent terms, nor to add counterterms by *fiat*, but to "trade" divergent terms for conditions on the scale-dependent couplings. The trade does not affect the approximation of the solution. The trade requires the couplings to satisfy ordinary differential equations in l. These equations are equivalent to the standard renormalization flow equations.

16.2 The $\lambda \phi^4$ system

The $\lambda \phi^4$ system is a self-interacting relativistic scalar field in $1+3$ dimensions described by the lagrangian density

$$\mathcal{L}(\phi(x)) = \frac{1}{2}\eta^{\mu\nu}\partial_\mu \phi(x)\partial_\nu \phi(x) - \frac{1}{2}m^2 \phi^2(x) - \frac{\lambda}{4!}\phi^4(x), \tag{16.41}$$

where λ is a dimensionless coupling constant when $\hbar = c = 1$, and the metric

$$\eta_{\mu\nu} = \text{diag}(1,-1,-1,-1). \tag{16.42}$$

There are many references for the study of the $\lambda\phi^4$ system since it is the simplest nontrivial example without gauge fields. The chapter "Relativistic scalar field theory" in Ashok Das' book [4] provides an excellent introduction to the system. In this section we apply the scale-evolution equation (16.40) for the effective action to this system as developed by Brydges *et al.*

- One minor difference: Brydges uses $\hbar = c = 1$ and gives physical dimensions in the mass dimension. We use $\hbar = c = 1$ and give physical dimensions in the length dimension because most of the work is done in spacetime rather than in 4-momentum space. The relationship

$$l \frac{\mathrm{d}}{\mathrm{d}l} = -m \frac{\mathrm{d}}{\mathrm{d}m} \tag{16.43}$$

provides the necessary correspondence.

- A nontrivial difference: Brydges investigates euclidean quantum field theory, whereas we investigate minkowskian field theory. The difference will appear in divergent expressions, which are singular at the origin in the euclidean case and are singular on the lightcone in the minkowskian case.

A normal-ordered lagrangian

The starting point of Brydges' work is the normal-ordered lagrangian

$$: \mathcal{L}(\phi(x)) : = \frac{1}{2}\eta^{\mu\nu} : \phi_{,\mu}(x)\phi_{,\nu}(x) : - \frac{1}{2}m^2 : \phi^2(x) : - \frac{\lambda}{4!} : \phi^4(x) : . \tag{16.44}$$

Originally, *operator* normal ordering was introduced by Gian-Carlo Wick to simplify calculations (it readily identifies vanishing terms in vacuum expectation values). It also eliminates the (infinite) zero-point energy of an assembly of harmonic oscillators. The normal ordering of a *functional* F of ϕ is, by definition,

$$: F(\phi) :_G := \exp\left(-\frac{1}{2}\Delta_G\right) F(\phi), \tag{16.45}$$

where Δ_G is the functional laplacian defined by the covariance G:

$$\Delta_G := \int \mathrm{dvol}x \int \mathrm{dvol}y \, G(x,y) \frac{\delta}{\delta\phi(x)} \frac{\delta}{\delta\phi(y)}. \tag{16.46}$$

In quantum physics, a normal-ordered lagrangian in a functional integral corresponds to a normal-ordered hamiltonian operator (see Appendix D).

Normal-ordered action functionals simplify calculations because integrals of normal-ordered monomials are eigenvalues of the coarse-graining operator P_l (2.96).

The effective action $S(l, \phi)$

We apply the procedure developed in Section 16.1 for constructing an effective action to the lagrangian (16.44).

The functional integral (16.22) can be written in terms of (16.27) with an effective action $S(l, \phi)$ defined by

$$\exp S(l, \phi) := P_l \exp\left(\frac{i}{\hbar} S_{\text{int}}(\phi)\right), \qquad (16.47)$$

where

$$\left\langle \mu_{[\Lambda,\infty[}, \exp\left(\frac{i}{\hbar} S_{\text{int}}\right) \right\rangle = \left\langle \mu_{[\Lambda,1[}, \mu_{[1,\infty[} * P_l \frac{i}{\hbar} S_{\text{int}}(\phi) \right\rangle. \qquad (16.48)$$

The integral with respect to $\mu_{[1,\infty[}$ is over polynomials that are normal-ordered according to the covariance

$$G_{[1,\infty[}(|x-y|) = \int_1^\infty d^\times l\, S_l u(|x-y|). \qquad (16.49)$$

Given an action

$$S(\phi) = \int \text{dvol} x\, \mathcal{L}(\phi(x)) = \frac{1}{2} Q(x) + S_{\text{int}}(\phi), \qquad (16.50)$$

one may often choose from among various ways of splitting the action into a quadratic term and an interaction term. Choosing Q amounts to choosing the concomitant gaussian μ_G. For the $\lambda \phi^4$ system there are three natural choices for Q:

$$Q(x) = \int \text{dvol} x\, \eta^{\mu\nu} : \partial_\mu \phi(x) \partial_\nu \phi(x) : \qquad \text{Brydges' choice}, \quad (16.51)$$

$$Q(x) = \int \text{dvol} x\, \eta^{\mu\nu} : \partial_\mu \phi(x) \partial_\nu \phi(x) : -m^2 : \phi^2(x) :,$$

$$Q(x) = S''(\psi_{\text{cl}}) \phi \phi, \qquad (16.52)$$

where $S''(\psi_{\text{cl}})$ is the value of the second variation in a classical solution of the Euler–Lagrange equation. Brydges' choice leads to mass and coupling-constant renormalization flow equations.

The second choice is not interesting for several reasons. The covariance $G(x,y)$ is, as in Brydges' choice, a function of $|x-y|$ but a more complicated one. The corresponding finite-dimensional case (Section 3.3) shows that the gaussian normalization in the second choice is awkward. Finally, and more importantly, the mass term must be included in the interaction in order to be renormalized.

The third choice (16.52) is challenging: the second variation as a quadratic form is very beneficial in the study of the anharmonic oscillator.[3]

We proceed with Brydges' choice (16.51). It follows from (16.44) that

$$S_{\text{int}}(\phi) = \int \mathrm{dvol}x \left(-\frac{1}{2}m^2 : \phi^2(x) :_{[1,\infty[} -\frac{\lambda}{4!} : \phi^4(x) :_{[1,\infty[}\right). \quad (16.53)$$

The effective action $S(l,\phi)$ at scales larger than l is defined by

$$P_l \exp\left(\frac{\mathrm{i}}{\hbar} S_{\text{int}}(\phi)\right) =: \exp S(l,\phi) \quad (16.54)$$

and satisfies the exact scale-evolution equation (16.40). Let us define $E(S)$ as the exact scale-evolution equation such that

$$E(S) := \left(\frac{\partial^\times}{\partial l} - \dot{S} - \frac{s}{4\pi}\dot\Delta\right) S(l,\phi) - \frac{s}{4\pi} S(l,\phi) \overset{\leftrightarrow}{\Delta} S(l,\phi) = 0. \quad (16.55)$$

An approximation $T(l,\phi)$ to $S(l,\phi)$ is called a solution at order $\mathcal{O}(\vec\lambda^k)$ if $E(T)$ is of order $\mathcal{O}(\vec\lambda^{k+1})$. We use $\vec\lambda$ to designate the set of coupling constants m, λ, and possibly others. In order to present the key issues as simply as possible, we consider a massless system: $m = 0$. Equation (16.53) becomes

$$S_{\text{int}}(\phi) = -\frac{\lambda}{4!} \int \mathrm{dvol}x : \phi^4(x) :_{[1,\infty[}. \quad (16.56)$$

First-order approximation to the effective action

The coarse-grained interaction, or naive scaling of the interaction,

$$S_{[l]}(\phi) = P_l S_{\text{int}}(\phi) = -\frac{\lambda}{4!} l^{4+4[\phi]} \int \mathrm{dvol}x : \phi^4(x) := S_{\text{int}}(\phi) \quad (16.57)$$

satisfies the evolution equation (16.38),

$$\left(\frac{\partial^\times}{\partial l} - \dot{S} - \frac{s}{4\pi}\dot\Delta\right) S_{[l]}(\phi) = 0. \quad (16.58)$$

$S_{[l]}$ is an approximate solution to $E(S)$ of order λ (16.55). The coupling constant λ is dimensionless and is not modified by the coarse-graining operator P_l.

---

[3] A first step towards using the second variation of $S(\phi)$ for computing the effective action of the $\lambda\phi^4$ system has been made by Xiaorong Wu Morrow; it is based on an effective action for the anharmonic oscillator (but remains unpublished).

Second-order approximation to the effective action

We expect the second-order approximation $T(l, \phi)$ to $S(l, \phi)$ to have the following structure:

$$T(l, \phi) = S_{[l]}(\phi) + \frac{1}{2} S_{[l]}(\phi) \overset{\leftrightarrow}{O} S_{[l]}(\phi), \qquad (16.59)$$

where the operator $\overset{\leftrightarrow}{O}$ is such that $E(T)$ is of order λ^3. The *Ansatz* proposed by Brydges is

$$\overset{\leftrightarrow}{O} = \exp\left(\frac{s}{2\pi} \overset{\leftrightarrow}{\Delta}_{[l-1,1[}\right) - 1. \qquad (16.60)$$

In the case of the $\lambda \phi^4$ system, the exponential terminates at $\overset{\leftrightarrow 4}{\Delta}$:

$$\overset{\leftrightarrow}{O} = \frac{s}{2\pi} \overset{\leftrightarrow}{\Delta}_{[l-1,1[} + \frac{1}{2!}\left(\frac{s}{2\pi}\right)^2 \overset{\leftrightarrow 2}{\Delta}_{[l-1,1[}$$

$$+ \frac{1}{3!}\left(\frac{s}{2\pi}\right)^3 \overset{\leftrightarrow 3}{\Delta}_{[l-1,1[} + \frac{1}{4!}\left(\frac{s}{2\pi}\right)^4 \overset{\leftrightarrow 4}{\Delta}_{[l-1,1[}. \qquad (16.61)$$

This provides proof that, with the *Ansatz* (16.60),

$$E(T) = \mathcal{O}(\lambda^3). \qquad (16.62)$$

A straight calculation of $E(T)$ is possible but far too long to be included here. We refer to the original literature [3, 6] and outline the key features of the proof. We shall show that

$$\left(\frac{\partial^{\times}}{\partial l} - \dot{S} - \frac{s}{4\pi}\dot{\Delta}\right) T(l, \phi) = \frac{1}{2} T(l, \phi) \frac{s}{2\pi} \overset{\leftrightarrow}{\Delta} T(l, \phi) + \mathcal{O}(\lambda^3). \qquad (16.63)$$

The approximation $T(l, \phi)$ is the sum of a term of order λ and four terms of order λ^2:

$$T(l, \phi) = S_{[l]}(\phi) + \frac{1}{2}\sum_{j=1}^{4} \frac{1}{j!}\left(\frac{s}{2\pi}\right)^j S_{[l]}(\phi) \overset{\leftrightarrow j}{\Delta}_{[l-1,1[} S_{[l]}(\phi) \qquad (16.64)$$

$$=: S_{[l]}(\phi) + \sum_{j=1}^{4} T_j(l, \phi).$$

- First one computes the left-hand side of (16.63) for $T_1(l, \phi)$:

$$\left(\frac{\partial^{\times}}{\partial l} - \dot{S} - \frac{s}{4\pi}\dot{\Delta}\right) S_{[l]}(\phi) \overset{\leftrightarrow}{\Delta}_{[l-1,1[} S_{[l]}(\phi)$$

$$= S_{[l]}(\phi)\left(\overset{\leftrightarrow}{\Delta} - \frac{s}{2\pi}\overset{\leftrightarrow}{\Delta}\overset{\leftrightarrow}{\Delta}_{[l-1,1[}\right) S_{[l]}(\phi). \qquad (16.65)$$

At first sight, this result is surprising. It seems that the Leibniz product rule has been applied to the left-hand side of (16.64) although it

16.2 The $\lambda\phi^4$ system

includes a scaling operator and a second-order differential operator. A quick handwaving argument runs as follows:

$$\dot{S} + \frac{s}{4\pi}\dot{\Delta}$$

is the generator H of the coarse-graining operator. When acting on $S_{[l]}$ (or any Wick ordered monomials) it acts as a first-order operator (16.58). Now let us see whether we can apply the Leibniz rule. Because

$$\left(\frac{\partial^\times}{\partial l} - H\right) S_{[l]}(\phi) = 0,$$

the only remaining term is

$$\left(\frac{\partial^\times}{\partial l} - \dot{S} - \frac{s}{4\pi}\dot{\Delta}\right) S_{[l]}(\phi) \overleftrightarrow{\Delta}_{[l-1,1[} S_{[l]}(\phi)$$

$$= S_{[l]}(\phi) \left(\frac{\partial^\times}{\partial l} - \dot{S} - \frac{s}{4\pi}\dot{\Delta}\right) \overleftrightarrow{\Delta}_{[l-1,1[} S_{[l]}(\phi). \quad (16.66)$$

What is the meaning of the right-hand side? How does the laplacian $\dot{\Delta}$ operate when sandwiched between two functionals? How can the right-hand side of (16.66) be equal to the right-hand side of (16.65)? It is easy to prove that

$$\left(\frac{\partial^\times}{\partial l} - \dot{S}\right) G_{[l-1,1[}(\xi) = \left(\frac{\partial^\times}{\partial l} - \dot{S}\right) \int_{l-1}^{1} d^\times s\, S_s u(\xi)$$

$$= S_{l-1} u(\xi) - \left.\frac{\partial^\times}{\partial t}\right|_{t=1} \int_{l-1 t-1}^{t-1} d^\times s\, S_s u(\xi)$$

$$= u(\xi).$$

It follows that

$$\left(\frac{\partial^\times}{\partial l} - \dot{S}\right) \overleftrightarrow{\Delta}_{[l-1,1[} = \overleftrightarrow{\dot{\Delta}}. \quad (16.67)$$

One cannot prove that $\frac{1}{2}\dot{\overleftrightarrow{\Delta}}\dot{\Delta}_{[l-1,1[}$ is equal to $\dot{\overleftrightarrow{\Delta}}\overleftrightarrow{\Delta}_{[l-1,1[}$. The reason lies in the improper use of the Leibniz rule. Indeed,

$$\frac{1}{2} \dot{\overleftrightarrow{\Delta}}\, S_{[l]}(\phi) \overleftrightarrow{\Delta}_{[l-1,1[} S_{[l]}(\phi)$$

contains, in addition to $S_{[l]}(\phi)\dot{\overleftrightarrow{\Delta}}\overleftrightarrow{\Delta}_{[l-1,1[} S_{[l]}(\phi)$, terms proportional to the third variation of $S_{[l]}(\phi)$, which are canceled out in the pedestrian computation of (16.65).
- Given (16.65) for T_1, we can record graphically the other terms necessary to prove that T satisfies (16.63). Let a straight line stand for $\overleftrightarrow{\Delta}_{[l-1,1[}$ and

let a dotted line stand for $\overset{\leftrightarrow}{\Delta}$,

$$\left(\frac{\partial^\times}{\partial l} - \dot{S} - \frac{s}{4\pi}\dot{\Delta}\right)\frac{s}{2\pi}S_{[l]}\text{------}S_{[l]}$$
$$= \frac{s}{2\pi}S_{[l]}\cdots\cdots S_{[l]} - \left(\frac{s}{2\pi}\right)^2 S_{[l]}\cdots\cdots S_{[l]}, \qquad (16.68)$$

$$\left(\frac{\partial^\times}{\partial l} - \dot{S} - \frac{s}{4\pi}\dot{\Delta}\right)\frac{1}{2}\left(\frac{s}{2\pi}\right)^2 S_{[l]}\text{=====}S_{[l]}$$
$$= \left(\frac{s}{2\pi}\right)^2 S_{[l]}\cdots\cdots S_{[l]} - \left(\frac{s}{2\pi}\right)^3 S_{[l]}\cdots\cdots S_{[l]} \qquad (16.69)$$

etc. The only remaining term on the right-hand side of the sum of these equations is

$$\frac{s}{2\pi}S_{[l]}\cdots\cdots S_{[l]};$$

therefore

$$\left(\frac{\partial^\times}{\partial l} - \dot{S} - \frac{s}{4\pi}\dot{\Delta}\right) S_{[l]}\overset{\leftrightarrow}{O} S_{[l]} = \frac{s}{2\pi}S_{[l]}\overset{\leftrightarrow}{\Delta} S_{[l]} \qquad (16.70)$$

and $T(l,\phi)$ satisfies (16.63). Brydges has proved for the case $s = 1$ that, in spite of the fact that the terms in $\mathcal{O}(\lambda^3)$ contain divergent terms when $l \to \infty$, these terms are uniformly bounded as $l \to \infty$.

The renormalization flow equation for λ

The approximate solution $T(l,\phi)$ of the effective action $S(l,\phi)$ contains one term of order λ and four terms of order λ^2 (16.64). We are interested in the divergent terms, since they are the ones which will be "traded" for conditions on the couplings. The other terms are of no particular interest in renormalization. In the massless $\lambda\phi^4$ system there are two terms, $T_2(l,\phi)$ and $T_3(l,\phi)$, each of which contains a singular part. Only $T_2(l,\phi)$ contributes to the flow of λ:

$$T_2(l,\phi) := \frac{1}{4}\left(\frac{s}{2\pi}\right)^2 \left(\frac{\lambda}{2}\right)^2 \int d\text{vol}x \int d\text{vol}y$$
$$\times :\phi^2(x): G^2_{[l-1,1[}(x-y) :\phi^2(y): . \qquad (16.71)$$

We can write $T_2(l,\phi)$ as the sum of a regular term and a singular term by subtracting from and adding to $T_2(l,\phi)$ the following term:

$$T_{2\text{sing}}(l,\phi) := \frac{1}{4}\left(\frac{s}{2\pi}\right)^2 \left(\frac{\lambda}{2}\right)^2 \int d\text{vol}x : \phi^4(x) :_{[l-1,1[} \int d\text{vol}y\, G^2_{[l-1,1[}(y). \qquad (16.72)$$

16.2 The $\lambda\phi^4$ system

The term $T_{2\text{sing}}$ is local and can be added to $S_{[l]}$ and simultaneously subtracted from T_2. The term $T_{2\text{sing}}$ regularizes T_2. Splitting $T_3(l,\phi)$ in a similar manner does not introduce terms including $:\phi^4(x):$, and need not be considered in the λ-flow. The λ-flow equation is obtained by imposing on $S_{[l]} + T_{2\text{sing}}$ the evolution equation (16.58) satisfied by $S_{[l]}$,

$$\frac{1}{4!}\left(\frac{\partial^\times}{\partial l} - H\right)\left(\lambda(l) + 36\left(\frac{s}{2\pi}\right)^2 \lambda^2(l) \int dvol y \, G^2_{[l^{-1},1[}(y)\right)$$
$$\times \int dvol x \, :\phi^4(x):_{[l]} = 0. \quad (16.73)$$

The irrelevant factor $1/4!$ is included for the convenience of the reader who may have used λ rather than $\lambda/4!$ in the interaction term. For the same reason we include $s/(2\pi)$ because it originates from the gaussian normalization (2.34).

The λ-flow equation reduces to

$$\frac{\partial^\times}{\partial l}\left(\lambda(l) + 36\left(\frac{s}{2\pi}\right)^2 \lambda^2(l)\right)\int dvol x \, G^2_{[l^{-1},1[}(x) = 0. \quad (16.74)$$

Proof. The operator H acts only on $\int dvol x :\phi^4(x):$ since the first factor does not depend on ϕ and is dimensionless. It follows that

$$\left(\frac{\partial^\times}{\partial l} - H\right)\int dvol x \, :\phi^4(x): = 0.$$

$\square$

The flow equation (16.74) plays the role of the β-function derived by loop expansion in perturbative quantum field theory to order λ^2 [5]:

$$\beta(\lambda) = -\frac{\partial^\times}{\partial l}\lambda(l) = \frac{3\lambda^2}{16\pi^2}. \quad (16.75)$$

The β-function is constructed from an effective action at a given loop-order built from one-particle-irreducible diagrams. This method is efficient because the set of one-particle-irreducible diagrams is much smaller than the set of all possible diagrams for the given system.

Remark. Dimensional regularization. Brydges uses a dimensional regularization of divergent integrals that is different from the standard regularization. Either method leads to the same final result, but Brydges' method works on the action functional rather than on the diagrams. We recall briefly the principle of diagram dimensional regularization (see also Section 15.4).

First consider the following integral over one momentum $k \in \mathbb{R}_D$:

$$I_D := \int_{\mathbb{R}_D} F(k, p^1, \ldots, p^r) \mathrm{d}^D k, \qquad k \equiv p^0, \qquad (16.76)$$

where the momenta $k, p^1, \ldots, p^r$ are linearly independent and the function F is invariant under rotation. Hence F is a function of the scalar products $p^i \cdot p^j$. Let $D \geq r + 1$ and decompose k in the basis $\{p^1, \ldots, p^r\}$:

$$k = k^\perp + c_1 p^1 + \cdots + c_r p^r, \qquad (16.77)$$

where $k^\perp$ is orthogonal to $p^1, \ldots, p^r$. Therefore

$$k \cdot p^i = \sum_1^r c_j p^i \cdot p^j.$$

The vector $k^\perp$ is in a space of dimension $D - r$, let κ be its length. The function F is a function of $\kappa, c_1, \ldots, c_r, p^1, \ldots, p^r$ and

$$\mathrm{d}^D k = \gamma(D, r) \kappa^{D-r-1} \, \mathrm{d}\kappa \, \mathrm{d}c_1 \ldots \mathrm{d}c_r \times \text{determinant}. \qquad (16.78)$$

$\gamma(D, r)$ is a constant depending on D and r, and the determinant depends on $p^1, \ldots, p^r$. The integral I_D is an integral not over D variables, but over $r + 1$ variables. The analytic continuation in D of the right-hand side of (16.78) is well defined. Applied to diagrams, the analysis of the integral I_D leads to integrals G_D of L 4-momenta $k^1, \ldots, k^L$, where each k is a loop momentum. The dimensional regularization of G_D consists of replacing $\mathrm{d}^4 k$ by $\mu^\epsilon \mathrm{d}^{4-\epsilon} k$, where the parameter μ has the same physical dimension as k. The volume element $\mathrm{d}^4 k^1 \ldots \mathrm{d}^4 k^L$ is replaced by

$$\mu^{\epsilon L} \, \mathrm{d}^{4-\epsilon} k^1 \ldots \mathrm{d}^{4-\epsilon} k^L = \left(\frac{\mu}{r}\right)^\epsilon r^3 \, \mathrm{d}r \, \mathrm{d}\omega, \qquad (16.79)$$

where r is the radial polar coordinate and $\mathrm{d}\omega$ is the volume element of the 3-sphere of radius 1. Equation (16.79) is justified by equation (16.78).

Brydges uses dimensional regularization by changing the dimensions of the field rather than the dimension of spacetime. The physical dimension of a field is chosen so that the action functional is dimensionless. When $\hbar = 1$ and $c = 1$, the remaining dimension is either a mass or a length dimension. For example, $\int \mathrm{d}^D x (\nabla \phi)^2$ is dimensionless if the

length dimension of ϕ is given by $2[\phi] = 2 - D$ \hfill (16.80)

and the mass dimension of ϕ is given by $2[\phi] = -2 + D$. \hfill (16.81)

Mass dimensions are appropriate for diagrams in momentum space; length dimensions are appropriate for fields defined on spacetimes. We use length dimensions. For $D = 4$, $2[\phi] = -2$. Brydges chooses

$$2[\phi] = -2 + \epsilon \qquad (16.82)$$

and works in $\mathbb{R}^4$. His works simulate works in $\mathbb{R}^{4-\epsilon}$. In Brydges' $\lambda\phi^4$ model

$$[\lambda] = -2\epsilon. \tag{16.83}$$

The goal of the $\lambda\phi^4$ application is only to introduce scaling renormalization. We refer to the original publications for an in-depth study of scaling renormalization [3] and to textbooks on quantum field theory for the use of β-functions.

The scaling method provides an exact equation (16.40) for the effective action $S(l,\phi)$. It has been used by Brydges, Dimock, and Hurd for computing fixed points of the flow equation (16.55). In the euclidean case they found that the gaussian fixed point is unstable, but that in dimension $4-\epsilon$ there is a hyperbolic non-gaussian fixed point a distance $\mathcal{O}(\epsilon)$ away. In a neighborhood of this fixed point they constructed the stable manifold. Brydges *et al.* work with euclidean fields. This method has been applied to minkowskian fields in [6].

The most attractive feature of the scaling method is that it works on the action functional, rather than on the lagrangian.

References

[1] D. J. Gross (1981). "Applications of the renormalization group to high energy physics," *Méthodes en Théorie des champs* (Les Houches, 1975), eds. R. Balian and J. Zinn-Justin (Amsterdam, North Holland), p. 144.

[2] K. G. Wilson (1975). "The renormalization group: critical phenomena and the Kondo problem," *Rev. Mod. Phys.* **47**, 773–840.
K. G. Wilson and J. Kogut (1974). "The renormalization group and the ϵ-expansion," *Phys. Rep.* **12**, 75–200.

[3] D. C. Brydges, J. Dimock, and T. R. Hurd (1998). "Estimates on renormalization group transformations," *Can. J. Math.* **50**, 756–793 and references therein.
D. C. Brydges, J. Dimock, and T. R. Hurd (1998). "A non-gaussian fixed point for ϕ^4 in $4-\epsilon$ dimensions," *Commun. Math. Phys.* **198**, 111–156.

[4] A. Das (1993). *Field Theory: A Path Integral Approach* (Singapore, World Scientific).

[5] B. Schwarzschild (2004), "Physics Nobel Prize goes to Gross, Politzer, and Wilczek for their discovery of asymptotic freedom," *Phys. Today*, December 2004, pp. 21–24.

[6] A. Wurm (2002). Renormalization group applications in area-preserving nontwist maps and relativistic quantum field theory, unpublished Ph.D. Dissertation, University of Texas at Austin.

17
Renormalization 3: combinatorics

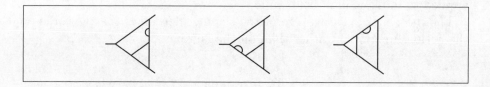

This chapter was contributed by M. Berg and is based on a collaboration with P. Cartier.

17.1 Introduction

It has been known for some time that the combinatorics of renormalization is governed by *Hopf algebras* (for a review, see Kreimer [30]). Unfortunately, since Hopf algebras are largely unfamiliar to physicists,[1] this intriguing fact has not been widely appreciated. In the paper [1] an attempt to help remedy this situation was made, by focusing instead on a related Lie algebra. This in principle allows one to perform the same renormalization procedure without introducing Hopf algebras at all, so we will not need to do so for the purposes of this chapter. (In fact, there is a theorem that guarantees the existence of this equivalent Lie algebra, given a Hopf algebra of renormalization.)

Still, we would like to draw attention to the "in principle" of the previous paragraph; the Lie algebra of renormalization is so intimately tied to the Hopf algebra of renormalization that any serious student needs to understand both. For the connection to Hopf algebras, see Project 19.7.3.

---

[1] Except in completely different and somewhat exotic contexts, e.g. as symmetries of some nonlinear sigma models in low-dimensional supergravity.

In addition to postponing the introduction of heavy mathematical machinery, the Lie-algebra point of view affords certain computational advantages compared with the Hopf-algebra methods; the Lie algebra can be expressed in terms of (in general infinite) matrices, so renormalization becomes a matter of matrix algebra. We will discuss the application to renormalization and give a number of examples.

There is also another, more speculative way in which this matrix Lie algebra may be applied, which is explained in the last section of this chapter, Section 17.10. First, let us step back and review some existing directions in the literature.

17.2 Background

Perturbative renormalization was regarded by most physicists as finished with a theorem by Bogoliubov, Parasiuk, Hepp, and Zimmermann (the BPHZ theorem), which was refined by Zimmermann in 1970. Original references for this theorem are [3], and a clear textbook version is given in [16]. Already hidden within the BPHZ theorem, however, was algebraic structure belonging to branches of mathematics that were largely unknown to physicists in the 1970s. The Hopf algebra of renormalization was first displayed by Kreimer in 1997, partially as a result of his excursions into knot theory [28]. In 1999, Broadhurst and Kreimer performed a Hopf-algebraic show of strength by computing contributions to anomalous dimensions in Yukawa theory to 30 loops [5].[2]

On a practical level, phenomenologists have been opening their eyes to Hopf algebra as a potential time-saver and organizing principle in massive computations, such as the five-loop computation of the beta function in quantum chromodynamics (QCD). Such high-precision analytic calculations are needed for comparisons with numerical calculations in lattice QCD, since the energy scales accessible in these lattice computations (and in many experiments) are often so low that perturbative QCD is only starting to become a valid approximation (see e.g. [2, 12]). Even when one is not seeking to push calculations to such heights of precision, savings in computer time through use of better algorithms could be welcome for routine calculations as well.

On the more mathematical side, a series of articles by Connes and Kreimer [19, 20] took the Hopf algebra of renormalization through some formal developments, amongst others the identification of the Lie algebra of the dual of the aforementioned Hopf algebra, through the Milnor–Moore theorem (see also [10] for related mathematical developments). This was

---

[2] This was later improved further [6].

later termed a study of *operads*,[3] which we will not go into here, but which can be seen as complementary to our discussion. Some geometric aspects were investigated in [14].

The point here is that representations of the dual Lie algebra can be useful in their own right. The Lie bracket is simply given by inserting one graph into another and subtracting the opposite insertion. The process of insertion may be represented by the multiplication of (in general infinite) matrices, hence matrix Lie algebra.

Although we give examples in scalar field theory, the representation of the algebra on graphs is independent of the specific nature of the action, except that we assume that the interactions are quartic. We could equally well consider fermions, vector fields, or higher-spin fields with quartic interactions. But surely Ward identities, ghosts, and so on make gauge-theory computations very different from those of scalar field theory? This question can be answered as follows. Keep in mind the distinction between Feynman graphs and their values. In performing field-theory computations, one routinely identifies graphs and their values. In the Connes–Kreimer approach one explicitly separates symmetries of the algebra of Feynman graphs from symmetries of Feynman integrals; the integrals are thought of as elements of the dual of the space of graphs. The dual has symmetries of its own, "quasi-shuffles" [29], which are obviously related but not identical to symmetries on graphs. The algebraic approach can help with computing Feynman integrals, but its greatest strength so far has been in the organization of counterterms, providing algebraic relations between seemingly unrelated quantities. The organization of counterterms is an essentially graph-theoretic problem, since the hierarchy of subdivergences can be read off from the topology of a graph without knowing precisely what fields are involved. In the extreme case of "rainbow" or "ladder" diagrams only, the Feynman integrals themselves are easily iterated to any loop order (as in [5]), but the BPHZ subtractions quickly become a combinatorial mess if the algebraic structure is ignored.

The attitude of separating the combinatorial and analytic problems (space of graphs and its dual) is also useful for computer implementation. Let us take one example: the Mathematica package *FeynArts* [25], which can automatically compute amplitudes up to one loop in any renormalizable theory, and, using additional software, up to two loops [26]. This particular software package calculates amplitudes by first writing down

---

[3] For instance, a simple example of an operad is a collection of maps from all tensor products of an algebra $\mathcal{A}$ into $\mathcal{A}$ itself, such as $\mathcal{A} \otimes \mathcal{A} \otimes \mathcal{A} \to \mathcal{A}$, with some compatibility requirements. See e.g. [27] for a precise definition.

scalar graphs with appropriate vertices, then it generates counterterm graphs, and then at a later stage it lets the user specify what fields are involved, namely gauge fields, spinors, or scalars. Since the number of graphs grows very quickly with loop order, it will become important to take advantage of algebraic structure if calculations at very high precision are to be feasible within a reasonable amount of computer time.

Apart from the direct application of the matrix Lie algebra to precision computations, there are other possible indirect applications. Such applications may be found in noncommutative geometry through the result of [20]: a homomorphism between the Hopf algebra of renormalization and the Hopf algebra of coordinates on the group of formal diffeomorphisms in noncommutative geometry [22], through which one may also be able to apply the matrix framework in noncommutative-geometry terms; this direction has not been developed in the literature, so we list it in the projects section of this book. (Reasons why noncommutative geometry could be relevant to physics abound [9, 18, 32, 36].)

Another promising application is to use the Hopf/Lie-algebraic techniques for functional integration. These directions are not discussed in detail in this book, but see Section 17.10 for further comments.

17.3 Graph summary

We consider ϕ^4 theory in four spacetime dimensions for ease of exposition (see the previous section for comments on adaptation to other theories). In ϕ^4 theory in four dimensions, we consider *graphs* (also called *diagrams*) with a quartic interaction vertex[4] and two and four external legs (see Section 17.9 for a reminder of why we do not need e.g. six external legs in this theory in four dimensions). All such graphs with up to three loops that are one-particle irreducible[5] (1PI) (for short 1PI graphs) are summarized in table 17.1.[6] In this table, "crossing" refers to graphs related to previously given graphs by crossing symmetry, such as the two graphs in parentheses for $L = 1$, $E = 4$; they are simply found by turning the first graph on its side (second graph), and then crossing the two lower external legs (third graph).[7] On a computer, graphs can be stored as lists. One example of such lists, adapted to the context of Feynman graphs, is given at the end of Section 17.4.

---

[4] That is, the interaction vertex is four-valent.
[5] That is, connected and which cannot be disconnected by removing one internal line.
[6] These graphs were drawn using FeynMF [34].
[7] The reader unfamiliar with Feynman diagrams may want to note that the third graph in $L = 1$, $E = 4$ still has only two vertices: the two lines crossing below the loop are going one above the other.

328 *Renormalization 3: combinatorics*

Table 17.1. *Graphs with L loops and E external legs*

L	E	1PI graphs
0	2	—
0	4	✕
1	2	◯
1	4	⊗ (⋈ ⊗)
2	2	⊖ ⊖
2	4	⊗ ⊗ ⊗ ⊗ + crossing
3	2	⊗ ⊗ ⊖ ⊖ ⊖
3	4	(many graphs) + crossing

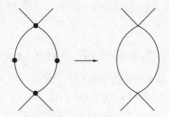

Fig. 17.1 Insertion points are implicit

17.4 The grafting operator

The Lie algebra of graphs comes from inserting (grafting) one Feynman graph into another. First a remark about terminology. In standard Feynman graphs, each internal line represents a propagator. We will work only with amputated graphs, where the external legs do not represent propagators. Now, in a theory with quartic interactions, one can consider inserting graphs with either two external legs (into an internal line) or four external legs (into an interaction vertex). Thus, the middle of an internal line and the interaction vertex itself are eligible as *insertion points*.[8] If desired, we could represent each insertion point by a dot, as in the graph on the left in figure 17.1. As in standard usage – the graph on the right in figure 17.1 – we leave those dots implicit. We alert the reader that the authors of [19] call the dots on lines "two-point vertices." To avoid confusion with

---

[8] By convention, the graph — has one insertion point.

17.4 The grafting operator

some of the physics terminology, we prefer to talk about insertion points instead of two-point vertices.

Here is a precise definition of the insertion, or grafting, operator. The practically minded reader can refer to the insertion tables in Appendix A2 of Section 17.10 to get a quick idea of the action of this operator. Enumerate the insertion points of a graph t by $p_i(t)$, the total number of insertion points of t by $N(t)$, and their valences (number of "prongs," i.e. lines sticking out from the vertex) by $v(p_i(t))$. Here, the valence is always either 2 or 4. If t_1 and t_2 are graphs, and the number $v(t_2)$ of external legs of t_2 is equal to the valence $v(p_i(t_1))$ of the insertion point $p_i(t_1)$ of t_1, then we define $s_{t_2}^{p_i(t_1)}$ to be insertion of the graph t_2 at $p_i(t_1)$ by summing over all permutations of the $v(p_i(t_1))$ propagators to be joined. Then *the grafting operator s_{t_2} is*

$$s_{t_2} t_1 = \frac{1}{v(t_2)!} \sum_{i=1}^{N(t_1)} s_{t_2}^{p_i(t_1)} t_1. \qquad (17.1)$$

The total number of graphs $\mathcal{N}$ (including multiplicities of the same graph) created by insertion of t_2 into t_1 is

$$\mathcal{N} = \sum_{i=1}^{N(t_1)} \delta_{v(p_i(t_1)), v(t_2)} v(t_2)!$$

We call t_1 the "object graph" and t_2 the "insertion graph." Often, we will use parentheses to delineate the insertion graph: $s(t_2)t_1$. Finally, we define $s(t)$ to be linear:

$$s(at_1 + bt_2) = as(t_1) + bs(t_2) \qquad a, b \in \mathbb{C}.$$

A remark about normalization: the normalization in (17.1) is a mean over insertions.

Let us denote by 1_E for $E = 2$ or 4 the graph with $L = 0$ loops and E external legs. Hence 1_2 is — an 1_4 is ✗. Our normalization gives

$$s(1_E) t = N_E(t) \cdot t,$$

where $N_E(t)$ is the number of insertion points of valence E in t. Note that 1_E has one insertion point, with valence E. The total number of insertion points $N(t) = N_2(t) + N_4(t)$ is given, when $L \neq 0$, by

$$N(t) = 3L \pm 1.$$

Here L is the number of loops of t, and the upper (lower) sign refers to graphs with four (two) external legs, respectively. Moreover, $s(t)1_E = t$ if $E = 2$ and t has two external legs. When $E = 4$ and t has four external legs, then $s(t)1_E$ is the mean of the various graphs obtained from t by permutation of the external legs. In the other cases $s(t)1_E$ is 0.

We can assign a grading (degree) to the algebra by loop number L, since then obviously
$$\deg(s_{t_2} t_1) = \deg t_2 + \deg t_1,$$
so within the set of graphs with a fixed number of external legs, s_1 acts via our grading operator. Since s_1 is the only s_t which is diagonal, we will exclude it from the algebra, but it can be adjoined whenever needed as in [20].

As an alternative to this construction of the grafting operator, one could consider a restricted grafting operator, for which permutations of the legs of the insertion that lead to the same topologies are represented by separate operators (see Project 19.7.3). In this work, we concentrate on exploring the full operator.

Nested divergences

It is of particular interest to know when a graph has nested divergences, i.e. when two divergent loops share the same propagator. For any graph, two lines are said to be indistinguishable if they are connected to the same vertex, and distinguishable if they are not. Whenever we insert a (loop) graph with four external legs at a certain vertex and two distinguishable external prongs of the insertion graph are connected to indistinguishable legs of another vertex, we create nested divergences. Of course, any nested divergences already present in the insertion graph are preserved.

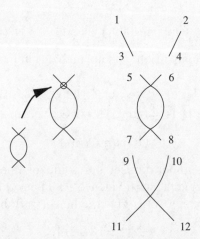

Fig. 17.2 The creation of nested divergences

Here is an example. Consider the insertion in figure 17.2. It is an insertion of ⌧ into the upper vertex of the same graph. With the above use of language, 5 and 6 are indistinguishable, and so are 9 and 10, but not e.g. 5 and 7. When we join 3–5, 4–6, 7–9, and 8–10, we create ⚹, which has only two disjoint divergent loops; but when we join 3–5, 4–8, 7–9, and 6–10, we create ⚹, which has nested divergences.

As an aside, we remark that figure 17.2 also lets us recall how ⚹ is easily represented in a format amenable to computer processing. The bottom vertex can be stored as $[9, 10, 11, 12]$, and the propagator going between 5 and 7 carrying loop momentum p can be stored as $[5, 7, p]$. This way, the list of vertices contains three sublists, and the list of propagators contains four sublists, which together completely specify the graph. This is also how graphs are stored in *FeynArts* [25]. Of course, there are many other equivalent representations, for instance using the relation between rooted trees and parenthesized words [4].

17.5 Lie algebra

We want to show that the grafting operators s_t form a Lie algebra $\mathcal{L}$. To this end, we first show a certain operator identity. Indeed, for graphs t_1 and t_2 define

$$[t_1, t_2] = s(t_1)t_2 - s(t_2)t_1; \qquad (17.2)$$

then we shall show that the commutator of the operators $s(t_1)$ and $s(t_2)$ is equal to $s([t_1, t_2])$:

$$[s(t_1), s(t_2)] = s([t_1, t_2]), \qquad (17.3)$$

as an operator identity acting on graphs. This identity is analogous to the Ihara bracket identity of the Magnus group, which is familiar to number theorists (see e.g. [35]).

Here is a sketch of the proof of (17.3). Writing that the operators in (17.3) give the same result when applied to a graph t_3, and using the definition (17.2), we see that the identity (17.3) amounts to the fact that

$$a(t_3|t_2, t_1) := s_{s_{t_1} t_2} t_3 - s_{t_1} s_{t_2} t_3$$

is symmetric in (t_1, t_2) (see equations (17.8) and (17.9)). When calculating $s_{t_1} s_{t_2} t_3$, we sum two kinds of insertions: (a) mutually nonlocal, where t_1 and t_2 are inserted at different vertices of t_3; and (b) mutually local, where t_1 is inserted into one new vertex created by an insertion of t_2 into t_3. This is illustrated in figure 17.3. The insertions of type (b) sum up to $s_{s_{t_1} t_2} t_3$, hence $-a(t_3|t_2, t_1)$ is the sum of contributions of type (a). However, insertions of type (a) clearly commute, since they modify different vertices. Hence $-a(t_3|t_2, t_1)$ is symmetric in t_1 and t_2. This concludes the outline of the proof.

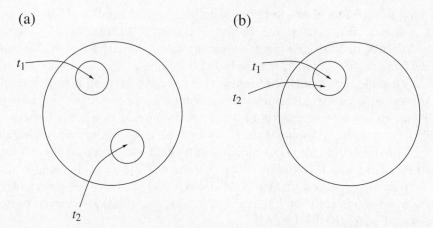

Fig. 17.3 The difference between mutually nonlocal (a) and mutually local (b) insertions

We remark that the identity trivially holds acting on 1:

$$[s(t_1), s(t_2)]1 = s(t_1)s(t_2)1 - s(t_2)s(t_1)1 = s(t_1)t_2 - s(t_2)t_1 = [t_1, t_2].$$

It is easy to check that the bracket $[t_1, t_2] = s(t_1)t_2 - s(t_2)t_1$ satisfies the Jacobi identity. By writing down three copies of the Ihara bracket identity (17.3) and adding them we find by linearity

$$s([t_1, t_2])t_3 + s(t_3)[t_2, t_1] + \text{cyclic} = 0$$

or

$$[[t_1, t_2], t_3] + \text{cyclic} = 0.$$

The Lie algebra is graded by loop order: $\mathcal{L} = \mathcal{L}^{(1)} \oplus \mathcal{L}^{(2)} \oplus \mathcal{L}^{(3)} \oplus \ldots$, and

$$[\mathcal{L}^{(m)}, \mathcal{L}^{(n)}] \subset \mathcal{L}^{(m+n)}.$$

Let us see what the Lie bracket looks like explicitly in our field theory with 4-vertices. To make the notation more succinct, we suppress any graph that is related by crossing symmetry to one that is already included. That is, since ⟨, ⟩, and ⟩ (see table 17.1) all represent the same function of the different Mandelstam variables[9] s, t, and u, we will display only ⟨ out of those three diagrams. We can always use a symmetric renormalization point, i.e. we can define the theory at some mass scale M, meaning $s = t = u = M^2$ at this scale, in which case counterterms are the same for all three diagrams mentioned previously.

---

[9] Here $s = (p_1 + p_2)^2$, $t = (p'_1 - p_1)^2$, and $u = (p'_2 - p_1)^2$, where p_1 and p_2 are incoming momenta and p'_1 and p'_2 are outgoing momenta.

17.5 Lie algebra

At the one-loop approximation, we have the following diagrams: —, X, ⊖, and ⊗. By consulting the insertion tables in the appendix, or equivalently the matrix representation which we introduce below, we find the commutators of tree-level (that is, zero-loop) diagrams with one-loop diagrams:

$[\mathcal{L}^{(0)}, \mathcal{L}^{(1)}] \subset \mathcal{L}^{(1)}$:

$$[-, \ominus] = s(-)\ominus - s(\ominus)-$$
$$= \ominus - \ominus$$
$$= 0.$$
$$[X, \otimes] = s(X)\otimes - s(\otimes)X$$
$$= 2\otimes - \otimes$$
$$= \otimes.$$

The only nonvanishing commutator of two one-loop graphs is

$[\mathcal{L}^{(1)}, \mathcal{L}^{(1)}] \subset \mathcal{L}^{(2)}$:

$$[\otimes, \ominus] = s(\otimes)\ominus - s(\ominus)\otimes$$
$$= \frac{1}{3}\text{⊗̄} + \frac{2}{3}\ominus\!\ominus - 2\otimes\!\!\!\otimes$$
$$\xrightarrow{I} \mathcal{O}(\hbar), \qquad (17.4)$$

where by $\xrightarrow{I}$ we mean evaluation of the graphs. Here we have restored $\hbar$ to connect with the semiclassical approximation; an L-loop graph is of order $\hbar^{L-1}$. At this point we note that the combinatorial factors $1/3$, $2/3$, and so on are *not* the usual symmetry factors of graphs, i.e. the multiplicity of operator contractions generating the same graph (this distinction is elaborated upon in Project 19.7.3).

Upon including two-loop graphs in the algebra, the following graphs come into play: ⊗, ⊗, ⊗, and ⊗. We can now consider a commutator of a one-loop and a two-loop graph, which creates a sequence of three-loop graphs:

$[\mathcal{L}^{(1)}, \mathcal{L}^{(2)}] \subset \mathcal{L}^{(3)}$:

$$[\otimes, \otimes] = s(\otimes)\otimes - s(\otimes)\otimes$$
$$= \frac{2}{3}\text{⊗} + \frac{2}{3}\text{⊗} + \frac{2}{3}\text{⊗}$$
$$\quad - \frac{2}{3}\text{⊗} - \text{⊗} - \frac{2}{3}\text{⊗} - \frac{2}{3}\text{⊗}$$
$$\xrightarrow{I} \mathcal{O}(\hbar^2),$$

which may be approximated by zero if we keep graphs only up to order $\hbar$. Thus, to this order, all the "new" graphs with two loops are central elements of the algebra.

In general, if we consider L loops (order $\hbar^{L-1}$), the commutator at order $\hbar^L$ is negligible, so only commutators of graphs of loop orders L_1 and L_2, where $L_1 + L_2 \leq L$, are nonvanishing. The effect of going from L to $L+1$ is then to add central graphs as new generators, which appear in lower-order commutators with one extra factor of $\hbar$. In other words, the new graphs *deform* the L-loop algebra by $\hbar$.

Now that we have displayed a few commutators, here are two examples of how the Ihara bracket identity works. Let us first consider a one-loop/one-loop commutator, acting on the bare propagator:

$$s([\mathord{\vcenter{\hbox{\includegraphics{}}}}, \mathord{\vcenter{\hbox{\includegraphics{}}}}]) \,\text{---}\, = s(s(\mathord{\vcenter{\hbox{\includegraphics{}}}})\mathord{\vcenter{\hbox{\includegraphics{}}}}) \,\text{---}\, - s(s(\mathord{\vcenter{\hbox{\includegraphics{}}}})\mathord{\vcenter{\hbox{\includegraphics{}}}}) \,\text{---}$$
$$= s\left(\frac{1}{3}\,\mathord{\vcenter{\hbox{\includegraphics{}}}} + \frac{2}{3}\,\mathord{\vcenter{\hbox{\includegraphics{}}}}\right) \,\text{---}$$
$$= \frac{1}{3}s(\mathord{\vcenter{\hbox{\includegraphics{}}}}) \,\text{---}\, + \frac{2}{3}s(\mathord{\vcenter{\hbox{\includegraphics{}}}}) \,\text{---}$$
$$= \frac{1}{3}\,\mathord{\vcenter{\hbox{\includegraphics{}}}} + \frac{2}{3}\,\mathord{\vcenter{\hbox{\includegraphics{}}}},$$

whereas

$$[s(\mathord{\vcenter{\hbox{\includegraphics{}}}}), s(\mathord{\vcenter{\hbox{\includegraphics{}}}})] \,\text{---}\, = s(\mathord{\vcenter{\hbox{\includegraphics{}}}})s(\mathord{\vcenter{\hbox{\includegraphics{}}}}) \,\text{---}\, - s(\mathord{\vcenter{\hbox{\includegraphics{}}}})s(\mathord{\vcenter{\hbox{\includegraphics{}}}}) \,\text{---}$$
$$= s(\mathord{\vcenter{\hbox{\includegraphics{}}}})\,\mathord{\vcenter{\hbox{\includegraphics{}}}}$$
$$= \frac{1}{3}\,\mathord{\vcenter{\hbox{\includegraphics{}}}} + \frac{2}{3}\,\mathord{\vcenter{\hbox{\includegraphics{}}}}.$$

At this level, the identity is quite trivial. Let us therefore also check the relation at the three-loop level:

$$[s(\mathord{\vcenter{\hbox{\includegraphics{}}}}), s(\mathord{\vcenter{\hbox{\includegraphics{}}}})]\,\mathord{\vcenter{\hbox{\includegraphics{}}}} = s(\mathord{\vcenter{\hbox{\includegraphics{}}}})s(\mathord{\vcenter{\hbox{\includegraphics{}}}})\,\mathord{\vcenter{\hbox{\includegraphics{}}}} - s(\mathord{\vcenter{\hbox{\includegraphics{}}}})s(\mathord{\vcenter{\hbox{\includegraphics{}}}})\,\mathord{\vcenter{\hbox{\includegraphics{}}}}$$
$$= s(\mathord{\vcenter{\hbox{\includegraphics{}}}})(2\,\mathord{\vcenter{\hbox{\includegraphics{}}}}) - s(\mathord{\vcenter{\hbox{\includegraphics{}}}})\left(\frac{2}{3}\,\mathord{\vcenter{\hbox{\includegraphics{}}}} + \frac{2}{3}\,\mathord{\vcenter{\hbox{\includegraphics{}}}} + \frac{2}{3}\,\mathord{\vcenter{\hbox{\includegraphics{}}}}\right)$$
$$= 2\left(\frac{2}{3}\,\mathord{\vcenter{\hbox{\includegraphics{}}}} + \frac{1}{3}\,\mathord{\vcenter{\hbox{\includegraphics{}}}} + \frac{2}{3}\,\mathord{\vcenter{\hbox{\includegraphics{}}}} + \frac{2}{3}\,\mathord{\vcenter{\hbox{\includegraphics{}}}} + \frac{2}{3}\,\mathord{\vcenter{\hbox{\includegraphics{}}}}\right)$$
$$- \frac{2}{3}\left(4\,\mathord{\vcenter{\hbox{\includegraphics{}}}} + 2\,\mathord{\vcenter{\hbox{\includegraphics{}}}} + 2\,\mathord{\vcenter{\hbox{\includegraphics{}}}} + 2\,\mathord{\vcenter{\hbox{\includegraphics{}}}} + 2\,\mathord{\vcenter{\hbox{\includegraphics{}}}}\right)$$
$$= \frac{4}{3}\,\mathord{\vcenter{\hbox{\includegraphics{}}}} + \frac{2}{3}\,\mathord{\vcenter{\hbox{\includegraphics{}}}} - \frac{4}{3}\,\mathord{\vcenter{\hbox{\includegraphics{}}}} - \frac{4}{3}\,\mathord{\vcenter{\hbox{\includegraphics{}}}} - \frac{4}{3}\,\mathord{\vcenter{\hbox{\includegraphics{}}}}, \quad (17.5)$$

but

$$s([\Gamma_1,\Gamma_2])\Gamma_3 = s\left(\frac{1}{3}⬡ + \frac{2}{3}⬡ - 2⬡\right)⬡$$

$$= \frac{1}{3}(2⬡) + \frac{2}{3}(2⬡) - 2\left(\frac{2}{3}⬡ + \frac{2}{3}⬡ + \frac{2}{3}⬡\right)$$

$$= \frac{4}{3}⬡ + \frac{2}{3}⬡ - \frac{4}{3}⬡ - \frac{4}{3}⬡ - \frac{4}{3}⬡.$$

Here we see that cancelation of graphs ⬡ and ⬡ was necessary in the first bracket, which is to be expected since these two graphs are generated by "mutually nonlocal" insertions, as defined above.

To summarize, if we know the commutator of two insertions, the identity (17.3) gives us the insertion of the commutator.

Going back to the question of normalization, not any combination of normalizations will preserve the Ihara bracket, of course, as can easily be seen from the example (17.5). One way to preserve it is to drop simultaneously overall normalization and multiplicities, but without multiplicities the grafting operator loses some of its combinatorial information (it ceases to count the number of ways a graph can be generated). Although from a physics point of view there is no freedom in the choice of normalization in the total sum of renormalized graphs – for given Feynman rules – it can be useful to consider different normalizations in these intermediate expressions.

Connection with the star operation

The Ihara bracket identity can be compared to the star product discussed by Kreimer [30].

Here is the connection between our operator s_t and Kreimer's star insertion $\star$:

$$t_1 \star t_2 \sim s(t_2)t_1,$$

where the $\sim$ alerts the reader that we use a different normalization, (17.1). Furthermore,

$$(t_1 \star t_2) \star t_3 \sim s(t_3)s(t_2)t_1, \qquad (17.6)$$
$$t_1 \star (t_2 \star t_3) \sim s(s(t_3)t_2)t_1. \qquad (17.7)$$

The coproduct Δ of the Connes–Kreimer Hopf algebra is coassociative, so the dual Hopf algebra should have an associative multiplication $\odot$. One would then naively expect the dual Lie algebra to have an associative multiplication as well. However, on comparing (17.6) and (17.7), one sees that the star operation is not associative. This is because the original Hopf algebra contains disconnected graphs, but only connected graphs are generated by the star product (or, equivalently, the grafting operator s_t). When the multiplication $\odot$ is truncated to connected graphs, it is no longer associative.

The deviation from associativity can be measured in terms of the *associativity defect*:

$$a(t_1|t_2, t_3) := t_1 \star (t_2 \star t_3) - (t_1 \star t_2) \star t_3. \qquad (17.8)$$

When this is nonzero, as in any nonassociative algebra, it is interesting to consider the "right-symmetric" algebra for which

$$a(t_1|t_2, t_3) = a(t_1|t_3, t_2) \qquad (17.9)$$

(a "left-symmetric" algebra can equivalently be considered). This leads naturally to an algebra of the star product [30], which is also common usage in differential geometry. On writing the right-symmetric condition (17.9) explicitly in terms of the associativity defect (17.8), the algebra is specified by

$$t_1 \star (t_2 \star t_3) - (t_1 \star t_2) \star t_3 = t_1 \star (t_3 \star t_2) - (t_1 \star t_3) \star t_2. \qquad (17.10)$$

This is known as a *pre-Lie* (or *Vinberg*) algebra. A pre-Lie algebra yields a bracket $[t_1, t_2] = t_1 \star t_2 - t_2 \star t_1$ that automatically satisfies the Jacobi identity, hence "pre-Lie." In our notation, using equations (17.6) and (17.7), the pre-Lie algebra identity (17.10) can be rewritten as the operator identity (acting on arbitrary t_3)

$$[s(t_1), s(t_2)] = s([t_1, t_2]), \qquad (17.11)$$

which we have already shown directly for the grafting operator s_t. Thus the two descriptions are equivalent, as expected.

Matrix representations

Graphs are graded by their number of loops, but there is no canonical ordering within each graded subspace. Given some ordering of graphs, the grafting operator s_t can be represented as a matrix. This matrix will be indexed by graphs. Now, graphs are inconvenient to use as subscripts, so we use a bra–ket notation instead: represent $s(t_1)t_2$ by

$$s(t_1)|t_2\rangle = \sum_{t_3} |t_3\rangle\langle t_3|s(t_1)|t_2\rangle,$$

where all graphs are to be thought of as orthogonal: $\langle t_1|t_2\rangle = 0$ for $t_1 \neq t_2$.

17.5 Lie algebra

An example: use of the insertion tables (consulting Appendix A2 in Section 17.10) gives

$$s(\graph{a})\,\graph{b} = \frac{2}{3}\graph{c} + \frac{2}{3}\graph{d} + \frac{2}{3}\graph{e},$$

which is to be compared with the expansion in a complete set

$$s(\graph{a})|\graph{b}\rangle = \cdots + \langle\graph{c}|s(\graph{a})|\graph{b}\rangle|\graph{c}\rangle + \langle\graph{d}|s(\graph{a})|\graph{b}\rangle|\graph{d}\rangle$$
$$+ \langle\graph{e}|s(\graph{a})|\graph{b}\rangle|\graph{e}\rangle + \cdots,$$

so we read off

$$\langle\graph{c}|s(\graph{a})|\graph{b}\rangle = 2/3, \qquad \langle\graph{d}|s(\graph{a})|\graph{b}\rangle = 2/3, \qquad \langle\graph{e}|s(\graph{a})|\graph{b}\rangle = 2/3,$$

giving the second column of the matrix:

$$s(\graph{a})\graph{b} \doteq \begin{pmatrix} 0 & 0 & 0 & 0 & 0 & 0 \\ 1/3 & 0 & 0 & 0 & 0 & 0 \\ 0 & 2/3 & 0 & 0 & 0 & 0 \\ 0 & 2/3 & 0 & 0 & 0 & 0 \\ 0 & 2/3 & 0 & 0 & 0 & 0 \\ 0 & 0 & 0 & 0 & 0 & 0 \end{pmatrix} \begin{pmatrix} 0 \\ 1 \\ 0 \\ 0 \\ 0 \\ 0 \end{pmatrix} = \begin{pmatrix} 0 \\ 0 \\ 2/3 \\ 2/3 \\ 2/3 \\ 0 \end{pmatrix} \doteq \frac{2}{3}\graph{c} + \frac{2}{3}\graph{e} + \frac{2}{3}\graph{d},$$

(17.12)

where we have ordered the graphs as $\graph{}, \graph{}, \graph{}, \graph{}, \graph{}, \graph{}$, and $\doteq$ denotes "represents." In Section 17.10 we give first insertion tables up to three-loop order (Appendix A2), then a few examples of matrices that represent the grafting operators (Appendix A3). These matrices are easily extracted from the insertion tables, and are indexed by graphs.

A word about practical implementation: because of our space-saving convention of suppressing graphs related by crossing symmetry, each number in the above 6×6 matrix is a 3×3 unit matrix. The exception is the first column, which has as second index the single tree-level diagram and so is a 3×1 matrix. A similar remark holds for the first row. Thus, in a completely explicit notation, the above matrix is 16×16.

A few remarks are in order. It is immediately clear that the matrices s_t are all lower triangular, since insertion can never decrease loop number. (Recall that s_1, which is represented by a diagonal matrix, is left out of the algebra.) Triangularity makes these matrices easy to exponentiate, since the series will cut off at finite order. This property will be crucial in Section 17.7. Triangular matrices are non-invertable, which makes sense – typically each application of s_t creates many different graphs. There is no unique inverse under s_t of a generic graph (but see Section 17.6). Finally, by a quick glance in the appendix we see that the matrices are very sparse, which is useful to know for computer storage; the size of the matrix is the number of relevant diagrams squared, which quickly becomes prohibitive if sparsity is not exploited.

It will be useful in the following to consider exponentiating matrices, for example the matrix representing $s(\lozenge)$ in (17.12):

$$\exp(s(\lozenge))\mathsf{X} = \left(1 + s(\lozenge) + \frac{1}{2}s(\lozenge)s(\lozenge)\right)\mathsf{X}$$

$$= \mathsf{X} + \frac{1}{3}\lozenge + \frac{1}{9}\lozenge + \frac{1}{9}\lozenge + \frac{1}{9}\lozenge + \text{crossing},$$

since the matrix (17.12) vanishes when cubed. On the other hand,

$$\exp(s(\varominus))\mathsf{X} = \mathsf{X}.$$

Thus, the exponential of the sum of all grafting operators acts as

$$e^{\sum_t s_t}\mathsf{X} = \mathsf{X} + \frac{1}{3}\lozenge + \frac{1}{9}\lozenge + \frac{1}{9}\lozenge + \frac{1}{9}\lozenge + \cdots + \text{crossing},$$

where the dots denote higher-order terms. Acting with the inverse is now a simple matter:

$$e^{-\sum_t s_t}\mathsf{X} = \mathsf{X} - \frac{1}{3}\lozenge + \frac{1}{9}\lozenge + \frac{1}{9}\lozenge + \frac{1}{9}\lozenge + \cdots + \text{crossing}. \quad (17.13)$$

With the three-loop matrices given in the appendix, one easily performs exponentiation to three-loop level as well (see Section 17.8). The alternating sign in equation (17.13) is already now suggestive for the reader who is familiar with Hopf algebra. The antipode of the Hopf algebra has a similar alternating sign – each shrinking of a subgraph comes with a sign change. In Section 17.7 we display the relation to the antipode, and thus the counterterms of the theory.

17.6 Other operations

Shrinking

When acting with the transpose of the matrix representation of a grafting operator, we shrink graphs. Here is an example:

$$s^{\mathrm{T}}(\lozenge)\lozenge \doteq \begin{pmatrix} 0 & 1/3 & 0 & 0 & 0 & 0 \\ 0 & 0 & 2/3 & 2/3 & 2/3 & 0 \\ 0 & 0 & 0 & 0 & 0 & 0 \\ 0 & 0 & 0 & 0 & 0 & 0 \\ 0 & 0 & 0 & 0 & 0 & 0 \\ 0 & 0 & 0 & 0 & 0 & 0 \end{pmatrix} \begin{pmatrix} 0 \\ 0 \\ 1 \\ 0 \\ 0 \\ 0 \end{pmatrix} = \begin{pmatrix} 0 \\ 2/3 \\ 0 \\ 0 \\ 0 \\ 0 \end{pmatrix} \doteq \frac{2}{3}\lozenge.$$

Gluing and external leg corrections

By insertion we create only one-particle-irreducible (1PI) diagrams. We can also define the operation of *gluing*, that is, joining any two external legs of two graphs with a new propagator. Under this operation, either the insertion graph becomes an external leg correction to the object graph (if the insertion has two external legs) or we add more external legs (if the insertion has n external legs with $n>2$). In the latter case, we obtain n-point functions with $n>4$, which are not superficially divergent in ϕ^4 theory, so we do not consider them. As for the external leg correction, it does not have to be considered here either, since we need only amputated graphs for the S-matrix.

Thus, the gluing operation will not be of much direct interest to us, but it does have one property we wish to emphasize. Define $t_1 \circ t_2$ to be gluing of t_1 onto t_2, e.g. ⚹ ∘ ⚭ is an external leg correction to ⚹. The operator s satisfies the Leibniz rule on these diagrams:

$$s(t_1)(t_2 \circ t_3) = (s(t_1)t_2) \circ t_3 + t_2 \circ (s(t_1)t_3). \quad (17.14)$$

It is easy to check that this relation holds for examples in ϕ^4 theory.

The structure of (17.14) also seems to indicate a possible interpretation as a coderivation. We will not investigate this direction in this work, since, as we have seen, the gluing operation is not needed for computing S-matrix elements in our model field theory.

17.7 Renormalization

Consider a square matrix M_1 depending on a parameter ϵ, with elements that diverge when $\epsilon \to 0$. By writing $M_1(1/\epsilon, \epsilon)$, we mean that both positive and negative powers of ϵ occur in the matrix elements. If only ϵ occurs in the argument and $1/\epsilon$ does not, there are no negative powers. We can decompose the divergent matrix M_1 into a divergent part M_2 and a finite part M_3 by the general procedure known as *Birkhoff decomposition*:

$$M_1(1/\epsilon, \epsilon) = M_2(1/\epsilon) M_3(\epsilon), \quad (17.15)$$

which is uniquely fixed by a boundary condition $M_2(0) = \mathbf{1}$. The matrix M_3 has a limit as $\epsilon \to 0$. It is important to notice that this is a *multiplicative decomposition*, which is not the same as an *additive decomposition* of each matrix element $m_1(1/\epsilon, \epsilon)$ as $m_2(1/\epsilon) + m_3(\epsilon)$.

We are considering here the procedure known as *dimensional regularization/minimal-subtraction*. Hence the matrices we encounter are triangular, with 1s on the diagonal. Such a matrix can be uniquely expressed as the exponential of a triangular matrix with 0s on the diagonal. Also our matrices $M_1(1/\varepsilon, \varepsilon)$ have elements $m_{ij}(1/\varepsilon, \varepsilon)$ which are power series in ε and $1/\varepsilon$ with finitely many negative powers $a_1/\varepsilon, a_2/\varepsilon^2, \cdots, a_N/\varepsilon_N$. In this case, the existence of the Birkhoff decomposition (17.15) is easily established where $M_2(1/\varepsilon)$ is a polynomial in $1/\varepsilon$ with constant term $M_2(0)$ equal to **1**. They are various algorithms available, but we shall not consider them here.[10]

Birkhoff decomposition was applied to renormalization by Connes and Kreimer [20]. We take a somewhat different approach; those authors treat the three matrices in the decomposition (17.15) as the same type of objects, whereas we will use the Lie algebra to reduce M_1 and M_3 to vectors, but keep M_2 as a matrix. The main point is that we choose M_2 in the group corresponding to the Lie algebra of matrices defined in Section 17.5. That is, the divergent matrix M_2 is a matrix C of the form

$$C = \exp\left(\sum_t C(t) s(t)\right)$$

with numerical coefficients $C(t)$ representing the overall divergence of the graphs t.[11] In dimensional regularization, ϵ represents $\epsilon = 4 - D$, where D is the (complex) spacetime dimension. The boundary condition $M_2(0) = \mathbf{1}$ is realized in the minimal-subtraction scheme, but other schemes can be accommodated by adapting other boundary conditions. The coefficient $C(t)$ will depend on ϵ so it should properly be denoted $C_\epsilon(t)$; however, we will suppress this dependence. The coefficient $C(t)$ is a polynomial in $1/\epsilon$ with no constant term (again, this is the boundary condition for (17.15)), but can depend on external momenta if t has two external legs, but not in the case of four. To calculate the complete set of counterterms, it suffices to know these overall-divergence coefficients $s(t)$ and the coefficients $C(t)$ which play the role of the overall divergence coefficients in the BPHZ scheme.

Let us describe the renormalization procedure in this framework, assuming that we know the matrix C, i.e. both the combinatorial matrices s_t. Denote the bare value of a graph t by $B(t)$ and the renormalized

---

[10] See for instance a recent paper by K. Ebrahilim-Fard and D. Manchan, **arXio:math-ph**/060639 02, as well as references therein.

[11] It should be stressed that our procedure *is not exactly* the BPHZ procedure, but gives the same final result for S-matrix elements. Hence, the overall divergence here is not the same as in the BPHZ procedure. Compare for instance formulas (17.18) and (17.19)!

17.7 Renormalization

value by $A(t)$ (in the graph-by-graph method [16]). Vectors containing these values, indexed by graphs, are denoted A and B, respectively. We Birkhoff-decompose the vector of bare values as in equation (17.15):

$$B_{1/\epsilon,\epsilon} = C_{1/\epsilon} A_\epsilon.$$

To find renormalized values from bare values, we have to invert the matrix C, which is a trivial matter when it is expressed as an exponential:

$$A_\epsilon = (C_{1/\epsilon})^{-1} B_{1/\epsilon,\epsilon}$$

$$= \exp\left(-\sum_t C(t)s(t)\right) B_{1/\epsilon,\epsilon}. \tag{17.16}$$

This is our main statement. As we have already pointed out, the sign in the exponential reproduces the sign of Zimmermann's forest formula [3], since every $s(t)$ factor in the expansion of the exponential corresponds to an insertion and a sign.

This is most easily understood in an example. Let us calculate the renormalized four-point function up to two loops:

$$\exp\left(-\sum_t C(t)s(t)\right) B =$$

$$\begin{pmatrix} 1 & 0 & 0 & 0 & 0 & 0 \\ -C(\mathfrak{X}) & 1 & 0 & 0 & 0 & 0 \\ -C(\mathfrak{X}) + C(\mathfrak{X})C(\mathfrak{X}) & -2C(\mathfrak{X}) & 1 & 0 & 0 & 0 \\ -C(\mathfrak{X}) + C(\mathfrak{X})C(\mathfrak{X}) & -2C(\mathfrak{X}) & 0 & 1 & 0 & 0 \\ -C(\mathfrak{X}) + C(\mathfrak{X})C(\mathfrak{X}) & -2C(\mathfrak{X}) & 0 & 0 & 1 & 0 \\ -C(\mathfrak{X}) + C(\mathfrak{X})C(\mathfrak{o}) & -2C(\mathfrak{o}) & 0 & 0 & 0 & 1 \end{pmatrix} \begin{pmatrix} B(\mathsf{X}) \\ B(\mathfrak{X}) \\ B(\mathfrak{X}) \\ B(\mathfrak{X}) \\ B(\mathfrak{X}) \\ B(\mathfrak{X}) \end{pmatrix},$$

where, as mentioned earlier, it is convenient to use a symmetric renormalization point so that $C(\mathfrak{X}) = C(\bowtie) = C(\mathfrak{X})$. In fact, these three graphs always appear in the same place in the expansion, since $s(\mathfrak{X}) = s(\bowtie) = s(\mathfrak{X})$.

We have, for the nontrivial elements,

$$\begin{pmatrix} A(\mathfrak{X}) \\ A(\mathfrak{X}) \\ A(\mathfrak{X}) \\ A(\mathfrak{X}) \\ A(\mathfrak{X}) \end{pmatrix} = \begin{pmatrix} -C(\mathfrak{X})B(\mathsf{X}) + B(\mathfrak{X}) \\ -(C(\mathfrak{X}) - C(\mathfrak{X})C(\mathfrak{X}))B(\mathsf{X}) - 2C(\mathfrak{X})B(\mathfrak{X}) + B(\mathfrak{X}) \\ -(C(\mathfrak{X}) - C(\bowtie)C(\mathfrak{X}))B(\mathsf{X}) - 2C(\mathfrak{X})B(\mathfrak{X}) + B(\mathfrak{X}) \\ -(C(\mathfrak{X}) - C(\mathfrak{X})C(\mathfrak{X}))B(\mathsf{X}) - 2C(\mathfrak{X})B(\mathfrak{X}) + B(\mathfrak{X}) \\ -(C(\mathfrak{X}) - C(\mathfrak{X})C(\mathfrak{o}))B(\mathsf{X}) - 2C(\mathfrak{o})B(\mathfrak{X}) + B(\mathfrak{X}) \end{pmatrix}.$$

To be specific, consider the third row:

$$A(\mathfrak{X}) = -(C(\mathfrak{X}) - C(\mathfrak{X})C(\mathfrak{X}))B(\mathsf{X}) - 2C(\mathfrak{X})B(\mathfrak{X}) + B(\mathfrak{X}). \tag{17.17}$$

The combination of graphs is clearly correct; the second counterterm (third term) cancels out the potentially nonlocal divergence in ⌘, and the first counterterm (first and second terms) takes care of the overall divergence after the nonlocal one has been canceled out. There is an even number of terms, as expected from the underlying Hopf-algebra structure. The matrix C "knew" the connection between a graph and its divergent subgraphs, since C was constructed from the grafting operators s_t. We can check the cancelation of nonlocal divergences in (17.17). Evaluating now in ϕ^4 theory, with f_1, f_2, g_1, g_2 containing no negative powers of ϵ and at most polynomial in p:

$$A(⌘) = \frac{\lambda^2}{2(4\pi)^2} \left\{ -\left[\left(\frac{2}{\epsilon^2} - \frac{2}{\epsilon}\ln p^2 + \frac{1}{\epsilon}f_1\right) - \left(\frac{2}{\epsilon^2} - \frac{2}{\epsilon}\ln p^2 + \frac{2}{\epsilon}g_1\right)\right] \right.$$

$$\left. -\left(\frac{2}{\epsilon^2} - \frac{2}{\epsilon}\ln p^2 + \frac{2}{\epsilon}g_1 + 2g_2\right) + \left(\frac{2}{\epsilon^2} - \frac{2}{\epsilon}\ln p^2 + \frac{1}{\epsilon}f_1 + f_2\right) \right\}$$

$$= \frac{\lambda^2}{2(4\pi)^2}(f_2 - 2g_2).$$

Here we included the symmetry factor (see Section 17.10) of 1/2. We see that the cancelation of $(1/\epsilon)\ln p^2$ is assured. The second check is that the second-order vertex counterterm $(C(⌘) - C(⌘)C(⌘)) \propto (2g_1 - f_1)/\epsilon$ is local.

The total contribution to the four-point function up to two-loop order is just the sum of all compatible (four-external-leg) elements of A, as in

$$\mathcal{A} = A(X) + A(⌘) + A(⌘) + A(⌘) + A(⌘) + A(⌘)(+ \text{ crossing}).$$

The total combination of graphs can be read off from the A vector above.

17.8 A three-loop example

All the diagrams at three-loop order in ϕ^4 theory have been computed, and clever techniques have been invented to perform the computations. Some techniques are reviewed in [13]. A useful reference for tables of divergent parts up to three loops is [15]. Therefore, we need worry only about the combinatorics, but for completeness we try to convey an idea of which integral computation techniques can be used to compute the relevant diagrams.

In one-loop computations, one usually employs the method of Feynman parameters. In fact, using Feynman parameters without additional techniques is, by some estimates, economical only for graphs with two lines. "Additional techniques" can include performing some series expansions before computing the integral (the Gegenbauer polynomials have been

17.8 A three-loop example

shown to be well suited for this) and imaginative use of integration by parts. In this context, integration by parts is applied to simplify integrals using the vanishing of integrals over total derivatives:

$$\int d^4 p \, \frac{\partial}{\partial p} I(p) = 0$$

for any integrand $I(p)$ depending on the loop momentum p and any other loop or external momenta. By applying integration by parts, the massive ladder diagram ⊠ at zero momentum transfer can be decomposed as [11]

$$⊠ \xrightarrow{I} \frac{1}{(4\pi)^4} \left(\frac{m^2}{4\pi M^2}\right)^{-\epsilon} \int \frac{d^D k}{(2\pi)^D} \frac{[F(k^2)]^2}{(k^2 + m^2)}$$

$$+ \frac{1}{\epsilon} \frac{4}{(4\pi)^2} \left(\frac{m^2}{4\pi M^2}\right)^{-\epsilon/2} W_6 - \frac{1}{\epsilon^2} \frac{4}{(4\pi)^2} \left(\frac{m^2}{4\pi M^2}\right)^{-\epsilon} S_3,$$

where F is a simpler genuine three-loop integral, i.e. it cannot be expressed in terms of integrals of lower loop order, W_6 is a two-loop integral (⊠ at zero momentum transfer), and S_3 is a one-loop integral (⊠ at zero momentum transfer). The result (given in [15]) is

$$⊠ \xrightarrow{I} \frac{m^4}{(4\pi)^6} \left(\frac{m^2}{4\pi M^2}\right)^{-3\epsilon/2}$$

$$\times \left[\frac{8}{3\epsilon^3} + \frac{1}{\epsilon^2}\left(\frac{8}{3} - 4\gamma\right) + \frac{1}{\epsilon}\left(\frac{4}{3} - 4a - 4\gamma + 3\gamma^2 + \frac{\pi^2}{6}\right) \right] + \text{(finite)},$$

where γ is Euler's constant, a is a numerical constant ($a = 1.17\ldots$) coming from integration over Feynman parameters, and M is the renormalization scale.

Now let us use the matrix Lie algebra to compute the renormalized graph (again working in the graph-by-graph context). From the previous section, we know that counterterms are generated by the exponential of a sum over grafting operators.

Using

$$A = \exp[-(C(⊠)s(⊠) + C(⊠)s(⊠) + C(⊠)s(⊠) + \cdots)] B,$$

we find the renormalized graphs. In particular, the relevant row of the vector A is

$A(⊠)$

$$= \left(-(2/81)C(⊠)^3 + (1/18)C(⊠)C(⊠) + (1/18)C(⊠)C(⊠) - (1/6)C(⊠) \right) B(✗)$$

$$+ (2/9)C(⊠)^2 B(⊠) - (1/3)C(⊠)B(⊠) - (1/3)C(⊠)B(⊠) + B(⊠).$$
(17.18)

This is to be compared with the known result for the renormalized graph:

$$\left.\vcenter{\hbox{⌘}}\right|_{\text{ren}} = \vcenter{\hbox{⌘}} + \left(-\langle\vcenter{\hbox{⌘}}\rangle + \langle\langle\mathsf{X}\rangle\mathsf{Y}\rangle + \langle\langle\mathsf{X}\rangle\mathsf{A}\rangle - \langle\langle\mathsf{X}\rangle^2\alpha\rangle\right)$$
$$- \langle\alpha\rangle\mathsf{Y} - \langle\alpha\rangle\mathsf{A} + \langle\alpha\rangle^2\alpha, \qquad (17.19)$$

where $\langle\,\rangle$ denotes the renormalization map (for example, in minimal subtraction we simply drop the finite part).

It is easy to establish a one-to-one correspondence between the different terms in (17.18) and in (17.19). For instance $\frac{1}{18}C(\mathsf{X})C(\mathsf{A})B(\mathsf{X})$ corresponds to $\langle\langle\mathsf{X}\rangle\mathsf{A}\rangle$. Here are the differences:

- In the BPHZ approach as exemplified by equation (17.19), all coefficients are $+1$ or -1, while we have rational coefficients, reminiscents of symmetry factors. We insist on the fact that, once we have calculated the insertion matrices, these coefficients are mechanically generated using the exponential of a matrix.
- The coefficients $C(t)$ are *not equal* to the overall divergence; for instance $C(\alpha)$ is not equal to $\langle\alpha\rangle$. Both are generated using different reversing procedures.
- In a term like $\frac{1}{18}C(\mathsf{X})C(\mathsf{A})B(\mathsf{X})$ above in formula (17.18), we take the product of the corresponding values, while in $\langle\langle\mathsf{X}\rangle\mathsf{A}\rangle$ we combine products and composition of the bracket operator $\langle\,\rangle$.
- The counterterms are generated from our coefficients $C(t)$ using an inverse exponential, while they are generated by the antipode in the BPHZ method revisited by using Hopf algebras.

We stress that, due to the uniqueness of Birkhoff factorization, both methods generate the same renormalized values for all graphs.

The Ihara bracket relation $s_{[t_1,t_2]} = [s_{t_1}, s_{t_2}]$ is still rather trivial in this example, as it turns out, because only $s(\mathsf{X})$ appears more than once. In the example of $\vcenter{\hbox{⌘}}$, we have $s(\mathsf{X})$, $s(\mathsf{X})$, and $s(\mathsf{Y})$ appearing, so there is potential for use of the relation here. However, the only nontrivial commutator, equation (17.4), yields graphs that do not appear as subgraphs of $\vcenter{\hbox{⌘}}$:

$$\frac{1}{3}s(\vcenter{\hbox{⊗}}) + \frac{2}{3}s(\ominus) - 2s(\mathsf{X}) = [s(\mathsf{X}), s(\mathsf{o})].$$

We have thus seen that it is not until the four-loop level that nontrivial application of the Lie-algebra relation $s_{[t_1,t_2]} = [s_{t_1}, s_{t_2}]$ appears. As mentioned in the introduction, new results are needed at five loops and higher loop order.

17.9 Renormalization-group flows and nonrenormalizable theories

In Wilson's approach to renormalization, short-distance (ultraviolet) fluctuations are integrated out of the functional integral to appear only as modifications of parameters of the Lagrangian for the remaining long-distance (infrared) degrees of freedom [37] (see also the previous chapter). The introduction of a mass scale M by renormalization parametrizes various renormalization schemes, and the change of M induces a flow of parameters in the Lagrangian, such that certain so-called irrelevant operators die away. In ϕ^4 theory in four dimensions, ϕ^6 is an example of such an operator; the size of this operator relative to other terms in the Lagrangian at a momentum scale p would be $(p/\Lambda)^{6-D} = (p/\Lambda)^2$ as $p \to 0$, where Λ is a cut-off.

Now, the grafting operator s_t may be thought of as a "magnifying glass"; by inserting graphs, we resolve details of graphs that were not visible at larger scales. In this specific sense, s_t induces scale change. In particular, in this chapter, we considered s_t only for graphs t with two or four external legs. We do not have to consider s_t for graphs with six external legs in ϕ^4 theory in four dimensions, precisely because ϕ^6 is an irrelevant operator in this theory. This shows how the matrix Lie algebra would appear in a nonrenormalizable theory; everything is the same as in the ϕ^4 example, except that there is an infinite number of insertion matrices. While this situation is certainly more cumbersome, it is not necessarily fatal.

According to Connes and Kreimer [20], renormalization-group flow boils down to the calculation of a matrix β, which in our notation becomes

$$\beta = \sum_t \beta(t) s_t,$$

and is independent of ϵ. Here $\beta(t)$ is the numerical contribution to the beta function due to a certain diagram t (again, in the graph-by-graph method). This matrix β generalizes the standard beta function of the renormalization group; there is now combinatorial information attached to each of the contributions $\beta(t)$ to the beta function.

See also the next section for some additional comments on Wilson's approach.

17.10 Conclusion

In this chapter, we displayed a matrix Lie algebra of operators acting on Feynman graphs, which, by exponentiation, yields group elements generating counterterms for diagrams with subdivergences. We discussed how to define a grafting operator s_t that inserts one Feynman graph t into another. The matrix representations of these operators satisfy a certain rule, similar to an Ihara bracket, that gives relations between them: $s_{[t_1,t_2]} = [s_{t_1}, s_{t_2}]$. In this way, not all matrices have to be computed separately, with potentially substantial savings in computation time. We displayed the relation to the star product of Kreimer, which is defined similarly to (but not in exactly the same way as) the grafting operator. A simple computation verifies that the right-symmetric (pre-Lie) nonassociative algebra of the star product is equivalent to the previously mentioned Ihara-bracket rule for our matrix representations.

We also gave a number of examples, mostly rather simple ones, and exhibited a three-loop check that the correct sequence of counterterms is provided by the exponential of Lie-algebra elements. Just as with Hopf algebra, the Lie-algebra rules are trivial at one-loop, almost trivial at two-loop, and just beginning to become interesting at three-loop order. At four loops, there is plenty of interesting structure to check, but it will require a fair amount of computation; this is an obvious direction in which future work could go (see the projects).

An equally obvious, but less direct, application is to noncommutative geometry. Using the Connes–Kreimer homomorphism between the Hopf algebra of renormalization and the Hopf algebra of coordinates on the group of formal diffeomorphisms [20, 22], and the Milnor–Moore theorem, the Lie algebra in this chapter has a corresponding Lie algebra in the context of noncommutative geometry. We hope that the matrix Lie algebra may eventually shed some light on certain aspects of noncommutative geometry, in this rather indirect but novel way.

In a less obvious direction, it is suggestive to note that the grafting operator s_t is an element of a Lie algebra, it satisfies a Leibniz rule (the gluing operator, Section 17.6), and its bracket (17.3) looks like that of a vector field X on a manifold. Loosely, the grafting operator is a "scale vector," which takes us into ever increasing magnification of graphs by including more and more subgraphs. If a Lie derivative $\mathcal{L}_{s_t}$ along this "vector field" could meaningfully be defined, that would open the following interesting speculative possibility.

It was proposed[12] in [8], on the basis of earlier work [7], that volume forms for functional integration be characterized in the following way: find a family of vector fields X on the space, define their divergence (this may be a nontrivial task in an infinite-dimensional space) and let the volume form ω be defined by the Koszul formula

$$\mathcal{L}_X \omega = (\operatorname{div} X)\omega.$$

---

[12] See also Chapter 11.

17.10 Conclusion

This definition reproduces familiar volume forms in some simple finite-dimensional examples (riemannian manifolds, symplectic manifolds), and gives some hope for generalization to infinite-dimensional spaces. Perhaps, if s_t is a vector field, it could be used in the role of X, in some extended sense, to characterize ω. In effect, this would be a perturbative definition, since including s_t up to a certain loop order will define different ω at different orders in the loop expansion.

An obvious problem with this idea is that s_t acts on the space of *graphs*, not the space of fields. This means that, even if a volume form ω could be defined, it would be a function on the space of graphs. In principle, it may be possible to exploit some analogy with the space of paths in quantum-mechanical functional integrals. This direction would be interesting to pursue in future work.

As a final note, there has been some dispute regarding the extent to which Wilson's picture is generally valid; there are some examples of (as yet) speculative theories in which infrared and ultraviolet divergences are connected [33]. Some of these examples naturally appear in connection with noncommutative geometry. In view of the relation between the Hopf/Lie-algebraic approach to renormalization and noncommutative geometry mentioned above, one could hope that these algebraic developments may eventually shed some light on the ultraviolet–infrared connection. Since there is so far no experimental evidence supporting speculative theories of the type discussed in [33], this development may seem premature, but it is of theoretical interest to consider whether there are limitations to Wilson's successful picture of renormalization as an organization of scales.

A1 Symmetry factors

Since conventions for how graphs carry operator-contraction symmetry factors vary, let us be precise.

The bare graph in our two-loop example (Section 17.7) contains in its bare value the nonlocal divergence $(1/\epsilon)\ln p^2$, including the $1/2$ due to the symmetry factor 2 appearing in the overall factor $\lambda^2/[2(4\pi)^2]$. The counterterms must subtract the $(1/\epsilon)\ln p^2$, nothing more, nothing less.

However, there is no consistent usage of notation for symmetry factors in the literature. The term $\langle\bowtie\rangle\chi$ taken literally looks as if it would contain two symmetry factors of 2, i.e. the forefactor would be $1/2 \times 1/2 = 1/4$. By itself, this is of course wrong – it subtracts only half the nonlocal divergence. Two possible resolutions are (i) to say that the graph with a cross (as in [16]) that represents $\langle\bowtie\rangle\chi$ has only symmetry factor 2, not 4 (which is not quite consistent notation on the non-graph level without a factor of 2, if $\langle t \rangle$ is really to mean the pole part of t); and

(ii) to posit, as we do here, that insertion can also create a term with $C(\mathcal{Q})$, which combines with $C(\mathsf{X})$ when we take the counterterms to be the same, to give a factor of 2.

It is a useful check to consider the s-channel counterterm as well. In contrast to the previous example, the graph with an s-channel insertion (the first $L = 2, E = 4$ graph in our table 17.1) has a forefactor of $1/4$. Each counterterm also has $1/4$, but now there are two of them, one for shrinking each of the two disjoint subdivergences. Thus they do not cancel out directly, but factor as usual. In the Hopf-algebra coproduct, factorization produces the factor of 2. The antipode then yields a factor of 2 similar to the above. With these conventions, finding a factor of 2 both in the s case and in the t/u case, the s-, t-, and u-channel graphs all have the same normalization after summing.

A2 Insertion tables

In this appendix, we provide tables of the action of the grafting operator s_t. We have defined s_t such that, on a bare diagram denoted by 1, we have $s_t 1 = t$. The matrix s_1 acts as $s_1 t = N(t)t$, where $N(t)$ is the number of insertion points of t. It is also true that the coefficients of each resulting graph in an insertion into a graph t should add up to the number of compatible insertion points of the graph t.

One-loop:
$$1 + 0 = 1$$

$$s(\mathcal{Q})\,\text{—} = \mathcal{Q}$$
$$s(\mathcal{Q})\mathsf{X} = 0$$
$$s(\mathsf{X})\,\text{—} = 0$$
$$s(\mathsf{X})\mathsf{X} = \frac{1}{3}\mathsf{X} + \text{crossing}.$$

Two-loop:
$$1 + 1 = 2$$

$$s(\mathcal{Q})\,\mathcal{Q} = \mathsf{8}$$
$$s(\mathcal{Q})\,\mathsf{X} = 2\mathsf{X}$$
$$s(\mathsf{X})\,\mathcal{Q} = \frac{1}{3}\mathsf{8} + \frac{2}{3}\ominus$$
$$s(\mathsf{X})\,\mathsf{X} = \frac{2}{3}\mathsf{X} + \frac{2}{3}\mathsf{X} + \frac{2}{3}\mathsf{X}.$$

17.10 Conclusion

Three-loop:

$1 + 2 = 3$

$$s(\text{◯})\,\text{8} = 2\,\text{⊗} + \text{8}$$

$$s(\text{◯})\,\ominus = 3\,\ominus$$

$$s(\text{◯})\,\text{⨉} = 4\,\text{⨉}$$

$$s(\text{◯})\,\text{⨉} = 2\,\text{⨉} + 2\,\text{⨉}$$

$$s(\text{◯})\,\text{⨉} = 2\,\text{⨉} + 2\,\text{⨉}$$

$$s(\text{◯})\,\text{⨉} = \frac{4}{3}\,\text{⨉} + \frac{4}{3}\,\text{⨉} + \frac{4}{3}\,\text{⨉}$$

$$s(\text{⨉})\,\text{8} = \frac{2}{3}\,\text{8} + \frac{2}{3}\,\text{⊖} + \frac{2}{3}\,\ominus$$

$$s(\text{⨉})\,\ominus = 2\,\infty$$

$$s(\text{⨉})\,\text{⨉} = \frac{2}{3}\,\text{⨉} + \text{⨉} + \frac{2}{3}\,\text{⨉} + \frac{2}{3}\,\text{⨉}$$

$$s(\text{⨉})\,\text{⨉} = \frac{1}{3}\,\text{⨉} + \frac{1}{3}\,\text{⨉} + \frac{2}{3}\,\text{⨉} + \frac{1}{3}\,\text{⨉} + \frac{2}{3}\,\text{⨉} + \frac{2}{3}\,\text{⨉}$$

$$s(\text{⨉})\,\text{⨉} = \frac{1}{3}\,\text{⨉} + \frac{1}{3}\,\text{⨉} + \frac{2}{3}\,\text{⨉} + \frac{1}{3}\,\text{⨉} + \frac{2}{3}\,\text{⨉} + \frac{2}{3}\,\text{⨉}$$

$$s(\text{⨉})\,\text{⨉} = \frac{2}{3}\,\text{⨉} + \frac{1}{3}\,\text{⨉} + \frac{2}{3}\,\text{⨉} + \frac{2}{3}\,\text{⨉} + \frac{2}{3}\,\text{⨉};$$

$2 + 1 = 3$

$$s(\,\text{8}\,)\,\text{◯} = \text{8}$$

$$s(\ominus)\,\text{◯} = \ominus$$

$$s(\,\text{8}\,)\,\text{⨉} = 2\,\text{⨉}$$

$$s(\ominus)\,\text{⨉} = 2\,\text{⨉}$$

$$s(\text{⨉})\,\text{◯} = \frac{1}{3}\,\text{8} + \frac{2}{3}\,\infty$$

$$s(\text{⨉})\,\text{◯} = \frac{2}{3}\,\infty + \frac{1}{6}\,\ominus + \frac{1}{6}\,\ominus$$

$$s(\text{⨉})\,\text{◯} = \frac{2}{3}\,\infty + \frac{1}{6}\,\ominus + \frac{1}{6}\,\ominus$$

Renormalization 3: combinatorics

$$s(\diagup\!\!\!\!\diagdown)\,\Omega = \tfrac{1}{3}\,\infty + \tfrac{2}{3}\,\theta$$

$$s(\diagup\!\!\!\!\diagdown)\,\diagup\!\!\!\!\diagdown = \tfrac{2}{3}\,\mathsf{X}_1 + \tfrac{2}{3}\,\mathsf{X}_2 + \tfrac{2}{3}\,\mathsf{X}_3$$

$$s(\diagup\!\!\!\!\diagdown)\,\diagup\!\!\!\!\diagdown = \tfrac{1}{3}\,\mathsf{Y}_1 + \tfrac{1}{6}\,\mathsf{Y}_2 + \tfrac{1}{6}\,\mathsf{Y}_3 + \tfrac{1}{3}\,\mathsf{Y}_4 + \tfrac{1}{3}\,\mathsf{Y}_5 + \tfrac{1}{3}\,\mathsf{Y}_6 + \tfrac{1}{3}\,\mathsf{Y}_7$$

$$s(\diagup\!\!\!\!\diagdown)\,\diagup\!\!\!\!\diagdown = \tfrac{1}{3}\,\mathsf{Z}_1 + \tfrac{1}{6}\,\mathsf{Z}_2 + \tfrac{1}{6}\,\mathsf{Z}_3 + \tfrac{1}{3}\,\mathsf{Z}_4 + \tfrac{1}{3}\,\mathsf{Z}_5 + \tfrac{1}{3}\,\mathsf{Z}_6 + \tfrac{1}{3}\,\mathsf{Z}_7.$$

$$s(\diagup\!\!\!\!\diagdown)\,\diagup\!\!\!\!\diagdown = \tfrac{2}{3}\,\mathsf{W}_1 + \tfrac{2}{3}\,\mathsf{W}_2 + \tfrac{2}{3}\,\mathsf{W}_3.$$

A3 Grafting matrices

In this appendix, we list matrix representations of the grafting operators $s(t)$. Since insertion into an n-point function can create only an n-point function, we consider submatrices of each separately, and denote them by subscripts. For example, if we put $s(\diagup\!\!\!\!\diagdown)_{\_\_} = A$ and $s(\diagup\!\!\!\!\diagdown)_X = B$, then the complete matrix representing $s(\diagup\!\!\!\!\diagdown)$ is the direct sum of A and B:

$$s(\diagup\!\!\!\!\diagdown) \stackrel{\bullet}{=} \left(\begin{array}{c|c} A & 0 \\ \hline 0 & B \end{array}\right).$$

Lines are drawn through the matrices to delineate loop order L.

One-loop insertions:

$$s(\Omega)_{\_\_} = \left(\begin{array}{c|cc|c} 0 & 0 & 0 & 0 \\ \hline 1 & 0 & 0 & 0 \\ 0 & 1 & 0 & 0 \\ \hline 0 & 0 & 0 & 0 \end{array}\right) \begin{array}{c} \_\_ \\ \Omega \\ \infty \\ \theta \end{array} \qquad s(\diagup\!\!\!\!\diagdown)_{\_\_} = \left(\begin{array}{c|cc|c} 0 & 0 & 0 & 0 \\ \hline 0 & 0 & 0 & 0 \\ 0 & \tfrac{1}{3} & 0 & 0 \\ \hline 0 & \tfrac{2}{3} & 0 & 0 \end{array}\right) \begin{array}{c} \_\_ \\ \Omega \\ \infty \\ \theta \end{array}$$

$$s(\Omega)_X = \left(\begin{array}{cc|cccc} 0 & 0 & 0 & 0 & 0 & 0 \\ 0 & 0 & 0 & 0 & 0 & 0 \\ \hline 0 & 0 & 0 & 0 & 0 & 0 \\ 0 & 0 & 0 & 0 & 0 & 0 \\ 0 & 0 & 0 & 0 & 0 & 0 \\ 0 & 2 & 0 & 0 & 0 & 0 \end{array}\right) \qquad s(\diagup\!\!\!\!\diagdown)_X = \left(\begin{array}{cc|cccc} 0 & 0 & 0 & 0 & 0 & 0 \\ \tfrac{1}{3} & 0 & 0 & 0 & 0 & 0 \\ \hline 0 & \tfrac{2}{3} & 0 & 0 & 0 & 0 \\ 0 & \tfrac{2}{3} & 0 & 0 & 0 & 0 \\ 0 & \tfrac{2}{3} & 0 & 0 & 0 & 0 \\ 0 & 0 & 0 & 0 & 0 & 0 \end{array}\right)$$

17.10 Conclusion

When it comes to three-loop matrices, it is typographically easier to give the transpose:

$$s^T(\lozenge)_\chi = \begin{pmatrix} 0 & 0 & 0 & 0 & 0 & 0 & 0 & 0 & 0 & 0 & 0 & 0 & 0 & 0 & 0 & 0 & 0 & 0 & 0 & 0 & 0 & 0 \\ 0 & 0 & 0 & 0 & 0 & 0 & 0 & 0 & 0 & 0 & 0 & 0 & 0 & 0 & 0 & 0 & 0 & 0 & 0 & 0 & 0 & 0 \\ 0 & 0 & 0 & \frac{2}{3} & 1 & 0 & 0 & \frac{2}{3} & \frac{2}{3} & 0 & 0 & 0 & 0 & 0 & 0 & 0 & 0 & 0 & 0 & 0 & 0 & 0 \\ 0 & 0 & \frac{1}{3} & \frac{1}{3} & 0 & \frac{2}{3} & 0 & \frac{1}{3} & 0 & \frac{2}{3} & 0 & \frac{2}{3} & 0 & 0 & 0 & 0 & 0 & 0 & 0 & 0 & 0 & 0 \\ 0 & 0 & \frac{1}{3} & \frac{1}{3} & 0 & 0 & \frac{2}{3} & 0 & \frac{1}{3} & 0 & \frac{2}{3} & 0 & \frac{2}{3} & 0 & 0 & 0 & 0 & 0 & 0 & 0 & 0 & 0 \\ 0 & 0 & 0 & 0 & 0 & 0 & 0 & 0 & 0 & 0 & 0 & 0 & \frac{2}{3} & \frac{1}{3} & 0 & 0 & \frac{2}{3} & \frac{2}{3} & \frac{2}{3} & 0 & 0 \end{pmatrix}$$

where all elements below those displayed are zero (it is, of course, a square matrix), and we have not displayed the zero-, one-, and two-loop submatrices already given. That is, the given matrix is the transpose of the 22×6 submatrix B in

$$\begin{pmatrix} A & 0 \\ \hline B & 0 \end{pmatrix},$$

where A is the 6×6 matrix given for $s(\lozenge)_\chi$ earlier. The ordering for the graphs with four external legs is as in table 17.1, summarized here:

7	8	9	10	11	12	13	14	15	16	17

18	19	20	21	22	23	24	25	26	27	28

Graphs numbered 1–6 are as in the rows of the matrix $s(\lozenge)_\chi$ above. We also give two two-loop insertions:

$$s^T(\lozenge)_\chi = \begin{pmatrix} 0 & 0 & 0 & 0 & 0 & 0 & 0 & 0 & 0 & 0 & 0 & 0 & 0 & 0 & 0 & 0 & 0 & 0 & 0 & 0 & 0 & 0 \\ 0 & 0 & 0 & 0 & \frac{2}{3} & \frac{2}{3} & \frac{2}{3} & 0 & 0 & 0 & 0 & 0 & 0 & 0 & 0 & 0 & 0 & 0 & 0 & 0 & 0 & 0 \\ 0 & 0 & 0 & 0 & 0 & 0 & 0 & 0 & 0 & 0 & 0 & 0 & 0 & 0 & 0 & 0 & 0 & 0 & 0 & 0 & 0 & 0 \\ 0 & 0 & 0 & 0 & 0 & 0 & 0 & 0 & 0 & 0 & 0 & 0 & 0 & 0 & 0 & 0 & 0 & 0 & 0 & 0 & 0 & 0 \\ 0 & 0 & 0 & 0 & 0 & 0 & 0 & 0 & 0 & 0 & 0 & 0 & 0 & 0 & 0 & 0 & 0 & 0 & 0 & 0 & 0 & 0 \\ 0 & 0 & 0 & 0 & 0 & 0 & 0 & 0 & 0 & 0 & 0 & 0 & 0 & 0 & 0 & 0 & 0 & 0 & 0 & 0 & 0 & 0 \end{pmatrix}$$

$$s^T(\lozenge)_\chi = \begin{pmatrix} 0 & 0 & 0 & 0 & 0 & 0 & 0 & 0 & 0 & 0 & 0 & 0 & 0 & 0 & 0 & 0 & 0 & 0 & 0 & 0 & 0 & 0 \\ 0 & 0 & 0 & \frac{1}{3} & 0 & 0 & 0 & \frac{1}{6} & \frac{1}{6} & \frac{1}{3} & \frac{1}{3} & \frac{1}{3} & \frac{1}{3} & 0 & 0 & 0 & 0 & 0 & 0 & 0 & 0 & 0 \\ 0 & 0 & 0 & 0 & 0 & 0 & 0 & 0 & 0 & 0 & 0 & 0 & 0 & 0 & 0 & 0 & 0 & 0 & 0 & 0 & 0 & 0 \\ 0 & 0 & 0 & 0 & 0 & 0 & 0 & 0 & 0 & 0 & 0 & 0 & 0 & 0 & 0 & 0 & 0 & 0 & 0 & 0 & 0 & 0 \\ 0 & 0 & 0 & 0 & 0 & 0 & 0 & 0 & 0 & 0 & 0 & 0 & 0 & 0 & 0 & 0 & 0 & 0 & 0 & 0 & 0 & 0 \\ 0 & 0 & 0 & 0 & 0 & 0 & 0 & 0 & 0 & 0 & 0 & 0 & 0 & 0 & 0 & 0 & 0 & 0 & 0 & 0 & 0 & 0 \end{pmatrix}$$

We note that, similarly to the previous case, there is also a top-left 6×6 matrix with entries in the first column only. The rest of the matrices are now trivial to extract from the insertion tables, so we shall not repeat them here.

References

[1] M. Berg and P. Cartier (2001). "Representations of the renormalization group as matrix Lie algebra," arXiv:hep-th/0105315, v2 14 February 2006.

[2] D. Becirevic, Ph. Boucaud, J. P. Leroy et al. (2000). "Asymptotic scaling of the gluon propagator on the lattice," *Phys. Rev.* **D61**, 114 508; arXiv:hep-ph/9910204.

[3] N. N. Bogoliubov and O. S. Parasiuk (1957). "Über die Multiplikation der Kausalfunktionen in der Quantentheorie der Felder," *Acta Math.* **97**, 227–266.
K. Hepp (1966). "Proof of the Bogoliubov–Parasiuk theorem of renormalization," *Commun. Math. Phys.* **2**, 301–326.
W. Zimmermann (1971). "Local operator products and renormalization in quantum field theory," in *Lectures on Elementary Particles and Quantum Field Theory, 1970 Brandeis University Summer Institute in Theoretical Physics*, Vol. 1, eds. S. Deser, M. Grisaru, and H. Pendleton (Cambridge, MA, MIT Press), pp. 301–306.

[4] D. J. Broadhurst and D. Kreimer (1999). "Renormalization automated by Hopf algebra," *J. Symb. Comput.* **27**, 581–600; arXiv:hep-th/9810087.

[5] D. J. Broadhurst and D. Kreimer (2000). "Combinatoric explosion of renormalization tamed by Hopf algebra: 30-loop Padé–Borel resummation," *Phys. Lett.* **B475**, 63–70; arXiv:hep-th/9912093.

[6] D. J. Broadhurst and D. Kreimer (2001). "Exact solutions of Dyson–Schwinger equations for iterated one-loop integrals and propagator-coupling duality," *Nucl. Phys.* **B600**, 403–422.

[7] P. Cartier and C. DeWitt-Morette (1995). "A new perspective on functional integration," *J. Math. Phys.* **36**, 2237–2312.
P. Cartier and C. DeWitt-Morette (2000). "Functional integration", *J. Math. Phys.* **41**, 4154–4187.

[8] P. Cartier, M. Berg, C. DeWitt-Morette, and A. Wurm (2001). "Characterizing volume forms," in *Fluctuating Paths and Fields*, ed. Axel Pelster (Singapore, World Scientific), pp. 139–156; arXiv:math-ph/0012009.

[9] A. H. Chamseddine and A. Connes (1997). "The spectral action principle," *Commun. Math. Phys.* **186**, 731–750.

[10] F. Chapoton (2000). "Un théorème de Cartier–Milnor–Moore–Quillen pour les bigèbres dendriformes et les algèbres braces," arXiv:math.QA/0005253.

[11] K. Chetyrkin, M. Misiak, and M. Münz (1998). "Beta functions and anomalous dimensions up to three loops," *Nucl. Phys.* **B518**, 473–494; arXiv:hep-ph/9711266.

[12] K. G. Chetyrkin and T. Seidensticker (2000). "Two loop QCD vertices and three loop MOM β functions," *Phys. Lett.* **B495**, 74–80; arXiv:hep-ph/0008094.

[13] K. G. Chetyrkin and F. V. Tkachov (1981). "Integration by parts: the algorithm to calculate beta functions in 4 loops," *Nucl. Phys.* **B192**, 159–204. K. G. Chetyrkin, A. H. Hoang, J. H. Kuhn, M. Steinhauser, and T. Teubner (1996). "QCD corrections to the e^+e^- cross section and the Z boson decay rate: concepts and results," *Phys. Rep.* **277**, 189–281.

[14] C. Chryssomalakos, H. Quevedo, M. Rosenbaum, and J. D. Vergara (2002). "Normal coordinates and primitive elements in the Hopf algebra of renormalization," *Commun. Math. Phys.* **225**, 465–485; arXiv:hep-th/0105259.

[15] J.-M. Chung and B. K. Chung (1999). "Calculation of a class of three-loop vacuum diagrams with two different mass values," *Phys. Rev.* **D59**, 105014; arXiv:hep-ph/9805432.

[16] J. C. Collins (1985). *Renormalization* (Cambridge, Cambridge University Press).

[17] A. Connes (2000). "A short survey of noncommutative geometry," *J. Math. Phys.* **41**, 3832–3866; arXiv:hep-th/0003006.

[18] A. Connes, M. R. Douglas, and A. Schwarz (1998). "Noncommutative geometry and matrix theory: compactification on tori," *J. High Energy Phys.* **02**, 003; arXiv:hep-th/9711162.

[19] A. Connes and D. Kreimer (2000). "Renormalization in quantum field theory and the Riemann–Hilbert problem I: the Hopf algebra structure of graphs and the main theorem," *Commun. Math. Phys.* **210**, 249–273; arXiv:hep-th/9912092.

[20] A. Connes and D. Kreimer (2001). "Renormalization in quantum field theory and the Riemann–Hilbert problem II: the β-function, diffeomorphisms and the renormalization group," *Commun. Math. Phys.* **216**, 215–241; arXiv:hep-th/0003188.

[21] A. Connes and D. Kreimer (1999). "Lessons from quantum field theory – Hopf algebras and spacetime geometries," *Lett. Math. Phys.* **48**, 85–96; arXiv:hep-th/9904044.

[22] A. Connes and H. Moscovici (2001). "Differentiable cyclic cohomology and Hopf algebraic structures in transverse geometry," arXiv:math.DG/0102167.

[23] J. M. Garcia-Bondía, J. C. Várilly, and H. Figueroa (2001). *Elements of Noncommutative Geometry* (Boston, Birkhäuser).

[24] R. Grossmann and R. G. Larson (1989). "Hopf-algebraic structure of families of trees," *J. Algebra* **126**, 184–210.

[25] T. Hahn (2000). "Generating Feynman diagrams and amplitudes with FeynArts 3," hep-ph/0012260. (The FeynArts web site is www.feynarts.de.)

[26] S. Heinemeyer (2001). "Two-loop calculations in the MSSM with FeynArts," arXiv:hep-ph/0102318.

[27] M. Kontsevich (1999). "Operads and motives in deformation quantization," *Lett. Math. Phys.* **48**, 35–72; arXiv:math.QA/9904055.

[28] D. Kreimer (1998). "On the Hopf algebra structure of perturbative quantum field theories," *Adv. Theor. Math. Phys.* **2**, 303–334; arXiv:q-alg/9707029.

[29] D. Kreimer (2000). "Shuffling quantum field theory," *Lett. Math. Phys.* **51**, 179–191; arXiv:hep-th/9912290.

[30] D. Kreimer (2002). "Combinatorics of (perturbative) quantum field theory," *Phys. Rep.* **363**, 387–424.; arXiv:hep-th/0010059.

[31] D. Kreimer (2000). *Knots and Feynman Diagrams* (Cambridge, Cambridge University Press).

[32] C. P. Martín, J. M. Gracia-Bondía, and J. C. Várilly (1998). "The Standard Model as a noncommutative geometry: the low energy regime," *Phys. Rep.* **294**, 363–406.

[33] S. Minwalla, M. Van Raamsdonk, and N. Seiberg (2000). "Noncommutative perturbative dynamics," *J. High Energy Phys.* **02**, 020, 30 pp.; arXiv:hep-th/9912072.

M. Van Raamsdonk and N. Seiberg (2000). "Comments on noncommutative perturbative dynamics," *J. High Energy Phys.* **03**, 035, 18 pp.; arXiv:hep-th/0002186.

L. Griguolo and M. Pietroni (2001). "Wilsonian renormalization group and the non-commutative IR/UV connection," *J. High Energy Phys.* **05**, 032, 28 pp.; arXiv:hep-th/0104217.

[34] T. Ohl (1995). "Drawing Feynman diagrams with LaTeX and Metafont," *Comput. Phys. Commun.* **90**, 340–354; arXiv:hep-ph/9505351.

[35] G. Racinet (2000). Séries génératrices non-commutatives de polyzêtas et associateurs de Drinfeld, unpublished Ph.D. Thesis, Université d'Amiens.

[36] N. Seiberg and E. Witten (1999). "String theory and noncommutative geometry," *J. High Energy Phys.* **09**, 032, 92 pp.; arXiv:hep-th/9908142.

[37] K. G. Wilson and J. Kogut (1974). "The renormalization group and the ϵ expansion," *Phys. Rep.* **12**, 75–200.

Note added in proof

A review of Hopf algebras, intended for physicist, P. Cartier, "A primer on Hopf algebras," will appear in *Frontiers in Number Theory, Physics and Geometry* II, eds. P. Cartier, B. Julia, P. Moussa, and P. Vunhour (Berlin, Springer, 2006). This book contains also a paper by A. Connes and M. Maveolli, and a paper by D. Kreimer, both about connections between renormalization and Hopf algebras. See also a forthcoming paper by P. Cartier and V. Ferag titled "Nonlinear transformations in lagrangians and Connes–Kreimer Hopf algebra."

18
Volume elements in quantum field theory

(contributed by Bryce DeWitt)

Remarkable properties of the "measure"

18.1 Introduction

The point of view of this contribution differs somewhat from that of the main text. Here whatever gaussian integrations are envisaged are with respect to general backgrounds and general boundary conditions. Statements of the boundary conditions, either explicit or implicit, must always be understood as accompanying every computation.

The general functional integral is to be understood as taken over the *space of field histories*. A "history" is a section of a fiber bundle having spacetime as its base space. Spacetime itself is assumed to have the structure $\mathbb{R} \times \Sigma$, where Σ is an $(n-1)$-dimensional manifold (compact or noncompact) that can carry a complete set of Cauchy data for whatever dynamical system is under consideration. The topology of Σ may be nontrivial and the bundle may be twisted. The typical fiber is a supermanifold (known as *configuration space*), which may itself be nontrivial; it need not be a vector space. When $n = 1$, the theory reduces to ordinary quantum mechanics, and the functional integral becomes a path integral.

The space of histories, denoted by Φ, will be regarded heuristically as an infinite-dimensional supermanifold. Spacetime will be thought of as a lattice, and both the bundle sections and the functional integral itself will be viewed as "continuum limits" of associated structures on this lattice. Such a view would seem to leave a certain amount of imprecision in the formalism, with regard, first, to how interpolations between lattice sites are to be performed and, second, to the manner in which field-theory infinities are to be handled. It turns out, however, that the formalism carries enough structure to guarantee its own self-consistency and, ultimately, its uniqueness.

The functional-integral volume element

Since the typical fiber is a general supermanifold, it must be described by an atlas of *charts* or *coordinate patches*, and one must be able to deal with the formal jacobians that arise under transformations of coordinates. For consistency the *volume element* of the functional integral must contain a *density functional* $\mu[\phi]$, often called the *measure*,[1] which is transformed, under changes of chart, by being multiplied by these jacobians. The volume element will be expressed in the form

$$\mu[\phi][\mathrm{d}\phi], \tag{18.1}$$

where $[\mathrm{d}\phi]$ is, formally,

$$[\mathrm{d}\phi] := \prod_{i,x} \mathrm{d}\phi^i(x). \tag{18.2}$$

Here the ϕ^i are the coordinates of a generic chart in configuration space and $\phi^i(x)$ denotes the point at which the field history ϕ intersects the fiber over x. The product in expression (18.2) is a formally infinite one, over the continuum of points x of spacetime and over the chart-coordinate label i. The ϕ^i are usually referred to as *field components*.

An approximate expression for $\mu[\phi]$

An approximate expression for $\mu[\phi]$ is derived in [1], namely

$$\mu[\phi] \approx \left(\mathrm{sdet}\, G^+[\phi]\right)^{-1/2}, \tag{18.3}$$

where "sdet" denotes the *superdeterminant* (or *berezinian*) and $G^+[\phi]$ is the advanced Green function of the *Jacobi field operator* evaluated at the history (or field) ϕ. The Jacobi field operator is the second functional derivative of the *action functional* $S : \Phi \longrightarrow \mathbb{R}_c$ of the dynamical system under consideration.[2] More precisely, $G^+[\phi]$ is defined by

$$_{i,}S_{,k}[\phi]G^{+kj}[\phi] = -_i\delta^j, \tag{18.4}$$

$$G^{+ij}[\phi] = 0 \quad \text{when } i \succ j, \tag{18.5}$$

---

[1] This is, of course, an abuse of language since it does not correspond to standard mathematical terminology.
[2] Here $\mathbb{R}_c$ denotes the space of commuting real supernumbers (see [1]). The ordinary real line is contained in $\mathbb{R}_c : \mathbb{R} \subset \mathbb{R}_c$.

where a condensed notation is being used, in which the spacetime points are absorbed into the indices i, j, k, and the summation over repeated indices includes an integration over spacetime. The symbol $_iS_{,k}[\phi]$ indicates that one of the two functional derivatives is from the left and the other is from the right, $_i\delta^j$ denotes a combined Kronecker delta and δ-function, and "$i \succ j$" means "the spacetime point associated with i lies to the future (relative to any foliation of spacetime into space-like hypersurfaces) of the spacetime point associated with j."

Remark. G^{+ij} is here being regarded as a matrix with indices ranging over a continuum of values. The superdeterminant in equation (18.3) is thus a formal one. Moreover, the times associated with the indices i and j are often restricted to lie within a certain range, namely the range over which the fields ϕ^i themselves are taken in the functional integral. Boundary conditions are usually imposed on the fields at the end points of this range, and G^+ in equation (18.3) is really a truncated Green function.

18.2 Cases in which equation (18.3) is exact

Equation (18.3) turns out to be exact in two important cases.

The first case is that of standard canonical systems having lagrangians of the form

$$L = \frac{1}{2}g_{ij}(x)\dot{x}^i\dot{x}^j + a_i(x)\dot{x}^i - v(x). \tag{18.6}$$

Here the coordinates in configuration space are denoted by x^i instead of ϕ^i, and the dot denotes differentiation with respect to the time. The quantum theory of these systems is ordinary quantum mechanics, and the associated functional integral is a path integral. Expression (18.6) may be regarded as the lagrangian for a nonrelativistic particle of unit mass and charge moving in a curved manifold with metric g_{ij}, in a magnetic field with vector potential a_i, and in an electric field with scalar potential v.[3]

For these systems the prefix "super" in "superdeterminant" becomes irrelevant and one finds (formally)

$$\det G^+[x] = \text{constant} \times g^{-1/2}(x(t_b))\left[\prod_{t_a < t < t_b} g^{-1}(x(t))\right]g^{-1/2}(x(t_a)), \tag{18.7}$$

---

[3] Expression (18.6) is easily generalized to the case in which g_{ij}, a_i, and v have explicit dependences on the time. None of the conclusions below are changed.

and hence

$$\mu[x] = \text{constant} \times g^{1/4}(x(t_b)) \left[\prod_{t_a<t<t_b} g^{1/2}(x(t)) \right] g^{1/4}(x(t_a)), \quad (18.8)$$

where

$$g := \det(g_{ij}) \quad (18.9)$$

and t_a and t_b are the times associated with the initial and final boundary conditions. Expression (18.8), which clearly transforms correctly under changes of chart, is what one would intuitively expect for the measure in these systems.

The second case is that of *linear*[4] *fields propagating in arbitrary backgrounds*. In the general case expression (18.3) is obtained (see [1]) by examining the difference between placing the factors in the quantum-dynamical equations in a symmetric (self-adjoint) order and placing them in chronological order. In the case of linear systems no factor-ordering problems arise, and the ϕ in equation (18.3) is replaced simply by the background (or external) field, which will be denoted by ϕ_{ex}. Thus

$$\mu[\phi_{\text{ex}}] = \left(\text{sdet}\, G^+[\phi_{\text{ex}}]\right)^{-1/2}, \quad (18.10)$$

and, although this measure does not vary over the domain of the functional integral (which becomes a simple gaussian), it does change when the background changes.

18.3 Loop expansions

It turns out that the measure (18.3) plays a host of remarkable roles. The first and simplest of these appears in the loop expansion for standard canonical systems. To one-loop order the point-to-point amplitude for these systems is given by[5]

$$\langle x, t | x', t' \rangle = \text{constant} \times \left(\frac{\det G[x_c]}{\det G^+[x_c]} \right)^{1/2} e^{iS(x,t|x',t')}, \quad (18.11)$$

where x_c is a dynamically allowed *classical* path between (x', t') and (x, t), and $S(x, t | x', t')$ is the associated action *function* (also known as

---

[4] Here "linear field" is synonymous with "free field," or field with a quadratic lagrangian.
[5] We take here $\hbar = 1$.

18.3 Loop expansions

Hamilton's principal function) between these points:

$$S(x,t|x',t') := S[x_c] := \int_{t'}^{t} L(x_c(t''), \dot{x}_c(t''), t'')dt''$$

with $x_c^i(t) = x^i$ and $x_c^i(t') = x'^i$. (18.12)

$G^+[x_c]$ in expression (18.11) comes from the measure (18.3), and $G[x_c]$, which is the point-to-point Green function of the Jacobi field operator, comes from the gaussian approximation to the functional integral

$$\int e^{iS[x]} \mu[x][dx].$$

The remarkable fact is that the ratio of the two formally infinite determinants in expression (18.11) is just the finite *Van Vleck–Morette determinant*:

$$\frac{\det G[x_c]}{\det G^+[x_c]} = \text{constant} \times D(x,t|x',t'), \quad (18.13)$$

where

$$D(x,t|x',t') := \det\left(-\frac{\partial^2 S(x,t|x',t')}{\partial x^i \, \partial x'^j}\right). \quad (18.14)$$

Therefore

$$\langle x,t|x',t'\rangle = \text{constant} \times D^{1/2}(x,t|x',t') e^{iS(x,t|x',t')}, \quad (18.15)$$

which is the well-known *semiclassical* or WKB approximation to the point-to-point amplitude. It should be noted that this result cannot be obtained from the functional integral without inclusion of the measure (18.3).

In higher loop orders the measure continues to play a remarkable role. For systems having lagrangians of the form (18.6), with non-flat g_{ij}, each vertex function contains two derivatives with respect to the time, and, when these derivatives act on the point-to-point Green functions G^{ij} in the loop expansion, terms containing factors $\delta(0)$ appear. These formally infinite terms are completely canceled out, *to all orders*, by corresponding terms coming from the measure.

Remarks.

1. Equations analogous to (18.11)–(18.15) hold for boundary conditions other than point-to-point, for example point-to-momentum, or point-to-eigenvalues of any other complete set of commuting observables. Each of these alternative boundary conditions has its own action function, its own "propagator" G^{ij}, and its own Van Vleck–Morette determinant.

2. If the configuration space has nontrivial topology, equation (18.11) is not strictly correct but should be replaced by

$$\langle x,t|x',t'\rangle = \text{constant} \times \sum_\alpha \chi(\alpha)\left(\frac{\det G[x_{c\alpha}]}{\det G^+[x_{c\alpha}]}\right)^{1/2} e^{iS_\alpha(x,t|x',t')}, \tag{18.16}$$

where the index α labels the homotopy classes of paths between (x',t') and (x,t), $x_{c\alpha}$ denotes the dynamically allowed classical path in the αth homotopy class, and $\chi(\alpha)$ is a one-dimensional character of the fundamental group of configuration space[6]. Similar generalizations are needed with boundary conditions other than point-to-point.

3. For systems with lagrangians of the form (18.6), if one evaluates the path integral to *two-loop order*, in the same (naive) way as the functional integrals for quantum field theories are evaluated, then one finds that the hamiltonian operator to which the path integral corresponds is given by the manifestly covariant expression

$$H = \frac{1}{2}g^{-1/4}(x)\left[-i\frac{\partial}{\partial x^i} - a_i(x) - \omega_i(x)\right]g^{1/2}(x)g^{ij}(x)$$
$$\times \left[-i\frac{\partial}{\partial x^j} - a_j(x) - \omega_j(x)\right]g^{-1/4}(x) + \frac{1}{8}R(x) + v(x), \tag{18.17}$$

where R is the curvature scalar corresponding to the matrix g_{ij} and $\boldsymbol{\omega}$ is a closed 1-form associated with the character χ (see [1]).

Linear fields

Consider a linear field, bosonic or fermionic. Suppose that the background field ϕ_{ex} in which it propagates has a stationary "in" region and a stationary "out" region. Since the "in" and "out" regions are stationary, Fock spaces associated with these regions can usually be set up. Each Fock space has its own mode functions $u^i_{\text{in }A}$, $u^i_{\text{out }X}$, each of which propagates from its defining region throughout the whole of spacetime. The field operators, denoted in boldface, can be decomposed in the alternative forms

$$\phi^i = \begin{cases} u^i_{\text{in }A}\mathbf{a}^A_{\text{in}} + u^{i\,*}_{\text{in }A}\mathbf{a}^{A\,*}_{\text{in}}, \\ u^i_{\text{out }X}\mathbf{a}^X_{\text{out}} + u^{i\,*}_{\text{out }X}\mathbf{a}^{X\,*}_{\text{out}}, \end{cases} \tag{18.18}$$

---
[6] If there is more than one $x_{c\alpha}$ in a given homotopy class, then a factor of i^μ must also be included, where μ is the *Morse index* of the path.

where, if the mode functions have been normalized correctly, the $\mathbf{a}^A_{\text{in}}$, $\mathbf{a}^{A\,*}_{\text{in}}$, $\mathbf{a}^X_{\text{out}}$, and $\mathbf{a}^{X\,*}_{\text{out}}$ satisfy the standard (anti)commutation relations for annihilation and creation operators.

The mode functions themselves are linearly related. One writes

$$u^i_{\text{out}\,X} = u^i_{\text{in}\,A}\alpha^A_X + u^{i\,*}_{\text{in}\,A}\beta^A_X, \tag{18.19}$$

or, suppressing indices,

$$u_{\text{out}} = u_{\text{in}}\alpha + u^*_{\text{in}}\beta. \tag{18.20}$$

The αs and βs are known as *Bogoliubov coefficients*. One can show (see [1]) that the transition amplitude between the initial and final vacuum states (also known as the *vacuum persistence amplitude*) is given by

$$\langle \text{out, vac} | \text{in, vac} \rangle = (\text{sdet}\,\alpha)^{-1/2}. \tag{18.21}$$

Using the well-known relations that the Bogoliubov coefficients satisfy (which are simple consequences of unitarity), one can also show that

$$\text{sdet}\,\alpha = \text{constant} \times \frac{\text{sdet}\,G^+[\phi_{\text{ex}}]}{\text{sdet}\,G[\phi_{\text{ex}}]}, \tag{18.22}$$

where $G[\phi_{\text{ex}}]$ is the *Feynman propagator* for the field ϕ^i in the background ϕ_{ex}. Equations (18.21) and (18.22) together imply

$$\langle \text{out, vac} | \text{in, vac} \rangle = \text{constant} \times \left(\frac{\text{sdet}\,G[\phi_{\text{ex}}]}{\text{sdet}\,G^+[\phi_{\text{ex}}]} \right)^{1/2}, \tag{18.23}$$

which bears a striking, and not at all accidental, resemblance to equation (18.11). Once again the measure (in this case expression (18.10)) is seen to play a fundamental role.

Anomalies

Suppose that the background ϕ_{ex} is a gauge field (e.g. Yang–Mills or gravitational) and that the action functional for the linear field ϕ^i is gauge-invariant. That is, suppose that, whenever the background suffers a gauge transformation, ϕ^i can be made to suffer a corresponding transformation so that the action remains invariant. Although the quantum theory of the field ϕ^i is fairly trivial, divergences generally occur in expressions (18.21) and (18.23) when the background itself is nontrivial. These divergences produce corresponding divergences in the vacuum currents (Yang–Mills current or stress-energy density) that the background induces. These divergences must be renormalized by adding counterterms to the action functional. However, the counterterms do not always remain invariant under gauge transformations of ϕ_{ex}, with the result that the standard conservation laws for the induced currents fail. The failure, which is always by a finite amount, is known as an *anomaly*. If anomalies are present, it is usually impossible to replace ϕ_{ex} by a dynamical quantized gauge field to which the ϕ^i can be coupled.

Instead of throwing the blame for anomalies onto the counterterms one sometimes tries to blame the volume element in the functional integral. (See [2].) In this case one mistakenly assumes the volume element to be given by expression (18.2), which indeed fails to be gauge-invariant. The correct volume element, however, is $\mu[\phi_{\text{ex}}][d\phi]$, which is fully gauge-invariant. The jacobian by which $[d\phi]$ is multiplied under gauge transformation is canceled out by a corresponding jacobian coming from $\mu[\phi_{\text{ex}}]$. One sees once again that the measure is an essential component of the formalism.

Wick rotation

In the case of general nonlinear quantum field theories expression (18.3) gives an approximation to the measure that is valid only up to one-loop order. Nonetheless, this approximation already suffices to reveal another remarkable property of the measure.

Let the background spacetime be taken as flat, so that we can work in momentum space. The generic one-loop graph then has a certain number, r, of external lines, which may represent either propagators for the nonlinear field ϕ^i itself or the effects (up to some power-series order) of inhomogeneities or non-flatness in the background. Delete the external lines, so that only vertex functions depending (possibly) on a number of incoming external momenta p_i satisfying $\sum_i p_i = 0$ are left. The value of the resulting graph has the generic form

$$I(C) := \text{constant} \times \int d^{r-1}y \int_C d^n k \, \frac{P(y, k, p, m)}{[k^2 + Q(y, p, m)]^r}, \qquad (18.24)$$

where $\int d^{r-1}y$ denotes schematically the integral over the auxiliary parameters y that are used to combine denominators, P and Q are polynomials in their indicated arguments (including particle masses denoted schematically by m), and C denotes a contour in the complex plane of the time component k^0 of the k-variable. The integral (18.24) will be typically divergent and is most easily dealt with by dimensional regularization.[7]

---

[7] Use of dimensional regularization generally requires the insertion, as an additional factor, of some power of an auxiliary mass μ (see Section 15.4).

18.3 Loop expansions

When only the gaussian contribution to the functional integral is kept, the contour C is that appropriate to the Feynman propagator (see e.g. [3]) and runs from $-\infty$ to 0 below the negative real axis (in the complex k^0-plane) and from 0 to ∞ above the positive real axis. If the integral (18.24) were convergent, the contour could be rotated so that it would run along the imaginary axis. One would set $k^0 = ik^n$, and (18.24) would become an integral over euclidean momentum-n-space. Generically, however, this rotation, which is known as *Wick rotation*, is not legitimate. Contributions from arcs at infinity, which themselves diverge or are nonvanishing, have to be included. *These contributions cannot be handled by dimensional regularization.*

When the measure is included it contributes to the generic one-loop graph an amount equal to the negative of the integral

$$I(C^+) := \text{constant} \times \int d^{r-1}y \int_{C^+} d^n k \, \frac{P(y,k,p,m)}{[k^2 + Q(y,p,m)]^r}, \qquad (18.25)$$

where C^+ is the contour (in the complex k^0-plane) appropriate to the advanced Green function; it runs from $-\infty$ to ∞ *below* the real axis. The gaussian and measure contributions, taken together, yield $I(C) - I(C^+)$ as the *correct* value of the graph. This corresponds to taking a contour that runs from ∞ to 0 below the positive real axis and then back to ∞ again above the positive real axis, and yields an integral that *can* be handled by dimensional regularization.

The remarkable fact is that $I(C) - I(C^+)$ is equal precisely to the value that is obtained by Wick rotation. This means that *the measure justifies the Wick-rotation procedure*. Although it has never been proved, one may speculate that the exact measure functional, whatever it is, will justify the Wick rotation to all orders and will establish a rigorous connection between quantum field theory in Minkowski spacetime and its corresponding euclideanized version.[8]

There is not enough space here to describe how the measure enters the study of non-abelian gauge fields, in which the functional integral for the quantum theory involves not only propagators for physical modes but also propagators for so-called *ghosts*. Suffice it to say that an approximate expression for the measure, valid to one-loop order, is again easily derived. It now involves not only the advanced Green function for the physical modes but also the advanced Green function for the ghosts, and in such a way that Wick rotation becomes justified for graphs containing physical loops as well as for graphs containing ghost loops.

---

[8] The present lack of rigor in establishing this connection stems entirely from the infinities of the theory.

Another satisfying feature of this measure is that it proves to be gauge-invariant, satisfying

$$\mathcal{L}_{\mathbf{Q}_\alpha}\mu = 0, \tag{18.26}$$

where the vector fields $\mathbf{Q}_\alpha$ (on Φ) are the generators of infinitesimal gauge transformations

$$\delta\phi^i = Q^i_\alpha\, \delta\xi^\alpha, \tag{18.27}$$
$$S_{,i}Q^i_\alpha \equiv 0. \tag{18.28}$$

References

[1] B. DeWitt (2003). *The Global Approach to Quantum Field Theory* (Oxford, Oxford University Press; with corrections, 2004), Chapter 10.
B. DeWitt (1992). *Supermanifolds*, 2nd edn. (Cambridge, Cambridge University Press), p. 387.
[2] K. Fujikawa (1979). "Path integral measure for gauge-invariant theories," *Phys. Rev. Lett.* **42**, 1195–1198.
K. Fujikawa (1980). "Path integral for gauge theories with fermions," *Phys. Rev.* **D21**, 2848–2858; erratum *Phys. Rev.* **D22**, 1499 (1980).
K. Fujikawa and H. Suzuki (2004). *Path Integrals and Quantum Anomalies* (Oxford, Oxford University Press).
[3] B.S. DeWitt (1965). *Dynamical Theory of Groups and Fields* (New York, Gordon and Breach), p. 31. First published in *Relativity, Groups and Topology*, eds. C. DeWitt and B.S. DeWitt (New York, Gordon and Breach, 1964), p. 617.

Part VI
Projects

19
Projects

The projects are grouped by topics in the same order as the book chapters. Some projects belong to several chapters. They are listed under the most relevant chapter and cross-referenced in the other chapters. Some projects do not cover new ground; but new approaches to the subject matter may shed light on issues whose familiarity has become a substitute for understanding.[1]

19.1 Gaussian integrals
(Chapter 2)

Project 19.1.1. Paths on group manifolds

Let $\mathbb{X}$ be the space of paths $x : \mathbb{T} \to \mathbb{M}^D$, and $\mathbb{X}'$ be its dual. In many cases volume elements on $\mathbb{X}$ are gaussians $d\Gamma(x)$ defined by their Fourier transforms on $\mathbb{X}'$. During the last forty years interesting functional integrals have been developed and computed when $\mathbb{M}^D$ is a group manifold. It has been proved that the WKB approximation for the propagator of free particles on a group manifold is exact.

On the other hand, the theory of Fourier transforms on commutative locally compact groups is well developed. It provides a natural framework for extending Chapter 2 to configuration spaces that are commutative locally compact groups. It is expected that old results will be simplified and new results will be obtained.

---

[1] As Feynman once said: "Every theoretical physicist who is any good knows six or seven [!] different theoretical representations for exactly the same physics" (*The Character of Physical Law*, Cambridge, MA, MIT Press, 1965, p. 168).

References

[1] J.S. Dowker (1971). "Quantum mechanics on group space and Huygens' principle," *Ann. Phys.* **62**, 361–382.
[2] W. Rudin (1962). *Fourier Analysis on Groups* (New York, Interscience).
[3] S.G. Low (1985). Path integration on spacetimes with symmetry, unpublished Ph.D. Dissertation, University of Texas at Austin.

Project 19.1.2. Delta-functional[2] and infinite-dimensional distributions

In Section 1.7, we hinted at *prodistributions*. One should develop a general theory along the following lines, suggested by the work of Albeverio and Høegh-Krohn [1]. We consider a Banach space $\mathbb{X}$ with dual $\mathbb{X}'$. We assume that both $\mathbb{X}$ and $\mathbb{X}'$ are separable. A space[3] of test functions $\mathcal{F}(\mathbb{X})$ on $\mathbb{X}$ should consist of the Fourier–Stieltjes transforms

$$f(x) = \int_{\mathbb{X}'} d\lambda(x') e^{2\pi i \langle x', x \rangle}, \tag{19.1}$$

where λ is a complex bounded measure on $\mathbb{X}'$ such that $\int_{\mathbb{X}'} d\lambda(x') ||x'||^N$ is finite for every integer $N = 0, 1, 2, \ldots$ A *distribution* T on $\mathbb{X}$ is defined by its Fourier transform $\Phi(x')$, a continuous function on $\mathbb{X}'$ with an estimate

$$|\Phi(x')| \leq C(1 + ||x'||)^N \tag{19.2}$$

for suitable constants $C > 0$ and $N \geq 0$. The distribution T is a linear form on $\mathcal{F}(\mathbb{X})$ of the form

$$\langle T, f \rangle = \int_{\mathbb{X}} d\lambda \cdot \Phi \tag{19.3}$$

for f given by (19.1). We also write $\int_{\mathbb{X}} T(x) f(x)$ for $\langle T, f \rangle$. In particular, for x' in $\mathbb{X}'$, the function $e(x) = e^{2\pi i \langle x', x \rangle}$ belongs to $\mathcal{F}(\mathbb{X})$ (take λ as a pointlike measure located at x') and

$$\Phi(x') = \int_{\mathbb{X}} T(x) e^{2\pi i \langle x', x \rangle}, \tag{19.4}$$

that is Φ is, as expected, the Fourier transform of T.

One can define various operations on such distributions, e.g. linear combinations with constant coefficients, translation, derivation, and convolution. The important point is that we can define the direct image $S = L_*(T)$ of a distribution T on $\mathbb{X}$ by a continuous linear map $L: \mathbb{X} \to \mathbb{Y}$;

---

[2] Based on notes by John LaChapelle.
[3] The letter $\mathcal{F}$ is a reminder of Feynman.

19.1 Gaussian integrals

it satisfies the integration formula

$$\int_Y S(y)g(y) = \int_X T(x)(g \circ L)(x) \qquad (19.5)$$

for any test function g in $\mathcal{F}(\mathbb{Y})$. Also the Fourier transform of $\mathcal{S}$ is the function $\Phi \circ \tilde{L}$ on $\mathbb{X}'$ where $\tilde{L} : \mathbb{Y}' \to \mathbb{X}'$ is the transposed map of $L : \mathbb{X} \to \mathbb{Y}$ defined by equation (2.58).

We list now some interesting particular cases.

(i) When $\mathbb{X}$ is finite-dimensional, a distribution in the sense given here is a tempered distribution in $\mathbb{X}$ (in Schwartz' definition) whose Fourier transform is a continuous function.

(ii) If $W(x')$ is any continuous quadratic form on $\mathbb{X}'$, with complex values, such that

$$\operatorname{Re} W(x') \geq 0 \qquad \text{for } x' \text{ in } \mathbb{X}', \qquad (19.6)$$

then the function $e^{-W(x')}$ on $\mathbb{X}'$ satisfies the estimate (19.2) for $N = 0$, and hence it is the Fourier transform of a distribution. This construction generalizes slightly our class of gaussian distributions.

(iii) For any a in $\mathbb{X}$, there exists a Dirac-like distribution δ_a on $\mathbb{X}$ such that $\langle \delta_a, f \rangle = f(a)$ for any test function f in $\mathcal{F}(\mathbb{X})$. Its Fourier transform is the function $x' \mapsto e^{2\pi i \langle x', a \rangle}$ on $\mathbb{X}'$. In particular $\delta := \delta_0$ is given by $\langle \delta, f \rangle = f(0)$ and the Fourier transform of δ is 1, but the situation is not symmetrical between $\mathbb{X}$ and $\mathbb{X}'$: there is no distribution like 1 on $\mathbb{X}$, with Fourier transform δ on $\mathbb{X}'$!

If W is an invertible positive real quadratic form on $\mathbb{X}'$, with inverse Q on $\mathbb{X}$, and s is a real positive parameter, we know the formula

$$\int \mathcal{D}_{s,Q} x \cdot \exp\left(-\frac{\pi}{s} Q(x)\right) \cdot \exp(-2\pi i \langle x', x \rangle) = \exp(-\pi s W(x')). \qquad (19.7)$$

On going to the limit $s = 0$, the right-hand side tends to 1, hence

$$\lim_{s=0} \mathcal{D}_{s,Q} x \cdot \exp\left(-\frac{\pi}{s} Q(x)\right) = \delta(x), \qquad (19.8)$$

which is a well-known construction of the Dirac "function" in the case $\mathbb{X} = \mathbb{R}$, $Q(x) = x^2$. Such formulas were developed by John LaChapelle in [2]. Notice also that, for $x'_1, \ldots, x'_n$ in $\mathbb{X}'$, the linear map

$$L : x \mapsto (\langle x'_1, x \rangle, \ldots, \langle x'_n, x \rangle)$$

from $\mathbb{X}$ to $\mathbb{R}^n$ enables one to define the *marginal* $T_{x'_1,\ldots,x'_n}$ of T as the direct image $L_* T$. Hence our distributions are prodistributions in the sense of Section 1.7.

One should search the literature for mathematical studies of distributions on test functionals (as opposed to test functions) and write up a summary of the results for the benefit of a wider readership (a user manual).

References

[1] S. A. Albeverio and R. J. Høegh-Krohn (1976). *Mathematical Theory of Feynman Path Integrals* (Berlin, Springer).

[2] J. LaChapelle (1997). "Path integral solution of the Dirichlet problem," *Ann. Phys.* **254**, 397–418.
J. LaChapelle (2004)."Path integral solution of linear second order partial differential equations I. The general construction," *Ann. Phys.* **314**, 262–295.

[3] P. Krée (1976). "Introduction aux théories des distributions en dimension infinie," *Bull. Soc. Math. France* **46**, 143–162, and references therein, in particular, *Seminar P. Lelong*, Springer Lecture Notes in Mathematics, Vols. **410** (1972) and **474** (1974).

19.2 Semiclassical expansions (Chapters 4 and 5)

Project 19.2.1. Dynamical Tunneling
(contributed by Mark Raizen)

A transition forbidden in classical physics but possible in quantum physics is called a tunneling transition. Tunnelings abound in physics, chemistry, and biology.

In Section 5.5, a simple path-integral technique is used for computing a simple tunneling effect, namely the propagation of a free particle whose classical motion is blocked by a knife edge. In principle one can use more powerful functional-integral techniques for computing more complex tunneling such as the dynamical tunneling observed by M. Raizen and his graduate students, Daniel Steck and Windell Oskay [1]. Tunneling is called "dynamical" when the classical transport is forbidden because of the system's dynamics rather than because of the presence of a potential barrier. The observed phenomenon is at the interface of classical and quantum physics: indeed, dynamical tunneling is influenced by the chaotic trajectories surrounding the islands of stability in the phase-space diagram of a classical two-dimensional system.

The observed quantum transitions between islands of stability are Poisson distributed. Path-integral techniques have been developed for studying Poisson processes (see Chapters 12 and 13).

References

[1] D. A. Steck, W. H. Oskay, and M. G. Raizen (2001). "Observation of chaos-assisted tunneling between islands of stability," *Science* **293**, 274–278.

[2] For a general introduction, see for instance "A chilly step into the quantum world," a feature article of the Winter 2002 edition of *Focus on Science*, published by UT College of Natural Sciences.

Project 19.2.2. Semiclassical expansion of a $\lambda\phi^4$ system

The background method (a functional Taylor expansion) is particularly useful when the action functional is expanded around a classical solution (semiclassical expansion) that is known explicitly. Indeed, the explicit solutions of the anharmonic oscillator have made it possible for Mizrahi to compute explicitly, to arbitrary order in $\hbar$, the quantum transitions and to dispel the "folkloric singularity at $\lambda = 0$." The solutions of the dynamical equation for the classical $\lambda\phi^4$ system, namely

$$(\Box + m^2)\phi + \lambda\phi^3/3! = 0,$$

are known explicitly, but have not been exploited. In Chapter 4, we examine the extent to which the anharmonic oscillator is a prototype for the $\lambda\phi^4$ model. A brief remark is made in Chapter 16 suggesting for the $\lambda\phi^4$ system the use of the hessian of the action functional which has been so beneficial for the anharmonic oscillator. It should be possible to adapt Brydges' scaling method to gaussians defined by the hessian of an action functional.

References

[1] G. Petiau (1960). "Les généralisations non linéaires des équations d'ondes de la mécanique ondulatoire," *Cahiers de Phys.* **14**, 5–24.
[2] D. F. Kurdgelaidze (1961). "Sur la solution des équations non linéaires de la théorie des champs physiques," *Cahiers de Phys.* **15**, 149–157.

19.3 Homotopy
(Chapter 8)

Project 19.3.1. Poisson-distributed impurities in $\mathbb{R}^2$
(contributed by Stéphane Ouvry)

In Section 8.5 a single Aharonov–Bohm (A–B) flux line piercing the plane at a given point was considered. A more complex system consists of several A–B lines piercing the plane at different points. A simplification arises when the locations of the punctures are random. Consider for example the random-magnetic-impurity model introduced in relation to the integer quantum Hall effect. It consists of an electron coupled to a poissonian distribution of A–B lines (the so-called magnetic impurities) having a mean density ρ and carrying a fraction α of the electron flux quantum

Φ_o. Periodicity and symmetry considerations allow us to take $\alpha \in [0, 1/2]$. It has been shown via path-integral random-walk simulations that, when $\alpha \to 0$, the average density of states of the electron narrows down to the Landau density of states for the average magnetic field $\langle B \rangle = \rho \alpha \Phi_o$ with broadened Landau levels (weak disorder). On the contrary, when $\alpha \to 1/2$, the density of states has no Landau level oscillations and rather exhibits a Lifschitz tail at the bottom of the spectrum (strong disorder).

In the path-integral formulation, one rewrites the average partition function as an average over C, the set of closed brownian curves of a given length t (the inverse temperature),

$$\langle Z \rangle = Z_o \left\langle \exp\left(\rho \sum_n S_n (e^{2\pi i \alpha n} - 1) \right) \right\rangle_{\{C\}},$$

where S_n is the arithmetic area of the n-winding sector of a given path in $\{C\}$ and Z_o is the free partition function. It amounts to saying that random Poisson A–B lines couple to the S_n, which is a different (intermediate) situation from that of the single A–B line, which couples to the angle spanned by the path around it, and from the homogeneous magnetic field, which couples to the algebraic area enclosed by the path. One rewrites the average partition function as

$$\langle Z \rangle = Z_o \int e^{-\rho t(S - iA)} P(S, A) \mathrm{d}S \mathrm{d}A,$$

where

$$S = \frac{2}{t} \sum_n S_n \sin^2(\pi \alpha n) \quad \text{and} \quad A = \frac{1}{t} \sum_n S_n \sin(2\pi \alpha n)$$

are random brownian loop variables and $P(S, A)$ is their joint probability density distribution. Since S_n scales like t – in fact $\langle S_n \rangle = t/(2\pi n^2)$, and, for n sufficiently large, $n^2 S_n \to \langle n^2 S_n \rangle = t/(2\pi)$ – the variables S and A are indeed t-independent.

Clearly, more precise analytic knowledge of the joint distribution function $P(S, A)$ is needed. In fact, very little is known:

(i) when $\alpha \to 0$,

$$\langle Z \rangle \to Z_o \left\langle \exp\left(\frac{2\pi i \langle B \rangle}{\Phi_0} \cdot \sum_n n S_n \right) \right\rangle_{\{C\}},$$

which is, as it should be, nothing other than the partition function for the homogeneous mean magnetic field $\langle B \rangle$, since $\sum n S_n$ is indeed the algebraic area enclosed by the path;

(ii) when $\alpha \to 1/2$, $\langle Z \rangle \to Z_o \langle \exp(-2\rho \sum_{n\,\text{odd}} S_n) \rangle_{\{C\}}$, implying that, for the average density of states (in terms of $\rho_o(E)$ the free density of states),

$$\langle \rho(E) \rangle = \rho_o(E) \int_0^{E/\rho} P(S') \mathrm{d}S',$$

where $P(S')$ is the probability density for the random variable $S' = \sum_n S_n$, n odd.

In this context, one should stress that important progress has been made in the determination of critical exponents related to planar random walk based on a family of conformally invariant stochastic processes, the stochastic Loewner evolution. It would be interesting to find an interpretation of the random variable S', and, in general, of the random variables S and A, in terms of conformally invariant stochastic processes.

References

[1] J. Desbois, C. Furtlehner, and S. Ouvry (1995). "Random magnetic impurities and the Landau problem," *Nucl. Phys.* B **453**, 759–776.
[2] A. Comtet, J. Desbois, and S. Ouvry (1990). "Winding of planar Brownian curves," *J. Phys.* A **23**, 3563–3572.
[3] W. Werner (1993). Sur l'ensemble des points autour desquels le mouvement brownien plan tourne beaucoup, unpublished Thèse, Université de Paris VII.
W. Werner (1994). "Sur les points autour desquels le mouvement brownien plan tourne beaucoup" [On the points around which plane Brownian motion winds many times], *Probability Theory and Related Fields* **99**, 111–144.
[4] W. Werner (2003). "SLEs as boundaries of clusters of Brownian loops," arXiv:math.PR/0308164.

19.4 Grassmann analysis
(Chapters 9 and 10)

Project 19.4.1. Berezin functional integrals.
Roepstorff's formulation

G. Roepstorff has proposed a formalism for Berezin functional integrals and applied it to euclidean Dirac fields. *It would be interesting to follow Roepstorff's approach and construct Berezin functional integrals for Dirac fields defined on (curved) spacetimes.* The essence of Roepstorff's formalism is to introduce evaluating functions or distributions η defined below. Let $\mathbb{E}$ be a complex vector space. In the construction of ordinary Berezin integrals, $\mathbb{E}$ is finite (or countably infinite) dimensional. In the construction of Berczin functional integrals, $\mathbb{E}$ is a space of test functions, and functionals on $\mathbb{E}$ are distributions. We summarize the finite-dimensional case as a point of departure for the infinite-dimensional case.

$\mathbb{E}$ is D-dimensional

An element $u \in \mathbb{E}$ in the basis $\{e_{(i)}\}$ is

$$u = \sum_{i=1}^{D} e_{(i)} u^i, \qquad u \in \mathbb{E}.$$

Let $A^p(\mathbb{E})$ be the space of p-linear antisymmetric functions on $\mathbb{E}$; let $\eta^{(i)} \in A^1(\mathbb{E})$ be the family of evaluating functions on $\mathbb{E}$

$$\eta^{(i)}(u) := u^i.$$

The evaluating functions generate a Grassmann algebra:

$$\eta^{(i)}\eta^{(k)} + \eta^{(k)}\eta^{(i)} = 0.$$

Each vector S in $A^p(\mathbb{E})$ can be represented by

$$S = \frac{1}{p!} \sum_{i_1 \ldots i_p} s_{i_1 \ldots i_p} \eta^{(i_1)} \ldots \eta^{(i_p)},$$

where $s_{i_1 \ldots i_p}$ are complex coefficients that are antisymmetric with respect to the indices. Basic operations on $S(u_{(1)}, \ldots, u_{(p)})$ can be formulated in terms of operations on the generators $\{\eta^{(i)}\}$. For instance, the Berezin integral follows from the fundamental rule $ID = 0$ (see Section 9.3), which here reads

$$\int \delta\eta \, \frac{\partial}{\partial \eta^{(i)}} S = 0.$$

$\mathbb{E}$ is a space of test functions f

Roepstorff's test functions map $\mathbb{R}^4$ into $\mathbb{C}^8$ because the distributions on $\mathbb{E}$ are euclidean Dirac bispinors. Bispinors serve two purposes: they provide

- a convenient formulation of charge conjugation; and
- a simple quadratic bispinor form in the action functional.

The space $A^p(\mathbb{E})$ is, by definition, the space of antisymmetric distributions defined on $\mathbb{E} \times \ldots \times \mathbb{E}$ (p factors). The evaluating distributions $\eta \in A^1(\mathbb{E})$ are defined by

$$\eta^a(x) f := f^a(x), \qquad f \in \mathbb{E}, \; \eta^a(x) \in A^1(\mathbb{E}).$$

They generate a Grassmann algebra of antisymmetric distributions:
$$\eta^a(x)\eta^b(y) + \eta^b(y)\eta^a(x) = 0,$$
$$\eta^a(x)\eta^b(y)(f,g) = f^a(x)g^b(y) - g^a(x)f^b(y).$$

Example. $S \in A^2(\mathbb{E})$. Let $S(x,y)$ be the two-point function of a Dirac field. It defines the following distribution **S**:
$$\mathbf{S} = \frac{1}{2}\int \mathrm{d}x \int \mathrm{d}y \, S(x,y)_{ab}\eta^a(x)\eta^b(y).$$

Therefore,
$$\langle f, g, \mathbf{S}\rangle = \mathbf{S}(f,g) = \int \mathrm{d}x \int \mathrm{d}y \, S(x,y)_{ab} f^a(x) g^b(y).$$

Reference

[1] G. Roepstorff (1994). *Path Integral Approach to Quantum Physics: An Introduction* (Berlin, Springer), pp. 329–330.

Project 19.4.2. Volume elements. Divergences. Gradients in superanalysis
(based on Chapters 9–11)

Volume elements in Grassmann analysis cannot be introduced via Berezin integrals because a Berezin integral is a derivation. On the other hand, the Koszul formula $\mathcal{L}_X \omega = \mathrm{Div}(X) \cdot \omega$ (11.1) applies to Grassmann volume elements [1]. This formula says that the divergence $\mathrm{Div}(X)$ measures the change $\mathcal{L}_X\omega$ of the volume element ω under the group of transformations generated by the vector field X.

Divergences are related to gradients by integration by parts. This relationship should apply to Grassmann divergences.

The triptych volume elements–divergences–gradients in finite and infinite dimensions is well established for bosonic variables; once established for fermionic variables it would provide a framework for superanalysis and supersymmetric analysis. In particular *the berezinian (determinant of matrices with odd and even entries) should follow from this framework.*

References

[1] T. Voronov (1992). "Geometric integration theory on supermanifolds," *Sov. Sci. Rev. C Math. Phys.* **9**, 1–138.
[2] Y. Choquet-Bruhat (1989). *Graded Bundles and Supermanifolds* (Naples, Bibliopolis).

[3] Y. Choquet-Bruhat and C. DeWitt-Morette (2000). *Analysis, Manifolds, and Physics Part II* (Amsterdam, North Holland), pp. 57–64.

19.5 Volume elements, divergences, gradients (Chapter 11)

Project 19.5.1. *Top forms as a subset of differential forms*

Differential forms form the subset of tensor fields that are totally antisymmetric covariant tensor fields. It is meaningful to identify this subset because it has its own properties that are not shared by arbitrary covariant tensor fields. Similarly D-forms on a D-dimensional manifold have their own properties that are not shared by other differential forms. They form a subset of choice for integration theory. An article on "top forms" would be useful and interesting.

Project 19.5.2. *Non-gaussian volume elements*
(contributed by John LaChapelle)

Although the gaussian volume element is prevalent, the general scheme of Chapter 11 allows the definition of other interesting volume elements. Three examples are the Dirac, Hermite, and gamma volume elements.

The Dirac volume element is the delta-functional presented in Project 19.1.2. It is the functional analog of an infinitely narrow gaussian: as such, it acts in functional integrals like the Dirac delta function in finite-dimensional integrals. The Dirac volume element has potential applications in quantum field theory, in particular (i) the Faddeev–Popov method for gauge field theories and (ii) equivariant localization in cohomological field theory.

The Hermite volume element is a generalization of the gaussian volume element – it includes the gaussian as a special case. As its name implies, it is defined in terms of weighted functional Hermite polynomials. It may offer a simplifying alternative to the gaussian volume element in quantum field theory, because it facilitates integration of Wick ordered monomials.

The gamma volume element is a scale-invariant volume element based on the gamma probability distribution. Avenues to be explored include (i) the functional-integral analog to the link between incomplete gamma integrals and sums of Poisson distributions

$$\frac{1}{\Gamma(\alpha)} \int_\lambda^\infty \mathrm{d}x\, x^{\alpha-1} \mathrm{e}^{-x} = \sum_{x=0}^{\alpha-1} \frac{\lambda^x \mathrm{e}^{-\lambda}}{x!} \tag{19.9}$$

for integer α and (ii) the functional analog of the central limit theorem which states that the average value of a random variable with a

19.5 Volume elements, divergences, gradients

gamma (or any other) probability distribution is asymptotically gaussian distributed.

The construction of the functional analog to (19.9) may serve as a pattern for the functional analog of another link, namely the link of the beta probability distribution to sums of binomial distributions:

$$\frac{\Gamma(\alpha+\beta)}{\Gamma(\alpha)\Gamma(\beta)} \int_0^p x^{\alpha-1}(1-x)^{\beta-1} \mathrm{d}x = \sum_{x=\alpha}^n \binom{n}{x} p^x (1-p)^{n-x} \qquad (19.10)$$

for $0 < p < 1$ and $n = \alpha + \beta - 1$.

References

[1] For definitions of the non-gaussian volume elements see Appendix B of J. LaChapelle (2004). "Path integral solution of linear second order partial differential equations I. The general construction," *Ann. Phys.* **314**, 362–395.
[2] A good review article on equivariant localization is R. J. Szabo (1996). "Equivariant localization of path integrals," arXiv:hep-th/9608068.

Project 19.5.3. The Schrödinger equation in a riemannian manifold

The title could equally well be "Does the Schrödinger equation include the Riemann scalar curvature?" Various answers to this question have been given since Pauli's remark in his 1951 lecture notes. Ultimately the answer will be given by observations because the Schrödinger equation is more phenomenological than fundamental, but the corresponding physical effects are very, very tiny. For the time being one can only say which formalism yields which Schrödinger equation.

In Chapter 7 we constructed a path integral on the frame bundle. The Schrödinger equation projected on the base space does not include the scalar curvature.

In Chapter 11 we characterized the riemannian volume element ω_g by the Koszul formula $\mathcal{L}_X \omega_g = \mathrm{Div}_g X \cdot \omega_g$. Lifting the Koszul formula on the frame bundle may provide a volume element different from the one used in Chapter 7, and hence a different Schrödinger equation.

In Chapter 18, Bryce DeWitt summarizes the properties of the "measure" $\mu(\phi)$ in functional integrals that represent the operator formalism of quantum physics based on the Peierls bracket. It follows that the path integral for a particle in a riemannian manifold satisfies a Schrödinger equation whose hamiltonian includes $\frac{1}{8}$ times the Riemann scalar curvature.

References

[1] For the history of the scalar curvature controversy, see C. DeWitt-Morette, K. D. Elworthy, B. L. Nelson, and G. S. Sammelman (1980). "A stochastic scheme for constructing solutions of the Schrödinger equations," *Ann. Inst. H. Poincaré* **A XXXII**, 327–341.

[2] H. Kleinert (2004). *Path Integrals in Quantum Mechanics, Statistics, and Polymer Physics*, 3rd edn. (Singapore, World Scientific).

[3] For the presence of a $\frac{1}{8}R$ term in the Schrödinger equation, see B. DeWitt (2003). *The Global Approach to Quantum Field Theory* (Oxford, Oxford University Press; with corrections, 2004), Chapters 15 and 16.

Project 19.5.4. The Koszul formula in gauge theories

The Koszul formula (11.1)

$$\mathcal{L}_X \omega = \mathrm{Div}_\omega(X) \cdot \omega$$

defines the divergence of a vector field X as the change of the volume element ω under the group of transformations generated by X; that is, as the Lie derivative $\mathcal{L}_X \omega$ of ω. This formula is well known when ω is a volume element ω_g in a riemannian manifold $(\mathbb{M}^{2N}, g)$. It is easy to derive it there because the volume element is a top form and $\mathcal{L}_X = \mathrm{d} i_X$. We have shown that the Koszul formula is valid for the volume element ω_Ω of a symplectic manifold $(\mathbb{M}^{2N}, \Omega)$. We have assumed that it remains valid in infinite-dimensional manifolds. One should be able to use the Koszul formula in gauge theories.

Let Φ be the space of field histories with gauge group of transformations $\mathcal{G}$. The space Φ is a principal fiber bundle having $\mathcal{G}$ as a typical fiber. Physics takes place in the phase space $\Phi/\mathcal{G}$. The typical fiber $\mathcal{G}$ is an infinite-dimensional group manifold. It admits an invariant (pseudo)riemannian metric. This metric can be extended in an infinity of ways to a group-invariant metric on Φ. However, if one imposes an ultra-locality requirement [1, p. 456] on the metric, then up to a scale factor it is unique in the case of the Yang–Mills field. Let γ be the preferred metric on Φ, it projects down to a metric g on $\Phi/\mathcal{G}$ [1, p. 459].

An explicit Koszul formula for Φ with metric γ and $\Phi/\mathcal{G}$ with metric g would be interesting.

Reference

[1] B. DeWitt (2003). *The Global Approach to Quantum Field Theory* (Oxford, Oxford University Press; with corrections, 2004).

19.6 Poisson processes
(Chapters 12 and 13)

Project 19.6.1. Feynman's checkerboard in 1 + 3 dimensions
(contributed by T. Jacobson)

In Chapter 12 we mention briefly the "Feynman's checkerboard" problem (figure 12.1): construction of the propagator for the Dirac equation in $1+1$ dimensions via a sum-over-paths method whereby the paths traverse a null lattice. The amplitude for each path is $(im\,\Delta t)^R$, where m is the particle mass, Δt is the time-step, and R is the number of reversals of direction. We give also references to works in $1+3$ dimensions inspired by Feynman's checkerboard.

Progress has been made recently in constructing a path integral for a massless two-component Weyl spinor in $1+3$ dimensions, preserving the simple and intriguing features of the Feynman's checkerboard path integral [1]. The path integral in [1] involves paths on a hypercubic lattice with null faces. The amplitude for a path with N steps and B bends is $\pm(1/2)^N(i/\sqrt{3})^B$.

It would be interesting to see whether the methods and results developed in Chapters 12 and 13 could be profitably applied to the path integral of [1].

We add a cautionary remark regarding convergence of the path integral. As in any discretization scheme for differential equations, convergence to the continuum limit is by no means guaranteed, but rather must be demonstrated directly. For the path integral of [1] convergence has not yet, at the time of the writing of this book, fully been established.

Reference

[1] B. Z. Foster and T. Jacobson (2003). "Propagating spinors on a tetrahedral spacetime lattice," arXiv:hep-th/0310166.

Project 19.6.2. Poisson processes on group manifolds

Path integrals based on Poisson processes on group manifolds have been defined and applied to several systems in [1]. The mathematical theory of Poisson processes developed in Chapter 13 applies to the issues discussed in [1], and may simplify and generalize the results derived in [1].

Reference

[1] Ph. Combe, R. Høegh-Krohn, R. Rodriguez, M. Sirugue, and M. Sirugue-Collin (1980). "Poisson processes on groups and Feynman path integrals," *Commun. Math. Phys.* **77**, 269–288.

(1982). "Feynman path integral and Poisson processes with piecewise classical paths," *J. Math. Phys.* **23**, 405–411.

19.7 Renormalization
(Chapters 15–17)

Project 19.7.1. *The principle of equivalence of inertial and gravitational masses in quantum field theory*

Einstein's theory of gravitation rests on two premises.

1. The first is an experimental observation: the numerical equality of the inertial mass, defined as the ratio of force over acceleration, and the gravitational mass defined by the intensity of the gravitational force. This fact is called the principle of equivalence.
2. The other premise is that there is no privileged frame of reference – which is a reasonable assumption.

In Section 15.3, it is shown that, for a pendulum immersed in a fluid, the renormalized inertial mass is different from the renormalized gravitational mass; cf. (15.51) and (15.52).

The principle of equivalence concerns pointlike masses. Related questions in classical field theory and in quantum physics are subtle. For instance, does an electron in free fall radiate? It does [1], but arguments based on the principle of equivalence have been used to claim that it does not. This complex issue has been clarified by D. Boulware [2].

Gravity in quantum mechanics and the phenomenon known as gravity-induced quantum interference, beautifully analyzed by J. J. Sakurai, bring forth the fact that at the quantum level gravity is not purely geometric; the mass term does not disappear from the Schrödinger equation. Nevertheless, the gravitational mass and the inertial mass are taken to be equal.

A preliminary investigation of the principle of equivalence in a quantized gravitational field [3] indicates that, to first order in radiative corrections, the principle of equivalence is satisfied. *One should study critically this one-loop-order calculation and reexamine its conclusion.*

Remark. Radiative corrections are computed nowadays with the help of Feynman diagrams in which at each vertex the sum of the 4-momenta vanishes [5, 6]. They used to be computed differently in the early days of quantum field theory [7]. Then the masses were real (on shell) and the sum of the 4-momenta did not vanish. Feynman's technique is a powerful labor-saving device, but some issues, such as unitarity, are simpler in the old-fashioned calculations. It is possible that, for the proposed calculation at one-loop order, the old-fashioned approach is preferable.

References

[1] C. Morette-DeWitt and B. S. DeWitt (1964). "Falling charges," *Physics* **1**, 3–20.
[2] D. G. Boulware (1980). "Radiation from a uniformly accelerated charge," *Ann. Phys.* **124**, 169–188 and references therein.
[3] J. J. Sakurai and San Fu Tuan (eds.) (1994). *Modern Quantum Mechanics, Revised Edition* (Reading, MA, Addison Wesley).
[4] C. DeWitt (1967). "Le principe d'équivalence dans la théorie du champ de gravitation quantifié," in *Fluides et champ gravitationnel en relativité générale* (Paris, CNRS), pp. 205–208.
[5] M. E. Peskin and D. V. Schroeder (1995). *An Introduction to Quantum Field Theory* (Reading, Perseus Books).
[6] N. D. Birrell and P. C. W. Davies (1982). *Quantum Fields in Curved Space* (Cambridge, Cambridge University Press).
[7] W. Heitler (1954). *The Quantum Theory of Radiation*, 3rd edn. (Oxford, Oxford University Press).

Project 19.7.2. Lattice field theory: a noncompact sigma model

Discretization of a functional integral is necessary for its numerical calculation, but discretization cannot be used for defining a functional integral without choosing a short-time propagator. Even with a carefully crafted short-time propagator, one has to analyze the continuum limit of the discretized functional integral. In the first paper in [1], two lattice decompositions of the noncompact sigma model $O(1,2)/(O(2) \times Z_2)$ are developed and computed by means of Monte Carlo simulations. In one of them, the differencing is done in the configuration space; in the other one it is done in the embedding space. They lead to totally different results. The first lattice action was chosen for further study. When the lattice cut-off is removed the theory becomes that of a pair of massless free fields. This continuum limit is not acceptable because it lacks the geometry and symmetries of the original model.

This model is worth further study. It has some features of quantum gravity: it is nonlinear, it is perturbatively nonrenormalizable, and it has a noncompact curved configuration space.

Instead of starting with a discretized model one could begin with a functional integral satisfying the criteria developed in this book, and then project it onto a lattice.

In [2], the path integral for a particle on a circle, a compact manifold, is constructed in two different ways: treating the motion on $\mathbb{R}^2$ with a constraint; and treating the motion on the U(1) group manifold. The resulting propagators are different. The authors use this example to illustrate methods applicable to less simple examples.

In [3], there are some results for path integrals when the paths take their values in noncompact group manifolds.

In [4], the short-time propagator obtained by projecting the functional integral onto a riemannian manifold given in Section 7.3 is compared with other short-time propagators introduced in discretized versions of path integrals.

The authors of [2–4] suggest methods for tackling the far more complex problem of the $O(1,2)/(O(2) \times Z_2)$ sigma model.

References

[1] J. L. de Lyra, B. DeWitt, T. E. Gallivan *et al.* (1992). "The quantized $O(1,2)/O(2) \times Z_2$ sigma model has no continuum limit in four dimensions. I. Theoretical framework," *Phys. Rev.* **D46**, 2527–2537.
J. L. de Lyra, B. DeWitt, T. E. Gallivan *et al.* (1992). "The quantized $O(1,2)/O(2) \times Z_2$ sigma model has no continuum limit in four dimensions. II. Lattice simulation," *Phys. Rev.* **D46**, 2538–2552.
[2] M. S. Marinov and M. V. Terentyev (1979). "Dynamics on the group manifold and path integral," *Fortschritte Phys.* **27**, 511–545.
[3] N. Krausz and M. S. Marinov (1997). "Exact evolution operator on noncompact group manifolds," avXiv:quant-ph/9709050.
[4] C. DeWitt-Morette, K. D. Elworthy, B. L. Nelson, and G. S. Sammelmann (1980). "A stochastic scheme for constructing solutions of the Schrödinger equation," *Ann. Inst. Henri Poincaré* **A XXXII**, 327–341.
[5] See Project 19.1.1, "Paths on group manifolds."

Project 19.7.3. Hopf, Lie, and Wick combinatorics of renormalization
(contributed by M. Berg)

As mentioned in Chapter 17, it is known that, given a graded Hopf algebra of renormalization, the existence of the related Lie algebra is guaranteed. This is the essence of the famous Milnor–Moore theorem (or Cartier–Milnor–Moore–Quillen theorem, see e. g. [1]), a fact that was exploited by Connes and Kreimer already in [2].

It is of interest to study complementary realizations of this Lie algebra, in particular ones that lie as close as possible to what is actually done in quantum field theory.

19.7 Renormalization

The grafting operator of Chapter 17 is simply the combinatorial action of inserting the given graph into another and counting the number of graphs of each type that are generated, and normalizing by an irrelevant overall factor.

Thus, this combinatorial grafting operator acts by "averaging over graphs." This is quite natural from a combinatorial/number-theoretic perspective, and this is how the Ihara bracket appeared in Chapter 17. However, this averaging is not exactly how one usually thinks of insertions in quantum field theory, even though the end result (e.g. the three-loop example considered in Chapter 17) is combinatorially the same.

In quantum field theory, one tries to avoid mixing graphs of different topologies, so one can consider a larger set of different grafting operators, where each operator is more restricted than the one considered in Chapter 17, in that permutations of the legs of the insertion are represented by separate operators.

For a four-point insertion, for example, there would a priori be 24 different grafting operators, instead of one grafting operator generating 24 graphs. In this way, one initially introduces a larger set of grafting operators, and then reduces the number of distinct ones by classifying them by the topologies they create.

In this construction, one could incorporate the natural "weighting" parameter in quantum field theory: the number of Wick contractions (in canonical quantization) that generate the same Feynman graph – essentially what is known as the *symmetry factor* of the Feynman graph. Clearly this factor cannot be reproduced by the multiplicity factors arising from the grafting operator of Chapter 17, since, when acting with a given grafting operator, graphs of different topologies (and hence different symmetry factors) can be generated with the same multiplicity. A grafting operator that more closely mimics quantum-field-theory intuition would recreate symmetry factors automatically.

It would be interesting to make this connection more precise, and compute grafting matrices for this "Wick-grafting" operator. Up to at least three-loop level, this could be done rather mechanically without much knowledge of Hopf algebras. Chapter 17 already lists all relevant graphs for ϕ^4 theory.

In addition, it would be interesting to carry out the program of matrix Lie-algebra renormalization in ϕ^4 theory to four-loop order and higher. For this, one needs to spell out the rule for the bracket structure (multiplication of $C(t)$) at higher loop order, as mentioned in Chapter 17. This requires the student to have a working understanding of the connection between the Lie and Hopf algebras. (A basic reference on the mathematical aspects of Hopf algebras is [3].)

References

[1] F. Chapoton (2000). "Un théorème de Cartier–Milnor–Moore–Quillen pour les bigèbres dendriformes et les algèbres braces," arXiv:math.QA/0005253.

[2] A. Connes and D. Kreimer (2000). "Renormalization in quantum field theory and the Riemann–Hilbert problem I: the Hopf algebra structure of graphs and the main theorem," *Commun. Math. Phys.* **210**, 249–273; hep-th/9912092.

[3] C. Kassel (1995). *Quantum Groups* (New York, Springer).

Appendices

Appendix A
Forward and backward integrals.[1]
Spaces of pointed paths

Classical random processes

Let $X = (X(t))_{t_a \leq t \leq t_b}$ be a random process. Usually, the initial position $X(t_a) = x_a$ is given, and the corresponding probability law over the space of paths $\mathcal{P}_a M^D$ is denoted by $\mathbb{P}_a$. An average over $\mathbb{P}_a$ of a functional $F(X)$ is denoted by $\mathbb{E}_a[F(X)]$. The probability of finding the particle at time t_b in a given domain U in the configuration space M^D (of dimension D) is given as the measure of *all paths beginning at* $a = (x_a, t_a)$ and with final position $X(t_b)$ in U (figure A.1).

For simplicity, assume that there exists a probability density $p(x_b, t_b; x_a, t_a)$ such that

$$\mathbb{P}_a[X(t_b) \in U] = \int_U d^D x_b \cdot p(x_b, t_b; x_a, t_a). \tag{A.1}$$

A general result[2] known as "disintegration of measures" ensures the existence of a measure $\mathbb{P}_{a,b}$ in the space $\mathcal{P}_{a,b}$ of all paths with boundary conditions

$$X(t_a) = x_a, \qquad X(t_b) = x_b \tag{A.2}$$

(where $a = (x_a, t_a)$ and $b = (x_b, t_b)$) such that[3]

$$\mathbb{E}_a[F(X)f(X(t_b))] = \int_{M^D} d^D x_b \cdot \mathbb{E}_{a,b}[F(X)] \cdot f(x_b) \tag{A.3}$$

---

[1] Here a "backward" integral is a function of (x_a, t_b); a "forward" integral is a function of (x_b, t_b).
[2] See for instance [1].
[3] We denote by $\mathbb{E}_{a,b}[F(X)]$ the integral of the functional $F(X)$ with respect to the measure $\mathbb{P}_{a,b}$.

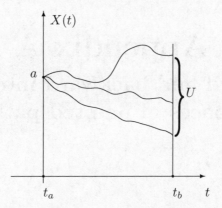

Fig. A.1 A backward integral. The space $\mathcal{P}_a \mathbb{M}^D$ of paths with a fixed initial point.

for any functional $F(X)$ of the process, and any function $f(x_b)$ on the configuration space $\mathbb{M}^D$. In particular, we see, by taking $F = 1$ and[4] $f = \chi_U$, that *the total mass of* $\mathbb{P}_{a,b}$ *is equal to*

$$p(b;a) = p(x_b, t_b; x_a, t_a) \tag{A.4}$$

(not to unity!). Hence, the *transition probability* $p(b;a)$ is a "sum over all paths going from a to b."

It is often the case that the initial position x_a is not fixed, but randomized according to a probability density $\pi(x_a, t_a)$. Then the probability density for the position at time t_b is given by the integral

$$\pi(x_b, t_b) = \int_{\mathbb{M}^D} \mathrm{d}^D x_a \cdot p(x_b, t_b; x_a, t_a) \pi(x_a, t_a). \tag{A.5}$$

This formula can be reinterpreted as a functional integral of $\pi(X(t_a), t_a)$ over the space *of all paths ending at b*. It also expresses how the probability density $\pi(x, t)$ moves in time.

Quantum mechanics

This situation is more symmetric in time than is the situation in classical probability theory. Here the main object is the *transition amplitude* $\langle b|a\rangle = \langle x_b t_b | x_a t_a \rangle$ represented as a functional integral,

$$\langle b|a\rangle = \int \mathcal{D}q \cdot \exp\left(\frac{\mathrm{i}}{\hbar} S(q)\right), \tag{A.6}$$

---

[4] χ_U is the characteristic function of U, that is

$$\chi_U(x_b) = \begin{cases} 1 & \text{for } x_b \text{ in } U, \\ 0 & \text{otherwise.} \end{cases}$$

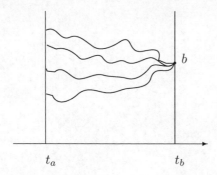

Fig. A.2 A forward integral. The space $\mathcal{P}_b M^D$ of paths with a fixed final point.

extended over all paths beginning at a and ending at b. If the wave function at time t_a, namely $\psi_a(x_a)$, is known then the wave function at time t_b dictated by the quantum dynamics is given by

$$\psi_b(x_b) = \int_{M^D} \mathrm{d}^D x_a \langle x_b\, t_b | x_a\, t_a \rangle \psi_a(x_a). \tag{A.7}$$

This can be interpreted as a *forward integral*

$$\psi_b(x_b) = \int \mathcal{D}q \cdot \psi_a(q(t_a)) \exp\left(\frac{\mathrm{i}}{\hbar} S(q)\right) \tag{A.8}$$

extended over the space of all paths q with $q(t_b) = x_b$ (see figure A.2). On the other hand, the amplitude of finding the system in a state Ψ_b at time t_b if it is known to be in a state Ψ_a at time t_a is given by

$$\langle \Psi_b | \Psi_a \rangle = \int_{M^D} \int_{M^D} \mathrm{d}^D x_b\, \mathrm{d}^D x_a\, \Psi_b(x_b)^* \langle x_b\, t_b | x_a\, t_a \rangle \Psi_a(x_a). \tag{A.9}$$

This can be expressed as a functional integral

$$\langle \Psi_b | \Psi_a \rangle = \int \mathcal{D}q\, \Psi_b(q(t_b))^* \Psi_a(q(t_a)) \exp\left(\frac{\mathrm{i}}{\hbar} S(q)\right), \tag{A.10}$$

extended over *all paths $q = (q(t))_{t_a \leq t \leq t_b}$ with free ends*.

Concluding remarks

Spaces of pointed paths (spaces of paths having one, and only one, common point) are particularly important (see for instance Chapter 7 in particular (7.33)) because they are contractible: there is a one-to-one mapping between a space $\mathcal{P}_a M^D$ of pointed paths on M^D and a space $\mathcal{P}_0 \mathbb{R}^D$ of pointed paths on $\mathbb{R}^D$. Volume elements on $\mathcal{P}_0 \mathbb{R}^D$ can be defined so that the total volume on $\mathcal{P}_0 \mathbb{R}^D$ is normalized to 1, in contrast to (A.4).

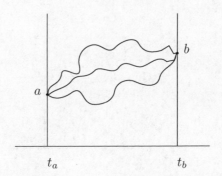

Fig. A.3 A tied-up integral

In this appendix, one can single out two important spaces of pointed paths: the space $\mathcal{P}_a\mathbb{R}^D$ of paths with fixed initial points (figure A.1) and the space $\mathcal{P}_b\mathbb{R}^D$ of paths with fixed final points (figure A.2). For the spaces $\mathcal{P}_a\mathbb{R}^D$ and $\mathcal{P}_b\mathbb{R}^D$ to be vector spaces, the fixed point is taken as the origin of the coordinates in the domains of the paths.

The space $\mathcal{P}_a\mathbb{M}^D$ is the space of interest in a diffusion problem. Equation (A.3) gives the probability distribution at time t_b of a system located at x_a at time t_a. It is a function of (x_a, t_b).

The space $\mathcal{P}_b\mathbb{M}^D$ is the space of interest in quantum mechanics. Equation (A.8) gives the wave function as a function of (x_b, t_b). A probabilist may think of (A.8) as a "backward integral" but a physicist cannot. The integral "moves forward" an initial wave function of (x_a, t_a) to a final wave function of (x_b, t_b).

Reference

[1] N. Bourbaki (1969). *Eléments de mathématiques, intégration* (Paris, Hermann), Chapter 9, p. 39. English translation by S. K. Berberian, *Integration II* (Berlin, Springer, 2004).

Appendix B
Product integrals

Product integrals were introduced by V. Volterra in 1896, and have been developed by J. D. Dollard and C. N. Friedman [1]. They have been used in a variety of problems – for instance by H. P. McKean Jr. in the study of brownian motions on a Lie group. They can be used to obtain path-integral solutions of the Schrödinger equation and representations of the Møller wave operators [2].

Product integration is to products what Riemann integration is to sums: the additive neutral element 0 becomes the multiplicative neutral element **1**; the additive inverse $-A$ becomes the multiplicative inverse A^{-1}, etc. Let A be a $D \times D$ matrix. The product-integral notation of

$$U(t, t_0) = \mathcal{T} \exp\left(\int_{t_0}^{t} A(s)\mathrm{d}s\right) \tag{B.1}$$

is

$$U(t, t_0) = \prod_{t_0}^{t} \exp(A(s)\mathrm{d}s); \tag{B.2}$$

the right-hand side is defined as the limit of a *time-ordered product* of N factors obtained from an N-partition of the time interval $[t_0, t]$:

$$\lim_{N=\infty} \prod_{i=1}^{N} \exp(A(s_i)\Delta s_i) = \lim_{N=\infty} \prod_{i=1}^{N} (\mathbf{1} + A(s_i)\Delta s_i). \tag{B.3}$$

We note that

$$\frac{\mathrm{d}}{\mathrm{d}t} \prod_{t_0}^{t} \exp(A(s)\mathrm{d}s) = A(t) \cdot \prod_{t_0}^{t} \exp(A(s)\mathrm{d}s), \qquad (B.4)$$

$$\frac{\mathrm{d}}{\mathrm{d}t_0} \prod_{t_0}^{t} \exp(A(s)\mathrm{d}s) = -\prod_{t_0}^{t} \exp(A(s)\mathrm{d}s) \cdot A(t_0), \qquad (B.5)$$

$$\prod_{t_0}^{t}(\ ) = \prod_{s}^{t}(\ ) \prod_{t_0}^{s}(\ ). \qquad (B.6)$$

Indefinite product integrals are defined by the multiplicative analog of

$$F(t) = \int_{t_0}^{t} \frac{\mathrm{d}F(s)}{\mathrm{d}s}\,\mathrm{d}s + F(t_0),$$

namely

$$P(t) = \prod_{t_0}^{t} \exp(A(s)\mathrm{d}s) P(t_0), \qquad P(t) \in \mathbb{R}^{D \times D}. \qquad (B.7)$$

It follows from (B.4) and (B.6) that

$$A(t) = P'(t)P^{-1}(t), \qquad P^{-1}(t)P(t) = \mathbf{1}. \qquad (B.8)$$

Therefore, the analog of the relationship (indefinite integral, derivative) is (indefinite product integral, logarithmic derivative), where the logarithmic derivative L is defined by

$$(LP)(t) = P'(t)P^{-1}(t); \qquad (B.9)$$

that is

$$P(t)P^{-1}(t_0) = \prod_{t_0}^{t} \exp((LP)(s)\mathrm{d}s). \qquad (B.10)$$

Note the following properties of logarithmic derivatives:

$$LP^{-1} = -P^{-1}P' \qquad (B.11)$$
$$L(PQ) = LP + P(LQ)P^{-1}. \qquad (B.12)$$

Comparing definite and indefinite product integrals yields

$$U(t, t_0) = P(t)P^{-1}(t_0). \qquad (B.13)$$

The sum rule. Let

$$U_A(t, t_0) = \prod_{t_0}^{t} \exp(A(s)\mathrm{d}s); \qquad (B.14)$$

then one can write either

$$\prod_{t_0}^{t} \exp(A(s)ds + B(s)ds) = U_A(t, t_0) \prod_{t_0}^{t} \exp(U_A(t_0, s)B(s)U_A(s, t_0)ds) \tag{B.15}$$

or

$$\prod_{t_0}^{t} \exp(A(s)ds + B(s)ds) = \prod_{t_0}^{t} \exp(U_A(t, s)B(s)U_A(s, t)ds)U_A(t, t_0). \tag{B.16}$$

Proof. Let

$$P(t) = \prod_{t_0}^{t} \exp(A(s)ds),$$

$$Q(t) = \prod_{t_0}^{t} \exp(P^{-1}(s)B(s)P(s)ds),$$

$$R(t) = \prod_{t_0}^{t} \exp(A(s)ds + B(s)ds);$$

then

$$LQ = P^{-1}BP$$

and, by the product rule (B.12),

$$L(PQ) = LP + P(P^{-1}BP)P^{-1} = A + B. \tag{B.17}$$

On the other hand,

$$LR = A + B. \tag{B.18}$$

The logarithmic derivatives of the two sides of (B.15) are equal, and both sides are equal to **1** for $t = t_0$. Hence they are equal. The proof of equation (B.16) is similar.

Remark. We recognize in the sum rule:
- a transformation to the interaction representation (see Section 6.3);
- an expression equivalent to the Trotter–Kato–Nelson formula; and
- an application of a method of variation of constants (see [1, p. 16]).

The similarity rule. Suppose that $P : [t_0, t] \longrightarrow \mathbb{R}^{D \times D}$ is nonsingular and has a continuous derivative P', then

$$P^{-1}(t) \prod_{t_0}^{t} \exp(A(s)\mathrm{d}s) P(t_0)$$
$$= \prod_{t_0}^{t} \exp(P^{-1}(s) A(s) P(s) \mathrm{d}s - P^{-1}(s) P'(s) \mathrm{d}s); \quad (B.19)$$

equivalently

$$P(t) \prod_{t_0}^{t} \exp(A(s)\mathrm{d}s) P^{-1}(t_0) = \prod_{t_0}^{t} \exp((P(s) A(s) P^{-1}(s) + (LP)(s))\mathrm{d}s). \quad (B.20)$$

Proof. Let

$$Q(t) = \prod_{t_0}^{t} \exp(A(s)\mathrm{d}s); \quad (B.21)$$

then, by (B.10),

$$P(t) Q(t) (P(t_0) Q(t_0))^{-1} = \prod_{t_0}^{t} \exp(L(PQ)(s)\mathrm{d}s). \quad (B.22)$$

Using equation (B.12), we get also

$$L(PQ) = LP + P(LQ)P^{-1}$$
$$= LP + PAP^{-1},$$

and hence (B.20). The proof of (B.19) is similar.

References

[1] J. D. Dollard and C. N. Friedman (1979). *Product Integration with Applications to Differential Equations* (Reading, MA, Addison-Wesley).
[2] C. DeWitt-Morette, A. Maheshwari, and B. Nelson (1979). "Path integration in nonrelativistic quantum mechanics," *Phys. Rep.* **50**, 266–372 and references therein.
[3] P. Cartier (2000). "Mathemagics," *Sém. Lothar. Combin.* **44** B44d, 71 pp.

Appendix C
A compendium of gaussian integrals

Gaussian random variables

A random variable X (see Section 1.6 for the basic definitions in probability theory) follows a normal law $N(0, \sigma)$ if its probability distribution is given by

$$\mathbb{P}[a \leq X \leq b] = \int_a^b \mathrm{d}\Gamma_\sigma(x), \tag{C.1}$$

where we use the definition

$$\mathrm{d}\Gamma_\sigma(x) \overset{f}{=} \frac{1}{\sigma\sqrt{2\pi}} e^{-x^2/(2\sigma^2)} \mathrm{d}x. \tag{C.2}$$

As a consequence, the expectation (or mean value) of any function $f(X)$ is given by

$$\mathbb{E}[f(X)] = \int_\mathbb{R} \mathrm{d}\Gamma_\sigma(x) \cdot f(x), \tag{C.3}$$

and, in particular,

$$\mathbb{E}[X] = \int_\mathbb{R} \mathrm{d}\Gamma_\sigma(x) \cdot x = 0, \tag{C.4}$$

$$\mathbb{E}[X^2] = \int_\mathbb{R} \mathrm{d}\Gamma_\sigma(x) \cdot x^2 = \sigma^2, \tag{C.5}$$

that is, the mean is $m = 0$ and the variance is σ^2.

The characteristic function and moments

The characteristic function of the probability law Γ_σ is defined as a Fourier transform, that is

$$\mathbb{E}[e^{iuX}] = \int_{\mathbb{R}} d\Gamma_\sigma(x) e^{iux}. \tag{C.6}$$

This is given by the *fundamental formula*

$$\mathbb{E}[e^{iuX}] = e^{-u^2\sigma^2/2}; \tag{C.7}$$

that is, we have the integral relation

$$\frac{1}{\sigma\sqrt{2\pi}} \int_{\mathbb{R}} dx\, e^{iux - x^2/(2\sigma^2)} = e^{-u^2\sigma^2/2}. \tag{C.8}$$

We shall make a few comments about this fundamental formula. The characteristic function is also known as the *moment-generating function*. Indeed, from the power-series expansion of the exponential, we immediately get

$$\mathbb{E}[e^{iuX}] = \sum_{n=0}^{\infty} \frac{(iu)^n}{n!} \mathbb{E}[X^n]. \tag{C.9}$$

By a similar argument we get

$$e^{-u^2\sigma^2/2} = \sum_{m=0}^{\infty} (-1)^m \frac{\sigma^{2m}}{2^m m!} u^{2m}. \tag{C.10}$$

This gives, on comparing the coefficients of the powers of u, the values of the *moments*

$$\mathbb{E}[X^{2m}] = \frac{(2m)!}{2^m m!} \sigma^{2m}, \tag{C.11}$$

$$\mathbb{E}[X^{2m+1}] = 0. \tag{C.12}$$

Equations (C.4) and (C.5) are particular cases of these formulas. The vanishing of the moments of odd order (equation (C.12)) is just a reflection of the invariance of the probability measure $d\Gamma_\sigma(x)$ under the symmetry $x \mapsto -x$.

The fundamental formula and the Γ-function

Conversely, a direct proof of equation (C.11) brings us back to the fundamental formula (C.7). By definition, we have

$$\mathbb{E}[X^{2m}] = \int_{\mathbb{R}} d\Gamma_\sigma(x) x^{2m} = \frac{1}{\sigma\sqrt{2\pi}} \int_{-\infty}^{+\infty} dx\, x^{2m} e^{-x^2/(2\sigma^2)}. \tag{C.13}$$

Using the change of integration variable $y = x^2/(2\sigma^2)$, hence $dy = x\,dx/\sigma^2$, and the definition of the gamma function

$$\Gamma(s) = \int_0^\infty dy\, y^{s-1} e^{-y}, \qquad (C.14)$$

we easily get

$$\mathbb{E}[X^{2m}] = 2^m \pi^{-1/2} \Gamma\left(m + \frac{1}{2}\right) \sigma^{2m}. \qquad (C.15)$$

Formula (C.11) amounts to

$$\Gamma\left(m + \frac{1}{2}\right) = \frac{(2m)!}{2^{2m} m!} \pi^{1/2}. \qquad (C.16)$$

This follows by induction on m from the particular case $m = 0$, that is $\Gamma(1/2) = \pi^{1/2}$, and the functional equation $\Gamma(s+1) = s\Gamma(s)$ for $s = m + \frac{1}{2}$.

Normalization

On setting $m = 0$ in formula (C.11), we get the normalization relation $\mathbb{E}[1] = 1$, that is $\int d\Gamma_\sigma(x) = 1$, or, explicitly,

$$\int_{-\infty}^{+\infty} dx\, e^{-x^2/(2\sigma^2)} = \sigma\sqrt{2\pi}. \qquad (C.17)$$

On putting $m = 0$ in formula (C.15), this normalization formula amounts to $\Gamma(1/2) = \pi^{1/2}$. We recall just two of the many proofs of this formula.

- From the Euler beta integral

$$\int_0^1 dx\, x^{\alpha-1}(1-x)^{\beta-1} = \frac{\Gamma(\alpha)\Gamma(\beta)}{\Gamma(\alpha+\beta)}, \qquad (C.18)$$

in the special case $\alpha = \beta = 1/2$, we get

$$\int_0^1 \frac{dx}{\sqrt{x(1-x)}} = \Gamma(1/2)^2, \qquad (C.19)$$

and, by standard trigonometry, the left-hand side is equal to π.
- The complement formula

$$\Gamma(s)\Gamma(1-s) = \frac{\pi}{\sin(\pi s)} \qquad (C.20)$$

for the special case $s = 1/2$ gives again $\Gamma(1/2)^2 = \pi$.

The fundamental formula and translation invariance

The fundamental formula (C.8) can be rewritten as

$$\int_{\mathbb{R}} \mathrm{d}x \exp\left(-\frac{(x - iu\sigma^2)^2}{2\sigma^2}\right) = \sigma\sqrt{2\pi}. \tag{C.21}$$

For $u = 0$, it reduces to the normalization formula (C.17), and the meaning of (C.21) is that its left-hand side is *independent of u*. More generally we consider the integral

$$\Phi(z) = \int_{\mathbb{R}} \mathrm{d}x \, \exp\left(-\frac{(x - z)^2}{2\sigma^2}\right), \tag{C.22}$$

for an arbitrary complex number $z = v + iu\sigma^2$ (with u and v real). When $z = v$ is real, $\Phi(v)$ is independent of v by virtue of the *invariance under translation* of the integral. However, it follows from general criteria that $\Phi(z)$ is an entire function of the complex variable z. By the principle of the analytic continuation, $\Phi(z)$ is a constant and, in particular, we get $\Phi(iu\sigma^2) = \Phi(0)$, that is formula (C.21).

Appendix D
Wick calculus

(contributed by A. Wurm)

Wick calculus began modestly in 1950 when G. C. Wick introduced operator normal ordering into quantum field theory: this technique is built on rearranging creation and annihilation operators, $a^\dagger$ and a, in order to bring as onto the right-hand side of the $a^\dagger$s. This arrangement simplifies calculations based on vacuum expectation values because

$$a|\text{vac}\rangle = 0.$$

It also removes very neatly the (infinite) zero-point energy of a free scalar field represented by an infinite collection of harmonic oscillators (the groundstate eigenvalue of each oscillator is equal to 1/2 in frequency units).

An example of operator ordering:

$$(a+a^\dagger)(a+a^\dagger) = :(a+a^\dagger)^2: + 1, \tag{D.1}$$

where the expression between colons is normal ordered. □

The Wick transform. Matrix elements can be represented by functional integrals – for instance, schematically

$$\left\langle B \middle| \exp\left(-\frac{i}{\hbar}\hat{H}t\right) \middle| A \right\rangle = \int_{\mathbb{X}_{AB}} \mathcal{D}x \exp\left(\frac{i}{\hbar}S(x)\right) \tag{D.2}$$

or[1]

$$\langle \text{out}|TF(\hat{\phi})|\text{in}\rangle = \int_\Phi \mathcal{D}\phi\, F(\phi) \exp\left(\frac{i}{\hbar}S(\phi)\right) \bigg/ \int_\Phi \mathcal{D}\phi\, \exp\left(\frac{i}{\hbar}S(\phi)\right). \tag{D.3}$$

---

[1] In this formula $TF(\hat{\phi})$ represents the time-ordered functional F of the field ϕ, quantized as the operator $\hat{\phi}$.

It follows from (D.3) that there must be a "Wick transform" that corresponds to the Wick operator ordering. The Wick transforms of polynomial and exponential functionals, also bracketed by colons, have been used profitably in functional integrals since the seventies. The simplest definition of Wick transforms is (2.65)

$$: F(\phi) :_G := \exp\left(-\frac{1}{2}\Delta_G\right) \cdot F(\phi), \qquad (D.4)$$

where Δ_G is the functional laplacian (2.63),

$$\Delta_G := \int_{\mathbb{R}^D} d^D x \int_{\mathbb{R}^D} d^D y \, G(x,y) \frac{\delta^2}{\delta\phi(x)\delta\phi(y)}, \qquad (D.5)$$

where, with our normalization, (2.32) and (2.36),

$$\int_\Phi d\mu_G(\phi) \phi(x)\phi(y) = \frac{s}{2\pi} G(x,y). \qquad (D.6)$$

Wick transforms have been used also when fields are treated as distributions, i.e. defined on test functions f. A gaussian on field distributions ϕ is usually defined by

$$\int_\Phi d\mu_G(\phi) \exp(-i\langle\phi, f\rangle) = \exp\left(-\frac{1}{2}C(f,f)\right), \qquad (D.7)$$

$$C(f,f) := \int_\Phi d\mu_G(\phi) \langle\phi, f\rangle^2. \qquad (D.8)$$

For comparing distribution gaussians with the gaussians (2.32) used in this book, let the distribution ϕ be equivalent to a function ϕ, i.e.

$$\langle\phi, f\rangle = \int_{\mathbb{R}^D} d^D x \, \phi(x) f(x), \qquad (D.9)$$

and compute the gaussian average w.r.t. μ_G of $\langle\phi, f\rangle^2$:

$$\int_\Phi d\mu_G(\phi) \int_{\mathbb{R}^D} d^D x \, \phi(x) f(x) \int_{\mathbb{R}^D} d^D y \, \phi(y) f(y)$$

$$= \frac{s}{2\pi} \int_{\mathbb{R}^D} d^D x \, f(x) \int_{\mathbb{R}^D} d^D y \, f(y) G(x,y)$$

$$= \frac{s}{2\pi} W(f,f). \qquad (D.10)$$

The variance W differs from, but has the same structure as, the covariance C, which is defined by

$$C(f,f) = \int_\Phi d\mu_G(\phi) \int_{\mathbb{R}^D} d^D x \, \phi(x) f(x) \int_{\mathbb{R}^D} d^D y \, \phi(y) f(y). \qquad (D.11)$$

The definition of μ_G on the space of distributions ϕ by the covariance $C(f, f)$ is similar to the definition of μ_G on the space of functions ϕ by the covariance $W(f, f)$. To sum up, use the relation

$$C(f, f) = \frac{s}{2\pi} W(f, f).$$

Representations of Wick ordering and Wick transforms by Hermite polynomials. Orthogonality, generating function recursion, λ. Let He_n be a scaled version of the usual Hermite polynomials H_n,

$$\mathrm{He}_n(x) := 2^{-n/2} H_n(x/\sqrt{2}). \tag{D.12}$$

The following properties have counterparts in Wick calculus:

orthogonality,

$$\int_{-\infty}^{\infty} dx \, e^{-x^2/2} \mathrm{He}_n(x) \mathrm{He}_m(x) = \delta_{mn} \sqrt{2\pi} n!; \tag{D.13}$$

the generating function,

$$\exp\left(x\alpha - \frac{1}{2}\alpha^2\right) = \sum_{n=0}^{\infty} \mathrm{He}_n(x) \frac{\alpha^n}{n!}; \tag{D.14}$$

and the recursion relation,

$$\mathrm{He}_{n+1}(x) = x\mathrm{He}_n(x) - n\mathrm{He}_{n-1}(x). \tag{D.15}$$

The normal-ordered product $:(a^\dagger + a)^n:$ is equal to $\mathrm{He}_n(a^\dagger + a)$.

Proof. Using the recursion formula (D.15), it suffices to establish the relation

$$(a + a^\dagger) : (a + a^\dagger)^n : = \, : (a + a^\dagger)^{n+1} : + n : (a + a^\dagger)^{n-1} : . \tag{D.16}$$

From the definition of the Wick ordering for operators,[2] the binomial formula takes here the form

$$: (a + a^\dagger)^n : \, = \sum_{i=0}^{n} \binom{n}{i} (a^\dagger)^i a^{n-i}. \tag{D.17}$$

Moreover, from the relation $aa^\dagger - a^\dagger a = 1$, we get

$$(a + a^\dagger)(a^\dagger)^i a^{n-i} = (a^\dagger)^{i+1} a^{n-i} + i(a^\dagger)^{i-1} a^{n-i+1} \tag{D.18}$$

and the proof of (D.16) results from a simple manipulation of binomial coefficients. □

---

[2] "Put the creation operators to the left of the annihilation operators."

The ordered polynomial $:\phi(x)^n:$ can also be represented by Hermite polynomials, provided that the covariance G is equal to 1 at coincidence points:[3]

$$:\phi(x)^n:_G = \text{He}_n(\phi(x)) \qquad \text{with } G(x,x) = 1. \qquad \text{(D.19)}$$

Proof. Expand the exponential in the definition (D.4) of Wick transforms. For example,

$$\begin{aligned} :\phi(y)^3:_G &= \left(1 - \frac{1}{2}\int_{\mathbb{R}^D} d^D x \int_{\mathbb{R}^D} d^D x'\, G(x,x')\frac{\delta^2}{\delta\phi(x)\delta\phi(x')}\right)\phi(y)^3 \\ &= \phi(y)^3 - 3G(y,y)\phi(y) \\ &= \phi(y)^3 - 3\phi(y). \end{aligned} \qquad \text{(D.20)}$$

□

In quantum field theory the covariance, i.e. the two-point function, is infinite at the coincidence points: $G(x,x) = \infty$. Two strategies have been developed for working with Wick transforms in spite of this difficulty:

- decomposing the covariance into scale-dependent covariances (see Chapter 16); and
- treating fields as distributions. It then follows from equations (D.7) and (D.8) that

$$:\phi(f)^n:_C = C(f,f)^{n/2}\,\text{He}_n\left(\frac{\phi(f)}{C(f,f)^{1/2}}\right). \qquad \text{(D.21)}$$

As remarked earlier, the covariance $C(f,f)$ has the same structure as the variance $W(f,f)$. The variance

$$W(f,f) = \langle f, Gf \rangle \qquad \text{(D.22)}$$

is equal to the covariance when f is a Dirac distribution, i.e. when J in equation (2.33) is a pointlike source. Pointlike sources are responsible for the infinity of covariances at coincidence points. Equation (D.21) conveniently eliminates pointlike sources.

The definition (D.21) is equivalent to the following recursive definition:

$$:\phi(f)^0:_C = 1,$$

$$\frac{\delta}{\delta\phi}:\phi(f)^n:_C = n:\phi(f)^{n-1}:_C, \qquad n = 1,2,\ldots \qquad \text{(D.23)}$$

$$\int d\mu_C(\phi):\phi(f)^n:_C = 0, \qquad n = 1,2,\ldots$$

---

[3] The general case can be reduced to this case by scaling. It suffices to apply formula (D.19) to the scaled field $\phi(x)/\sqrt{G(x,x)}$.

Recall that a distribution ϕ is a linear map on the space of test functions f, and therefore

$$\frac{\delta}{\delta\phi}\phi(f) = f.$$

Wick ordered polynomials as eigenfunctions of the Brydges coarse-graining operator. See (2.94) and Section 16.2 on the $\lambda\phi^4$ system. Let P_L be the coarse-graining operator (2.83),

$$P_L F = S_{L/\Lambda}\bigl(\mu_{[\Lambda,L[} * F\bigr),$$

which is obtained by rescaling the average over the range $[\Lambda, L[$ of scale-dependent functionals

$$P_L \int_{\mathbb{R}^D} d^D x \; :\phi^n(x): \;:= L^{4+n[\phi]} \int_{\mathbb{R}^D} d^D x \; :\phi^n(x):,$$

where $[\phi]$ is the physical length dimension of the field, $[\phi] = -1$.

For other properties of Wick transforms and their application in calculating the specific heat in the two-dimensional Ising model with random bonds, see [6].

References

[1] G. C. Wick (1950). "The evaluation of the collision matrix," *Phys. Rev.* **80**, 268–272.
[2] B. Simon (1974). *The $P(\phi)_2$ Euclidean (Quantum) Field Theory* (Princeton, Princeton University Press).
[3] J. Glimm and A. Jaffe (1981). *Quantum Physics*, 2nd edn. (New York, Springer).
[4] L. H. Ryder (1996). *Quantum Field Theory*, 2nd edn. (Cambridge, Cambridge University Press).
[5] M. Peskin and D. V. Schroeder (1995). *An Introduction to Quantum Field Theory* (Reading, Perseus Books).
[6] A. Wurm and M. Berg (2002). "Wick calculus," arXiv:Physics/0212061. (This appendix is based on parts of this reference.)

Appendix E
The Jacobi operator

Introduction: expanding the action functional
(See also Chapters 4 and 5)

Consider a space $\mathcal{P}\mathrm{M}^D$ of paths $x: \mathbb{T} \to \mathrm{M}^D$ on a D-dimensional manifold M^D (or simply M). The action functional S is a function

$$S: \mathcal{P}\mathrm{M}^D \to \mathbf{R}, \qquad (\text{E.1})$$

defined by a lagrangian L,

$$S(x) = \int_{\mathbb{T}} dt\ L(x(t), \dot{x}(t), t). \qquad (\text{E.2})$$

Two subspaces of $\mathcal{P}\mathrm{M}^D$ dominate the semiclassical expansion:

- $U \subset \mathcal{P}\mathrm{M}^D$, the space of critical points of S

$$q \in U \Leftrightarrow S'(q) = 0, \qquad (\text{E.3})$$

i.e. q is a solution of the Euler–Lagrange equations; it follows that U is $2D$-dimensional.
- $\mathcal{P}_{\mu,\nu}\mathrm{M}^D \subset \mathcal{P}\mathrm{M}^D$, the space of paths satisfying D initial conditions (μ) and D final conditions (ν).

Let $U_{\mu,\nu}$ be their intersection,

$$U_{\mu,\nu} := U \cap \mathcal{P}_{\mu,\nu}\mathrm{M}^D. \qquad (\text{E.4})$$

In Section 4.2 the intersection $U_{\mu,\nu}$ consists of only one point q, or several isolated points $\{q_i\}$. In Section 4.3 the intersection $U_{\mu,\nu}$ is of dimension $\ell > 0$. In Section 5.2 the intersection $U_{\mu,\nu}$ is a multiple root of $S'(q) \cdot \xi = 0$. In Section 5.3 the intersection $U_{\mu,\nu}$ is an empty set.

The hessian and the Jacobi operator

The action functional S is defined on the infinite-dimensional space $\mathcal{P}_{\mu,\nu}\mathbb{M}^D$. Its expansion near a path q, namely

$$S(x) = S(q) + S'(q) \cdot \xi + \frac{1}{2!}S''(q) \cdot \xi\xi + \Sigma(q;\xi), \tag{E.5}$$

is greatly simplified by introducing a one-parameter family of paths $\bar{\alpha}(u) \in \mathcal{P}_{\mu,\nu}\mathbb{M}^D$ such that

$$\bar{\alpha}(u)(t) = \alpha(u,t), \qquad \bar{\alpha}(0) = q, \text{ and } \bar{\alpha}(1) = x. \tag{E.6}$$

Set

$$\frac{\partial \alpha(u,t)}{\partial u} = \bar{\alpha}'(u)(t), \qquad \text{then } \xi := \bar{\alpha}'(0). \tag{E.7}$$

The functional expansion (E.5) is then the ordinary Taylor expansion of a function of u around its value at $u = 0$; it reads[1]

$$(S \circ \bar{\alpha})(1) = \sum_{n=0}^{\infty} \frac{1}{n!}(S \circ \bar{\alpha})^{(n)}(0), \tag{E.8}$$

$$(S \circ \bar{\alpha})'(u) = S'(\bar{\alpha}(u)) \cdot \bar{\alpha}'(u), \tag{E.9}$$

$$(S \circ \bar{\alpha})''(u) = S''(\bar{\alpha}(u)) \cdot \bar{\alpha}'(u)\bar{\alpha}'(u) + S'(\bar{\alpha}(u)) \cdot \bar{\alpha}''(u) \text{ etc.} \tag{E.10}$$

The derivatives of the map $S \circ \bar{\alpha} : [0,1] \to \mathbb{R}$ can be understood either as ordinary derivatives or as covariant derivatives according to one's convenience. We consider them as covariant derivatives (with respect to a linear connection on the manifold $\mathbb{M}^D$).

The term $\xi := \bar{\alpha}'(0)$ is a vector field along q, i.e.

$$\xi \in T_q \mathcal{P}_{\mu,\nu}\mathbb{M}^D. \tag{E.11}$$

The second variation, namely the hessian, defines the (differential) Jacobi operator $\mathcal{J}(q)$ on $T_q \mathcal{P}_{\mu,\nu}\mathbb{M}^D$,

$$S''(q) \cdot \xi\xi = \langle \xi, \mathcal{J}(q)\xi \rangle \text{ in the } L^{2,1} \text{ duality}$$
$$= \int_T dt\, \xi^\alpha(t) \mathcal{J}_{\alpha\beta}(q) \xi^\beta(t). \tag{E.12}$$

---

[1] An important consequence of equation (E.10) is the following. Denote by Ξ the tangent space $T_q \mathcal{P}\mathbb{M}^D$ of the path space $\mathcal{P}\mathbb{M}^D$ at the point q. The first variation $S'(q)$ of S is a linear form $\xi \mapsto S'(q) \cdot \xi$ on Ξ. Let Ξ_0 be the set of all ξ with $S'(q) \cdot \xi = 0$; when q is a critical point of S, then $\Xi_0 = \Xi$. Then $S''(q)$ is defined invariantly as a quadratic form on Ξ_0.

The integral kernel of the hessian of S defines the (functional) Jacobi operator

$$\mathcal{J}_{\alpha\beta}(q,s,t) = \frac{1}{2} \frac{\delta^2}{\delta\xi^\alpha(s)\delta\xi^\beta(t)} S''(q) \cdot \xi\xi. \tag{E.13}$$

A Jacobi field $h(q)$ is a vector field along q in the space TU, where U is the set of critical points of S. A Jacobi field is a solution of the Jacobi equation

$$\mathcal{J}_{\alpha\beta}(q)(h^\beta(q))(t) = 0, \tag{E.14}$$

which is often abbreviated to

$$\mathcal{J}_{\alpha\beta}(q)h^\beta(t) = 0. \tag{E.15}$$

The Jacobi operator, its Green function, and its eigenvectors are powerful tools for gaussian functional integrals. For the Jacobi operator on phase space, see Section 3.4.

Jacobi fields and Green's functions

Let

$$S(x) = \int_\mathbb{T} dt\, L(x(t), \dot{x}(t), t), \qquad \mathbb{T} = [t_a, t_b].$$

Let $x_{\text{cl}}(u)$ be a $2D$-parameter family of classical paths with values in a manifold $\mathbb{M}^D$. Set

$$(x_{\text{cl}}(u))(t) =: x_{\text{cl}}(t; u),$$
$$\frac{\partial x_{\text{cl}}(t; u)}{\partial t} =: \dot{x}_{\text{cl}}(t; u),$$
$$\frac{\partial x_{\text{cl}}(t; u)}{\partial u} =: x'_{\text{cl}}(t; u). \tag{E.16}$$

Varying initial conditions (x_a, p_a)

We choose $u = (u^1, \ldots, u^{2D})$ to stand for $2D$ initial conditions that characterize the classical paths $x_{\text{cl}}(u)$, which are assumed to be unique for the time being,

$$S'(x_{\text{cl}}(u)) = 0, \qquad \text{for all } u.$$

The Jacobi operator

By varying successively the $2D$ initial conditions, one obtains $2D$ Jacobi fields[2]

$$\left\{\frac{\partial x_{\mathrm{cl}}}{\partial u^\alpha}\right\}.$$

Indeed,

$$0 = \frac{\partial}{\partial u^\alpha} S'(x_{\mathrm{cl}}(u)) = S''(x_{\mathrm{cl}}(u)) \frac{\partial x_{\mathrm{cl}}(u)}{\partial u^\alpha}.$$

If the hessian of the action is not degenerate, the $2D$ Jacobi fields are linearly independent.

Let the initial conditions be

$$u = (x_a, p_a),$$

where $x_a := x(t_a)$ and

$$p_a := \left.\frac{\partial L}{\partial \dot{x}(t)}\right|_{t=t_a}.$$

The corresponding Jacobi fields are[3]

$$j^{\bullet\beta}(t) = \frac{\partial x_{\mathrm{cl}}^\bullet(t; u)}{\partial p_{a,\beta}},$$

$$k^\bullet{}_\beta(t) = \frac{\partial x_{\mathrm{cl}}^\bullet(t; u)}{\partial x_a^\beta},$$

$$\tilde{k}_\bullet{}^\beta(t) = \frac{\partial p_{\mathrm{cl}\bullet}(t; u)}{\partial p_{a,\beta}},$$

$$\ell_{\bullet\beta}(t) = \frac{\partial p_{\mathrm{cl}\bullet}(t; u)}{\partial x_a^\beta}. \tag{E.17}$$

The Jacobi fields can be used to construct Jacobi matrices:

$$J^{\alpha\beta}(t, t_a) := j^{\alpha\beta}(t),$$
$$K^\alpha{}_\beta(t, t_a) := k^\alpha{}_\beta(t),$$
$$\tilde{K}_\alpha{}^\beta(t, t_a) := \tilde{k}_\alpha{}^\beta(t),$$
$$L_{\alpha\beta}(t, t_a) := \ell_{\alpha\beta}(t). \tag{E.18}$$

The $2D \times 2D$ matrix constructed from the four $D \times D$ blocks $J, K, \tilde{K}, L$ is a solution of the equation defined by the Jacobi operator in phase space.

---

[2] Jacobi fields have been constructed as needed in [1, 2, 8] and references therein.
[3] In these formulas $p_{\mathrm{cl}}(t; u)$ denotes the momentum along the classical trajectory $t \mapsto x_{\mathrm{cl}}(t; u)$.

Let $M, N, \tilde{N}, P$ be the matrix inverses of $J, K, \tilde{K}, L$, respectively, in the following sense:

$$J^{\alpha\beta}(t, t_a) M_{\beta\gamma}(t_a, t) = \delta^\alpha{}_\gamma, \qquad \text{abbreviated to } JM = \mathbf{1}. \qquad (\text{E.19})$$

The Jacobi matrices and their inverses have numerous properties. We note only that the inverse matrices $M, N, \tilde{N}, P$ are the hessians of the corresponding action function $\mathcal{S}$ – i.e. the Van Vleck matrices. For example,

$$M_{\alpha\beta}(t_b, t_a) = \frac{\partial^2 \mathcal{S}(x_{\text{cl}}(t_b), x_{\text{cl}}(t_a))}{\partial x_{\text{cl}}^\beta(t_b) \partial x_{\text{cl}}^\alpha(t_a)}. \qquad (\text{E.20})$$

The matrices N and $\tilde{N}$ are, respectively, the hessians of $\mathcal{S}(x_{\text{cl}}(t_b), p_{\text{cl}}(t_a))$ and $\mathcal{S}(p_{\text{cl}}(t_b), x_{\text{cl}}(t_a))$. The matrix P is the hessian of $\mathcal{S}(p_{\text{cl}}(t_b), p_{\text{cl}}(t_a))$.

Green's functions of Jacobi operators are solutions of

$$\mathcal{J}_{\alpha\beta}(x_{\text{cl}}(t)) G^{\beta\gamma}(t, s) = \delta^\gamma_\alpha \delta(t - s), \qquad (\text{E.21})$$

where x_{cl} is a solution of the Euler–Lagrange equation characterized by $2D$ boundary conditions. We list below the Green functions when x_{cl} is characterized by position and/or momentum.

The Green functions are also two-point functions [1, 2, 4, 8]:

$$\int_{\Xi_{\mu,\nu}} d\gamma(\xi) \xi^\alpha(t) \xi^\beta(s) = \frac{s}{2\pi} G^{\alpha\beta}(t, s), \qquad (\text{E.22})$$

$$\Xi_{\mu,\nu} := T_{x_{\text{cl}}} \mathcal{P}_{\mu,\nu} \mathbb{M}^D, \qquad \text{often abbreviated to } \Xi, \qquad (\text{E.23})$$

where γ is the gaussian of covariance G and Ξ is defined by (E.11). The boundary conditions of the Green functions G are dictated by the boundary conditions of ξ – i.e. by the pair (μ, ν). They can be read off (E.22).

We list below the Green functions of Jacobi operators for classical paths characterized by various boundary conditions. See [1, 2, 4, 8].

(i) The classical path is characterized by $p_{\text{cl}}(t_a) = p_a$, $x_{\text{cl}}(t_b) = x_b$; the boundary values of $\xi \in \Xi$ are $\dot{\xi}(t_a) = 0$, $\xi(t_b) = 0$,

$$\begin{aligned} G(t, s) = {}& \theta(s - t) K(t, t_a) N(t_a, t_b) J(t_b, s) \\ & - \theta(t - s) J(t, t_b) \tilde{N}(t_b, t_a) \tilde{K}(t_a, s). \end{aligned} \qquad (\text{E.24a})$$

(ii) The classical path is characterized by $x_{\text{cl}}(t_a) = x_a$, $p_{\text{cl}}(t_b) = p_b$; the boundary values of $\xi \in \Xi$ are $\xi(t_a) = 0$, $\dot{\xi}(t_b) = 0$,

$$\begin{aligned} G(t, s) = {}& \theta(s - t) J(t, t_a) \tilde{N}(t_a, t_b) \tilde{K}(t_b, s) \\ & - \theta(t - s) K(t, t_b) N(t_b, t_a) J(t_a, s). \end{aligned} \qquad (\text{E.24b})$$

(iii) The classical path is characterized by $x_{cl}(t_a) = x_a$, $x_{cl}(t_b) = x_b$; the boundary values of $\xi \in \Xi$ are $\xi(t_a) = 0$, $\xi(t_b) = 0$,

$$G(t,s) = \theta(s-t)J(t,t_a)M(t_a,t_b)J(t_b,s)$$
$$\quad - \theta(t-s)J(t,t_b)\widetilde{M}(t_b,t_a)J(t_a,s), \quad (E.24c)$$

where $\widetilde{M}$ is the transpose of M.

(iv) The classical path is characterized by $p_{cl}(t_a) = p_a$, $p_{cl}(t_b) = p_b$; the boundary values of $\xi \in \Xi$ are $\dot\xi(t_a) = 0$, $\dot\xi(t_b) = 0$,

$$G(t,s) = \theta(s-t)K(t,t_a)P(t_a,t_b)\tilde{K}(t_b,s)$$
$$\quad - \theta(t-s)K(t,t_b)\tilde{P}(t_b,t_a)\tilde{K}(t_a,s), \quad (E.24d)$$

where $\tilde{P}$ is the transpose of P.

Example: a simple harmonic oscillator (SHO)

The action functional of the SHO is

$$S(x) = \frac{m}{2}\int_\mathbb{T} dt(\dot x^2(t) - \omega^2 x^2(t)) \quad (E.25)$$

$$= S(q) + \frac{1}{2}S''(q)\cdot \xi\xi \quad \text{with } S'(q)\cdot\xi = 0$$

$$= S(q) + \frac{m}{2}\int_\mathbb{T} dt(\dot\xi^2(t) - \omega^2\xi^2(t)). \quad (E.26)$$

In order to identify the differential Jacobi operator $\mathcal{J}_{\alpha\beta}(q)$ defined by (E.12) we integrate (E.26) by parts

$$S''(q)\cdot\xi\xi = m\int_\mathbb{T} dt\,\xi(t)\left(-\frac{d^2}{dt^2} - \omega^2\right)\xi(t) + m\xi(t)\dot\xi(t)\Big|_{t_a}^{t_b} \quad (E.27)$$

$$= m\int_\mathbb{T} dt\,\xi(t)\left(-\frac{d^2}{dt^2} - \omega^2 + (\delta(t-t_b) - \delta(t-t_a))\frac{d}{dt}\right)\xi(t)$$
$$\quad (E.28)$$

$$= \int_\mathbb{T} dt\,\xi(t)\mathcal{J}(q)\xi(t). \quad (E.29)$$

We note that the argument q of the Jacobi operator consists only of the pair (m,ω). The Green functions[4] corresponding to the four cases (E.24a)–(E.24d) are, respectively, as follows.

---

[4] Note that the Green functions of the Jacobi operator defined by (E.29) differ by a constant factor from the Green functions of the operator defined by (3.75). In Chapter 3 the physical constants of the forced harmonic oscillator (FHO) are displayed explicitly in order to prove that the FHO is not singular at $\lambda = 0$.

(i) Momentum-to-position transitions, $\dot{\xi}(t_a) = 0$, $\xi(t_b) = 0$,

$$G(t,s) = \theta(s-t)\cos[\omega(t-t_a)](\cos[\omega(t_b-t_a)])^{-1}\frac{1}{\omega}\sin[\omega(t_b-s)]$$
$$- \theta(t-s)\frac{1}{\omega}\sin[\omega(t-t_b)](\cos[\omega(t_a-t_b)])^{-1}\cos[\omega(t_a-s)].$$
(E.30a)

(ii) Position-to-momentum transitions, $\xi(t_a) = 0$, $\dot{\xi}(t_b) = 0$,

$$G(t,s) = \theta(s-t)\frac{1}{\omega}\sin[\omega(t-t_a)](\cos[\omega(t_b-t_a)])^{-1}\cos[\omega(t_b-s)]$$
$$- \theta(t-s)\cos[\omega(t-t_b)](\cos[\omega(t_a-t_b)])^{-1}\frac{1}{\omega}\sin[\omega(t_a-s)].$$
(E.30b)

(iii) Position-to-position transitions, $\xi(t_a) = 0$, $\xi(t_b) = 0$,

$$G(t,s) = \theta(s-t)\frac{1}{\omega}\sin[\omega(t-t_a)](\sin[\omega(t_b-t_a)])^{-1}\sin[\omega(t_b-s)]$$
$$- \theta(t-s)\frac{1}{\omega}\sin[\omega(t-t_b)](\sin[\omega(t_a-t_b)])^{-1}\sin[\omega(t_a-s)].$$
(E.30c)

(iv) Momentum-to-momentum transitions, $\dot{\xi}(t_a) = 0$, $\dot{\xi}(t_b) = 0$,

$$G(t,s) = \theta(s-t)\cos[\omega(t-t_a)]\frac{-1}{\omega\sin[\omega(t_b-t_a)]}\cos[\omega(t_b-s)]$$
$$- \theta(t-s)\cos[\omega(t-t_b)]\frac{-1}{\omega\sin[\omega(t_a-t_b)]}\cos[\omega(t_a-s)].$$
(E.30d)

Varying initial and final positions (x_a, x_b)

Alternatively, $x_{\text{cl}}(u)$, $u \in \mathbb{R}^{2D}$, can be specified by the D components of $x_a = x_{\text{cl}}(t_a; u)$ and the D components of $x_b = x_{\text{cl}}(t_b; u)$. See the references in this appendix for expressing one set of Jacobi fields in terms of another set. Here we state only the Green function of the Jacobi operator for a classical path characterized by (x_a, x_b), namely

$$G^{\alpha\beta}(t,s) = \theta(t-s)\mathcal{K}^{\alpha\beta}(t,s) + \theta(s-t)\mathcal{K}^{\beta\alpha}(s,t), \qquad \text{(E.31)}$$

where

$$\mathcal{K}^{\alpha\beta}(t,s) := -\frac{\partial x_{\text{cl}}^\alpha(t;u)}{\partial x_a^\gamma} J^{\gamma\delta}(t_a, t_b) \frac{\partial x_{\text{cl}}^\beta(s;u)}{\partial x_b^\delta}, \qquad u = (x_a, x_b). \quad \text{(E.32)}$$

Remark. We have given detailed results obtained by varying (x_a, p_a), in equations (E.17)–(E.30), and a few results obtained by varying (x_a, x_b). Varying (x_a, p_a) or (x_b, p_b) is useful in the study of caustics; varying

(x_a, x_b) is useful when correlating action functional and action function; varying (p_a, p_b) is useful when conservation laws restrict the momentum states.

Determinants

(See Sections 4.2 and 11.2)

The finite-dimensional formula

$$d \ln \det A = \operatorname{Tr} A^{-1} dA = -\operatorname{Tr} A \, dA^{-1} \tag{E.33}$$

is valid in the infinite-dimensional case for Green's functions satisfying

$$S'' G = 1, \tag{E.34}$$

where S'' is defined by (E.12).

Note that the Green function G expressed by (E.31) can be reexpressed in terms of the advanced Green function G^{adv} of the same Jacobi operator; in terms of components:

$$G^{\text{adv}\,\alpha\beta}(t,s) := \theta(s-t)(-\mathcal{K}^{\alpha\beta}(t,s) + \mathcal{K}^{\beta\alpha}(s,t)), \tag{E.35}$$

namely

$$G(t,s) = \mathcal{K}(t,s) + G^{\text{adv}}(t,s). \tag{E.36}$$

Equation (E.36) follows from (E.31) on using the qualified (valid when integrated) equality

$$\theta(t-s) \stackrel{\int}{=} 1 - \theta(s-t).$$

Given two Green functions of the same Jacobi operator, one can use the basic equation (E.33) for the ratio of their determinants:

$$\delta \ln \frac{\operatorname{Det} G}{\operatorname{Det} G^{\text{adv}}} = -\operatorname{Tr}(G - G^{\text{adv}}) \delta S'' \tag{E.37}$$
$$= -\operatorname{Tr} \mathcal{K} \, \delta S''.$$

At this point, one can either follow Bryce DeWitt's calculations[5] and compute the variation $\delta S''$ of the hessian of S by varying the action functional S, or one can use the relationship between the action functional $S(x_{\text{cl}}(u))$ and the action function $\mathcal{S}(x_{\text{cl}}(t_b; u), x_{\text{cl}}(t_a; u))$:

$$S(x_{\text{cl}}(u)) = \mathcal{S}(x_{\text{cl}}(t_b; u), x_{\text{cl}}(t_a; u)) \equiv \mathcal{S}(x_b, t_b; x_a, t_a), \tag{E.38}$$

which is sometimes abbreviated to $\mathcal{S}(t_b, t_a)$ or $\mathcal{S}(x_b, x_a)$.

---

[5] Advanced Green functions (labeled G^+) have been used extensively by Bryce DeWitt for constructing volume elements (see Chapter 18).

Remark. The letters S, for the action functional, and $\mathcal{S}$, for the action function, are unfortunately hard to distinguish. They are easily identified by their arguments.

Remark. When the path is characterized by boundary values other than (x_a, x_b), equation (E.38) is modified by end-point contributions (see e.g. [3], p. 2504); the following argument remains valid but is more clumsy to spell out.

Taking the derivatives of (E.38) with respect to $u \in \mathbb{R}^{2D}$, and then setting $u = 0$, gives

$$\frac{\partial x_{\text{cl}}^{\alpha}}{\partial x_a^{\gamma}} S_{,\alpha\beta}(x_{\text{cl}}) \frac{\partial x_{\text{cl}}^{\beta}}{\partial x_b^{\delta}} = \frac{\partial^2}{\partial x_a^{\gamma} \partial x_b^{\delta}} \mathcal{S}(x_b, x_a). \tag{E.39}$$

We have

$$\delta(S(x_{\text{cl}})) = (\delta S)(x_{\text{cl}}) + S_{,\alpha}(x_{\text{cl}}) \delta x_{\text{cl}}^{\alpha}$$
$$= (\delta S)(x_{\text{cl}}).$$

After some straightforward algebraic manipulations, it follows from (E.38), (E.39), and (E.32) that

$$\operatorname{Tr} \mathcal{K} \delta S'' = -\mathrm{r} J^{\gamma\delta}(t_a, t_b) \delta \frac{\partial^2 \mathcal{S}}{\partial x_b^{\delta} \partial x_a^{\gamma}} \tag{E.40}$$

$$= -\mathrm{r} J^{\gamma\beta}(t_a, t_b) \delta M_{\beta\gamma}(t_b, t_a). \tag{E.41}$$

According to (E.20) and to [1], p. 373,

$$M_{\alpha\beta}(t_a, t_b) = -M_{\beta\alpha}(t_b, t_a), \tag{E.42}$$
$$J^{\alpha\beta}(t_a, t_b) M_{\beta\gamma}(t_b, t_a) = \delta^{\alpha}_{\gamma}. \tag{E.43}$$

Therefore, using the basic formulas relating traces and determinants,

$$\mathrm{r} J^{\gamma\beta} \delta M_{\beta\gamma} = (M^{-1})^{\gamma\beta} \delta M_{\beta\gamma}$$
$$= \delta \ln \det M.$$

where we denote by M the matrix with entries $M_{\beta\gamma} = M_{\beta\gamma}(t_b, t_a)$.

By equating (E.41) and (E.37) one obtains

$$\delta \ln \det M = \delta \ln \left(\frac{\operatorname{Det} G}{\operatorname{Det} G^{\text{adv}}} \right).$$

Finally, up to a multiplicative constant,

$$\det M = \frac{\operatorname{Det} G}{\operatorname{Det} G^{\text{adv}}}. \tag{E.44}$$

Eigenvectors of the Jacobi operator

The eigenvectors of the Jacobi operator diagonalize the hessian of the action functional and make a convenient basis for $T_q \mathcal{P}_{\mu,\nu} \mathrm{M}^D$, $q \in U^{2D}$.

Let $\{\Psi_k\}_k$ be a complete set of orthonormal eigenvectors of $\mathcal{J}(q)$:

$$\mathcal{J}(q) \cdot \Psi_k = \alpha_k \Psi_k, \qquad k \in \{0, 1, \ldots\}. \tag{E.45}$$

Let $\{u^k\}$ be the coordinates of ξ in the $\{\Psi_k\}$ basis:

$$\xi^\alpha(t) = \sum_{k=0}^\infty u^k \Psi_k^\alpha(t), \tag{E.46}$$

$$u^k = \int_{t_a}^{t_b} dt (\xi(t) | \Psi_k(t)). \tag{E.47}$$

The $\{\Psi_k\}$ basis diagonalizes the hessian

$$S''(q) \cdot \xi\xi = \sum_{k=0}^\infty \alpha_k (u^k)^2; \tag{E.48}$$

therefore the space of the $\{u^k\}$ is the (hessian) space l^2 of points u such that $\sum \alpha_k (u^k)^2$ is finite.

Vanishing eigenvalues of the Jacobi operator

If one (or several) eigenvalues of the Jacobi operator vanishes – assume only one, say $\alpha_0 = 0$, for simplicity – then Ψ_0 is a Jacobi field with vanishing boundary conditions,

$$\mathcal{J}(q) \cdot \Psi_0 = 0, \qquad \Psi_0 \in T_q \mathcal{P}_{\mu,\nu} M^d. \tag{E.49}$$

The $2D$ Jacobi fields are not linearly independent, $S''(q) \cdot \xi\xi$ is degenerate.

The various types of hessian degeneracies are analyzed in Sections 5.1 and 5.2.

References

The properties of Jacobi operators, fields, matrices, Green's functions, and determinants have been developed as needed in the following publications listed in chronological order.

[1] C. DeWitt-Morette (1976). "The semi-classical expansion," *Ann. Phys.* **97**, 367–399 (see p. 373 and Appendix A); erratum, *Ann. Phys.* **101**, 682–683.

[2] C. DeWitt-Morette, A. Maheshwari, and B. Nelson (1979). "Path integration in non relativistic quantum mechanics," *Phys. Rep.* **50**, 266–372 (see Appendix B and p. 278).

[3] C. DeWitt-Morette and T.-R. Zhang (1983). "Path integrals and conservation laws," *Phys. Rev.* **D28**, 2503–2516 (see the appendix).
[4] C. DeWitt-Morette and T.-R. Zhang (1983). "A Feynman–Kac formula in phase space with application to coherent state transitions," *Phys. Rev.* **D28**, 2517–2525.
[5] C. DeWitt-Morette, B. Nelson, and T.-R. Zhang (1983). "Caustic problems in quantum mechanics with applications to scattering theory," *Phys. Rev.* **D28**, 2526–2546.
[6] C. DeWitt-Morette (1984). "Feynman path integrals. From the prodistribution definition to the calculation of glory scattering," in *Stochastic Methods and Computer Techniques in Quantum Dynamics*, eds. H. Mitter and L. Pittner, *Acta Phys. Austriaca* Suppl. **26**, 101–170 (see Appendix A).
[7] P. Cartier (1989). "A course on determinants," in *Conformal Invariance and String Theory*, eds. P. Dita and V. Georgescu (New York, Academic Press).
[8] P. Cartier and C. DeWitt-Morette (1995). "A new perspective on functional integration," *J. Math. Phys.* **36**, 2237–2312 (located on the World Wide Web at http://godel.ph.utexas.edu/Center/Papers.html and http://babbage.sissa.it/list/funct-an/9602) (see Appendix B and p. 2255).
[9] J. LaChapelle (1995). Functional integration on symplectic manifolds, unpublished Ph.D. Dissertation, University of Texas at Austin (see pp. 68–81, Appendix C).
[10] B. DeWitt (2003). *The Global Approach to Quantum Field Theory* (Oxford, Oxford University Press; with corrections, 2004), pp. 216–217.

Appendix F
Change of variables of integration

The purpose of this appendix is to show how the standard formula for change of variables of integration can be derived from the formulas for integration by parts. It is intended as an elementary introduction to the general methods described in Chapter 11. A similar method could also be applied to the Berezin integral of Grassmann variables.

The set up

Let $\mathbb{X}$ and $\mathbb{Y}$ be open sets in $\mathbb{R}^D$, and let f be a differentiable map, not necessarily invertible:

$$f : \mathbb{X} \to \mathbb{Y} \qquad \text{by } y = f(x). \tag{F.1}$$

The derivative mapping of f at x maps the tangent space $T_x \mathbb{X}$ into $T_y \mathbb{Y}$:

$$f'(x) : T_x \mathbb{X} \to T_y \mathbb{Y} \qquad \text{by } V(x) \mapsto W(y). \tag{F.2}$$

The pull-back $f^*(y)$ maps the dual $T_y^* \mathbb{Y}$ of $T_y \mathbb{Y}$ into $T_x^* \mathbb{X}$:

$$f^*(y) : T_y^* \mathbb{Y} \to T_x^* \mathbb{X} \qquad \text{by } \theta(y) \mapsto \eta(x). \tag{F.3}$$

In coordinates, $y = f(x)$ is the vector with coordinates $y^j = f^j(x)$,

$$x = \{x^\alpha\}, \qquad y = \{y^j\}; \tag{F.4}$$

$f'(x)$ is the jacobian matrix $J(x)$,

$$f'(x) = J(x), \qquad \text{with } J^j_\alpha(x) = \frac{\partial f^j(x)}{\partial x^\alpha}. \tag{F.5}$$

We do not use inverse mappings so that this scheme can be applied when f is not invertible. It also bypasses the burden of inverting the jacobian. This scheme has been applied to Grassmann variables in Section 9.3. See in (9.54) the change of variable of integration; it introduces naturally the

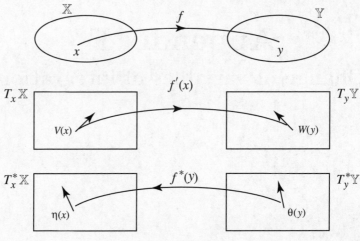

Fig. F.1

jacobian, not its inverse as usually claimed. In a subsequent step one can move the determinant from one side of the equation to the other side; it then enters the equation as an inverse determinant!

The following equations give the transformation laws, under the application $f : \mathbb{X} \to \mathbb{Y}$, of basis and components of 1-forms and contravariant vectors (see figure F.1).

Basis transformation laws

Action of f on 1-forms:

$$\mathrm{d}y^{(j)} = \frac{\partial f^j(x)}{\partial x^\alpha} \mathrm{d}x^{(\alpha)} = J^j_\alpha(x)\mathrm{d}x^{(\alpha)}. \tag{F.6}$$

Components of 1-forms[1] $\theta(y) = \theta_j \mathrm{d}y^{(j)}$:

$$\eta_\alpha \, \mathrm{d}x^{(\alpha)} = (\theta_j \circ f)\mathrm{d}y^{(j)} \Leftrightarrow \eta_\alpha = (\theta_j \circ f) J^j_\alpha. \tag{F.7}$$

Components of vector fields[2] $\langle \eta, V \rangle = \langle \theta, W \rangle \circ f$, hence

$$W^j \circ f = J^j_\alpha V^\alpha. \tag{F.8}$$

Vector basis $e_{(\alpha)} V^\alpha = e_{(j)} W^j$, hence

$$e_{(\alpha)} = e_{(j)} J^j_\alpha. \tag{F.9}$$

---

[1] Given a 1-form θ on $\mathbb{Y}$, there always exists a unique 1-form η on $\mathbb{X}$ satisfying (F.7).
[2] Given the vector field V on $\mathbb{X}$, the existence and uniqueness of a vector field W on $\mathbb{Y}$ satisfying (F.8) is granted if f is invertible, but not in general.

Change of variables of integration

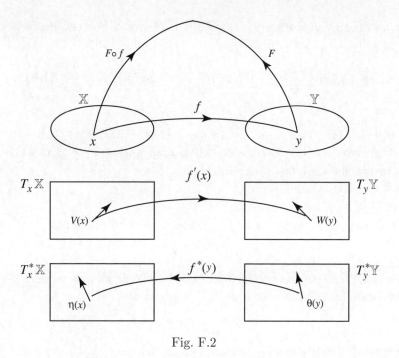

Fig. F.2

Note that, given the multiplication rules for column and row vectors with matrices, we place the basis elements $\{e_{(j)}\}$ to the left of the transformation matrix J_α^j.

The main formulas

From now on, we assume that f is *invertible*. Hence, the matrix $J(x) = (J_\alpha^j(x))$ is invertible for all x in $\mathbb{X}$. Furthermore, given any vector field V on $\mathbb{X}$, with components $V^\alpha(x)$, there exists by virtue of formula (F.8) a unique vector field W on $\mathbb{Y}$ related to V by f with components $W^j(y) = J_\alpha^j(x) V^\alpha(x)$ for $x = f^{-1}(y)$.

For integration purposes, we need change-of-variable formulas for Lie derivatives and for divergences, and the basic formula $\int_\mathbb{X} f^*\theta = \int_\mathbb{Y} \theta$. We shall prove the following formulas (where F is a smooth function with compact support on $\mathbb{Y}$, hence $F \circ f$ is a smooth function with compact support on $\mathbb{X}$; see figure F.2):

$$\mathcal{L}_V (F \circ f) = (\mathcal{L}_W F) \circ f, \tag{F.10}$$

$$\operatorname{div} V = (\operatorname{div} W) \circ f - \mathcal{L}_V \mathbf{J}/\mathbf{J}, \quad \mathbf{J} := \det J \neq 0, \tag{F.11}$$

$$\int_\mathbb{X} (F \circ f) |\mathbf{J}| d^D x = \int_\mathbb{Y} F \, d^D y. \tag{F.12}$$

Proving (F.10). Geometrically (F.10) is obvious. For the proof in components, use the chain rule, equations (F.5) and (F.8):

$$\mathcal{L}_V(F \circ f) = \frac{\partial(F \circ f)}{\partial x^\alpha} V^\alpha = \left(\frac{\partial F}{\partial y^j} \circ f\right) J_\alpha^j V^\alpha = \left(\frac{\partial F}{\partial y^j} W^j\right) \circ f$$
$$= (\mathcal{L}_W F) \circ f. \tag{F.13}$$

Proving (F.11). This equation applies to invertible maps $f : \mathbb{X} \to \mathbb{Y}$ having, therefore, an invertible derivative mapping $J(x) := f'(x)$. We can then invert the matrix $(J_\alpha^j(x)) = J(x)$,

$$L_k^\alpha(x) J_\alpha^j(x) = \delta_k^j, \tag{F.14}$$

and introduce the vector field $L_{(k)}$ on $\mathbb{X}$ with α-components

$$(L_{(k)}(x))^\alpha = L_k^\alpha(x). \tag{F.15}$$

It is related by f to the vector field $\partial/\partial y^k$ on $\mathbb{Y}$. It follows from (F.8) that in the basis $\{L_{(j)}\}$ the vector field V is given by

$$V = L_{(j)} \cdot (W^j \circ f). \tag{F.16}$$

The r.h.s. of (F.16) has the structure $U\varphi$, where U is the vector $L_{(j)}$, φ is the function $W^j \circ f$, and[3]

$$\mathrm{div}(U\varphi) = \partial_\alpha(U^\alpha \varphi) = \mathrm{div}\, U \cdot \varphi + \mathcal{L}_U \varphi. \tag{F.17}$$

Applying the rule (F.17) to (F.16) gives

$$\mathrm{div}\, V = \mathrm{div}\, L_{(j)} \cdot (W^j \circ f) + \mathcal{L}_{L_{(j)}}(W^j \circ f), \tag{F.18}$$

where

$$\mathrm{div}\, L_{(j)} \cdot (W^j \circ f) = \mathrm{div}\, L_{(j)} \cdot J_\alpha^j V^\alpha \quad \text{by (F.8)}, \tag{F.19}$$

$$\mathcal{L}_{L_{(j)}}(W^j \circ f) = \left(\mathcal{L}_{\partial/\partial y^j} W^j\right) \circ f = (\mathrm{div}\, W) \circ f \quad \text{by (F.10)}. \tag{F.20}$$

It remains to prove that $\mathrm{div}\, L_{(j)} \cdot J_\alpha^j V^\alpha$ is equal to $-\mathcal{L}_V \mathbf{J}/\mathbf{J}$. For this purpose we apply the matrix rules

$$\mathrm{d}(\det A)/\det A = \mathrm{Tr}(A^{-1} \cdot \mathrm{d}A), \tag{F.21}$$

$$\mathrm{d}(A^{-1}) = -A^{-1} \cdot \mathrm{d}A \cdot A^{-1} \tag{F.22}$$

to the matrix $J_\alpha^j(x) = \partial_\alpha f^j(x)$ and its inverse $L_j^\alpha(x) = (L_{(j)}(x))^\alpha$. We begin by computing

$$-\mathcal{L}_V \mathbf{J}/\mathbf{J} = -V^\alpha \partial_\alpha \mathbf{J}/\mathbf{J} = -V^\alpha L_j^\beta \partial_\alpha J_\beta^j \quad \text{by (F.21)}. \tag{F.23}$$

---

[3] We use the standard abbreviation ∂_α for $\partial/\partial x^\alpha$.

On the other hand,
$$\begin{aligned}
\operatorname{div} L_{(j)} \cdot J^j_\alpha V^\alpha &= \partial_\beta L^\beta_j \cdot J^j_\alpha V^\alpha \\
&= -L^\beta_k (\partial_\beta J^k_\gamma) L^\gamma_j \cdot J^j_\alpha V^\alpha \quad \text{by (F.22)} \\
&= -L^\beta_k (\partial_\beta J^k_\alpha) V^\alpha. \quad \text{(F.24)}
\end{aligned}$$

Since
$$\partial_\beta J^k_\alpha = \partial_\beta \partial_\alpha f^k = \partial_\alpha \partial_\beta f^k = \partial_\alpha J^k_\beta, \quad \text{(F.25)}$$
the r.h.s. of equations (F.24) and (F.23) are identical and the proof of (F.11) is completed. □

The main result

Proving (F.12). This equation follows readily from[4]
$$\int_\mathbb{X} f^* \theta = \int_\mathbb{Y} \theta \quad \text{(F.26)}$$
and from (F.6) when $\theta(y) = F(y) \mathrm{d}^D y$. We present another proof of (F.12) that involves formulas of interest in the theory of integration. Let F be a test function on $\mathbb{Y}$, that is a smooth function with compact support, and T be the distribution based[5] on $\mathbb{Y}$ defined by
$$\langle T, F \rangle := \int_\mathbb{X} (F \circ f) |\mathbf{J}| \mathrm{d}^D x. \quad \text{(F.27)}$$
We shall prove that
$$\langle T, F \rangle = C_f \int_\mathbb{Y} F \mathrm{d}^D y, \quad \text{(F.28)}$$
where C_f is a number depending, a priori, on the transformation f. It is then proved that C_f is invariant under deformations of f and it is found equal to 1.

What we have to prove is that the distribution T on $\mathbb{Y}$ is a constant, that is its derivatives $\partial_j T := \partial T / \partial y^j$ are 0. More explicitly, we need to prove the relation
$$\langle T, \partial_j F \rangle = 0 \quad \text{(F.29)}$$
for every test function F on $\mathbb{Y}$. By replacing F by $F W^j$ and summing over j, we derive the equation $\langle T, H \rangle = 0$ from (F.29), where $H = \partial_j (F W^j)$.

---

[4] This formula requires that f does not change the orientation, that is $\mathbf{J} > 0$. Taking the absolute value $|\mathbf{J}|$ in (F.12) bypasses this (slight) difficulty.
[5] A distribution defined on test functions F on $\mathbb{Y}$ is said to be based on $\mathbb{Y}$.

Conversely, (F.29) follows from this identity by the specialization $W^k = \delta_j^k$. Notice that H is the divergence of the vector field FW, hence

$$H := \mathcal{L}_W F + F \operatorname{div} W. \tag{F.30}$$

It is therefore enough to establish the identity

$$0 = \langle T, H \rangle = \langle T, \mathcal{L}_W F + F \operatorname{div} W \rangle, \tag{F.31}$$

where F is a test function on $\mathbb{Y}$ and W a vector field on $\mathbb{Y}$. On the other hand, the defining equation (F.27) says that

$$\langle T, H \rangle = \int_\mathbb{X} (H \circ f) |\mathbf{J}| \mathrm{d}^D x. \tag{F.32}$$

Using (F.30), (F.10), and (F.11), one gets[6]

$$\langle T, H \rangle = \int_\mathbb{X} ((\mathcal{L}_W F + F \operatorname{div} W) \circ f) |\mathbf{J}| \mathrm{d}^D x$$
$$= \int_\mathbb{X} (\mathcal{L}_V (F \circ f) \cdot |\mathbf{J}| + (F \circ f) |\mathbf{J}| \operatorname{div} V + (F \circ f) \mathcal{L}_V |\mathbf{J}|) \mathrm{d}^D x. \tag{F.33}$$

Set $G := (F \circ f) |\mathbf{J}|$, hence

$$\langle T, H \rangle = \int_\mathbb{X} (\mathcal{L}_V G + G \operatorname{div} V) \mathrm{d}^D x$$
$$= \int_\mathbb{X} \operatorname{div}(GV) \mathrm{d}^D x = \int_\mathbb{X} \partial_\alpha (GV^\alpha) \mathrm{d}^D x$$

and finally $\langle T, H \rangle = 0$ by integration by parts.

To prove that the constant C_f in (F.28) is invariant under deformation of f, one introduces a deformation $\{f_\lambda\}$, with λ a parameter,[7] $\lambda \in I$. Let φ be a diffeomorphism

$$\varphi : \mathbb{X} \times I \to \mathbb{Y} \times I \quad \text{by } \varphi(x, \lambda) = (f_\lambda(x), \lambda).$$

Let $J_\lambda(x)$ be the jacobian matrix of f_λ with determinant $\mathbf{J}_\lambda(x)$. Then the jacobian of φ is the determinant of the matrix

$$\begin{pmatrix} J_\lambda(x) & 0 \\ * & 1 \end{pmatrix},$$

that is $\mathbf{J}(x, \lambda) = \mathbf{J}_\lambda(x)$. Apply the transformation rule on $\mathbb{Y} \times I$ analogous to

$$\int_\mathbb{X} (F \circ f) \cdot |\mathbf{J}| \mathrm{d}^D x = C_f \int_\mathbb{Y} F \cdot \mathrm{d}^D y$$

---
[6] Notice that $\mathcal{L}_V \mathbf{J}/\mathbf{J}$ is equal to $\mathcal{L}_V |\mathbf{J}|/|\mathbf{J}|$.
[7] I is an open interval in $\mathbb{R}$.

to a function on $\mathbb{Y} \times I$ of the form $F(y)u(\lambda)$. One gets

$$C_\varphi \int_{\mathbb{Y} \times I} F(y)u(\lambda) \mathrm{d}^D y\, \mathrm{d}\lambda = \int_{\mathbb{Y} \times I} C_{f_\lambda} F(y)u(\lambda) \mathrm{d}^D y\, \mathrm{d}\lambda.$$

This equation is valid for F and u arbitrary; hence, for all λ in I,

$$C_\varphi = C_{f_\lambda}.$$

Thus C_{f_λ} is independent of λ as promised.

To conclude the proof, notice that one can deform locally a smooth transformation f to its linear part f^0, hence $C_f = C_{f^0}$. A linear transformation can be deformed either to the identity map or to the map given by $y^1 = -x^1$, $y^2 = x^2, \ldots, y^D = x^D$ according to the sign of its determinant. The last particular case is easily settled. □

Appendix G
Analytic properties of covariances

Generalities

Covariances occur in two different situations, classical and quantum.

- Suppose that $(x(t))_{t\in\mathbb{T}}$ is a family of real random variables, indexed say by a time interval $\mathbb{T}$. Put

$$G(t,t') = \mathbb{E}[x(t)x(t')] \tag{G.1}$$

for t, t' in $\mathbb{T}$.

- Let $(\hat{x}(t))_{t\in\mathbb{T}}$ be a family of self-adjoint operators in some (complex) Hilbert $\mathcal{H}$, and $\Omega \in \mathcal{H}$ a state vector. Put

$$\hat{G}(t,t') = \langle\Omega|\hat{x}(t)\hat{x}(t')|\Omega\rangle \tag{G.2}$$

for t, t' in $\mathbb{T}$.

The first case is subsumed into the second one by taking for $\mathcal{H}$ the space of complex random variables with finite second moment $\mathbb{E}[|X|^2] < +\infty$, and the scalar product $\mathbb{E}[X^*Y]$, with $\hat{x}(t)$ the operator of multiplication by $x(t)$, and $\Omega \equiv 1$. Conversely, *if the operators $\hat{x}(t)$ commute in pairwise manner*, then the quantum case is obtained from the classical. This is one form of the spectral theorem, and constitutes the basis of Born's statistical interpretation of wave functions.

We summarize the main properties.

- *The classical case:* the function G is real-valued, symmetric $G(t',t) = G(t,t')$, and satisfies the *positivity condition*

$$\sum_{i,j=1}^{N} c_i c_j G(t_i, t_j) \geq 0 \tag{G.3}$$

for $t_1, \ldots, t_N$ in $\mathbb{T}$, and real constants $c_1, \ldots, c_N$.

- *The quantum case:* the function $\hat{G}$ is complex-valued, hermitian $\hat{G}(t',t) = \hat{G}(t,t')^*$, and satisfies the *positivity condition*

$$\sum_{i,j=1}^{N} c_i^* c_j \hat{G}(t_i, t_j) \geq 0 \qquad \text{(G.4)}$$

for $t_1, \ldots, t_N$ in $\mathbb{T}$, and complex constants $c_1, \ldots, c_N$.

The classical case is the special case of the quantum one when the function $\hat{G}$ is real-valued. In that which follows, we treat in detail the real case and call a symmetric function G satisfying (G.3) a *positive kernel*; it is the generalization of a symmetric positive-semidefinite matrix.

The reconstruction theorem

In what follows, $\mathbb{T}$ may be an arbitrary space. We consider *a real continuous symmetric function G on $\mathbb{T} \times \mathbb{T}$ satisfying the inequalities* (G.3). We proceed to construct a (real) Hilbert space $\mathcal{H}$ and a family of vectors $(v(t))_{t \in \mathbb{T}}$ in $\mathcal{H}$ such that

$$G(t,t') = \langle v(t) | v(t') \rangle \qquad \text{(G.5)}$$

for t, t' in $\mathbb{T}$. This is a particular case of the so-called *Gelfand–Naïmark–Segal* (GNS) *reconstruction theorem*.

The construction is in three steps.[1]

(1) We first construct a real vector space E with a basis $(e(t))_{t \in \mathbb{T}}$ (see any textbook on algebra). Then define a bilinear form B on $E \times E$ by the requirement

$$B(e(t), e(t')) = G(t, t'). \qquad \text{(G.6)}$$

For any vector $e = \sum_{i=1}^{N} c_i e(t_i)$ in E, we obtain

$$B(e, e) = \sum_{i,j=1}^{N} c_i c_j G(t_i, t_j)$$

by linearity, hence $B(e,e) \geq 0$ for all e in E. Furthermore, the symmetry of G entails the symmetry of B, namely $B(e, e') = B(e', e)$.

(2) Let N be the set of vectors e in E with $B(e,e) = 0$. From the positivity of the form B, one derives in the usual way the Cauchy–Schwarz inequality

$$B(e, e')^2 \leq B(e, e) B(e', e') \qquad \text{(G.7)}$$

for e, e' in E. Hence N is also the vector subspace of E consisting of the vectors e such that $B(e, e') = 0$ for all e' in E. Let $V = E/N$; it is easily

---

[1] See [1], p. V.8 for more details.

seen that on V there exists a bilinear scalar product $\langle v|v'\rangle$, that is

$$\langle v|v'\rangle = \langle v'|v\rangle,$$
$$\langle v|v\rangle > 0 \quad \text{for } v \neq 0$$

such that $B(e,e') = \langle \bar{e}|\bar{e}'\rangle$ if $\bar{e}(\bar{e}')$ is the class of $e(e')$ modulo N.

(3) On V, one defines a norm by $\|v\| = \langle v|v\rangle^{1/2}$. Complete V w.r.t. this norm to get a Banach space $\mathcal{H}$. The scalar product on V extends by continuity to a scalar product on $\mathcal{H}$, hence $\mathcal{H}$ is the Hilbert space sought. It remains to define $v(t)$ as the class of $e(t)$ modulo N in $V \subset \mathcal{H}$.

Notice that, by construction, the linear combinations of the vectors $v(t)$ are dense in $\mathcal{H}$.

Reproducing kernels

We want a representation of $\mathcal{H}$ by functions. Let $\mathcal{H}'$ be the Banach space dual to $\mathcal{H}$. By Riesz' theorem, one may identify $\mathcal{H}$ with $\mathcal{H}'$, but we refrain from doing so. By definition an element of $\mathcal{H}'$ is a linear form L on $\mathcal{H}$ for which there exists a constant $C > 0$ satisfying the inequality $|L(v)| \leq C\|v\|$ for all v in $\mathcal{H}$. Since the finite linear combinations $v = \sum_{i=1}^{N} c_i v(t_i)$ are dense in $\mathcal{H}$, the inequality on L becomes

$$\left|\sum_{i=1}^{N} c_i L(v(t_i))\right|^2 \leq C^2 \sum_{i,j=1}^{N} c_i c_j \langle v(t_i)|v(t_j)\rangle. \tag{G.8}$$

By associating L with the function $t \mapsto L(v(t))$ on $\mathbb{T}$, we obtain an isomorphism of $\mathcal{H}'$ (hence of $\mathcal{H}$) with the space $\mathcal{R}$ of *functions f on $\mathbb{T}$ such that, for some $C > 0$, the kernel $C^2 G(t,t') - f(t)f(t')$ be positive*. This space is said to be defined by the *reproducing kernel G*. We quote the properties coming from this definition.

- For each t in $\mathbb{T}$, the function $K_t : t' \mapsto G(t,t')$ on $\mathbb{T}$ belongs to $\mathcal{R}$. Furthermore, we have the identities

$$\langle K_t|K_{t'}\rangle = G(t',t), \tag{G.9}$$
$$\langle K_t|f\rangle = f(t) \tag{G.10}$$

for f in $\mathcal{R}$ and t,t' in $\mathbb{T}$. The scalar product on $\mathcal{R}$ is derived from the scalar product on $\mathcal{H}'$ via the isomorphism of $\mathcal{H}'$ with $\mathcal{R}$.

- Every function f in $\mathcal{R}$ is continuous on $\mathbb{T}$. Indeed, for t_0, t in $\mathbb{T}$ we have

$$|f(t) - f(t_0)|^2 \leq C^2[G(t,t) + G(t_0,t_0) - 2G(t,t_0)] \tag{G.11}$$

as a particular case of (G.8). For t converging to t_0, the right-hand side converges to 0 by virtue of the continuity of G; hence $f(t)$ converges to $f(t_0)$.

- Since $\mathcal{R}$ is a Hilbert space (like $\mathcal{H}$ and $\mathcal{H}'$), there exists in $\mathcal{R}$ an orthonormal basis (e_n) (countable in all applications) and from (G.9) and (G.10) one derives the representation

$$G(t,t') = \sum_n e_n(t)e_n(t') \qquad (G.12)$$

(in the complex case, replace $e_n(t)$ by $e_n(t)^*$).
- For any function f in $\mathcal{R}$, the norm $\|f\| = \langle f|f\rangle^{1/2}$ is the smallest constant $C \geq 0$ such that the kernel $C^2 G(t,t') - f(t)f(t')$ is positive.

Example: the $L^{2,1}_-$ space

We illustrate this on the example[2]

$$\mathbb{T} = [0,T], \qquad G(t,t') = \inf(t,t'). \qquad (G.13)$$

The positivity of G comes from the integral representation

$$G(t,t') = \int_0^T d\tau\, I_t(\tau) I_{t'}(\tau), \qquad (G.14)$$

where

$$I_t(\tau) = \begin{cases} 1 & \text{if } 0 \leq \tau \leq t, \\ 0 & \text{otherwise.} \end{cases} \qquad (G.15)$$

Let us introduce the Sobolev space $L^{2,1}_-$. A function $f : [0,T] \to \mathbb{R}$ is in $L^{2,1}_-$ if it can be written as a primitive

$$f(t) = \int_0^t d\tau\, u(\tau) \qquad \text{for } 0 \leq t \leq T, \qquad (G.16)$$

with u in L^2, that is $\int_0^T d\tau |u(\tau)|^2 < +\infty$. Stated otherwise, f is continuous, $f(0) = 0$, and the derivative (in the sense of distributions) of f is the function u in L^2. Using the classical results of Lebesgue, one can also characterize the functions in $L^{2,1}_-$ as the absolutely continuous functions f, such that $f(0) = 0$, whose derivative $u = \dot{f}$, which exists almost everywhere, is square-integrable. The norm in $L^{2,1}_-$ is the L^2-norm of the derivative, or, more explicitly, the scalar product is given by

$$\langle f_1 | f_2 \rangle = \int_0^T d\tau\, \dot{f}_1(\tau) \dot{f}_2(\tau). \qquad (G.17)$$

---

[2] To simplify the notation: if $\mathbb{T} = [t_a, t_b]$, when t runs over $\mathbb{T}$, then $t - t_a$ runs over $[0,T]$ with $T = t_b - t_a$.

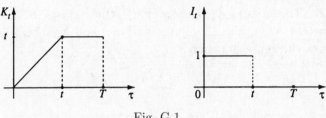

Fig. G.1

The function G is the reproducing kernel of the space $L_-^{2,1}$. To prove this, we have to check formulas (G.9) and (G.10). Notice that the derivative of K_t is I_t (see figure G.1), hence (G.9) follows from the integral representation (G.14). Moreover,

$$\langle K_t | f \rangle = \int_0^T d\tau\, \dot{K}_t(\tau) \dot{f}(\tau) = \int_0^T d\tau\, I_t(\tau) \dot{f}(\tau) = \int_0^t d\tau\, \dot{f}(\tau) = f(t), \tag{G.18}$$

hence (G.10). Notice also that

$$-\frac{d^2}{dt^2} G(t, t') = -\frac{d}{dt} I_{t'}(t) = \delta(t - t'). \tag{G.19}$$

This corresponds to the property $DG = \mathbf{1}$ of Section 2.3, formula (2.27).

From formula (G.17), the norm in $L_-^{2,1}$ is given by

$$\|f\|^2 = \int_0^T dt\, \dot{f}(t)^2 = \int_0^T \frac{(df)^2}{dt}. \tag{G.20}$$

We show how to calculate this by *Riemann sums*. Indeed, from the definition of the norm in a space with reproducing kernel, $\|f\|^2$ is the lowest upper bound (l.u.b.) of quantities

$$\left(\sum_{i=1}^N c_i f(t_i) \right)^2 \bigg/ \sum_{i,j=1}^N c_i c_j \inf(t_i, t_j), \tag{G.21}$$

where we take arbitrary real constants $c_1, \ldots, c_N$ (not all 0) and an arbitrary subdivision $\Delta : 0 < t_1 < \ldots < t_N = T$ of the interval $[0, T]$. Diagonalize the quadratic form in the denominator as $\gamma_1^2 + \cdots + \gamma_N^2$, with the following definitions:

$$\begin{aligned} t_0 &= 0, \\ \Delta t_j &= t_j - t_{j-1} & \text{for } 1 \leq j \leq N, \\ \gamma_j &= (\Delta t_j)^{1/2} \cdot \sum_{i=j}^N c_i & \text{for } 1 \leq j \leq N. \end{aligned} \tag{G.22}$$

The numerator in (G.21) becomes

$$\sum_{i=1}^{N} \gamma_i \frac{\Delta f_i}{(\Delta t_i)^{1/2}}$$

with $\Delta f_i = f(t_i) - f(t_{i-1})$ for $1 \le i \le N$. By taking the l.u.b. first over $\gamma_1, \ldots, \gamma_N$ we can conclude that

$$\|f\|^2 = \text{l.u.b.} \sum_{i=1}^{N} \frac{(\Delta f_i)^2}{\Delta t_i}, \qquad (G.23)$$

where the l.u.b. is now over all subdivisions

$$\Delta : 0 = t_0 < t_1 < \cdots < t_N = T.$$

We can characterize the functions f in $L_-^{2,1}$ by the property that the quantity $\|f\|$ defined by (G.23) is finite. Moreover, by standard arguments in Riemann sums,[3] one proves that, given any $\varepsilon > 0$, there exists a constant $\eta > 0$ such that, for all subdivisions with mesh $\text{l.u.b.}_{1 \le i \le N} \Delta t_i$ smaller than η, the following inequalities hold:

$$\|f\|^2 \ge \sum_{i=1}^{N} \frac{(\Delta f_i)^2}{\Delta t_i} \ge \|f\|^2 - \varepsilon. \qquad (G.24)$$

Exercise. Given a subdivision Δ as before, calculate the orthogonal projection f_Δ of $f \in L_-^{2,1}$ onto the vector space consisting of the functions that are linearly affine in each of the subintervals $[t_{i-1}, t_i]$. Prove the formula $\|f_\Delta\|^2 = \sum_{i=1}^{N} (\Delta f_i)^2 / \Delta t_i$. Explain formula (G.24).

Example: coherent states

We give an example of a *complex* positive kernel. In this subsection, $\mathbb{T}$ is the plane $\mathbb{C}$ of complex numbers z. We introduce the volume element

$$d\gamma(z) = \pi^{-1} e^{-|z|^2} d^2 z \qquad (G.25)$$

in $\mathbb{C}$, where $d^2 z = dx\, dy$ for $z = x + iy$. The Hilbert space $\mathcal{H}$ consists of the entire functions $f(z)$ such that $\int_\mathbb{C} d\gamma(z) |f(z)|^2$ is finite, with the scalar product

$$\langle f_1 | f_2 \rangle = \int_\mathbb{C} d\gamma(z) f_1(z)^* f_2(z). \qquad (G.26)$$

---

[3] For f in $L_-^{2,1}$, the derivative $\dot{f}$ can be approximated in L^2-norm by a continuous function u, and u is uniformly continuous.

It is immediately checked that the functions $e_n(z) = z^n/(n!)^{1/2}$ for $n = 0, 1, 2, \ldots$ form an orthonormal basis of $\mathcal{H}$; that is, $\mathcal{H}$ consists of the functions $f(z) = \sum_{n=0}^{\infty} c_n z^n$ with $\|f\|^2 := \sum_{n=0}^{\infty} n! |c_n|^2$ finite.

The reproducing kernel is given by

$$K(z, z') = \sum_{n=0}^{\infty} e_n(z)^* e_n(z'), \qquad (G.27)$$

that is

$$K(z, z') = e^{z^* z'}. \qquad (G.28)$$

For each z in $\mathbb{C}$, the function $K_z : z' \mapsto e^{z^* z'}$ belongs to the Hilbert space $\mathcal{H}$, and from the orthonormality of the functions $e_n(z)$, one derives immediately the characteristic properties (G.9) and (G.10) for the *reproducing kernel* K.

Remarks.

(1) The function $\Omega = e_0$ is the *vacuum*, the operator a of derivation $\partial/\partial z$ is the *annihilation* operator, with adjoint $a^\dagger$ (the *creation* operator) the multiplication by z. Hence $a e_n = n^{1/2} e_{n-1}, a\Omega = 0, e_n = (a^\dagger)^n \Omega/(n!)^{1/2}$, and $[a, a^\dagger] = \mathbf{1}$. This is the original Fock space!
(2) We can extend the results to the case of entire functions of N complex variables $z_1, \ldots, z_N$, with the volume element

$$d\gamma_N(z) = d\gamma(z_1) \ldots d\gamma(z_N) \qquad (G.29)$$

for $z = (z_1, \ldots, z_N)$ in $\mathbb{C}^N$.

The case of quantum fields

In general, quantum fields are operator-valued distributions. Let $\mathbb{M}^D$ be the configuration space, or the spacetime. A quantum field is a family of self-adjoint operators $\hat{\varphi}(u)$ in some complex Hilbert space $\mathcal{H}$, depending linearly on a (real) test function u on $\mathbb{M}^D$. We use the integral notation

$$\int d^D x \, \hat{\varphi}(x) u(x)$$

for $\hat{\varphi}(u)$. If Ω is any state vector, we define the *covariance* by

$$\hat{G}(u, u') = \langle \Omega | \hat{\varphi}(u) \hat{\varphi}(u') | \Omega \rangle. \qquad (G.30)$$

According to the kernel theorem of L. Schwartz [2], there exists a distribution $\hat{\mathbf{G}}$ on $\mathbb{M}^D \times \mathbb{M}^D$ such that

$$\hat{G}(u, u') = \iint d^D x \, d^D x' \, u(x) u'(x') \hat{\mathbf{G}}(x, x'). \qquad (G.31)$$

We can write symbolically

$$\hat{G}(x,x') = \langle \Omega | \hat{\varphi}(x) \hat{\varphi}(x') | \Omega \rangle, \tag{G.32}$$

by not distinguishing between $\hat{G}$ and $\hat{\mathbf{G}}$.

Let $\mathcal{H}^0$ be the closed subspace of $\mathcal{H}$ spanned by the vectors $\hat{\varphi}(u)\Omega$, for u running over the (real or complex) test functions on $\mathbb{M}^D$. Let $\mathcal{H}'_0$ be the dual space of $\mathcal{H}^0$. It can be identified with the space of complex distributions L on $\mathbb{M}^D$ such that, for some constant $C > 0$, the distribution $C^2\hat{G}(x,x') - L(x)^*L(x')$ is a (generalized) hermitian positive kernel (*exercise:* define precisely this notion). We say that $\mathcal{H}'_0$ is the *space with* (generalized) *reproducing kernel* $\hat{G}(x,x')$. Again, if L_n is an orthonormal basis of $\mathcal{H}'_0$, we have the representation

$$\hat{G}(x,x') = \sum_n L_n(x)^* L_n(x'), \tag{G.33}$$

or, more rigorously,

$$\hat{G}(u^*, u') = \sum_n \langle L_n, u \rangle^* \langle L_n, u' \rangle \tag{G.34}$$

for any two *complex* test functions u, u' on $\mathbb{M}^D$.

If u is a test function on $\mathbb{M}^D$, acting on u by the kernel $\hat{G}$ defines the distribution $u \cdot \hat{G}$ on $\mathbb{M}^D$; symbolically

$$(u \cdot \hat{G})(x') = \int d^D x \, u(x) \hat{G}(x, x'). \tag{G.35}$$

Any such distribution belongs to $\mathcal{H}'_0$; these elements are dense in $\mathcal{H}'_0$ and the scalar product in $\mathcal{H}'_0$ is given by the following generalization of (G.9)

$$\langle u \cdot \hat{G} | u' \cdot \hat{G} \rangle = \hat{G}(u', u^*) \tag{G.36}$$

for any two complex test functions u, u' on $\mathbb{M}^D$.

Since $u \cdot \hat{G} = \int d^D x \, u(x) K_x$ and

$$\hat{G}(u', u^*) = \iint d^D x \, d^D x' \, u^*(x) \hat{G}(x', x) u'(x'), \tag{G.37}$$

formula (G.36) is the integrated form of (G.9), namely $\langle K_x | K_{x'} \rangle = \hat{G}(x', x)$.

Example: the Feynman propagator

The purpose of this example is to illustrate three points:

- how to dispense with positivity;
- the role of boundary conditions; and
- the need for distributions in reproducing kernels.

The space $\mathbb{X}$ consists of the smooth (C^∞) functions $u(t)$ of a real variable t, with the boundary conditions

$$\dot{u}(t) + i\omega u(t) = 0,$$
$$\dot{u}(-t) - i\omega u(-t) = 0, \tag{G.38}$$

for large positive t. This follows the advice of Feynman: "propagate positive (negative) frequencies to the future (past)." The space $\mathbb{X}'$ consists of the smooth functions $U(t)$ with a compact support, and the duality between $\mathbb{X}$ and $\mathbb{X}'$ is given by

$$\langle U, u \rangle = \int dt\, U(t) u(t). \tag{G.39}$$

We introduce a pair of operators $\mathbb{X} \underset{G}{\overset{D}{\rightleftarrows}} \mathbb{X}'$, by writing the formulas

$$Du(t) = -\ddot{u}(t) - \omega^2 u(t), \tag{G.40}$$

$$GU(t) = \int dt'\, G(t - t') U(t'), \tag{G.41}$$

with the kernel

$$G(t) = e^{-i\omega|t|}/(2i\omega). \tag{G.42}$$

Notice that, by necessity, $\mathbb{X}$ and $\mathbb{X}'$ consist of *complex functions* since the kernel G is complex. Notice also that $G(t - t') = G(t' - t)$ is symmetric, not hermitian!

That D maps $\mathbb{X}$ into $\mathbb{X}'$ comes from the fact that, for large $|t|$, $u(t)$ is of the form $c \cdot e^{-i\omega|t|}$ for some constant c, hence $\ddot{u}(t) + \omega^2 u(t)$ vanishes for large $|t|$. To prove that GU satisfies the "radiative" conditions (G.38) is an easy exercise. That $DG = \mathbf{1}_{\mathbb{X}'}$ on $\mathbb{X}'$ follows from the relation

$$\ddot{G}(t) + \omega^2 G(t) = -\delta(t). \tag{G.43}$$

Furthermore, D is *injective* since $Du = 0$ implies $u(t) = ae^{i\omega t} + be^{-i\omega t}$ with complex constants a and b, and the boundary conditions (G.38) imply $a = b = 0$, hence $u = 0$. The calculation

$$D(GD - \mathbf{1}_{\mathbb{X}}) = DGD - D = (DG - \mathbf{1}_{\mathbb{X}'})D = 0$$

and the injectivity of D imply $GD = \mathbf{1}_{\mathbb{X}}$. Finally, $G : \mathbb{X}' \to \mathbb{X}$ is symmetric, that is $\langle U, GU' \rangle = \langle U', GU \rangle$, since the kernel G is symmetric. That D is symmetric follows by integration by parts, since from (G.38) it follows that $\dot{u}u' - u\dot{u}'$ vanishes off a bounded interval for u, u' in $\mathbb{X}$. Hence *all assumptions made in Section 2.3 are fulfilled, except that $\mathbb{X}$ and $\mathbb{X}'$ are not Banach spaces!*

We introduce now in $\mathbb{X}$ the symmetric bilinear (not hermitian!) scalar product

$$\langle u|u'\rangle := \langle Du, u'\rangle = \int dt(-\ddot{u}u' - \omega^2 uu'). \tag{G.44}$$

An integration by parts would give an obviously symmetric definition[4]

$$\langle u|u'\rangle = \int dt(\dot{u}\dot{u}' - \omega^2 uu'), \tag{G.45}$$

but the boundary conditions (G.38) show that the integral oscillates at infinity like $e^{\pm 2i\omega t}$, hence the integral needs a regularization (for instance, a damping factor $e^{-\varepsilon |t|}$ with $\varepsilon \to 0$). Defining $K_{t_0}(t) = G(t - t_0)$, we have $GD = \mathbf{1}_\mathbb{X}$, that is

$$u(t_0) = \int dt\, Du(t) K_{t_0}(t),$$

for u in $\mathbb{X}$. But K_{t_0} is not smooth, hence it is not in the space $\mathbb{X}$ (and $DK_{t_0}(t) = \delta(t - t_0)$ is not in $\mathbb{X}'$!). So we cannot write directly $u(t_0) = \langle u|K_{t_0}\rangle$ and $\langle K_t|K_{t'}\rangle = G(t' - t)$, but, after smoothing with a test function U in $\mathbb{X}'$, these relations become true (see equation (G.36)).

Exercise.

(a) Replace $\mathbb{X}'$ by the Schwartz space $\mathcal{S}(\mathbb{R})$ and describe the boundary conditions satisfied by the functions GU for U in $\mathcal{S}(\mathbb{R})$.
(b) The Fourier transform $\mathcal{F}G$ of G is given by

$$\mathcal{F}G(p) = \lim_{\varepsilon \to 0} \frac{1}{p^2 - \omega^2 + i\varepsilon}.$$

Using (a), give the Fourier transform version of $\mathbb{X}, \mathbb{X}'$ etc.

References

[1] N. Bourbaki (1981). *Espaces vectoriels topologiques* (Paris, Masson).
[2] J. Dieudonné (1977). *Eléments d'analyse* (Paris, Gauthier-Villars), Vol. VII, p. 52.

---

[4] We remind the reader that u and u' are *complex-valued* functions. Notice that, for $u = u'$ real, there can be *no positivity* due to the negative term $-\omega^2 u^2$ in the integrand: the operator $-d^2/dt^2 - \omega^2$ does not have a positive spectrum.

Appendix H
Feynman's checkerboard

We deal with the Dirac equation in a spacetime of dimension $1+1$, and we calculate the corresponding retarded propagator (not the Feynman propagator) as a sum over paths. For the relationships between the Feynman propagator and other propagators, see for instance [1].

The Dirac propagator

We consider a spacetime with two coordinates x and t and introduce the (dimensionless) lightcone coordinates

$$x_\pm = \frac{mc}{2\hbar}(ct \pm x).$$

The corresponding derivations are given by

$$\partial_\pm = \frac{\hbar}{mc}(c^{-1}\partial_t \pm \partial_x).$$

The Klein–Gordon equation for a particle of mass m, namely

$$\left(c^{-2}\partial_t^2 - \partial_x^2 + \frac{m^2 c^2}{\hbar^2}\right)\psi = 0, \tag{H.1}$$

becomes in these new coordinates

$$(\partial_+ \partial_- + 1)\psi = 0. \tag{H.2}$$

The *retarded propagator*[1] is $G(x'-x)$, where $G(x)$ (or $G(x_+, x_-)$) is given by

$$G(x) = J_0(2\sqrt{x_+ x_-})\theta(x_+)\theta(x_-) \tag{H.3}$$

---
[1] From now on, we write x for (x_+, x_-).

and satisfies

$$(\partial_+\partial_- + 1)G(x) = \delta(x_+)\delta(x_-), \tag{H.4}$$

$$G(x) = 0 \quad \text{outside the domain } x_+ \geq 0, x_- \geq 0. \tag{H.5}$$

We recall that $\theta(u)$ is the Heaviside function and that the Bessel function $J_k(z)$ with integral order k is defined by

$$J_k(z) = \left(\frac{z}{2}\right)^k \sum_{n\geq 0} \frac{(-z^2/4)^n}{n!(n+k)!}. \tag{H.6}$$

For the corresponding Dirac equation we introduce the Dirac matrices

$$\gamma_+ = \begin{pmatrix} 0 & 1 \\ 0 & 0 \end{pmatrix}, \qquad \gamma_- = \begin{pmatrix} 0 & 0 \\ 1 & 0 \end{pmatrix} \tag{H.7}$$

satisfying

$$(\gamma_+)^2 = (\gamma_-)^2 = 0, \qquad \gamma_+\gamma_- + \gamma_-\gamma_+ = 1_2. \tag{H.8}$$

The Dirac equation is written as

$$(\gamma_+ \partial_+ + \gamma_- \partial_- + i)\psi = 0, \tag{H.9}$$

whose retarded propagator is $S(x' - x)$ with

$$S(x) = (\gamma_+ \partial_+ + \gamma_- \partial_- - i)G(x) \tag{H.10}$$

or, more explicitly,

$$S(x) = -\begin{pmatrix} iJ_0(2\sqrt{x_+x_-}) & \sqrt{\frac{x_-}{x_+}}J_1(2\sqrt{x_+x_-}) \\ \sqrt{\frac{x_+}{x_-}}J_1(2\sqrt{x_+x_-}) & iJ_0(2\sqrt{x_+x_-}) \end{pmatrix}\theta(x_+)\theta(x_-)$$

$$+ \begin{pmatrix} 0 & \delta(x_+)\theta(x_-) \\ \theta(x_+)\delta(x_-) & 0 \end{pmatrix}. \tag{H.11}$$

Summing over paths

In the plane with coordinates x_+, x_-, denote by D the domain defined by $x_+ > 0, x_- > 0$, with boundary $L_+ \cup L_-$, where $L_\pm$ is the positive part of the $x_\pm$-axis. For a given point $x = (x_+, x_-)$ in D, we denote by $\mathcal{P}_x^p$ the set of polygonal paths beginning at 0 and ending at x, composed of p segments, alternately parallel to L_+ and L_- and oriented in the same direction. The space of paths $\mathcal{P}_x$ is the disjoint union of the spaces $\mathcal{P}_x^p$ for $p = 2, 3, \ldots$ The space $\mathcal{P}_x$ is subdivided into four classes $\mathcal{P}_x^{++}, \mathcal{P}_x^{+-}, \mathcal{P}_x^{-+}$, and $\mathcal{P}_x^{--}$, where for instance $\mathcal{P}_x^{+-}$ consists of the paths beginning parallel to L_+ and ending parallel to L_-. For a path ξ in $\mathcal{P}_x^p$ we denote by $n_+(\xi)$

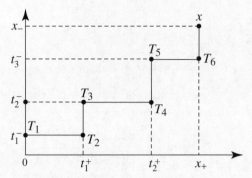

Fig. H.1 A path in the class $\mathcal{P}_x^{--}$ with $p = 7, n_+ = 3, n_- = 4$

the number of segments parallel to L_+ and similarly for $n_-(\xi)$. We have the following table:

$$\begin{aligned} \xi \text{ in } \mathcal{P}_x^{++} & \qquad n_+(\xi) = n_-(\xi) + 1, \\ \xi \text{ in } \mathcal{P}_x^{+-} \cup \mathcal{P}_x^{-+} & \qquad n_+(\xi) = n_-(\xi), \\ \xi \text{ in } \mathcal{P}_x^{--} & \qquad n_+(\xi) = n_-(\xi) - 1. \end{aligned} \qquad (\text{H}.12)$$

In every case, we have $p = n_+(\xi) + n_-(\xi)$, hence p is even (odd) for ξ in $\mathcal{P}_x^{+-} \cup \mathcal{P}_x^{-+} (\mathcal{P}_x^{++} \cup \mathcal{P}_x^{--})$.

We now give a description of a volume element in the space $\mathcal{P}_x$. Once we know that the first segment is parallel to L_+ or L_-, a path is fully parametrized by the lengths $\ell_i^+ (1 \leq i \leq n_+)$ of the segments parallel to L_+ and the lengths $\ell_i^- (1 \leq i \leq n_-)$ of the segments parallel to L_-. They satisfy the equations

$$\begin{aligned} \ell_1^+ + \cdots + \ell_{n_+}^+ &= x_+, \\ \ell_1^- + \cdots + \ell_{n_-}^- &= x_-. \end{aligned} \qquad (\text{H}.13)$$

Hence, we can take as independent parameters the numbers $\ell_i^+ (1 \leq i < n_+)$ and $\ell_i^- (1 \leq i < n_-)$ and define the volume element

$$\mathcal{D}\xi = d\ell_1^+ \ldots d\ell_{n_+-1}^+ d\ell_1^- \ldots d\ell_{n_--1}^-. \qquad (\text{H}.14)$$

Altogether, there are $(n_+ - 1) + (n_- - 1) = p - 2$ independent parameters.

Remark. Feynman, in his original construction, chose a mesh $\varepsilon > 0$ and assumed that the particle is traveling on the checkerboard (see Section 12.2). That is, x_+, x_- and $\ell_1^+, \ldots, \ell_{n_+}^+, \ell_1^-, \ldots, \ell_{n_-}^-$ are integral multiples of ε. He then assigned the same weight ε^p to every such path, and let ε go to 0. It is easy to check that our definition is equivalent to this limiting process.

We analyze in more detail this volume element in $\mathcal{P}_x^p \cap \mathcal{P}_x^{++}$ and leave to the reader the details of the other combinations of signs. A path in $\mathcal{P}_x^p$ has $p-1$ turning points $T_1, \ldots, T_{p-1}$. In our case $p = n_+ + n_-$ is equal to $2n_- + 1$; hence it is odd. We introduce new parameters satisfying the inequalities

$$0 < t_1^+ < t_2^+ < \cdots < t_{n_-}^+ < x_+,$$
$$0 < t_1^- < t_2^- < \cdots < t_{n_--1}^- < x_-,$$
(H.15)

as the partial sums

$$t_1^+ = \ell_1^+ \qquad\qquad t_1^- = \ell_1^-$$
$$t_2^+ = \ell_1^+ + \ell_2^+ \qquad\qquad t_2^- = \ell_1^- + \ell_2^-$$
$$t_3^+ = \ell_1^+ + \ell_2^+ + \ell_3^+ \qquad\qquad t_3^- = \ell_1^- + \ell_2^- + \ell_3^-$$

The parameters t_i^+, t_i^- are independent and the volume element in $\mathcal{P}_x^p \cap \mathcal{P}_x^{++}$ is given by

$$\mathcal{D}\xi = dt_1^+ \ldots dt_{n_-}^+ dt_1^- \ldots dt_{n_--1}^-.$$
(H.16)

A path ξ has $p-1$ turning points $T_1, \ldots, T_{p-1}$ with coordinates

$$(t_1^+, 0), (t_1^+, t_1^-), (t_2^+, t_1^-), (t_2^+, t_2^-), \ldots, (t_{n_-}^+, t_{n_--1}^-), (t_{n_-}^+, x_-).$$

In general we denote by $T(\xi)$ the number of turning points in the path ξ.

Path-integral representation of the Dirac propagator

According to (H.10), for $x = (x_+, x_-)$ in D, that is $x_+ > 0, x_- > 0$, the Dirac propagator is given by

$$S(x) = \begin{pmatrix} -iG(x_+, x_-) & \partial_+ G(x_+, x_-) \\ \partial_- G(x_+, x_-) & -iG(x_+, x_-) \end{pmatrix}.$$
(H.17)

Hence we have to give path-integral representations for $G(x_+, x_-)$ and its first derivatives $\partial_+ G$ and $\partial_- G$. We know that

$$G(x_+, x_-) = \sum_{n=0}^{\infty} (-i)^{2n} \frac{x_+^n x_-^n}{n! \, n!},$$
(H.18)

hence

$$\partial_+ G(x_+, x_-) = \sum_{n \geq 1} (-i)^{2n} \frac{x_+^{n-1} x_-^n}{(n-1)! \, n!},$$
(H.19)

$$\partial_- G(x_+, x_-) = \sum_{n \geq 1} (-i)^{2n} \frac{x_+^n x_-^{n-1}}{n! \, (n-1)!}.$$
(H.20)

Notice that paths ξ in $\mathcal{P}_x^{2n+1} \cap \mathcal{P}_x^{++}$ have $2n \geq 2$ turning points, and that according to (H.15) and (H.16) the volume of the space $\mathcal{P}_x^{2n+1} \cap \mathcal{P}_x^{++}$ is equal to

$$\frac{x_+^n}{n!} \frac{x_-^{n-1}}{(n-1)!}$$

(here $n_- = n$ and $n_+ = n+1$). Hence formula (H.20) is equivalent to

$$\partial_- G(x_+, x_-) = \int_{\mathcal{P}_x^{++}} \mathcal{D}\xi \cdot (-\mathrm{i})^{T(\xi)}. \tag{H.21}$$

We leave to the reader the derivation of similar expressions for the other elements of the matrix $S(x)$ in (H.17), namely

$$\partial_+ G(x_+, x_-) = \int_{\mathcal{P}_x^{--}} \mathcal{D}\xi \cdot (-\mathrm{i})^{T(\xi)}, \tag{H.22}$$

$$-\mathrm{i}G(x_+, x_-) = \int_{\mathcal{P}_x^{+-}} \mathcal{D}\xi \cdot (-\mathrm{i})^{T(\xi)}, \tag{H.23}$$

$$-\mathrm{i}G(x_+, x_-) = \int_{\mathcal{P}_x^{-+}} \mathcal{D}\xi \cdot (-\mathrm{i})^{T(\xi)}. \tag{H.24}$$

Notice that the integrand is always the same, only the boundary conditions for the path ξ vary. The expression $(-\mathrm{i})^{T(\xi)}$ for the phase factor associated with the path ξ is the one given originally by Feynman.

Reference

[1] C. DeWitt and B. DeWitt (eds.) (1964). *Relativity, Groups and Topology* (New York, Gordon and Breach), pp. 615–624.

Bibliography

R. Abraham and J. E. Marsden (1985). *Foundations of Mechanics* (New York, Addison-Wesley), cited in Chapter 14.

M. Abramowitz and I. A. Stegun (1970). *Handbook of Mathematical Functions* 9th edn. (New York, Dover Publications), cited in Chapters 12 and 13.

S. A. Albeverio and R. J. Høegh-Krohn (1976). *Mathematical Theory of Feynman Path Integrals* (Berlin, Springer), cited in Chapters 1, 11, and Project 19.1.2.

R. Balian and C. Bloch (1971). "Analytical evaluation of the Green function for large quantum numbers," *Ann. Phys.* **63**, 592–600, cited in Chapter 14.
(1974). "Solution of the Schrödinger equation in terms of classical paths," *Ann. Phys.* **85**, 514–545, cited in Chapter 14.

R. Balian and J. Zinn-Justin (eds.) (1976). *Methods in Field Theory* (Amsterdam, North-Holland), cited in Chapter 15.

R. Balian, G. Parisi, and A. Voros (1978). "Discrepancies from asymptotic series and their relation to complex classical trajectories," *Phys. Rev. Lett.* **41**, 1141–1144; erratum, *Phys. Rev. Lett.* **41**, 1627, cited in Chapter 14.

D. Bar-Moshe and M. S. Marinov (1994). "Realization of compact Lie algebras in Kähler manifolds," *J. Phys. A* **27**, 6287–6298, cited in Chapter 7.
(1996). "Berezin quantization and unitary representations of Lie groups," in *Topics in Statistical and Theoretical Physics: F. A. Berezin Memorial Volume*, eds. R. L. Dobrushin, A. A. Minlos, M. A. Shubin, and A. M. Vershik (Providence, RI, American Mathematical Society), pp. 1–21. also hep-th/9407093, cited in Chapter 7.

H. Bateman and A. Erdelyi (1953). *Higher Transcendental Functions*, Vol. II, (New York, McGraw-Hill), cited in Chapter 13.

D. Becirevic, Ph. Boucaud, J. P. Leroy *et al.* (2000). "Asymptotic scaling of the gluon propagator on the lattice," *Phys. Rev.* **D61**, 114508, cited in Chapter 17.

M. E. Bell (2002). Introduction to supersymmetry, unpublished Master's Thesis, University of Texas at Austin, cited in Chapter 9.

F. A. Berezin (1965). *The Method of Second Quantization* (Moscow, Nauka) [in Russian]. English translation: Academic Press, New York, 1966, cited in Chapter 9.

(1974). "Quantization," *Izv. Math. USSR* **8**, 1109–1165, cited in Chapter 7.

F. A. Berezin and M. S. Marinov (1975). "Classical spin and Grassmann algebra," *JETP Lett.* **21**, 320–321, cited in Chapter 9.

(1977). "Particle spin dynamics as the Grassmann variant of classical mechanics," *Ann. Phys.* **104**, 336–362, cited in Chapter 9.

M. Berg (2001). Geometry, renormalization, and supersymmetry, unpublished Ph. D. Dissertation, University of Texas at Austin, cited in Chapter 15.

M. Berg and P. Cartier (2001). "Representations of the renormalization group as matrix Lie algebra," arXiv:hep-th/0105315, cited in Chapter 17.

N. Berline, E. Getzler, and M. Vergne (1992). *Heat Kernels and Dirac Operators* (Berlin, Springer), cited in Chapter 10.

M. V. Berry (1985). "Scaling and nongaussian fluctuations in the catastophe theory of waves," *Prometheus* **1**, 41–79 (A Unesco publication, eds. Paulo Bisigno and Augusto Forti) [in Italian]; and in *Wave Propagation and Scattering*, ed. B. J. Uscinski (Oxford, Clarendon Press, 1986), pp. 11–35 [in English], cited in Chapter 4.

N. D. Birrell and P. C. W. Davies (1982). *Quantum Fields in Curved Space* (Cambridge, Cambridge University Press), cited in Project 19.7.1.

Ph. Blanchard and M. Sirugue (1981). "Treatment of some singular potentials by change of variables in Wiener integrals," *J. Math. Phys.* **22**, 1372–1376, cited in Chapter 14.

S. Bochner (1955). *Harmonic Analysis and the Theory of Probability* (Berkeley, CA, University of California Press), cited in Chapter 1.

N. N. Bogoliubov and O. S. Parasiuk (1957). "Über die Multiplikation der Kausalfunktionen in der Quantentheorie der Felder," *Acta Math.* **97**, 227–266, cited in Chapter 17.

B. Booss and D. D. Bleecker (1985). *Topology and Analysis, the Atiyah–Singer Index Formula and Gauge-Theoretic Physics* (Berlin, Springer), Part IV, cited in Chapter 10.

M. Born (1927). *The Mechanics of The Atom* (London, G. Bell and Sons), cited in Chapter 14.

D. G. Boulware (1980). "Radiation from a uniformly accelerated charge," *Ann. Phys.* **124**, 169–188, cited in Project 19.7.1.

N. Bourbaki (1969). *Eléments de mathématiques, intégration*, chapitre 9 (Paris, Hermann). English translation by S. K. Berberian, *Integration II* (Berlin, Springer, 2004). See in particular "Note historique," pp. 113–125, cited in Chapters 1, 10, 13, and Appendix A.

N. Bourbaki (1981). *Espaces vectoriels topologiques* (Paris, Masson), cited in Appendix G.

L. Brillouin (1938). *Les tenseurs en mécanique et en élasticité* (Paris, Masson), cited in Chapter 9.

D. J. Broadhurst and D. Kreimer (1999). "Renormalization automated by Hopf algebra," *J. Symb. Comput.* **27**, 581–600, cited in Chapter 17.

(2000). "Combinatoric explosion of renormalization tamed by Hopf algebra: 30-loop Padé–Borel resummation," *Phys. Lett.* **B475** 63–70, cited in Chapter 17.

(2001). "Exact solutions of Dyson–Schwinger equations for iterated one-loop integrals and propagator-coupling duality," *Nucl. Phys.* **B600** 403–422, cited in Chapter 17.

D. C. Brydges, J. Dimock, and T. R. Hurd (1998). "Estimates on renormalization group transformations," *Can. J. Math.* **50**, 756–793 and references therein, cited in Chapters 2 and 16.

(1998). "A non-gaussian fixed point for ϕ^4 in $4 - \epsilon$ dimensions," *Commun. Math. Phys.* **198**, 111–156, cited in Chapters 2 and 16.

P. F. Byrd and M. D. Friedman (1954). *Handbook of Elliptic Integrals for Engineers and Physicists* (Berlin, Springer), cited in Chapter 4.

D. M. Cannell (1993). *George Green Mathematician and Physicist 1793–1841* (London, The Athlone Press Ltd.), cited in Chapter 15.

H. Cartan (1979). *Œuvres* (Berlin, Springer), Vol. III, cited in Chapter 10.

P. Cartier (1989). "A course on determinants," in *Conformal Invariance and String Theory*, eds. P. Dita and V. Georgescu (New York, Academic Press), cited in Chapter 11 and Appendix E.

(2000). "Mathemagics" (a tribute to L. Euler and R. Feynman), *Sém. Lothar. Combin.* **44** B44d. 71 pp., cited in Chapter 1 and Appendix B.

P. Cartier and C. DeWitt-Morette (1993). "Intégration fonctionnelle; éléments d'axiomatique," *C. R. Acad. Sci. Paris* **316 Série II**, 733–738, cited in Chapter 11.

(1995). "A new perspective on functional integration," *J. Math. Phys.* **36**, 2237–2312, cited in Chapters 2, 4, 7, 8, 11, and 17 and Appendix E.

(1997). "Physics on and near caustics," in *NATO-ASI Proceedings, Functional Integration: Basics and Applications* (Cargèse 1996) (New York, Plenum), cited in Chapter 4.

(1999). "Physics on and near caustics. A simpler version," in *Mathematical Methods of Quantum Physics*, eds C. C. Bernido, M. V. Carpio-Bernido, K. Nakamura, and K. Watanabe (New York, Gordon and Breach), pp. 131–143, cited in Chapter 4.

(2000). "Functional integration," *J. Math. Phys.* **41**, 4154–4187, cited in Chapter 17.

P. Cartier, M. Berg, C. DeWitt-Morette, and A. Wurm (2001). "Characterizing volume forms," in *Fluctuating Paths and Fields*, ed. A. Pelster (Singapore, World Scientific), cited in Chapters 11 and 17.

P. Cartier, C. DeWitt-Morette, M. Ihl, and Ch. Sämann, with an appendix by M. E. Bell (2002). "Supermanifolds – applications to supersymmetry," arXiv:math-ph/0202026; and in

(2002). Michael Marinov memorial volume *Multiple Facets of Quantization and Supersymmetry*, eds. M. Olshanetsky and A. Vainshtein (River Edge, NJ, World Scientific), pp. 412–457, cited in Chapters 9 and 10.

M. Chaichian and A. Demichev (2001). *Path Integrals in Physics*, Vols. I and II (Bristol, Institute of Physics), cited in Chapter 1.

A. H. Chamseddine and A. Connes (1997). "The spectral action principle," *Commun. Math. Phys.* **186**, 731–750, cited in Chapter 17.

F. Chapoton (2000). "Un théorème de Cartier–Milnor–Moore–Quillen pour les bigèbres dendriformes et les algèbres braces," arXiv:math.QA/0005253, cited in Chapters 17 and Project 19.7.3.

S. S. Chern (1979). "From triangles to manifolds," *Am. Math. Monthly* **86**, 339–349, cited in Chapter 10.

K. G. Chetyrkin and T. Seidensticker (2000). "Two loop QCD vertices and three loop MOM β functions," *Phys. Lett.* **B495** 74–80, cited in Chapter 17.

K. G. Chetyrkin and F. V. Tkachov (1981). "Integration by parts: the algorithm to calculate beta functions in 4 loops," *Nucl. Phys.* **B192**, 159–204, cited in Chapter 17.

K. G. Chetyrkin, A. H. Hoang, J. H. Kuhn, M. Steinhauser, and T. Teubner (1996). "QCD corrections to the e^+e^- cross section and the Z boson decay rate: concepts and results," *Phys. Rep.* **277**, 189–281, cited in Chapter 17.

K. Chetyrkin, M. Misiak, and M. Münz (1998). "Beta functions and anomalous dimensions up to three loops," *Nucl. Phys.* **B518**, 473–494, cited in Chapter 17.

Ph. Choquard and F. Steiner (1996). "The story of Van Vleck's and Morette–Van Hove's determinants," *Helv. Phys. Acta* **69**, 636–654, cited in Chapter 4.

Y. Choquet-Bruhat (1989). *Graded Bundles and Supermanifolds* (Naples, Bibliopolis), cited in Chapters 9 and Project 19.4.2.

Y. Choquet-Bruhat and C. DeWitt-Morette (1996 and 2000). *Analysis, Manifolds, and Physics*, Part I.: Basics (with M. Dillard–Bleick), revised edn.; Part II, revised and enlarged edn. (Amsterdam, North Holland), cited in Chapters 5, 7–11, and Project 19.4.2.

C. Chryssomalakos, H. Quevedo, M. Rosenbaum, and J. D. Vergara (2002). "Normal coordinates and primitive elements in the Hopf algebra of renormalization," *Commun. Math. Phys.* **225**, 465–485, cited in Chapter 17.

J.-M. Chung and B. K. Chung (1999). "Calculation of a class of three-loop vacuum diagrams with two different mass values," *Phys. Rev.* **D59**, 105014, cited in Chapter 17.

C. Cohen-Tannoudji, B. Diu, and F. Laloe (1977). *Quantum Mechanics*, Vol. 1 (New York, Wiley-Interscience), cited in Chapter 12.

D. Collins (1997). Two-state quantum systems interacting with their environments: a functional integral approach, unpublished: Ph.D. Dissertation, University of Texas at Austin, cited in Chapter 12.

 (1997). "Functional integration over complex Poisson processes," appendix to "A rigorous mathematical foundation of functional integration," by P. Cartier and C. DeWitt-Morette, in *Functional Integration: Basics and Applications* eds. C. DeWitt-Morette, P. Cartier, and A. Folacci (New York, Plenum), cited in Chapter 12.

J. C. Collins (1985). *Renormalization* (Cambridge, Cambridge University Press), cited in Chapter 17.

Ph. Combe, R. Høegh-Krohn, R. Rodriguez, M. Sirugue, and M. Sirugue-Collin (1980). "Poisson processes on groups and Feynman path integrals," *Commun. Math. Phys.* **77**, 269–288, cited in Project 19.6.2.

(1982). "Feynman path integral and Poisson processes with piecewise classical paths," *J. Math. Phys.* **23**, 405–411, cited in Project 19.6.2.

A. Comtet, J. Desbois, and S. Ouvry (1990). "Winding of planar Brownian curves," *J. Phys. A* **23**, 3563–3572, cited in Project 19.3.1.

A. Connes (1994). *Noncommutative Geometry* (San Diego, CA, Academic Press), cited in Chapter 10.

(2000). "A short survey of noncommutative geometry," *J. Math. Phys.* **41**, 3832–3866, cited in Chapter 17.

A. Connes and D. Kreimer (1999). "Lessons from quantum field theory – Hopf algebras and spacetime geometries," *Lett. Math. Phys.* **48**, 85–96, cited in Chapter 17.

(2000). "Renormalization in quantum field theory and the Riemann–Hilbert problem I: the Hopf algebra structure of graphs and the main theorem," *Commun. Math. Phys.* **210**, 249–273, cited in Chapters 17 and Project 19.7.8.

(2001). "Renormalization in quantum field theory and the Riemann–Hilbert problem II: the β-function, diffeomorphisms and the renormalization group," *Commun. Math. Phys.* **216**, 215–241, cited in Chapter 17.

A. Connes and H. Moscovici (2001). "Differentiable cyclic cohomology and Hopf algebraic structures in transverse geometry," cited in Chapter 17.

A. Connes, M. R. Douglas, and A. Schwarz (1998). "Noncommutative geometry and matrix theory: compactification on tori," *J. High Energy Phys.* **02**, 003, cited in Chapter 17.

A. Das (1993). *Field Theory, A Path Integral Approach* (Singapore, World Scientific), cited in Chapters 1, 15, and 16.

J. L. de Lyra, B. DeWitt, T. E. Gallivan *et al.* (1992). "The quantized $O(1,2)/O(2) \times Z_2$ sigma model has no continuum limit in four dimensions. I. Theoretical framework," *Phys. Rev.* **D46**, 2527–2537, cited in Project 19.7.2.

(1992). "The quantized $O(1,2)/O(2) \times Z_2$ sigma model has no continuum limit in four dimensions. II. Lattice simulation," *Phys. Rev.* **D46**, 2538–2552, cited in Project 19.7.2.

J. Desbois, C. Furtlehner, and S. Ouvry (1995). "Random magnetic impurities and the Landau problem," *Nucl. Phys.* **B 453**, 759–776, cited in Project 19.3.1.

B. DeWitt (1992). *Supermanifolds*, 2nd edn. (Cambridge, Cambridge University Press), cited in Chapters 9, 10, and 18.

(2003). *The Global Approach to Quantum Field Theory* (Oxford, Oxford University Press; with corrections 2004), cited in Chapters 1, 4, 6, 10, 15, 18, Projects 19.5.3 and 19.5.4, and Appendix E.

B. S. DeWitt (1965). *Dynamical Theory of Groups and Fields* (New York, Gordon and Breach), cited in Chapter 18.

C. DeWitt (1967). "Le principe d'équivalence dans la théorie du champ de gravitation quantifié," in *Fluides et champ gravitationnel en relativité générale*, cited in Project 19.7.1.

C. DeWitt and B. DeWitt (eds.) (1964). *Relativity Groups and Topology* (New York, Gordon and Breach), cited in Appendix H.

C. M. DeWitt (1969). "L'intégrale fonctionnelle de Feynman. Une introduction," *Ann. Inst. Henri Poincaré* **XI**, 153–206, cited in Chapter 8.

C. DeWitt-Morette (1972). "Feynman's path integral; definition without limiting procedure," *Commun. Math. Phys.* **28**, 47–67, cited in Chapter 1.

(1974). "Feynman path integrals; I. Linear and affine techniques; II. The Feynman–Green function," *Commun. Math. Phys.* **37**, 63–81, cited in Chapters 1, 3, and 4.

(1976). "The Semi-Classical Expansion," *Ann. Phys.* **97**, 367–399 and **101**, 682–683, cited in Chapter 4 and Appendix E.

(1976). "Catastrophes in Lagrangian systems," and its appendix C. DeWitt-Morette and P. Tshumi, "Catastrophes in Lagrangian systems. An example," in *Long Time Prediction in Dynamics*, eds. V. Szebehely and B. D. Tapley (Dordrecht, D. Reidel), pp. 57–69, cited in Chapter 5.

(1984). "Feynman path integrals. From the prodistribution definition to the calculation of glory scattering," in *Stochastic Methods and Computer Techniques in Quantum Dynamics*, eds. H. Mitter and L. Pittner, *Acta Phys. Austriaca Suppl.* **26**, 101–170. Cited in Chapters 1, 4, and 5 and Appendix E.

(1993). "Stochastic processes on fibre bundles; their uses in path integration," in *Lectures on Path Integration, Trieste 1991*, eds. H. A. Cerdeira, S. Lundqvist, D. Mugnai et al. (Singapore, World Scientific) (includes a bibliography by topics), cited in Chapter 1.

(1995). "Functional integration; a semihistorical perspective," in *Symposia Gaussiana*, eds. M. Behara, R. Fritsch, and R. Lintz (Berlin, W. de Gruyter and Co.), pp. 17–24, cited in Chapter 1.

C. DeWitt-Morette and K. D. Elworthy (1981). "A stepping stone to stochastic analysis," *Phys. Rep.* **77**, 125–167, cited in Chapter 6.

C. DeWitt-Morette and S.-K. Foong (1989). "Path integral solutions of wave equation with dissipation," *Phys. Rev. Lett.* **62**, 2201–2204, cited in Chapters 12 and 13.

(1990). "Kac's solution of the telegrapher's equation revisited: part I," *Annals Israel Physical Society*, **9**, 351–366 (1990), Nathan Rosen Jubilee Volume, eds. F. Cooperstock, L. P. Horowitz, and J. Rosen (Jerusalem, Israel Physical Society), cited in Chapters 12 and 13.

C. DeWitt-Morette and B. Nelson (1984). "Glories – and other degenerate critical points of the action," *Phys. Rev.* **D29**, 1663–1668, cited in Chapters 1, 4, and 5.

C. DeWitt-Morette and T.-R. Zhang (1983). "Path integrals and conservation laws," *Phys. Rev.* **D28**, 2503–2516, cited in Appendix E.

(1983). "A Feynman–Kac formula in phase space with application to coherent state transitions," *Phys. Rev.* **D28**, 2517–2525, cited in Chapters 3 and 5 and Appendix E.

(1984). "WKB cross section for polarized glories," *Phys. Rev. Lett.* **52**, 2313–2316, cited in Chapters 1 and 4.

C. DeWitt-Morette, K. D. Elworthy, B. L. Nelson, and G. S. Sammelman (1980). "A stochastic scheme for constructing solutions of the Schrödinger equations," *Ann. Inst. Henri Poincaré* **A XXXII**, 327–341, cited in Projects 19.5.3 and 19.7.2.

C. DeWitt-Morette, S. G. Low, L. S. Schulman, and A. Y. Shiekh (1986). "Wedges I," *Foundations Phys.* **16**, 311–349, cited in Chapter 5.

C. DeWitt-Morette, A. Maheshwari, and B. Nelson (1977). "Path integration in phase space," *General Relativity and Gravitation* **8**, 581–593, cited in Chapter 3.

(1979). "Path integration in non relativistic quantum mechanics," *Phys. Rep.* **50**, 266–372, cited in Chapters 1, 3, 4, 6, 7, 11, and 14 and Appendices B and E.

C. DeWitt-Morette, B. Nelson, and T.-R. Zhang (1983). "Caustic problems in quantum mechanics with applications to scattering theory," *Phys. Rev.* **D28**, 2526–2546, cited in Chapter 5 and Appendix E.

P. A. M. Dirac (1933). "The Lagrangian in quantum mechanics," *Phys. Z. Sowjetunion* **3**, 64–72. Also in the collected works of P. A. M. Dirac, ed. R. H. Dalitz (Cambridge, Cambridge University Press, 1995), cited in Chapter 1.

(1947). *The Principles of Quantum Mechanics* (Oxford, Clarendon Press), cited in Chapters 1, 11, and 15.

(1977). "The relativistic electron wave equation," *Europhys. News* **8**, 1–4. (This is abbreviated from "The relativistic wave equation of the electron," *Fiz. Szle* **27**, 443–445 (1977).) Cited in Chapter 1.

J. Dieudonné (1977). *Eléments d'analyse*, Vol. VII (Paris, Gauthier-Villars), cited in Appendix G.

J. D. Dollard and C. N. Friedman (1979). *Product Integration with Applications to Differential Equations* (Reading, MA, Addison-Wesley), cited in Appendix B.

J. S. Dowker (1971). "Quantum mechanics on group space and Huygens' principle," *Ann. Phys.* **62**, 361–382, cited in Project 19.1.1.

(1972). "Quantum mechanics and field theory on multiply connected and on homogenous spaces," *J. Phys. A: Gen. Phys.* **5**, 936–943, cited in Chapter 8.

I. H. Duru and H. Kleinert (1979). "Solution of the path integral for the H-atom," *Phys. Lett. B* **84**, 185–188, cited in Chapter 14.

(1982). "Quantum mechanics of H-atoms from path integrals," *Fortschr. Phys.* **30**, 401–435, cited in Chapter 14.

F. J. Dyson (1949). "The radiation theories of Tomonaga, Schwinger, and Feynman," *Phys. Rev.* **75**, 486–502, cited in Chapter 6.

K. D. Elworthy (1982). *Stochastic Differential Equations on Manifolds* (Cambridge, Cambridge University Press), cited in Chapter 1.

L. D. Faddeev and A. A. Slavnov (1980). *Gauge Fields, Introduction to Quantum Theory* (Reading, MA, Benjamin Cummings), cited in Chapter 4.

R. P. Feynman (1942). *The Principle of Least Action in Quantum Mechanics* (Ann Arbor, MI, Princeton University Publication No. 2948. Doctoral Dissertation Series), cited in Chapter 1.

(1951). "An operator calculus having applications in quantum electrodynamics," *Phys. Rev.* **84**, 108–128, cited in Chapter 6.

(1966). "The development of the space-time view of quantum electrodynamics," *Phys. Today* (August), 31–44, cited in Chapter 15.

R. P. Feynman and A. R. Hibbs (1965). *Quantum Mechanics and Path Integrals* (New York, McGraw-Hill), cited in Chapters 1, 12, and 14.

R. P. Feynman and F. L. Vernon (1963). "Theory of a general quantum system interacting with a linear dissipative system," *Ann. Phys.* **24**, 118–173, cited in Chapter 12.

S.-K. Foong (1990). "Kac's solution of the telegrapher's equation revisited: part II," *Annals Israel Physical Society* **9**, 367–377, cited in Chapters 12 and 13.

(1993). "Path integral solution for telegrapher's equation," in *Lectures on Path Integration, Trieste 1991*, eds. H. A. Cerdeira, S. Lundqvist, D. Mugnai *et al.* (Singapore, World Scientific), cited in Chapters 12 and 13.

B. Z. Foster and T. Jacobson (2003). "Propagating spinors on a tetrahedral spacetime lattice," arXiv:hep-th/0310166, cited in Project 19.6.1.

K. Fujikawa (1979). "Path integral measure for gauge-invariant theories," *Phys. Rev. Lett.* **42**, 1195–1198, cited in Chapter 18.

(1980). "Path integral for gauge theories with fermions," *Phys. Rev.* **D21**, 2848–2858; erratum *Phys. Rev.* **D22**, 1499, cited in Chapter 18.

K. Fujikawa and H. Suzuki (2004). *Path Integrals and Quantum Anomalies* (Oxford, Oxford University Press), cited in Chapter 18.

A. B. Gaina (1980). "Scattering and Absorption of Scalar Particles and Fermions in Reissner–Nordstrom Field" [in Russian], VINITI (All-Union Institute for Scientific and Technical Information) 1970–80 Dep. 20 pp., cited in Chapter 5.

C. Garrod (1966). "Hamiltonian path integral methods," *Rev. Mod. Phys.* **38**, 483–494, cited in Chapter 14.

S. Gasiorowicz (1974). *Quantum Physics* (New York, John Wiley and Sons Inc.), cited in Chapter 5.

B. Gaveau, T. Jacobson, M. Kac, and L. S. Schulman (1984). "Relativistic extension of the analogy between quantum mechanics and Brownian motion," *Phys. Rev. Lett.* **53**, 419–422, cited in Chapter 12.

E. Getzler (1985). "Atiyah–Singer index theorem," in Les Houches Proceedings *Critical Phenomena, Random Systems, Gauge Theories*, eds. K. Osterwalder and R. Stora (Amsterdam, North Holland), cited in Chapter 10.

J. Glimm and A. Jaffe (1981). *Quantum Physics*, 2nd edn. (New York, Springer), cited in Chapter 1 and Appendix D.

S. Goldstein (1951). "On diffusion by discontinuous movements, and the telegraph equation," *Quart. J. Mech. Appl. Math.* **4**, 129–156, cited in Chapter 12.

J. M. Garcia-Bondía, J. C. Várilly, and H. Figueroa (2001). *Elements of Noncommutative Geometry* (Boston, Birkhäuser), cited in Chapter 17.

G. Green (1836). "Researches on the vibration of pendulums in fluid media," *Royal Society of Edinburgh Transactions*, 9 pp.. Reprinted in *Mathematical Papers* (Paris, Hermann, 1903), pp. 313–324. Cited in Chapter 15.

W. Greub, S. Halperin, and R. Vanstone (1973). *Connections, Curvature, and Cohomology*, Vol. II (New York, Academic Press), cited in Chapter 10.

L. Griguolo and M. Pietroni (2001). "Wilsonian renormalization group and the noncommutative IR/UV connection," *J. High Energy Phys.* **05**, 032, cited in Chapter 17.

C. Grosche and F. Steiner (1998). *Handbook of Feynman Path Integrals* (Berlin, Springer), cited in Chapter 1.

D. J. Gross (1981). "Applications of the renormalization group to high energy physics," *Méthodes en théorie des champs* (Les Houches, 1975), eds. R. Balian and J. Zinn-Justin (Amsterdam, North Holland), p. 144, cited in Chapter 16.

R. Grossmann and R. G. Larson (1989). "Hopf-algebraic structure of families of trees," *J. Algebra* **126**, 184–210, cited in Chapter 17.

M. C. Gutzwiller (1967). "Phase-integral approximation in momentum space and the bound state of an atom," *J. Math. Phys.* **8**, 1979–2000, cited in Chapter 14.
(1995). *Chaos in Classical and Quantum Mechanics (Interdisciplinary Applied Mathematics, Vol. 1)* (Berlin, Springer), cited in Chapter 14.

T. Hahn (2000). "Generating Feynman diagrams and amplitudes with FeynArts 3," arXiv:hep-ph/0012260. (The FeynArts web site is www.feynarts.de.) Cited in Chapter 17.

S. Heinemeyer (2001). "Two-loop calculations in the MSSM with FeynArts," arXiv:hep-ph/0102318, cited in chapter 17.

W. Heitler (1954). *The Quantum Theory of Radiation*, 3rd edn. (Oxford, Oxford University Press), cited in Project 19.7.1.

K. Hepp (1966). "Proof of the Bogoliubov–Parasiuk theorem of renormalization," *Commun. Math. Phys.* **2**, 301–326, cited in Chapter 17.

Y. Jack Ng and H. van Dam (2004). "Neutrix calculus and finite quantum field theory," arXiv:hep-th/0410285, v3 30 November 2004, cited in Chapter 15.

T. Jacobson (1984). "Spinor chain path integral for the Dirac equation," *J. Phys. A* **17**, 2433–2451, cited in Chapter 12.

T. Jacobson and L. S. Schulman (1984). "Quantum stochastics: the passage from a relativistic to a non-relativistic path integral," *J. Phys. A* **17**, 375–383, cited in Chapter 12.

G. W. Johnson and M. L. Lapidus (2000). *The Feynman Integral and Feynman's Operational Calculus* (Oxford, Oxford University Press, paperback 2002), cited in Chapter 1.

M. Kac (1956). "Some stochastic problems in physics and mathematics," Magnolia Petroleum Co. Colloquium Lectures, cited in Chapters 4 and 12.

C. Kassel (1995). *Quantum Groups* (New York, Springer), cited in Project 19.7.3.

J. B. Keller (1958). "Corrected Bohr–Sommerfeld quantum conditions for non-separable systems," *Ann. Phys.* **4**, 180–188, cited in Chapter 14.
(1958). "A geometrical theory of diffraction," in *Calculus of Variations and its Applications* (New York, McGraw-Hill), pp. 27–52, cited in Chapter 14.

J. B. Keller and D. W. McLaughlin (1975). "The Feynman integral," *Am. Math. Monthly* **82**, 457–465, cited in Chapter 14.

H. Kleinert *Festschrift*: (2001). *Fluctuating Paths and Fields*, eds. W. Janke, A. Pelster, H.-J. Schmidt, and M. Bachmann (Singapore, World Scientific), cited in Chapter 1.

H. Kleinert (2004). *Path Integrals in Quantum Mechanics, Statistics, and Polymer Physics*, 3rd edn. (Singapore, World Scientific), cited in Chapters 1, 14, and Project 19.5.3.

M. Kontsevich (1999). "Operads and motives in deformation quantization," *Lett. Math. Phys.* **48**, 35–72, cited in Chapter 17.

J. L. Koszul (1985). "Gochet de Schouten–Nijenhuis et cohomologie," in *The Mathematical Heritage of Elie Cartan, Astérisque*, Numéro Hors Série, 257–271, cited in Chapter 11.

N. Krausz and M. S. Marinov (1997). "Exact evolution operator on noncompact group manifolds," arXiv:quant-ph/9709050, cited in Project 19.7.2.

P. Krée (1976). "Introduction aux théories des distributions en dimension infinie," *Bull. Soc. Math. France* **46**, 143–162, cited in Project 19.1.2.

D. Kreimer (1998). "On the Hopf algebra structure of perturbative quantum field theories," *Adv. Theor. Math. Phys.* **2**, 303–334, cited in Chapter 17.

(2000). "Shuffling quantum field theory," *Lett. Math. Phys.* **51**, 179–191, cited in Chapter 17.

(2000). *Knots and Feynman Diagrams* (Cambridge, Cambridge University Press), cited in Chapter 17.

(2002). "Combinatorics of (perturbative) quantum field theory," *Phys. Rep.* **363**, 387–424, cited in Chapter 17.

D. F. Kurdgelaidzé (1961). "Sur la solution des équations non linéaires de la théorie des champs physiques," *Cahiers de Phys.* **15**, 149–157, cited in Chapters 4 and Project 19.2.2.

P. Kustaanheimo and E. Stiefel (1965). "Perturbation theory of Kepler motion based on spinor regularization," *J. Reine Angew. Math.* **218**, 204–219, cited in Chapter 14.

J. LaChapelle (1995). Functional integration on symplectic manifolds, unpublished Ph. D. Dissertation, University of Texas at Austin, cited in Chapters 3, 7, and 14 and Appendix E.

(1997). "Path integral solution of the Dirichlet problem," *Ann. Phys.* **254**, 397–418, cited in Chapters 11, 14, and Project 19.1.2.

(2004). "Path integral solution of linear second order partial differential equations, I. The general construction," *Ann. Phys.* **314**, 362–395, cited in Chapters 11, 14, and Project 19.1.2.

(2004). "Path integral solution of linear second order partial differential equations, II. Elliptic, parabolic, and hyperbolic cases," *Ann. Phys.* **314**, 396–424, cited in Chapter 14.

M. G. G. Laidlaw (1971). Quantum mechanics in multiply connected spaces, unpublished Ph. D. Dissertation, University of North Carolina, Chapel Hill, cited in Chapter 8.

M. G. G. Laidlaw and C. M. DeWitt (1971). "Feynman functional integrals for systems of indistinguishable particles," *Phys. Rev.* **D3**, 1375–1378, cited in Chapter 8.

D. A. Leites (1980). "Introduction to the theory of supermanifolds," *Russian Mathematical Surveys* **35**, 1–64, cited in Chapter 10.

Seminar on Supermanifolds, known as SoS, over 2000 pages written by D. Leites and his colleagues, students, and collaborators from 1977 to 2000 (expected to be available electronically from arXiv), cited in Chapters 9 and 10.

P. Lévy (1952). *Problèmes d'analyse fonctionnelle*, 2nd edn. (Paris, Gauthier-Villars), cited in Chapter 1.

P. Libermann and C.-M. Marle (1987). *Symplectic Geometry and Analytic Mechanics* (Dordrecht, D. Reidel), cited in Chapter 7.

S. G. Low (1985). Path integration on spacetimes with symmetry, unpublished Ph.D. Dissertation, University of Texas at Austin, cited in Project 19.1.1.

A. Maheshwari (1976). "The generalized Wiener–Feynman path integrals," *J. Math. Phys.* **17**, 33–36, cited in Chapters 3 and 4.

P. M. Malliavin (1997). *Stochastic Analysis* (Berlin, Springer), cited in Chapter 15.

M. S. Marinov (1980). "Path integrals in quantum theory," *Phys. Rep.* **60**, 1–57, cited in Chapter 9.
(1993). "Path integrals in phase space," in *Lectures on Path Integration, Trieste 1991*, eds. H. A. Cerdeira, S. Lundqvist, D. Mugnai *et al.* (Singapore, World Scientific), pp. 84–108, cited in Chapter 1.

M. S. Marinov and M. V. Terentyev (1979). "Dynamics on the group manifold and path integral," *Fortschritte Phys.* **27**, 511–545, cited in Project 19.7.2.

C. P. Martín, J. M. Gracia-Bondía, and J. C. Várilly (1998). "The Standard Model as a noncommutative geometry: the low energy regime," *Phys. Rep.* **294**, 363–406, cited in Chapter 17.

R. A. Matzner, C. DeWitt-Morette, B. Nelson, and T.-R. Zhang (1985). "Glory scattering by black holes," *Phys. Rev.* **D31**, 1869–1878, cited in Chapter 5.

D. McDuff (1998). "Symplectic structures," *Notices AMS* **45**, 952–960, cited in Chapter 11.

H. P. McKean Jr. (1969). *Stochastic Integrals* (New York, Academic Press), cited in Chapter 14.

P. A. Meyer (1966). *Probabilités et potentiel* (Paris, Hermann), cited in Chapter 13.

W. J. Miller (1975). "Semi-classical quantization of non-separable systems: a new look at periodic orbit theory," *J. Chem. Phys.* **63**, 996–999, cited in Chapter 14.

S. Minwalla, M. Van Raamsdonk, and N. Seiberg (2000). "Noncommutative perturbative dynamics," *J. High Energy Phys.* **02**, 020, cited in Chapter 17.

M. M. Mizrahi (1975). An investigation of the Feynman path integral formulation of quantum mechanics, unpublished Ph.D. Dissertation, University of Texas at Austin, cited in Chapter 4.
(1978). "Phase space path integrals, without limiting procedure," *J. Math. Phys.* **19**, 298–307, cited in Chapter 3.
(1979). "The semiclassical expansion of the anharmonic-oscillator propagator," *J. Math. Phys.* **20**, 844–855, cited in Chapter 4.
(1981). "Correspondence rules and path integrals," *Il Nuovo Cimento* **61B**, 81–98, cited in Chapter 6.

C. Morette (1951). "On the definition and approximation of Feynman's path integral," *Phys. Rev.* **81**, 848–852, cited in Chapters 2 and 4.

C. Morette-DeWitt and B. S. DeWitt (1964). "Falling charges," *Physics* **1**, 3–20, cited in Project 19.7.1.

A. Mostafazadeh (1994). "Supersymmetry and the Atiyah–Singer index theorem. I. Peierls brackets, Green's functions, and a proof of the index theorem via Gaussian superdeterminants," *J. Math. Phys.* **35**, 1095–1124, cited in Chapter 10.

(1994). "Supersymmetry and the Atiyah–Singer index theorem. II. The scalar curvature factor in the Schrödinger equation," *J. Math. Phys.* **35**, 1125–1138, cited in Chapter 10.

D. Nualart (1995). *The Malliavin Calculus and Related Topics (Probability and its Applications)* (Berlin, Springer), cited in Chapter 11.

T. Ohl (1995). "Drawing Feynman diagrams with LaTeX and Metafont," *Comput. Phys. Commun.* **90**, 340–354, cited in Chapter 17.

W. Pauli (1958). *Theory of Relativity* (New York, Pergamon Press); translated with supplementary notes by the author from "Relativitätstheorie" in *Enzyclopädie der mathematischen Wissenschaften* (Leipzig, B. G. Teubner, 1921), Vol. 5, Part 2, pp. 539–775. Cited in Chapter 9.

R. E. Peierls (1952). "The commutation laws of relativistic field theory," *Proc. Roy. Soc. (London)* **A214**, 143–157, cited in Chapter 1.

M. E. Peskin and D. V. Schroeder (1995). *An Introduction to Quantum Field Theory* (Reading, Perseus Books), cited in Chapters 10, 15, Project 19.7.1 and Appendix D.

G. Petiau (1960). "Les généralisations non-linéaires des équations d'ondes de la mécanique ondulatoire," *Cahiers de Phys.* **14**, 5–24, cited in Chapters 4 and Project 19.2.2.

G. Racinet (2000). Séries génératrices non-commutatives de polyzêtas et associateurs de Drinfeld, unpublished Ph. D. Dissertation, Université d'Amiens, cited in Chapter 17.

M. Van Raamsdonk and N. Seiberg (2000). "Comments on noncommutative perturbative dynamics," *J. High Energy Phys.* **03**, 035, cited in Chapter 17.

F. Ravndal (1976). *Scaling and Renormalization Groups* (Copenhagen, Nordita), cited in Chapter 15.

G. Roepstorff (1994). *Path Integral Approach to Quantum Physics: An Introduction* (Berlin, Springer). Original German edition *Pfadintegrale in der Quantenphysik* (Braunschweig, Friedrich Vieweg & Sohn, 1991). Cited in Chapters 1, 9, and Project 19.4.1.

W. Rudin (1962). *Fourier Analysis on Groups* (New York, Interscience), cited in Project 19.1.1.

L. H. Ryder (1996). *Quantum Field Theory*, 2nd edn. (Cambridge, Cambridge University Press), cited in Chapter 15 and Appendix D.

J. J. Sakurai (1985). *Modern Quantum Mechanics* (Menlo Park, CA, Benjamin/Cummings), cited in Chapters 3 and Project 19.7.1.

A. Salam and J. Strathdee (1974). "Super-gauge transformations," *Nucl. Phys.* **B76**, 477–482. Reprinted in *Supersymmetry*, ed. S. Ferrara (Amsterdam/Singapore, North Holland/World Scientific, 1987). Cited in Chapter 10.

L. S. Schulman (1968). "A path integral for spin," *Phys. Rev.* **176**, 1558–1569, cited in Chapter 8.

(1971). "Approximate topologies," *J. Math. Phys.* **12**, 304–308, cited in Chapter 8.

(1981). *Techniques and Applications of Path Integration* (New York, John Wiley), cited in Chapters 1 and 3.

(1982). "Exact time-dependent Green's function for the half plane barrier," *Phys. Rev. Lett.* **49**, 559–601, cited in Chapter 5.

(1984). "Ray optics for diffraction: a useful paradox in a path-integral context," in *The Wave, Particle Dualism* (Conference in honor of Louis de Broglie) (Dordrecht, Reidel), cited in Chapter 5.

B. Schwarzschild (2004). "Physics Nobel Prize goes to Gross, Politzer and Wilczek for their discovery of asymptotic freedom," *Phys. Today*, December, 21–24, cited in Chapter 16.

N. Seiberg and E. Witten (1999). "String theory and noncommutative geometry," *J. High Energy Phys.* **09**, 032, cited in Chapter 17.

B. Simon (1974). *The $P(\phi)_2$ Euclidean (Quantum) Field Theory* (Princeton, Princeton University Press), cited in Appendix D.

(1979). *Functional Integration and Quantum Physics* (New York, Academic Press), cited in Chapter 1.

D. A. Steck, W. H. Oskay, and M. G. Raizen (2001). "Observation of chaos-assisted tunneling between islands of stability," *Science* **293**, 274–278, cited in Project 19.2.1.

S. N. Storchak (1989). "Rheonomic homogeneous point transformation and reparametrization in the path integrals," *Phys. Lett. A* **135**, 77–85, cited in Chapter 14.

R. J. Szabo (1996). "Equivariant localization of path integrals," arXiv:hep-th/9608068, cited in Project 19.5.2.

G. 't Hooft and M. J. G. Veltmann (1973). "Diagrammar," CERN Preprint 73-9, 1–114, cited in Chapter 1.

A. Voros (1974). "The WKB–Maslov method for non-separable systems," in *Géométrie symplectique et physique mathématique* (Paris, CNRS), pp. 277–287. Cited in Chapter 14.

T. Voronov (1992). "Geometric integration theory on supermanifolds," *Sov. Sci. Rev. C. Math. Phys.* **9**, 1–138, cited in Chapters 9, 11, and Project 19.4.2.

S. Weinberg (1995, 1996, and 2000). *The Quantum Theory of Fields* (Cambridge, Cambridge University Press), Vols. I, II, and III, cited in Chapter 10.

W. Werner (1993). Sur l'ensemble des points autour desquels le mouvement brownien plan tourne beaucoup, unpublished Thèse, Université de Paris VII, cited in Project 19.3.1.

(1994). "Sur les points autour desquels le mouvement brownien plan tourne beaucoup [On the points around which plane Brownian motion winds many times]," *Probability Theory and Related Fields* **99**, 111–144, cited in Project 19.3.1.

(2003). "SLEs as boundaries of clusters of Brownian loops," arXiv:math.PR/0308164, cited in Project 19.3.1.

H. Weyl (1921). *Raum–Zeit–Materie* (Berlin, Springer). English translation *Space–Time–Matter* (New York, Dover, 1950). Cited in Chapter 9.

(1952). *Symmetry* (Princeton, NJ, Princeton University Press), cited in Chapter 7.

G. C. Wick (1950). "The evaluation of the collision matrix," *Phys. Rev.* **80**, 268–272, cited in Appendix D.

F. W. Wiegel and J. Boersma (1983). "The Green function for the half plane barrier: derivation from polymer entanglement probabilities," *Physica A* **122**, 325–333, cited in Chapter 5.

N. Wiener (1923). "Differential space," *J. Math. Phys.* **2**, 131–174, cited in Chapter 1.

K. G. Wilson (1975). "The renormalization group: critical phenomena and the Kondo problem," *Rev. Mod. Phys.* **47**, 773–840, cited in Chapter 16.

K. G. Wilson and J. Kogut (1974). "The renormalization group and the ϵ-expansion," *Phys. Rep.* **12**, 75–200, cited in Chapters 16 and 17.

E. Witten (1982). "Supersymmetry and Morse theory," *J. Diff. Geom.* **17**, 661–692, cited in Chapter 10.

N. M. J. Woodhouse (1992). *Geometric Quantization*, 2nd edn. (Oxford, Clarendon Press), cited in Chapter 7.

A. Wurm (1997). The Cartier/DeWitt path integral formalism and its extension to fixed energy Green's function, unpublished Diplomarbeit, Julius-Maximilians-Universität, Würzburg, cited in Chapter 14.

(2002). Renormalization group applications in area-preserving nontwist maps and relativistic quantum field theory, unpublished Ph. D. Dissertation, University of Texas at Austin, cited in Chapters 2, 15, and 16.

A. Wurm and M. Berg (2002). "Wick calculus," arXiv:Physics/0212061, cited in Appendix D.

A. Young and C. DeWitt-Morette (1986). "Time substitutions in stochastic processes as a tool in path integration," *Ann. Phys.* **169**, 140–166, cited in Chapter 14.

T.-R. Zhang and C. DeWitt-Morette (1984). "WKB cross-section for polarized glories of massless waves in curved space-times," *Phys. Rev. Lett.* **52**, 2313–2316, cited in Chapter 5.

W. Zimmermann (1970). "Local operator products and renormalization in quantum field theory," in *Lectures on Elementary Particles and Quantum Field Theory*, eds. S. Deser *et al.* (Cambridge, MA, MIT Press), Vol. 1, pp. 300–306, cited in Chapter 17.

J. Zinn-Justin (2003). *Intégrale de chemin en mécanique quantique: introduction* (Les Ulis, EDP Sciences and Paris, CNRS), cited in Chapter 1.

J. Zinn–Justin (2004). *Path Integrals in Quantum Mechanics* (Oxford, Oxford University Press), cited in Chapter 1.

Index

$L^{2,1}_-$ space, 425
n-point function, 301
SO(3)
   geodesics, 148
   propagators, 147
SU(2)
   geodesics, 148
   propagators, 147
Γ-function, 396

$\lambda\phi^4$ model, 91, 300, 314, 327, 371
ℤ, group of integers, 153

action functional, 79, 356, 404
   critical points, 98, 404
   effective action, 311
   fixed-energy systems, 270
   relative critical points, 97
   Taylor expansion, 405
affine transformation, 64, 65
Aharanov–Bohm system, 155
analytic continuation, 92
analytic properties of covariances, 422
anharmonic oscillator, 88
annihilation operator, 428
   bosonic, 163
   fermionic, 163
anomalies, 361
anyons, 151
Archimedes' principle, 297
Arzela's theorem, 19

background method, 64
backward integral, 387
Baire hierarchy, 10

Bargmann–Siegel transform, 47
Berezin functional integral, 373
Berezin integral, 164
berezinian, 355
Bogoliubov coefficients, 361
Boltzmann constant, 25
Borel function, 13
Borel subsets, 10
bound-state energy spectrum, 276
BPHZ theorem, 325
brownian motion
   Lie group (on), 391
brownian path, 57
brownian process
   intrinsic time, 282
Brydges coarse-graining operator, 50, 403

Cameron–Martin subspace, 62
Cameron–Martin formula, 85
canonical gaussians in L^2, 59
canonical gaussians in $L^{2,1}$, 59
Cartan development map, 141
causal ordering, 301
caustics, 97, 101
change of variables of integration, 415
characteristic class, 175
characteristic function, 20, 396
characteristic functional, 20
chronological ordering, 6
coarse-grained interaction, 314
coarse-graining, 46, 312
   Brydges, 312
   operator, 50, 312, 403
   transformation, 312
coherent states, 427

452   Index

combinatorics
   Hopf, Lie, Wick, 382
   symmetry factors, 347
configuration space, 355
Connes–Kreimer Hopf algebra
   associativity defect, 336
   coproduct Δ, 335
   right-symmetric algebra, 336
constants of motion, 97, 100
convolution, 47, 50
covariance, 42, 76
   analytic properties, 422
   scaled, 49
   self-similar contributions, 310
creation operator, 428
   bosonic, 163
   fermionic, 163
cylindrical measure, 18

Darboux' theorem, 196
degeneracy, 97
delta-functional, 274, 275, 368
densities
   Grassmann, weight -1, 163, 168, 172
   ordinary, weight 1, 162
disintegration of measure, 387
determinant
   Van Vleck–Morette, 87
determinants, 411
   finite-dimensional, 84
   traces, 412
diagrams, 5, 44, 304
differential equations
   interaction representation, 252
   stochastic solutions, 251
differential space, 3, 57
dimensional regularization, 300, 321, 363
Dirac δ-function, 166
Dirac equation, 220, 241
   four-component, 224
   spinor-chain path-integral solution, 224
   two-dimensional, 221, 260
Dirac matrices, 433
Dirac propagator, 432
Dirac volume element, 274
distribution, 368
   Dirac, 369
   Fourier transform, 368
   infinite-dimensional, 368
   tempered, 369
divergence, 191, 203, 375
   in function spaces, 205
dominated convergence theorem, 14
dynamical tunneling, 370
dynamical vector fields, 135, 140

Dyson series, 120
Dyson's time-ordered exponential formula, 252

effective action, 40, 316
   first-order approximation, 317
   second-order approximation, 318
elementary solution, 264
energy constraint, 272
energy function, 30
energy states
   density, 276
engineering dimension, 314
Euler beta integral, 397
Euler class, 175
Euler number, 175
Euler–Lagrange equation, 100, 404
Euler–Poincaré characteristic, 175
external-leg correction, 339

Feynman formula, 27
Feynman functional integral, 7
Feynman propagator, 361, 429
Feynman's checkerboard, 223, 379, 432
Feynman's integral, 31
Feynman–Kac formula, 119
Feynman–Vernon influence functional, 231
first exit time, 268, 270
first-order perturbation, 128
fixed-energy amplitude, 272
Fock space, 360
forced harmonic oscillator, 63, 70
form
   symplectic, 195
forms
   Grassmann, 163, 168, 172
   ordinary, 162
forward integral, 389
Fourier transform, 19, 39, 166
free field, 302
functional integral
   density functional, 356
   volume element, 356
functional Laplacian, 47

gaussian
   distribution, 400
   on a Banach space, 38
   on field distributions, 400
   scale-dependent, 311
gaussian integral, 35, 367
gaussian random variable, 395
   normalization, 397
   translation invariance, 398
gaussian volume element, 291
Gelfand–Naïmark–Segal, 423

Index

ghosts, 326
glory scattering, 6, 97, 104
gluing operation, 339
GNS, 423
graded
  algebra, 158
  anticommutative products, 157
  anticommutator, 158
  antisymmetry, 159
  commutative products, 157
  commutator, 158
  exterior product, 162
  Leibniz rule, 159
  Lie derivative, 159
  matrices, 160
  operators on Hilbert space, 161
  symmetry, 159
gradient, 191, 203, 375
  in function spaces, 205
grafting operator, 328
  matrix representation, 336, 350
graph, 327
  insertion, 329
  object, 329
Grassmann analysis, 157, 175
  berezinian, 161
  complex conjugation, 159
  generators, 159
  superdeterminant, 161
  superhermitian conjugate, 161
  supernumbers, 159
  superpoints, 160
  supertrace, 161
  supertranspose, 161
  supervector space, 160
Green's example, 297
Green's function
  fixed energy, 268
  Jacobi operator, 406
  spectral decomposition, 269
group-invariant symbols, 209
group manifold, 367
groups of transformations, 135, 141

half-edges, 304
Hamilton–Cayley equation, 260
hamiltonian operator, 29
hamiltonian vector field, 196
heat equation, 28
Hermite polynomial, 401
hessian, 79, 405
Hilbert space ℓ^2, 10
homotopy, 146, 371
homotopy theorem for path integration, 150
Hopf algebra, 324, 383

Ihara bracket, 335, 344
immersion in a fluid, 298
indistinguishable particles, 151
influence functional, 225
influence operator, 231
injection, 62
insertion
  insertion points, 328
  operator, 329
  table, 337, 348
integral, 13
  locally bounded, 13
integration by parts, 295
intrinsic time, 281
Ito integral, 127

Jacobi equation, 279, 406
Jacobi field, 99, 279, 406
Jacobi field operator, 356
Jacobi matrix, 99
Jacobi operator, 74, 98, 404
  eigenvalues, 413
  eigenvectors, 413
  functional, 406
  in phase space, 279
joint probability density, 237

Kac's formula, 29
Kac's solution, 242
kernel theorem of L. Schwartz, 428
Klein–Gordon equation, 220, 241, 263, 432
Koszul formula, 192, 196, 346, 375, 377, 378
  in function space, 205
Koszul's parity rule, 157
Kustaanheimo–Stiefel transformations, 283

lagrangian, 30
Laplace–Beltrami operator, 140
lattice field theory, 381
Lebesgue measure, 12
Lebesgue integration, 9
length dimension, 309
Lie algebra, 324, 383
Lie derivative
  in function space, 205
linear maps, 46, 56
logarithmic derivative, 392
loop expansion, 73, 358
  three-loop, 342

marginal, 16, 369
markovian principle, 123
mass dimension, 315
mean value, 16

measure, 7, 377
measure in a Polish space, 11
  complex, 11
  locally bounded, 11
  product measure, 12
  regular, 11
  variation of, 11
Mellin transform, 311
Milnor–Moore theorem, 325
moment-generating function, 396
moments, 42
momentum-to-position transition, 82
Morse index, 278
Morse lemma, 101
multiplicative functional, 254
multistep method, 119

nested divergences, 330
nonrenormalizable theories, 345
normal-ordered lagrangian, 315
normal-ordered product, 401
nuclear space, 199

operator causal ordering, 292
operator formalism, 114
operator ordering, 399
operator quantization rule, 7
orbits
  periodic, 276
  quasiperiodic, 276

parity
  forms, 162
  matrix parity, 158, 160
  vector parity, 158
particles in a scalar potential, 118
particles in a vector potential, 126
partition function, 26
path integral, 4
path-integral representation of the Dirac propagator, 435
Peierls bracket, 7
pfaffian, 195
phase space, 271
phase-space path integrals, 73
physical dimension, 114
Planck constant, 25
Poincaré map, 279
Poisson bracket, 7
Poisson functional integral, 228
Poisson law, 239
  probability measure, 236
Poisson process, 379
  basic properties of, 234
  counting process, 233, 237
  decay constant, 233, 236
  generating functional, 240
  group manifolds, 379
  hits, 234
  mathematical theory, 233
  unbounded domain, 234
  waiting time, 233, 236
Poisson-distributed impurities, 371
polarization, 43, 44
Polish space, 9
  closed, 10
  complete, 9
  open, 10
  separability, 9
position-to-position transition, 88
positive kernel, 423
pre-Lie algebra, 336
probability density, 17
  compatibility, 18
  symmetry, 18
probability law, 15, 236
probability measure, 15
probability space, 271
prodistribution, 6, 19, 22
product integrals, 119, 391
  indefinite, 392
  similarity, 394
  sum rule, 392
projective system, 85
promeasure, 6, 18, 243
pseudomeasure, 6, 22
pulsation, 23

quantum dynamics, 114
quantum field theory
  measure, 356
  volume elements, 355
quantum fields, 428
quantum mechanics, 388
quantum partition function, 27

radiative corrections, 380
rainbow scattering, 102
random hits, 234, 244
random process
  classical, 387
random times, 219
random variable, 16
reconstruction theorem, 423
regularization, 290, 305
renormalization, 289, 308, 324, 339, 380
  combinatorics, 324
  flow, 320
  scale, 311

renormalization group, 308, 344
reproducing kernel, 424, 428
resolvent operator, 268
Riemann sums, 426
riemannian geometry, 193
Riesz–Markoff theorem, 15
Roepstorff's formulation, 373

sample space, 15
scale evolution, 313, 314, 315
scale variable, 48, 309
scaling, 46, 308
scaling operator, 309, 312
scattering matrix (S-matrix), 301
scattering of particles by a repulsive
    Coulomb potential, 101
Schrödinger equation, 28, 114, 121, 377
  free particle, 115, 116
    in momentum space, 116
    time-independent, 268
Schwinger variational principle, 7
Schwinger's variational method, 124
semiclassical approximation, 359
semiclassical expansion, 78, 96, 370
short-time propagator, 255
sigma model
  noncompact, 381
simple harmonic oscillator, 409
soap-bubble problem, 101
space L^2, 14
space of field histories, 355
spaces of pointed paths, 389
spaces of Poisson paths, 241
  complex bounded measure, 247
  complex measure, 243
  volume element, 250
spectral energy density, 24
spinning top, 146
star operation, 335
Stefan's law, 23
stochastic independence, 16
stochastic Loewner distribution, 373
stochastic process
  conformally invariant, 373
Stratonovich integral, 127
sum over histories, 254
superdeterminant, 356
supermanifold, 355
supersymmetry, 177
supertrace, 175
supervector spaces $\mathbb{R}^{n|\nu}$, 196
symmetries, 135
symplectic geometry, 193
symplectic manifold, 141
symplectomorphism, 141

telegraph equation, 215, 261
  analytic continuation, 223, 234
  four-dimensional, 223
  Kac's solution, 218, 219, 261
  Monte Carlo calculation, 217
thermal equilibrium distribution, 26
time-dependent potential, 72
time–frequency diagram, 23
time-ordered exponentials, 120
time-ordered operator, 129
time-ordering, 4, 129
  time-ordered product, 391
Tomonaga–Schwinger theory, 124
top form, 376
trace/determinant relation, 198
transition amplitude, 388
  fixed energy, 269
transition probability, 388
transitions
  momentum-to-momentum, 410
  momentum-to-position, 410
  position-to-momentum, 410
  position-to-position, 410
translation invariance, 295
Trotter–Kato–Nelson formula, 119, 393
tuned time, 281
tunneling, 97, 106
two-loop order, 360
two-point function, 408
two-state system
  interacting with its environment,
    225, 227
  isolated, 225

vacuum, 428
vacuum persistence amplitude, 361
variance, 42
variational methods, 293
Vinberg algebra, 336
volume element, 172, 191, 375
  Dirac, 376
  gamma, 376
  group-invariant, 209
  Hermite, 376
  non-gaussian, 376
  riemannian, 192
  symplectic, 195
  translation-invariant, 194, 206

Ward identities, 326
wave equation, 242
  Cauchy data, 262
Wick calculus, 399
Wick-grafting operator, 383
Wick ordered monomials, 51

Wick rotation, 362, 363
Wick transform, 47, 399, 400
Wien's displacement law, 23
Wiener measure, 57
   marginals of, 58

Wiener's integral, 30
Wiener–Kac integrals, 31
WKB, 78
WKB approximation, 273, 277, 359
   physical interpretation, 93